AGRICULTURAL GLOBALIZATION TRADE AND THE ENVIRONMENT

NATURAL RESOURCE MANAGEMENT AND POLICY

Editors:

Ariel Dinar
Rural Development Department
The World Bank
1818 H Street, NW
Washington, DC 20433

David Zilberman
Dept. of Agricultural and
Resource Economics
Univ. of California, Berkeley
Berkeley, CA 94720

EDITORIAL STATEMENT

There is a growing awareness to the role that natural resources such as water, land, forests and environmental amenities play in our lives. There are many competing uses for natural resources, and society is challenged to manage them for improving social well being. Furthermore, there may be dire consequences to natural resources mismanagement. Renewable resources such as water, land and the environment are linked, and decisions made with regard to one may affect the others. Policy and management of natural resources now require interdisciplinary approach including natural and social sciences to correctly address our society preferences.

This series provides a collection of works containing most recent findings on economics, management and policy of renewable biological resources such as water, land, crop protection, sustainable agriculture, technology, and environmental health. It incorporates modern thinking and techniques of economics and management. Books in this series will incorporate knowledge and models of natural phenomena with economics and managerial decision frameworks to assess alternative options for managing natural resources and environment.

Concerns about the social and environmental effects of agricultural globalization and trade have grown substantially in recent years. Countries whose economies depend on production and export of agricultural commodities and minerals are particularly vulnerable to changing conditions in the global economy, notably to declining terms of trade, as witnessed in their steadily falling standards of living. An additional concern is the steady decline of the global environment. "Agricultural Globalization, Trade, and the Environment" addresses these concerns via a series of approaches, and helps the reader approach a conclusion on the likely direction of these effects.

The Series Editors

Recently Published Books in the Series

Bauer, Carl J.:
Against the Current: Privatization, Water Markets, and the State in Chile
Easter, K. William, Rosegrant, Mark W., and Dinar, Ariel:
Markets for Water: Potential and Performance
Smale, Melinda:
Farmers, Gene Banks, and Crop Breeding: Economic Analyses of Diversity in Wheat, Maize, and Rice
Casey, Frank, Schmitz, Andrew, Swinton, Scott, and Zilberman, David:
Flexible Incentives for the Adoption of Environmental Technologies in Agriculture
Feitelson, Eran and Haddad, Marwan
Management of Shared Groundwater Resources: the Israeli-Palestinian Case with an International Perspective
Wolf, Steven and Zilberman, David
Knowledge Generation and Technical Change: Institutional Innovation in Agriculture

AGRICULTURAL GLOBALIZATION TRADE AND THE ENVIRONMENT

EDITED BY

Charles B. Moss
Gordon C. Rausser
Andrew Schmitz
Timothy G. Taylor
David Zilberman

KLUWER ACADEMIC PUBLISHERS
Boston / Dordrecht / London

Distributors for North, Central and South America:
Kluwer Academic Publishers
101 Philip Drive
Assinippi Park
Norwell, Massachusetts 02061 USA
Telephone (781) 871-6600
Fax (781) 871-6528
E-Mail <kluwer@wkap.com>

Distributors for all other countries:
Kluwer Academic Publishers Group
Distribution Centre
Post Office Box 322
3300 AH Dordrecht, THE NETHERLANDS
Telephone 31 78 6392 392
Fax 31 78 6546 474
E-Mail <services@wkap.nl>

Electronic Services <http://www.wkap.nl>

Library of Congress Cataloging-in-Publication Data

Agricultural globalization, trade, and the environment / edited by Charles B. Moss...[et al].
p. cm. - (Natural resource management and policy)
ISBN 0-7923-7472-X (alk. paper)
1. Agriculture—Environmental aspects. 2. Globalization—Environmental aspects. 3. International trade—Environmental aspects. I. Moss, Charles B. (Charles Britt) II. Series.

S589.75 .A363 2001
338.1'873—dc21 **2001038670**

Printed on acid-free paper.

Printed in Great Britain by IBT Global, London

Contents

List of Tables

List of Figures

Acknowledgements

The editors thank all those who participated in and attended the symposium entitled "Agricultural Globalization, Trade, and the Environment," held at the University of California at Berkeley on March 8 and 9 of 1999. We thank all the participants who presented papers at the conference, and especially those who contributed chapters for the present book. We also thank the authors who contributed additional chapters to this volume after the conference was completed.

We particularly thank the agencies that contributed financially to both the conference and to this volume's publication. These agencies include the College of Natural Resources and the Department of Agricultural and Resource Economics at the University of California at Berkeley; the Office of the Vice President for Agriculture and Natural Resources; the Office of the Dean for Research for the Institute of Food and Agricultural Sciences; and the Department of Food and Resource Economics at the University of Florida. Additional funding was provided by the Ben Hill Griffin, Jr. Endowed Chair at the University of Florida and by the U.S. Department of Agriculture through its National Research Initiative.

Special thanks to H. Carole Schmitz for her tireless efforts organizing the Berkeley conference, as well as for her invaluable administrative work editing this volume and managing its production for publication. Special thanks also to Jason Snart from the University of Florida Department of English for his work as editor, research specialist, and technical advisor in the production of this volume. Invaluable assistance with manuscript preparation and layout was also provided by Carol Fountain, Rose Herman, and Tom Stevens.

About the Contributors

Ballayram is a Guyanese national and currently a Post-doctoral Research Associate in the Sea Grant College, University of Florida, Gainesville, Florida. His research interests are in the areas of trade, agriculture and the environment.

Chris de Bodisco is currently working at the Food and Resource Economics Department at the University of Florida, while writing his dissertation for the Department of Economics at Vanderbilt University. His primary interest is comparing implicit values of irrigation across regions.

Charlene M. Brewster is a Post-doctoral Research Associate in the Food and Resource Economics Department at the University of Florida, Gainesville, Florida.

Jung-Sup Choi has been working, since 1981, with the Korea Rural Economic Institute. His research interests include agricultural policy and trade. He received his Ph.D degree in agricultural economics from the University of Wisconsin at Madison.

Carlton G. Davis is a distinguished professor in the Food and Resource Economics Department at the University of Florida, Gainesville, Florida. His area of research is agricultural trade and development, with emphasis on the Caribbean Basin.

Chris Dumas received his Ph.D degree in Agricultural and Resource Economics in 1997 from the University of California at Berkeley. He is currently Assistant Professor of Economics at the University of North Carolina at Wilmington.

Edward A. Evans is a Visiting Assistant Professor in the Department of Food and Resource Economics, Gainesville, Florida.

Bruce L. Gardner is the Distinguished University Professor and Chair, Department of Agricultural and Resource Economics, University of Maryland, where he has been since 1981 except for a period during 1989-1992 when he was Assistant Secretary for Economics at the U.S. Department of Agriculture. He is a Fellow of the American Agricultural Economics Association.

Richard Gray is a Professor in the Department of Agricultural Economics, and Director of the Centre for Studies in Agriculture Law and Environment at the University of Saskatchewan. He received his Ph.D from the University of California at Berkeley in 1991.

Harry de Gorter is Associate Professor in the Department of Agricultural, Resource and Managerial Economics at Cornell University. Before 1987, he taught at the University of Guelph and was policy advisor to the government of Canada on trade issues.

Hamish R. Gow is Assistant Professor in International Agribusiness Management in the Department of Agricultural and Consumer Economics and Department of Business Administration at the University of Illinois Urbana-Champaign, and he is Senior Research Associate at the Policy Research Group, Katholieke Universiteit Leuven, Belgium.

Hartley Furtan is a Professor of Agricultural Economics at the University of Saskatchewan. He has published a number of papers on agricultural policy, technological change, and agricultural trade issues. His Ph.D is from Purdue.

Faycal Haggui is a Ph.D graduate student in the Department of Agricultural Economics at the University of Saskatchewan.

Roland Herrmann is a Professor at the Institute of Agricultural Policy and Market Research and currently Acting Director of the Center for International Development and Environmental Research (CIDER), both at the University of Giessen, Germany.

Jill E. Hobbs is an Assistant Professor in the Department of Agricultural Economics at the University of Saskatchewan. Her research interests include agri-food marketing, supply-chain management, and the economics of food safety and biotechnology.

Tim Josling is Professor and Senior Fellow at the Institute for International Studies, Stanford University. The focus of his research measures distortions within the realm of international agricultural trade.

Richard E. Just is Distinguished University Professor at the University of Maryland and was formerly a professor at the University of California at Berkeley, where he received his Ph.D. He is a Fellow of the American Agricultural Economics Association.

Larry Karp is Professor of Agricultural and Resource Economics at the University of California at Berkeley.

Bonwoo Koo is a Post-doctoral Fellow at the International Food Policy Research Institute (IFPRI). His main research area is the economics of intellectual property rights, biotechnology, and genebank management.

John G. Lee is a Professor in the Department of Agricultural Economics at Purdue University. His research program focuses on environmental and natural resource economics as well as applied production economics.

Xuemei Liu is a Ph.D Candidate in the Department of Agricultural and Resource Economics at the University of California at Berkeley.

Charles B. Moss is a Professor in the Department of Food and Resource Economics at the University of Florida. He received his Ph.D in Agricultural Economics from Purdue University in 1987.

Sinaia Netanyahu is Lecturer in the Department of Industrial Engineering and Management at the Ben-Gurion University of the Negev.

David Orden is Professor of Agricultural and Applied Economics at Virginia Polytechnic Institute and State University (Virginia Tech) and Chairman of the department's graduate program.

Phillip L. Paarlberg has been a Professor of Agricultural Economics at Purdue University since 1985. His areas of research include agricultural trade and trade policy as well as agricultural reform in transition economies.

Peter W.B. Phillips is Professor of Agricultural Economics in the NSERC/SSHRC Chair in Managing Knowledge-based Agri-food Development at the University of Saskatchewan.

Gordon C. Rausser is the Robert Gordon Sproul Distinguished Professor at the University of California at Berkeley. He is a former dean of the College of Natural Resources (1994-2000) and served as chair on two occasions of Berkeley's Department of Agricultural and Resource Economics. He is a Fellow of the American Agricultural Economics Association.

Robert Romain is Professor in the Department of Agricultural Economics and Consumer Sciences at Laval University, and Director of the Centre for Research in the Economics of Agrifood.

Andrew Schmitz, Eminent Scholar, is the Ben Hill Griffin, Jr., Endowed Chair in the Department of Food and Resource Economics at the University of Florida in Gainesville, Florida. He served as Chairman of Berkeley's Agricultural and Resource Economics Department. He is a Fellow of the American Agricultural Economics Association.

Troy G. Schmitz received his M.Sc and Ph.D in Agricultural and Resource Economics from University of California, Berkeley in 1995. He is currently Assistant Professor in the Morrison School of Agribusiness and Resource Management at Arizona State University.

James L. Seale, Jr. is a Professor in the Food and Resource Economics Department at the University of Florida, Gainesville, Florida. His research areas include International Agriculture, Trade/Development, Econometrics, and Macroeconomics.

Richard Sexton is a Professor in the Department of Agricultural and Resource Economics at the University of California at Davis, and Director of the Giannini Foundation of Agricultural Economics.

Thomas H. Spreen is Professor, Food and Resource Economics Department, University of Florida, Gainesville, FL. Dr. Spreen holds a research and teaching appointment specializing in quantitative methods and marketing.

Daniel A. Sumner is the Frank H. Buck, Jr. Chair Professor in the Department of Agricultural and Resource Economics at the University of California at Davis, and Director of the University of California Agricultural Issues Center. His Ph.D is from the University of Chicago.

David Sunding is Director of the Center for Sustainable Resource Development, College of Natural Resources, University of California, Berkeley. He is also an Associate Cooperative Extension Specialist in the Department of Agricultural and Resource Economics at Berkeley.

Johan F. M. Swinnen is Economic Advisor at the European Union Commission, and Associate Professor of Agricultural and Food Policy and Director of the Policy Research Group in the Department of Agricultural Economics, Katholieke Universiteit Leuven, Belgium.

Timothy G. Taylor is a Professor in the Food and Resource Economics Department at the University of Florida in Gainesville, Florida. He is Co-Director of the Center for Agribusiness. His research areas include International Economics and Agribusiness.

Clive Y. Thomas is Professor of Economics and Director of the Institute of Development Studies, University of Guyana. His area of interest is development economics.

Suzanne D. Thornsbury is an Assistant Professor in the Food and Resource Economics Department at the University of Florida. Her research interests include food policy and trade.

Luther Tweeten is Professor Emeritus of Agricultural Policy and Trade at Ohio State University. He is a former President and Fellow of the American Agricultural Economics Association.

John J. VanSickle is a Professor of Food and Resource Economics and is Director of the International Agricultural Trade and Development Center. His research focuses on international trade and agricultural policy.

Brian D. Wright is Professor in the Department of Agricultural and Resource Economics at the University of California at Berkeley. Before joining the faculty at Berkeley, he was in the Department of Economics at Yale University.

David Zilberman is Professor at the University of California at Berkeley's Department of Agricultural and Resource Economics, where he holds the Robinson Chair. He is also the Director of the Center for Sustainable Resource Development in the U.C. Berkeley College of Natural Resources, and he is a Fellow of the American Agricultural Economics Association.

Preface

The relative prosperity in U.S. agriculture that attended the passage of the Federal Agriculture Improvement and Reform Act of 1996 was followed by a general decline in U.S. agricultural prices from 1998 to 2000. This trend in declining prices continues through the year 2001, despite the movement toward more liberalized agricultural trade. Trade liberalization has been the result of a variety of factors, including the implementation of the Uruguay Round Agreement, and the establishment of a variety of regional trade agreements, such as the North America Free Trade Agreement. Needless to say, in the face of falling agricultural prices and increasingly liberalized agricultural trade, the agricultural policy scene is an extremely complex one, both locally and globally.

The chapters in this volume look to understand this complexity by addressing the interaction between trade, the economic well-being of the farm sector, and the possibilities for future policy reform. The chapters collected here explore a number of different issues, including the operation of the tariff-rate quotas established under the Uruguay Round Agreement, the implications of sanitary and phytosanitary restrictions on trade, and the growing controversy over genetically modified organisms. In addition, several chapters analyze the interaction between agricultural trade and environmental concerns.

This volume does not pretend to offer a single, systematic prescription for what the next agricultural policy should be. Rather, the arguments and analyses contained herein are intended to highlight several issues that must be considered in the continuing debates on agricultural policy. This volume will be useful to students and practicing academics interested in the areas of agricultural trade and resource policy in the United States and abroad. In addition, we hope this volume will provide useful background and insight to policymakers, producer groups, and to all others involved in the debates over agricultural policy.

Chapter 1

Agricultural Globalization, Trade, and the Environment: Introduction

Charles B. Moss and Timothy G. Taylor
University of Florida

Since the 1980s, world agriculture has become increasingly integrated. Some of the impetus for this global integration had its origin at the end of the Cold War, and stemmed from the continuing philosophical debate over the merits of planned economies versus market economies. Commensurate with this, many developing economies began replacing inward-looking, import-industrialization-substitution policies with outward-looking policies and with export-led growth. Many countries also undertook unilateral liberalization of their trade and investment policies.

Trade liberalization occurred on both regional and global bases. Regional-trade agreements began appearing throughout the globe. For example, in the Western Hemisphere, the Caribbean Basin Initiative (CBI) and the Caribbean-Canada Trade Agreement (CARIBCAN) provided preferential access to the North American market for Caribbean and Latin American countries, and the North American Free Trade Agreement (NAFTA) laid the groundwork for a free-trade area among Canada, the United States, and Mexico. Finally, the completion of the Uruguay Round of the General Agreement on Tariffs and Trade (GATT) and the creation of the World Trade Organization (WTO), resulted in the first comprehensive, multilateral, rules-based trade agreement.

The Uruguay Round of GATT was especially notable in that it represented the first global trade accord to explicitly address the complex and politically charged issues associated with trade in agricultural products. Despite the many limitations of the Agreement on Agriculture, it nonetheless represented a significant step toward rationalizing the global food system.

At the turn of the 21st century, it appears that the rush toward further trade policy liberalization is slowing down. Within the western hemisphere, the failure of the U.S. administration to gain fast-track authority has slowed progress toward completing the proposed Free-trade Area of the Americas. It has, no doubt, hindered the United States' role in WTO discussions related to further agricultural trade liberalization. However, a more important factor slowing the trade-liberalization process is the desire of many interest groups to link environmental and labor issues with trade policy. Indeed, the Seattle WTO Ministerial meetings suggest that labor and environmental issues, and their linkage to trade policy, will likely prove pivotal in further trade-liberalization discussions.

Whether environmental and labor policy issues should be linked to trade policy is a normative question, the answer to which might be beyond the scope of strictly empirical analysis. However, the implications of linking environmental and labor policies to various trade policies is of great significance. This is especially true for agriculture, given its strong and unique relationship to the environment, in addition to the fact that much of world agriculture remains labor intensive. Further, agriculture accounts for a large proportion of low-wage employment.

This book contains twenty-three chapters that focus on a number of critical issues associated with agricultural globalization, trade, and the environment. The majority of these manuscripts were presented at a conference held at the University of California, Berkeley, in March of 1999. The book is divided into five parts: (1) farm programs and trade liberalization; (2) tariffs, quotas, and rent-seeking behavior; (3) foreign direct investment (FDI), trade, and vertical contracting; (4) trade and the environment; and (5) trade and biotechnology.

The six chapters of Part 1 focus on the evolution of farm programs under trade liberalization. Orden takes a politico-economic perspective in his evaluation of the establishment of U.S. farm policies, in general, and the Federal Agriculture Improvement and Reform (FAIR) Act of 1996, in particular. One stated purpose of FAIR was to introduce a freer market for agricultural products in the United States. Attempts to move toward greater market freedom subjected the agricultural sector to increased pressure from agricultural globalization. The reduction of subsidies, however, implied significant reductions in economic rents to farmers. As a result, through the political process, agricultural interests increased pressure to maintain some level of price supports. Based on the emerging difficulties of permanent-subsidy reductions through FAIR, Orden analyzes four possible scenarios for their elimination.

Tweeten's chapter provides a natural extension of Orden's chapter by suggesting that instability is the principle problem facing commercial agriculture. He demonstrates how trade could reduce instability, and contrasts the effect of a more open trading system with the effects of policy instruments (such as loan rates, deficiency payments, and income insurance) as methods for stabilizing the agricultural sector. He argues that international trade agreements may reduce the willingness of governments to undertake trade-distorting policies that contribute to instability. Thus, the emergence of international trade agreements may provide a self-enforcing mechanism through decreased distortions and a reduction of instability.

Gardner focuses on the effects of increased trade on agricultural income in the United States. He argues that export markets have been important to the economic health of U.S. agriculture, but that this importance may not be reflected directly in net farm income. Gardner discusses several explanations for this missing linkage. One argument is that export markets provide a backstop for agricultural prices by reducing the downside risk. However, in his empirical model, Gardner focuses on the possibility that the increased returns from exports accrue through increased agricultural asset values.

Departing from U.S. policy questions, Davis, Ballayram, Evans and Thomas analyze the effects of trade liberalization on small countries. They focus on the Caribbean Community (CARICOM) countries that are characterized by small economies with limited resources in which agriculture contributes significantly to the gross domestic product (GDP). Agriculture in the Caribbean depends heavily upon preferential trade agreements (such as the various Lomé Conventions, and the Caribbean Basin Economic Recovery Act). Given their dependence upon protected markets, these small countries stand to suffer significant losses under freer trade.

The significance of agriculture to GDP and to export earnings may imply significant sectional adjustments for Caribbean countries if they are to keep pace with processes of globalization. These kinds of adjustments will likely be required of most developing nations, and thus will serve to balance the significant gains to agricultural globalization described by Tweeten and Gardner. In addition, the complicated array of preferential trade agreements and strategies described by Davis et al. echo the difficulties, cited by Orden, associated with implementing institutional change in agricultural policies.

Moss and de Bodisco present another dimension of agricultural globalization by analyzing the implication of changes in agricultural policy for water management. Whereas Gardner suggests the value accruing to agriculture may be found in returns to fixed factors, such as land, Moss and de Bodisco explore another dimension of the returns to agriculture: specifically, returns that can be found in factors that have alternative uses in the long run, such as water. Moss and de Bodisco analyze the implications of the agricultural demand for water due to changes in agricultural and trade policy. They find that both FAIR and NAFTA have significant consequences for the agricultural demand for water in Florida. Further, the extent of this impact depends on the relative price changes for export-oriented agricultural commodities (such as corn and soybeans) as opposed to crops that compete domestically with imports.

Sumner and Choi examine the issue of tariff-rate quotas (TRQs). Under the Uruguay agreement, quotas are to be converted to tariff equivalents that are subject to minimum-access agreements. Several countries have exploited this to maintain specific commodity prices (for example, the U.S. Sugar Policy). A TRQ allows the importation of a specific quantity of a commodity at a nominal tariff. Once the minimum-access requirement is filled, additional imports are allowed, but they are subject to higher, and sometimes prohibitive, tariffs.

Sumner and Choi demonstrate how manipulating commodity definitions can extend policy rents under a TRQ. They analyze Korean imports of rice and show that by defining rice as either high-quality table rice or as low-quality rice, and stipulating that the government will import rice from the lowest bidder under the TRQ, Korea extracts an additional rent. The low-quality rice wins each bid to the exclusion of the high-quality table rice.

The six chapters comprising Part 2 focus on economic rent-seeking behavior through tariffs and quotas. Herrmann and Sexton decompose the general economic effect of TRQs into a tariff component and a quota component. This allows them to demonstrate how prices are determined, given that imports are constrained within the quota portion of the TRQ. Herrmann and Sexton show the economic gains and losses of a TRQ compared to a free-trade solution. Their empirical application focuses on the German banana market.

The new institutional economics of rent-seeking behavior developed in the de Gorter, Rausser, and Schmitz chapter can be used to further uncover costs to the Korean rice policy as discussed by Sumner and Choi. De Gorter et al. focus on the fact that TRQs can be established in a variety of ways depending on who receives the commodity in the country. In the case of the Korean rice policy, the right to import rice is maintained by the Korean government. The economic rents accruing to the higher import prices are captured domestically by the Korean government. De Gorter et al. then develop more complicated rent allocations focusing on the New Zealand Dairy Board's (NZDB) use of TRQs within the world dairy market. Their general conclusion is that TRQs, and other provisions introduced in the Uruguay Round of GATT, have led to rampant rent-seeking behavior within the agricultural trade sector.

Dumas and Schmitz analyze the case of upland cotton to determine whether or not the provisions of the FAIR Act are internally consistent. That is, the policy goals should be the same when they are viewed from the standpoint of producers, consumers, and taxpayers. Orden suggests that observed high commodity prices in 1996 made the passage of FAIR possible. Given high commodity prices, producer payments under the preceding agricultural program would have been limited. However, under FAIR, farmers received production flexibility contract (PFC) payments. After the fact, the PFC payments in 1996 were higher than would have been observed under the previous regime, while the payments in 1997, 1998, and 1999 were lower. Since the cotton market is highly dependent on storage, Dumas and Schmitz construct a stochastic model of the cotton market, which includes random supply shocks in the domestic and foreign market in addition to storage. Their results indicate that gains *ex ante* to FAIR were U.S. $57.6 million.

Paarlberg and Lee examine the implications of import restrictions intended to control foot-and-mouth disease (FMD) outbreaks in beef cattle in the United States. Such restrictions open up the opportunity for rent-seeking behavior, since producers have an incentive to overvalue the loss to the industry that may result from an FMD outbreak due to the importation of infected animals. Intuitively, domestic producers can obtain additional profits through the use of phytosanitary restrictions. Paarlberg and Lee suggest that tariffs imposed on beef imports should include a risk premium component, which would offset the cost of remediation should an FMD outbreak occur. It is interesting that this optimal policy includes a prohibitive tariff for some countries. Thus, a prohibitive tariff, by itself, is not an indication of excessive rent-seeking behavior.

Spreen, VanSickle, and Brewster expand the discussion of sanitary and phytosanitary (SPS) restrictions to include possible health concerns. Focusing on SPS restrictions within the citrus trade, they discuss a certification program that is used by Florida citrus producers to gain access to the European market. Florida is subject to sporadic infestations from the Mediterranean fruit fly. Since the fruit-fly pests can be transported in fresh fruit, and given that fresh fruit is far more profitable than processed fruit, Florida producers have entered into a certification process whereby pest traps are set in orchards to verify that their produce is pest free. One obvious possible SPS restriction is the prohibition of fresh fruit into the state to control the possibility of a pest infestation. However, the outright prohibition of the importation of fruit is unfeasible. Due to the seasonal nature of the citrus trade, it may be necessary to import citrus into Florida during certain times of the year. While this importation poses little economic threat from a market-share perspective to Florida's citrus production, it does imply risk of fruit-fly infestation.

Thornsbury provides a retrospective of five years of experience with SPS restrictions, including perceived expectations or goals of the SPS agreement, by focusing on four goals: (1) the establishment of scientific criteria for the evaluation of SPS restrictions; (2) the establishment of the transparency of the implementation process; (3) the establishment of restrictions under the SPS agreement that would be subject to geographic regions that are smaller than national boundaries; and (4) the establishment of efficient and credible settlement procedures that are not necessarily limited to SPS restrictions.

The number of SPS notifications filed, and the responses to these filings, suggest that the process is becoming increasingly transparent. The relaxation of the geographical restrictions on cattle imports from Argentina appears to represent a move toward smaller quarantine regions by regulatory authorities. However, questions regarding the establishment of efficient and credible dispute settlement mechanisms can be pursued in terms of the U.S.-E.U. beef-hormone controversy. Thornsbury concludes that the SPS agreement has resulted in freer agricultural trade, but that the process of liberalization has been sporadic.

The six chapters comprising Part 3 investigate the implications of the growing trends in vertical integration, and of multinationals in the coordination and globalization of the agribusiness sector.

Gow and Swinnen examine the role of vertical integration and of foreign direct investment in Eastern European agriculture, and they describe the trend toward FDI in the agribusiness sector. Western European businesses, because of their geographical proximity to Eastern Europe, have invested in the new markets of Eastern Europe. However, they have encountered difficulties in contract enforcement. The inability to enforce contracts has lead to the classic hold-up problem in which one party has extracted excessive economic rents from the other.

According to the new institutional economics literature, the solution to hold-up problems that stem from costly enforcement is vertical integration. Gow and Swinnen present a case study of a sugar processor (Juhocukor a.s.) and demonstrate that rents were extracted from farmers through delayed payment, which resulted in a decreased supply of sugar beets, and, hence, in decreased profitability for agribusiness.

Reforms, such as contract design, intended to mitigate the processor hold-up problem, were required to counter a downward cycle in sugar-beet production. The linkage between FDI and institutional design can be made in two dimensions. First, without efficient contracting, FDI is limited because of potential hold-up problems. Second, given past difficulties, FDI may provide capital and ground rules to overcome institutional problems.

Josling, who focuses on the emerging agricultural and trade-policy dimensions of agribusiness in the global economy, takes a slightly different perspective. First, he notes that the role of agribusiness in the formulation of agricultural and trade policy has never been well documented. Thus, he contemplates whether agribusiness gains from lobbying with the producers, or lobbying with the consumers, when it comes to agricultural and trade policy. Josling suggests that, historically, the net effect of such policies within the food industry has been relatively small. However, given the increasingly global nature of agribusiness, it may be affected by agricultural and trade policy to a greater degree.

Taylor and Seale sharpen Josling's discussion of agribusiness by examining the potential behavior of multinational firms in light of trade-policy liberalization. They note that 39,000 parent firms with 279,000 foreign subsidiaries controlled an FDI stock of U.S. $2.7 trillion in 1995. These firms conducted transactions across international boundaries that resulted in considerable intra-firm trade. These transactions could have been motivated by considerations different than the arms-length transactions assumed for the economic firms that are contained in the classical trade-theory models. This suggests that trade policy may work in ways considerably different than those generally posited by classical, or even by the new trade theory. When explaining the motivation for multinationals to trade across country boundaries, Taylor and Seale present three distinct models of firm behavior that result in the creation of multinational enterprises.

The four chapters in Part 4 look at the effects of increased globalization on the interaction between agriculture and the environment. Gray, Romain, and Furtan examine three potential environmental policies in Quebec, Canada. The WTO rules allow for the subsidization of production if subsidies offset pollution-abatement requirements. While such policies may be well intended, the net effect of such an abatement offset depends on the nature of the technology adopted. If adoption of the environmental technology improves efficiency, the total effect of price supports and technological adoption may distort the terms of trade.

Gray, Romain, and Furtan's first scenario assumes that world price is such that hog production does not lead to an externality or to a price support. The second scenario hypothesizes that hog production externalities are significant enough to be regulated, and sets a maximum manure-application rate. Under this scenario and its assumptions, the increased cost is offset with a subsidy that is countervailed by the United States. The authors note, however, that environmental regulations vary among the different provinces within Canada. Thus, hog producers in western Canada are not subject to the environmental requirements that are effective in Quebec, hence, they receive no subsidy to offset the cost of environmental compliance. Hog producers in western Canada are, however, subject to countervailing duties. Thus, the environmental regulation and the subsidy offset would have real regional effects on the price received.

A third scenario proposes that the government build a waste-disposal plant for hog-waste remediation. If Quebec constructed a waste-disposal plant instead of issuing the producers a government subsidy, the United States would have fewer grounds for countervailing. Hence, this policy alternative would not bring about a regional price.

Just and Netanyahu extend the discussion of environmental and trade policy by developing a game-theoretic model of linked bargaining. Taken separately, trade and environmental policies may be characterized as a zero-sum game in which mutually beneficial trade does not exist. One way to break the impasse involves linked games. By linking the games, it is possible to generate cooperation. A country may offset losses in the environmental dimension with gains in trade. One such linkage the authors discuss is tradable carbon permits.

Karp and Liu analyze the economic value of tradable emission permits. Their model estimates the effect of carbon emissions on the GDP of various countries. The value of tradable permits is then computed as a reduction in GDP, given that each country is forced individually to meet the carbon-emission standards. Tradable emission permits are compared to the scenario in which carbon emissions in each country are traded, and the marginal value of carbon is considered equal. Their results indicate that an additional eight percent reduction in carbon emissions can be achieved without income loss by using these tradable permits.

Zilberman, Liu and Sunding expand the discussion of the effects of global warming beyond the effect of temperature to consider the possible effects of daylight hours, fertilization, pests, geography and resettlement costs. In addition, their paper develops the mechanisms of environmental damage including rising sea levels, saltwater intrusion, changes in precipitation and other impacts on natural resources. They analyse several potential policy responses such as increased research, migration, and food supply policies.

Dumas, Moss, and Schmitz explore the contentious issue of global warming from an agricultural perspective. The general aim of the Kyoto Agreement on agricultural trade can be achieved in two ways: (1) Given that fuel represents a significant input in agricultural production, agriculture would be adversely affected by the imposition of a carbon tax. This carbon tax, however, may have additional trade implications. Agriculture in developed countries relies more heavily on technology than is the case in developing countries. In addition, under the Kyoto Agreement, developing nations are generally given higher targets, which imply lower relative carbon taxes. Both of these imply less-favorable terms of trade for developed nations. (2) Given that atmospheric carbon can be reduced through carbon sinks, forests can be planted to sequester carbon. And while not explicitly addressed in the Kyoto Agreement, another possibility for the sequestration of carbon involves the re-establishment of permanent grasslands. Dumas et al. examine the agricultural implications of these methods to reduce carbon in the atmosphere.

The advent of genetically modified organisms (GMOs) is intimately linked to the biotech revolution. Advances in biotechnology have radically changed the timeline for the production of new plant varieties when GMOs are used to meet emerging challenges to production. However, the new technologies also raise a host of questions related to the management of the agricultural research agenda. The three chapters comprising Part 5, the final part of the book, explore the potential effects of GMOs on agriculture and on agricultural trade.

Koo and Wright investigate the effects of the biotech revolution on the management of gene banks by examining a scenario in which a disease or pest emerges to cause significant economic damage to agricultural production. Under this scenario, the disease or pest can be controlled by genetically modifying local cultivars. The genetic source of the modification is contained in a gene bank of varieties.

The challenge for policymakers is then to design an optimal strategy for the structure of the gene bank. One strategy would be to begin testing cultivars in the gene bank for potential solutions before the outbreak. Another solution is to wait until an outbreak actually occurs before starting the search. Koo and Wright derive rules indicating that testing before the pest or disease actually occurs is optimal when the probability of an outbreak is high. The results presented in this chapter have significant implications for international research centers, such as the International Maize and Wheat Improvement Center (CIMMYT) and the International Food Policy Research Institute (IFPRI), and thus for the global nature of commercial agriculture.

Gray, Hobbs, and Haggui explore one possibility for the management of GMOs in terms of a market-based approach. Under this scenario, consumers weigh the risks and benefits of GMOs in their consumption decisions. The problem with the market-based system is the possibility of non-market externalities. Gray, Hobbs, and Haggui divide these potential externalities into six groups: (1) human health; (2) ecosystems; (3) reduced demand for undifferentiated non-GM products; (4) marketing and market segmentation; (5) production costs; and (6) informational cost.

Some question exists as to whether human health effects are externalities or part of the market. If the product is labeled, consumers can make rational decisions regarding the potential health consequences of GMOs. However, it could be argued that the long-run nature of most health risks binds the rationality of personal decisions.

The remaining externalities are much less controversial. The use of GMOs may imply positive and negative externalities for the ecosystem. On the positive side, GMOs may allow for the reduction of pesticide use; on the negative side, cross-fertilization between GMOs and the related wild varieties may have dramatic consequences for genetic diversity.

Phillips analyzes the political economy of internationally traded GMOs. He notes that, in 2001, GMOs are produced primarily in 12 countries. The potential effect of this concentration is heightened in that the crop/country combinations are focused on production for export markets. For example, the United States represents 72 percent of the acreage planted globally in transgenic crops. Soybeans account for 54 percent of this area, 28 percent of the area is corn, and 28 percent is cotton. This distribution is important given the export-oriented nature of the soybean, corn, and cotton markets in the United States. Further compounding the implications of GMOs is the fact that the objective of many of the technologies is to minimize cost. Taken together, these characteristics point to lower commodity prices worldwide.

Counteracting these supply pressures on the adoption of GMOs are regulatory efforts to ensure that the new products do not constitute a danger to human health or to ecosystems. Phillips points out that most countries have institutions charged with the maintenance of food safety. However, these institutions often have trouble keeping up with advances in biotechnology. They may be inadequately designed to regulate GMOs.

For international organizations, the provisions of the WTO are the most likely to be sustainable. Under the SPS Agreement, countries may prevent importation of commodities if the country can demonstrate a scientifically based suspicion of increased risk. Undoubtedly, these criteria may be difficult to satisfy in the case of GMOs. Specifically, health risks linked to GMOs may only emerge in the long-run, such as cancer. As such, scientifically establishing the linkage between GMOs and increased risk requires long-term study.

Private sector responses to GMOs include the branding or labeling of food as GMO-free. This market-based approach allows for consumers to form their own preferences about the risk implied by GMOs. However, as pointed out in Gray, Hobbs, and Haggui, this marketing-based approach is subject to potential externalities. As such, market-based approaches may still require additional agricultural policy responses.

PART I

FARM PROGRAMS AND TRADE LIBERALIZATION

Chapter 2

Farm Policy Reform in the United States

David Orden

Virginia Polytechnic Institute and State University

1. INTRODUCTION

Efforts to reform farm policy (that is, efforts both to reduce interventions in markets and to reduce the fiscal cost of transfers to the agricultural sector) have a long and informative history in the United States. This chapter assesses recent policy developments in their historical context and draws inferences about limits to and prospects for future reforms. When the first Republican-controlled Congress in 40 years wrote the Federal Agriculture Improvement and Reform (FAIR) Act of 1996, it was thought that a new era had been launched in agricultural policy, one that marked a substantial departure from the scope and (eventually) cost of past farm programs. Analysis and subsequent events have proven otherwise. Changes to farm policy made in 1996 were expensive, incomplete, and not necessarily permanent. The Republican members of Congress who authored the FAIR Act were among the strongest defenders of farm support policies, not their most outspoken critics. They were dissatisfied with some aspects of past program implementation, and they initiated market-liberalizing changes in several policy instruments. But Congress failed in 1996 to repeal the permanent agricultural acts that authorized intrusive government interventions. To avoid reversion to the terms of the antiquated permanent legislation, Congress will have to write yet another farm bill by 2002, under conditions that may, or may not, be propitious for reform. Thus, the central policy-reform problem in American agriculture has not been resolved, but strategies and circumstances can be imagined under which reform progress will occur that builds upon the FAIR Act.

In the next section of this chapter, I summarize the historical record of agricultural subsidies and supply interventions, and I review arguments concerning the difficulty of achieving reform. In the third section I posit four alternative paths along which reform might proceed, and characterize the dominant policy-driving forces and principal policy results of past decades within this framework. I present arguments concerning determinants of the FAIR Act outcomes in the fourth section, and in the fifth I look at policy reform setbacks that occurred when market prices fell after the FAIR Act was adopted. In a concluding section I consider future reform prospects. The chapter confirms that dismantling intervention and subsidy policies that are no longer needed or effective remains a formidable task in the face of organized interests prepared to defend them.[i]

2. DIFFICULTY OF REFORMING FARM POLICY

Reasons for the persistence of farm programs abound. As summarized in Table 2.1, neither the percentage of net farm income derived from government payments nor the available crop acreage idled under government programs shows a consistent decline from the 1950s through adoption of the FAIR Act; nor does their sum—an intervention index akin to the macroeconomic misery index of the inflation plus unemployment rates. The life span of a government program will depend on an evolving balance of political power between those who support the program (typically, those who benefit from it most directly) and those who oppose the program (typically, those who must pay for it, or who would like public resources to be used in some other way). In this struggle over the content of public policy, the most important determinant of the outcome will often be political organization. Organization is emphasized in equilibrium models of government price

Table 2.1 Forty-five Years of Farm Program Interventions, 1956-2000

Years	*Farm income from government payments*	*Idled to planted acreage*	*Intervention index (sum)*
		percent	
1956-1959	7.2	8.6	15.8
1960-1969	20.2	21.9	42.1
1970-1979	9.5	8.3	17.8
1980-1989	29.9	15.2	45.1
1990-1995	21.4	23.1	44.5
1996-2000	29.9	13.5	43.4

Sources: Orden, Paarlberg and Roe (1999: 5), and author's calculations from USDA data.

interventions along the lines of Grossman and Helpman [(1994, 1995); see also Dixit, Grossman and Helpman (1997) and Schleich (1999)], which combine differences in organizational strength among industries with a distinction between mobile factors (which benefit little from the interventions) and sector-specific fixed factors (which capture the benefits). The farm commodity programs that were first initiated in the United States during the 1930s have proved difficult to reform primarily because the defenders of these programs have earned rents and have remained better organized and more effective than the critics over the ensuing years. As agriculture modernizes in the context of a growing economy, the incentive for farmers to organize for political action tends to increase. When industrial productivity and incomes increase, those farmers who fail to make adjustments (by postponing adoption of new technologies or movement of their labor out of farming) tend to experience slower income growth than the rest of the economy. These slow-adjusting farmers, feeling themselves caught on what Willard Cochrane (1958) dubs a treadmill, will have a heightened motive to organize for political action. They will seek policy protection as an escape from the adjustment imperative. This will bring them into an alliance with the larger low-cost farmers who are not feeling adjustment pain, but whose asset values rise if protection is enacted. Farmers will have increased capabilities to organize for political action as industrialization occurs, because they will be better educated and more wealthy as time goes on, and, paradoxically, because they will be fewer in number.

In their efforts to understand the difficulty of policy reform, political scientists point also to the role of ideas and institutions. Analysts of farm policy such as Bonnen and Browne (1989) have adopted the argument that ideas can be influential long beyond their original moment of formation. They claim that farm groups remain powerful politically in the United States, in part, because of their ability to mobilize a badly out-of-date, but still potent, mythology of agrarian values. On institutions, Peter Hall (1993) argues that first-order policy reforms, in which only the settings of policy instruments change, can be relatively easy to achieve, but that higher-order reforms will be difficult because existing institutions and long-held ideas will have to be challenged. The only way to get the most powerful third-order reform (for example, an end to the farm support programs) is to adopt an entirely new theoretical or ideological vantage point. In Hall's view, movement toward this third order of change will have to come from more than just a routine political contest between organized lobby interests. It will require some form of society-wide contestation over ideas.

Because ideas and institutions both matter, and because both tend to resist change, it is that much easier to understand why farm policies already in place become difficult to reform. Paul Pierson (1994) argues further that welfare-state programs will tend to endure, in part because the interest groups that want those programs will be strengthened by their operation through feedback effects. A second type of feedback making programs already in existence easier for lobby groups to defend arises because the governmental apparatus that grows up around those benefit programs will come to share the lobby groups' interests in program perpetuation.

Political parties also play a role in the complex process of entrenchment of established entitlement programs. Political parties are organizations that seek to promote the material interests of their most active members and leaders by winning competitive elections. Parties respond to more than just material interests; they also carry and give expression to embedded ideas, which can help them retain and recruit activists. In the United States, with farm numbers quite low, the policy positions taken on agriculture by the two parties will more likely be responsive to the financial or organizational contributions of their respective farm constituents than to numerical clout in the voting booth. The interests of these farm constituents can change over time (for example, as the structure of agriculture changes) and, even when interests do not change, links to parties can sometimes be reversed (for example, when the Republican Party—mostly for reasons linked to race policy issues—makes gains in the once solid Democratic south). For these reasons, but also because parties with different farm group constituencies can suddenly replace one another in their control of government, parties are sometimes a source of agricultural policy change, occasionally of the third-order magnitude (in Hall's terms), but mostly of the first or second order.

3. ALTERNATIVE REFORM STRATEGIES

How might farm policy reform proceed—if it can proceed at all—in the face of organized interests prepared to defend established interventions? I have argued with Robert Paarlberg and Terry Roe (1999) that there are four alternative strategies available to advocates of policy reform in agriculture. As shown in Figure 2.1, farm program interventions can either be dismantled slowly or quickly, either with or without compensation. A slow, compensated transformation of programs is a cash out. A quick, compensated termination is a buyout. A slow, uncompensated diminution is a squeeze out, while a quick, uncompensated elimination is a cutout.

Compensation	*Speed of implementation*	
	Slow	*Fast*
Yes	Cash out	Buyout
No	Squeeze out	Cutout

Figure 2.1 Alternative Reform Strategies

3.1 Cash Out

A cash out is only a partial approach to reform, since it leaves market intervention policies in place for some time. It also imposes a potentially open-ended burden on taxpayers, while allowing farm lobby groups and their supporters in Congress and at the U.S. Department of Agriculture to retain dominance over the policy process. The cash-out strategy gives markets more room to operate, and has been particularly favored by export-oriented farmers and agribusinesses as a way to attain sales abroad without sacrificing government support for farmers at home. An early proposal to cash out farm price guarantees was made by the Secretary of Agriculture in 1949, under the Brannan Plan. Congress rejected this early proposal as too broad, too associated with supply control, and too costly. Not until the 1960s, when it became clear that high price support levels were hurting U.S. farm exports, did a belated shift begin toward letting market prices fall and instead using cash payments to support farm income. This cash out has provided only a partial solution to the larger policy reform problem. The gains from pursuing a cash out in the meantime are fewer market distortions, fewer production restraints, and more competitive export pricing.

3.2 Buyout

This approach would bring a quick end to farm support programs, with political resistance from farmers overcome through an up-front windfall of temporary compensatory payments. A short-run increase in subsidies might be a cost-effective way to proceed if it could be linked to a guarantee of permanent program demise, and might be justified as compensation for any drop in farm asset values associated with an end to farm subsidies. This reform strategy is not easily accomplished in farm legislation, in which no Congress has ever considered itself held to account by the decisions of a predecessor. A few buyouts have been attempted in the past, most notably in the 1985 farm-bill debate when Senators Boschwitz (Republican, Minnesota) and Boren (Democrat, Oklahoma) offered a proposal to drop traditional farm supports and replace them with a generous, but temporary, stream of decoupled transition payments. This buyout failed to catch hold in 1985 when it was rejected simultaneously by budget officials, who thought it was too expensive, and by farmers, who thought it was not generous enough. The substantial cash payments made to farmers under the 1996 FAIR Act were occasionally represented at the time as a buyout, but they bought little assurance of final program termination.

3.3 Squeeze Out

A slow squeeze-out strategy would reduce the market-distorting effects of farm programs without cash compensation. If the costs of program participation increase over time while the growth of benefits is constrained, farmers might be induced to quit the programs voluntarily, thus avoiding a difficult final legislative confrontation. This strategy can be pursued explicitly, through a legislated or administered withdrawal of market interventions or cash payments, or it can be pursued tacitly, by fixing benefits in nominal terms, then allowing inflation to erode their real value. A slow squeeze out also can be accomplished by targeting benefits only to certain categories of farmers, or by tying those benefits to a progressively more onerous set of performance requirements, such as stringent environmental regulations.

It may seem surprising with such a variety of options available that little reform has been attained through squeeze outs. Attempts made to lower farm support prices during the 1950s had the intent of a squeeze out, but farm-production costs were falling so rapidly that program participation remained attractive to most farmers. An implicit squeeze out seemed underway in the inflationary 1970s, but farm program defenders were able to ratchet up intervention levels, which itself proved costly when inflation was later squeezed out of the economy. The squeeze-out approach was unattractive in the past, partly because of the reliance of commodity support programs on production controls; in order for these controls to be effective, rates of participation in the programs had to be kept high.

3.4 Cutout

The abrupt cutout strategy would produce a quick and explicit end to farm support programs, without cash compensation. Among other things, under this approach the permanent legislation for these programs would have to be repealed. This approach, which would leave markets free to function, and taxpayers free from the burden of paying cash subsidies to farmers, has been favored by the most radical farm program critics. Virtual cutouts were proposed several times by Republican presidents in the 1970s and 1980s, but in each case a bipartisan coalition of agriculturalists in Congress rejected the administration's plan immediately and then wrote their own farm bill to keep existing support levels mostly intact. Farmers are likely to view any sharp termination of cash payments as an unacceptable cutout, despite any windfall of decoupled payments they earlier received when market prices were high.

The historical record of circumstances that have dominated farm policy and the associated outcomes are characterized in summary form in Table 2.2 using the framework of alternative reform strategies. Depressed world agricultural markets in the 1920s did not evoke substantive intervention until the world economic collapse and Democratic election victories in the 1930s brought a third-order change in government's role in the economy. Once

Table 2.2 Dominant Forces and Policy Outcomes

Years	*Dominant force(s)*	*Policy outcome(s)*
1920s	Collapse of world agricultural markets	Minimal intervention
1930s	Collapse of world economy	New Deal, Agricultural Adjustment Act (AAA)
1940s	World War II and consolidation of farm lobby	High price supports
1950s	Structural change and contested Congress	Thwarted squeeze out
1960s	Structural change and Democratic control	Partial cash out
1970s	Inflation devaluation	Failed squeeze out
1980s	Inflation suppression and budget pressure	Costly cash out
1990-97	Relative stability, high price interlude, Republican control	Minimal GATT effects, FAIR Act cash out
1998-2000	Falling prices, contested Congress	Costly supplemental cash-out payments

Source: Author.

farm programs were established, powerful constituencies formed to defend them, and they attained high wartime support prices. A squeeze out of farm programs was thwarted in the 1950s; reform proceeded along the lines of a cash out in the 1960s; and farm programs again proved impossible to squeeze out in the inflationary era of the 1970s. In the 1980s, a farm crisis occurred partly because of cash-out miscalculations, and commodity prices and land values decreased. Even so, the cash-out reform path was pursued, potentially setting the stage for a modest squeeze out by fixing nominal target prices, albeit at great expense and backed up with a high level of market-intrusive supply control. International trade negotiations subsequently proved ineffective in achieving basic reform during the General Agreement on Tariffs and Trade (GATT) Uruguay Round of 1986-94, but when control of Congress shifted into Republican hands and market prices temporarily soared, the FAIR act pushed U.S. policy further than ever toward a cash out.

The basic features of the FAIR Act are well known: decoupling of payments to export-crop producers from market prices and planting decisions; planting flexibility; elimination of annual acreage-reduction programs (ARPs); and capping of loan rates at what seemed to be low levels at the time. These policy moves were accompanied by others that achieved less reform: higher decoupled payments to farmers in the short run than payments would have been under previously enacted law, continuation of export subsidy authority, and renewal of protective programs for dairy, sugar, and peanuts. The FAIR Act did not put farm policy on a new strategic path to reform. It did not achieve a cutout because permanent legislation was left in place and generous cash subsidies were provided to farmers in the short run. Just as clearly, Congress was not advancing a squeeze out with the FAIR Act because the generous new decoupled payments it authorized were much greater than expected and were so unencumbered by regulations and easy for farmers to obtain that voluntary program participation increased. The FAIR Act also failed to ensure a buyout of farm programs because it maintained permanent legislation and a baseline for continued farm-program spending.

The reform path Congress took in the 1996 FAIR Act (another heavily compensated cash out of farm programs) shares all of the benefits and risks that the cash-out strategy has historically entailed. Under the FAIR Act, a market-oriented cash out was advanced for export crops, with payments more fully decoupled than ever before—but at high budget cost. The farm subsidy budget baseline was extended, and provision of additional subsidies (as occurred in 1998, 1999, and 2000) was never precluded. As long as the farm lobby remains powerful, and the agriculture committees in Congress retain control of the policy process, there are ready venues under the FAIR Act either for traditional program defense or for the design of new support programs to benefit a new generation of organized farm-lobby constituents.

4. EXPLAINING THE FAIR ACT OUTCOMES

Prior to the 1996 FAIR Act, most policy analysts neglected the influence of party control in Congress on the content of U.S. farm policy. Gardner's (1995) account of the politics of U.S. agriculture placed emphasis on changing structures in Congress which had diminished the role of political parties and had weakened the influence of party leadership. Likewise, Browne (1988) acknowledged an early history of party influence in U.S. farm policy debates, but concluded that the influence of parties within the agricultural committees of Congress had long since been replaced by the influence of bipartisan coalition building, based largely on issue expertise commodity by commodity.

As suggested by the policy characterization in Table 2.2, Browne was correct to recognize the importance for farm policy of party control in Congress. Republicans resisted farm-program enactment in the 1920s, before Democratic ascendancy helped bring the Agricultural Adjustment Act of 1933. After the Second World War, Republican capture of Congress in the mid-term elections of 1946 helped facilitate passage of Title II of the 1948 Agricultural Act, a measure which—if it had survived—was designed to put post-war U.S. farm policy on a market-oriented track by lowering support prices. The Democratic recapture of control of Congress in the November 1948 election led, by late 1949, to an extension of high price supports fixed at 90 percent of parity. When the Republicans regained control of Congress in the 1952 election, they successfully replaced these high price supports with a lower minimum-parity level. In the 1954 mid-term elections, the Democratic Party recaptured Congress once again, and farm programs reverted to heavier dependence on supply controls. Thus, during this important early period, the fluctuating fortunes of market-oriented reforms corresponded closely to changes of party control in Congress.

After 1954, the Democrats' control of the House of Representatives was maintained for the next four decades, from January 1955 through December 1994. Because it became fixed during this period, the latent importance of party control in Congress was neglected.

Forty years later, in the election campaign of 1994, the Republican House candidates used their unified support for the Contract with America to drive home where they differed as a party from the Democrats. Party differences on agricultural policy were not presented in this 1994 House Republican manifesto. Whether there were such differences and, if so, whether the new Republican majority would make them an issue in 1995-96 was open to question.

An argument can be made along the lines of those by Gardner (1995) and Browne (1988) that party does not matter anymore for farm policy. Inside the agriculture committees of Congress, whenever narrow commodity issues are at stake, party differences tend to be trumped by regional, or commodity, loyalties. The durable strength of regional- and commodity-based preferences, separate from party label, is vividly illustrated by the changed party control in the south. The deep south is now a more solid region for Republicans than for Democrats, and a key part of their party power base. Yet, House and Senate agriculture committee endorsement for highly protective policies for southern commodities has scarcely been altered by the shift in party alliance. What matters most about the south for agricultural policy is that it still produces cotton, rice, sugar, peanuts, and tobacco, commodities for which there is a long history of market intervention.

The apparent power of commodity and regional loyalties over party differences also reflects successful recruitment and cultivation of members of Congress by narrow commodity-based political action committees (PACs). Across all issues (not just agriculture), members of the House of Representatives tend to rely heavily on PACs for campaign funding; those who win elections receive on average more than 40 percent of their campaign funds from PACs. Agricultural PACs gave a total of U.S. $15.5 million to members of Congress in 1993-94, making them the nation's third largest source of PAC money overall.

Having conceded this much, party control still matters for farm policy. In debates among producer groups, party loyalties are easily trumped, but for farm program legislation to get through Congress, a broader coalition must be formed. Farm commodity PACs and lobby groups alone lack the political resources to deliver floor majorities in the House and Senate. Commodity-based agricultural interests are salient only to a small fraction of the members of Congress, well short of a minimum winning floor coalition. To be enacted, farm bills must be attractive, or at least acceptable, to a mix of constituencies with broader interests. These supporters are drawn from various constituencies, among them are farm organizations that are not commodity based, agribusiness, environmentalists, food consumers, or advocates for the rural and urban poor. When this larger set of non-commodity farm-bill constituencies is surveyed, policy preferences divide sharply along party lines.

A measure of such divergent party leanings between Congressional Republicans and Democrats can be derived by comparing the favorability ratings assigned to key members of Congress by a sample of constituency-linked advocacy groups. The favorability ratings (from 0 to 100) made by six different groups—the American Farm Bureau Federation (AFBF), the National Farmers' Union (NFU), the League of Conservation Voters (LCV), the Libertarian Party (LP), the National Taxpayers' Union (NTU), and Bread for the World (BFW)—are shown in Table 2.3. For each of these six lobby groups, the ratings given to Republicans tend to differ sharply from ratings given to Democrats; the Republican committee chairs who took over in the 104th Congress were viewed as ideologically distinct from the Democratic chairs they replaced along every dimension. Key farm-bill constituencies perceive substantial differences between Republicans and Democrats.

4.1 Republican Congress Ends Annual Supply Controls

The fact that there are significant differences between Republicans and Democrats on farm-bill issues does not mean one party is more favorably inclined toward farmers than the other. Both parties have traditionally advocated helping farmers, and, in the end, both parties tend to support generous farm bills. The capture of Congress by Republicans in November 1994 did not mean farm programs were soon going to disappear. But the Republican-controlled 104th Congress was more inclined than any Democratic Congress

Table 2.3 Advocacy Group Ratings of House and Senate Leaders in the 101st, 103rd, and 104th Congress

Group	*101st Congress 1989-90 (Democratic majority)*		*103rd Congress 1993-94 (Democratic majority)*		*104th Congress 1995-96[a] (Republican majority)*	
	House	*Senate*	*House*	*Senate*	*House*	*Senate*
	average ratings for all committee chairs					
American Farm Bureau Federation	28.7	44.3	21.6	41.6	89.5	92.0
National Farmers Union	79.4	78.3	73.3	85.7	29.2	30.8
League of Conservation Voters	68.5	58.5	73.3	72.1	12.4	16.3
Libertarian Party	11.6	—	9.6	9.4	64.8	74.0
National Taxpayers Union	18.0	28.7	18.5	20.8	82.5	83.6
Bread for the World	74.9	82.7	62.6	79.2	12.9	18.6

[a]Rankings based on 1995 only.
Source: Orden, Paarlberg and Roe (1999: 180).

had been toward full production agriculture and agribusiness interests. The new Republican agricultural leadership set as one of its key objectives the elimination of unpaid annual land diversions under ARPs.

In the early stage of the farm-bill debate, another set of preferences brought to Congress by the Republican leadership first had to be addressed. This was the Republican preference (clearly visible in the NTU ratings) for greater federal budget discipline. In this contest, the new House agriculture committee chairman, Pat Roberts of Kansas, used the fact that 33 of the 76 freshman Republicans were dependent on agriculture in their districts to convince the House leadership not to cut farm program budgets too severely. His "posse" of Republican agriculturalists who opposed cutting farm-program spending emerged as the winners in the farm-program budget battle.

Once the Republican agriculturalists had contained the threat of excessive budget cuts, they were faced with the next step in the process, which was building a coalition among their majority on the agriculture committee for specific policies affordable within the budget space they had worked so hard to win. Over most of the summer of 1995, this presented itself as a nearly insurmountable political problem. Because of near unanimous Democratic opposition to the Republican budget agenda, the Republicans knew they needed to craft a farm bill that could gain the support of almost every party member on the agriculture committee. A consensus would have to be reached among old-style agriculturalists and new-style budget hawks, and would have to include not only midwestern grain constituents, who wanted ARP elimination and greater planting flexibility, but also the southern crop advocates on the committee who were generally opposed to any radical policy change.

4.2 High Commodity Prices Induce Decoupled Subsidy Payments

Without steadily rising commodity prices in late 1995, Congress never would have been persuaded to adopt the decoupled payments approach eventually included in the 1996 FAIR Act. Without high prices, in fact, this approach likely never would have been proposed in the first place. Full decoupling of support payments from market prices and farmers' planting decisions had not been considered seriously by Congress when it rewrote farm policy in 1990, and it remained essentially off the table when the 1995 farm-bill debate began. New policy ideas were being generated just before the 1995 debate, but the new concept attracting the greatest attention was the revenue assurance idea pushed by a coalition of Iowa corn growers. This Iowa Plan represented a sharp break from the traditional method of sending payments to farmers, but it was not decoupling. A different market-oriented proposal prepared for the Democratic Leadership Council endorsed a budgetary cap, yet explicitly rejected decoupling as an option, arguing instead for safety-net payments that would be made only when farm revenues fell below levels set by new federal standards (Runge, Schnittker, and Penny, 1995). Fully decoupled payments (payments to farmers for being farmers) were also rejected as a policy disaster at a conference of the market-oriented American Enterprise Institute (Wright, 1995).

Early efforts by the Republican agricultural leadership to sell the radical idea of fully decoupled payments were unsuccessful. The initial argument revolved too much around a speculative and exaggerated warning of how vulnerable farm spending might be to budget cuts in the future, unless the monies negotiated into the Republicans' 1995 budget resolution were locked in through fixed-payment contracts, which would turn an intended budget ceiling into a budget floor. This argument for using decoupling to protect agriculture from future budget cuts was not enough to persuade cotton and rice interests in the south, who remained firmly attached to payments tied to existing base acreages. It was not until crop prices rose sharply in late 1995 that the idea of decoupling became sufficiently attractive, first to southern Republicans in Congress and eventually even to a sufficient number of Democrats, to ensure its final passage. By November 1995, market prices were well above the target-price levels set by the 1990 farm bill. At this point it became clear that decoupling would not just protect farm spending from future budget cuts, it would also bring an immediate increase in cash payments to farmers. Once decoupling could be advocated as a way to increase spending on farmers in the short run, it received a more positive response. In just six months, a farm-bill debate that had begun by making no reference at all to decoupled payments was able to coalesce around the desirability of this radical new choice. An unacceptable option had been transformed into an irresistible choice by rising market prices.

4.3 Alternative Explanations for FAIR Act Outcomes

Are there convincing explanations for the FAIR Act outcomes other than the change in party control of Congress in 1994 and unexpected high-market prices the following year? One possibility, the implementation of the final 1994 Uruguay Round Agreement on Agriculture, does not provide a convincing explanation for U.S. farm policy change in 1995-96, since the terms of the international agreement were carefully written to require no support-level reductions beyond those already undertaken in the 1985 and 1990 farm bills. A reassurance that the Uruguay Round would require no immediate changes in U.S. farm policy was also built into the implementing legislation negotiations in late 1994 between the congressional farm lobby and the Clinton administration.

Nor can the unexpected shift toward decoupled payments under the FAIR Act be attributed to the structural modernization that has occurred in the agricultural sector. The modernization of agricultural production since the 1950s has continued to improve U.S. competitiveness. This systemic change helps explain the progressive movement toward a cash out of farm programs for export crops since the mid 1960s, and the impetus to eliminate the remaining annual acreage set asides in 1995-96, a move which enjoyed wide (though not universal) support among farm groups. However, the structural transformation that had occurred in agriculture did not lead to widespread calls for decoupling when Congress debated the 1995 farm-bill.

Perhaps the FAIR Act resulted from a long-term systemic change in the demography of the farm sector in a different sense: the slow erosion of farmers' political clout in Washington. The 1996 FAIR Act suggests little in the way of such a diminished political clout for farm organizations. For export crops, the most important policy changes—the end to unpaid ARP land diversions and decoupling of support payments—provided more taxpayer money to farmers, not less. For import-competing crops (sugar and peanuts) farm organizations continued to win head-on battles against well-funded and widely publicized reform initiatives, and dairy policy changes in the FAIR Act added protection for high-cost producers in New England by imposing higher market prices on consumers.

Federal budget pressures are also unconvincing as an explanation for the key reforms in the FAIR Act. When the farm-bill debate began, Republicans in Congress thought they had won a mandate to cut government spending sharply. They learned the limits of that mandate as the year wore on. Actual budget disciplines proved relatively weak in the end, but the budget rules used by Congress were influential to the final FAIR Act outcome. If there had not been a pay-as-you-go rule (which put a high budget cost on the option of ARP elimination, assuming traditional deficiency payments), and if there had not been an outdated Congressional Budget Office (CBO) budget baseline projection (making it possible to capture higher spending with decoupled payments, once market prices started to go up), the move toward decoupling in 1995 would have been less compelling.

A final alternative explanation for the FAIR Act is one based on the power of ideas. Perhaps the shift made in the 1996 FAIR Act toward ARP elimination and decoupled payments was more than a product of clashing interests; perhaps it represented a victory for market-oriented ideas, a suggestion made, for example, by Coleman, Skogstad, and Atkinson (1997) and Tweeten and Zulauf (1997).

Several features of the 1995-96 farm-bill debate suggest that new ideas about the role of government in agricultural policy were not being embraced. First, traditional use of supply controls for export crops was not fully abandoned, as paid land diversion (of nearly 15 percent of planted acreage) was continued under the Conservation Reserve Program (CRP). Second, if the idea of market orientation had been driving the outcome, reform would have been felt by all crops, not just by the deficiency-payment crops, where the budget baseline-capture option existed. Moreover, if third-order policy changes are unlikely to occur without some kind of society-wide contestation and engagement, as Hall (1993) argues, then there is little to suggest such a fundamental change in the 1990s. There was actually far more social contestation and engagement over agricultural policy at the peak of the inflationary commodity boom of the 1970s (when labor and consumer interests were mobilized against high food prices), or at the depths of the farm crisis of the 1980s (when populist farm interests were mobilized against high debts and low crop prices) than there was during the 1995-96 farm-bill debate.

Had the 1994 Republican election mandate been somewhat stronger, and had a corresponding third-order change taken place in U.S. farm policy, one can imagine what it might have looked like from the policy ledger briefly touted in 1995 by House agriculture committee chairman Roberts. This ledger surfaced at a time when ideas of budget discipline and free-market fundamentalism did seem to reign, just after the Republican congressional election victory. The policy ledger, which Roberts offered as common ground on which both Republican agriculturalists and the new House leadership could stand, promised to lower federal-support expenditures and to reduce market interventions. Offsetting this, farmers were to benefit from greater planting flexibility and elimination of ARPs; from deficit-reductions and lower taxes that the Republican Party would champion; from specific tax breaks; from lessening of burdensome regulations; and from strengthening of private property rights. If the market-oriented, small government Republican revolution (as the leaders, themselves, called it) had not been so easily turned back (by Roberts and his supporters for agriculture, among others), the ledger might have emerged as the essential outline of a new American farm policy: a loss of traditional producer supports compensated by tax breaks and deregulation. This would have marked a sharp break from the fundamental assumptions of the New Deal regarding both the role of government and the vulnerability of the farm sector to market forces. Farmers would no longer be protected from the dangers of the market by a benevolent government; they would be protected not from the market but from the tax collectors and regulators of the government itself, and they would be protected not as farmers but as citizens with property rights. The farm bill that passed in 1996 never quite incorporated this bold reformulation of basic assumptions.

5. POST-FAIR ACT REFORM SETBACKS

At the time the FAIR Act became law in 1996, buoyant agricultural forecasts suggested that a number of circumstances would continue to generate high crop prices for some years into the future. The market boom in 1995-96 reduced worldwide levels of grain stock. A tapering off of global yield trends, world population growth, and income growth in China, India, and other populous developing countries were all cited as pointing toward continued high prices. The USDA and other forecasters were projecting much stronger price trends by 1997 and early 1998 than they had at the outset of the farm-bill debate in February 1995, but the projections for strong farm prices soon proved erroneous.

Export-crop prices had already fallen sharply from their spring peak by late 1996, and they moved persistently downward again in the spring and summer of 1998. Commodity prices were driven down by a recovery of U.S. production from low 1996 levels, by worldwide supply responses induced from the high prices of the previous years, by appreciation of the U.S. dollar of nearly 25 percent between 1995 and 1998, and by slowed growth and recession in Asia. With prices falling from strong forecast levels in 1997-98, continuation of the traditional deficiency-payment programs, not decoupling, would have created an opportunity to capture fiscal outlays for farm programs. Had deficiency payments remained in effect, falling prices also would have undermined arguments to abandon annual supply-control authority. The most important reforms in the FAIR Act simply would not have been enacted had the farm-bill debate taken place in 1997-98 instead of 1995-96.

With the 1996 FAIR Act already in place, the importance of party-based farm policy differences was again in full evidence as market prices fell in 1998. During the summer, congressional Democrats called for a virtual rewriting of farm policy to restore the safety net, particularly through higher loan rates and off-budget emergency assistance to farmers facing crop or income losses. Republicans argued for staying the course with the FAIR Act and offered help in the form of measures to boost foreign demand for U.S. farm products plus tax relief targeted toward farmers.

As prices continued to move down through the autumn of 1998, an election year bidding war ensued between Democratic and Republican agriculturalists over the granting of new benefits to farmers, and in this bidding war the budget discipline features of the FAIR Act were ignored. Congressional

Democrats failed to win endorsement for higher loan rates from the Clinton administration, but the president supported emergency crop disaster relief expenditures of U.S. $500 million (not subject to pay-as-you-go budget rules). He also instructed the secretary of agriculture to use existing Commodity Credit Corporation (CCC) authority to buy U.S. $250 million of wheat for food-aid donations. Republicans countered with calls for even higher emergency disaster relief and a speed-up of scheduled tax breaks. Bipartisan support arose for various indirect aids to farmers: enlarged use of export credit guarantees, exemption of farm products from economic embargoes imposed after India and Pakistan carried out underground nuclear tests, and an allowance permitting farmers to take one-half of their scheduled 1999 decoupled payments in advance. Then, with mid-term elections approaching and prices still low, the bidding heightened. Senate Minority Leader Thomas Daschle continued to press for higher loan rates—despite being rebuffed three times by floor votes. Removing the FAIR Act loan-rate caps would have raised the price support levels (per bushel) from U.S. $1.89 to U.S. $2.19 for corn, and from U.S. $2.58 to U.S. $3.18 for wheat, at a projected cost for a single year of between U.S. $4 billion and U.S. $8 billion. Republicans, in their own way, sought to be equally generous; they unveiled an alternative plan for special market-loss payments to supplement the income support provided under the FAIR Act. When the Republican Congress sent the president an agricultural appropriations bill authorizing U.S. $3.9 billion in disaster relief and new cash payments, he vetoed it as not generous enough. Only when the new aid was pushed up to nearly U.S. $7 billion were the Democrats willing to abandon their effort to raise loan rates and agree to the farm-spending package.[ii]

Thus with low prices in 1998 and with Republicans holding a slim 22-seat majority in the House of Representatives, Congress and the administration delivered substantial additional support to farmers in various ways, enough to keep net cash-farm income near its 1996 level, despite low prices. With Republicans still in the congressional majority, the structural components of the FAIR Act reforms were sustained: farmers retained planting flexibility, annual supply controls were not re-imposed (indeed planted acreage remained near its 1997-98 peak), the loan-rate caps held, and cash payments remained nominally decoupled from market prices. But the budget discipline once touted by Republican leaders as a virtue of the fixed payments adopted under the FAIR Act proved illusory; and the increased decoupled payments and loan deficiency payments made as market prices fell were a *defacto* abandonment of full decoupling. This situation worsened in 1999 and 2000 when the originally scheduled FAIR-Act decoupled payments were doubled and loan deficiency payments rose to more than U.S. $5 billion annually.

6. PROSPECTS FOR FURTHER REFORM

This chapter has described cash-out, buyout, squeeze-out, and cutout strategies to achieve farm-policy reform. The multi-decade history of policy reform efforts confirms some of the reasons traditionally given to explain why the reform process is so difficult and clarifies the actual promise of various hypothetical reform paths. The political power of farm organizations persists despite declining numbers of farmers, confirming the expectations of various public-choice theorists of policy change. The political power of farm organizations also persists consistent with those that have stressed the importance of legislative logrolling coalitions, including coalitions internal to the farm sector and also coalitions with interests from beyond the sector. The expectation of political theorists that the abrupt cutout approach is likely to fail in the face of strong resistance from defenders of entrenched programs is confirmed by the demise of various unilateral cutout proposals. Explicit options to buy out farm program entitlements have seldom been proposed, and have likewise fallen short. The 1996 FAIR Act failed to qualify as a clear buyout of traditional farm-support programs, even for export crops, because a budget baseline remained in place indefinitely, and because permanent legislation was never repealed.

The squeeze-out strategy has also been historically quite difficult to follow. The original squeeze-out attempt launched by the Eisenhower administration in the 1950s ran afoul of a Democratic Congress and was thwarted as well by high productivity growth in the sector. A different kind of squeeze-out opportunity emerged during the inflationary period of the 1970s; this opportunity for a squeeze out was lost when Congress ratcheted nominal support prices upward. Some progress toward an eventual squeeze out, via inflation, was finally made when the 1985 and 1990 farm bills imposed a freeze on nominal target-price levels, but the 1996 FAIR Act was not a squeeze out. It resulted in more program spending and higher program participation rates than would have been the case under a simple extension of the 1990 law.

The strategic option that has served reform best, so far, has been the cash out. Moves toward a cash out are always precarious reform victories. Cashed out farm programs remain costly to taxpayers and leave the traditional agricultural policy process establishment in place, ever ready to seek expanded benefits whenever circumstances offer the political opportunity, and all the while searching for new programs to deliver and then to manage. Market conditions and party control of Congress have been shown to be two key determinants of cash-out reform progress, and the 1996 FAIR Act was a substantial move down the cash-out path. The unexpected reform outcomes in 1996 were reflections not only of momentarily higher crop prices and the

CBO budget baseline capture option but just as much of how the first Republican Congress in four decades decided to exercise that option. The partisan wrangling and subsequent reform setbacks that followed in 1998-2000, when prices fell and the Republican majority in Congress shrank, demonstrate again that there is no end to the policy-reform problem in American agriculture as long as only a cash out is underway.

Despite reform setbacks, partial steps toward supporting farm income with cash payments (eventually decoupled cash payments) rather than with market interventions are better than no reform steps at all. Once income support entitlements have been cashed out and decoupled, under propitious circumstances perhaps they can be subjected to a second-stage squeeze out. Under a successful post-FAIR-Act reform scenario, the market-oriented features of the 1996 bill will be retained, and subsequent legislation could even establish a schedule for ending the remaining cash payments to export-crop producers. The import-competing commodities that escaped reform in 1996 could also be obliged to face market disciplines. Under this reform scenario, the 1996 FAIR Act would deserve to be remembered as a critical point of change. The generous FAIR Act payments that were made in 1996 and 1997 under the FAIR Act and the subsequent payments made in 1998-2000 under emergency approved legislation would, in retrospect, be components of a successful buyout of the U.S. farm policies that were established during the New Deal. At this juncture, with political feedback from the operation of the FAIR Act weakening the traditional agricultural process establishment, the idea of trusting markets would emerge as more clearly dominant. The farm economy would come to be seen as not much different from the rest of the U.S. economy, and for this reason no longer in need of its distinctive support programs.

The evolution of farm-policy reform from a cash out to a squeeze out is more improbable with market prices low, after the reform setbacks that have been enacted since 1998, and with control of Congress more uncertain than it was when the FAIR Act became law under the new Republican majority's rhetoric of being the first step in a transition to less direct intervention or support. Still, several new factors may work in favor of a squeeze out under the 1996 FAIR Act, at least for export crops.

First, the caps on per-unit commodity loan rates imposed under the FAIR Act prevented those rates from moving up in response to the 1995-96 market boom. A political battle had to be fought to retain the loan-rate caps in 1998 (and the rates fixed in 1996 proved too high), but reform effort was not needed just to bring rates back down that increased automatically, as they had in the 1980s. Second, with ARPs no longer in use as a policy instrument, it is no longer necessary to make program benefits sufficiently attractive to large commercial farmers to ensure their participation. This opens a potentially interesting option of reducing program costs and squeezing out program participants through a targeting or means testing of benefits. Farm groups have successfully resisted, or found legal grounds to avoid, most benefit targeting and means-testing efforts in the past and, no doubt, they will continue to resist any such restrictions on farm-program eligibility. But the argument that targeting could drive the largest farmers out of participating in annual supply-control programs can no longer be used as a cover for this resistance. The transparency of decoupled cashed-out income supports to farmers may make them more susceptible to a political squeeze out over the long run.

For import-competing commodities, simply getting to a cash out will be the challenge. A cash out of these programs might not seem an attractive policy option, given the implication of larger rather than smaller budget outlays in the short run, yet it may be the essential prelude to a squeeze out and more feasible than either a cutout or a costly buyout. Cash outs are half measures but they give markets more room to operate; they are politically palatable; and they can always lead to more complete squeeze outs, buyouts, or cutouts later on. Perhaps the emergence of surplus budget circumstances at the turn of the 21st century, which might pose a risk to further reform progress, can, instead, be turned to an advantage for a legitimate policy-reform objective. For sugar, pressure is arising from an international agreement that liberalizes agricultural trade regionally (the clock is ticking on Mexican access to the U.S. sugar market under NAFTA), and by 2000 the domestic support program was unable to maintain market prices above legislated loan rates without purchase of nearly 1 million tons of sugar by the CCC. Under this pressure, or with a dramatic political opening to Cuba, perhaps a partial cash out of the U.S. sugar or peanut program will not be an impossible option for reformers to imagine in the years ahead (Orden, 2000). Even further down the road, the next step might be to decouple this cash out for import-competing crops just as the 1996 FAIR Act decoupled cash payments for export crops. This step would encourage producers of these crops to undertake more rapid adjustments to international market realities, perhaps creating economic and political space for an eventual squeeze out.

7. ENDNOTES

[i] The historical record and some of the arguments described herein are laid out in greater detail in Orden, Paarlberg, and Roe (1999).

[ii] The 1998 farm spending package included U.S. $3.1 billion in one-time payments to offset low prices, U.S. $2.4 billion to compensate farmers for crop disaster losses, and U.S. $1 billion in accelerated tax relief. In addition, direct payments of more than U.S. $2 billion resulted in market price levels falling below the FAIR Act loan rates.

8. REFERENCES

Bonnen, J.T. and W.P. Browne. 1989. "Why Is Agricultural Policy so Difficult to Reform?" in Carol S. Kramer, ed., pp. 7–33. *The Political Economy of U.S. Agriculture: Challenges for the 1990s.* Washington, DC: National Center for Food and Agricultural Policy.

Browne, W.P. 1988. *Private Interests, Public Policy, and American Agriculture.* Lawrence, KS: University Press of Kansas.

Coleman, W., D. Skogstad, and M.M. Atkinson. 1997. "Paradigm Shifts and Policy Networks: Cumulative Change in Agriculture." *Journal of Public Policy* 16: 273–301.

Cochrane, W.W. 1958. *Farm Prices, Myth, and Reality.* Minneapolis, MN: University of Minnesota Press.

Dixit, A., G.M. Grossman, and E. Helpman. 1997. "Common Agency and Coordination: General Theory and Application to Government Policymaking." *Journal of Political Economy* 105: 752–769.

Gardner, B.D. 1995. *Plowing Ground in Washington: The Political Economy of U.S. Agriculture.* San Francisco, CA: Pacific Research Institute for Public Policy.

Grossman, G.M. and E. Helpman. 1995. "Trade Wars and Trade Talks." *Journal of Political Economy* 103: 675–708.

_____. 1994. "Protection for Sale." *American Economic Review* 84: 675–708.

Hall, P.A. 1993. "Policy Paradigms, Social Learning and the State: The Case of Economic Policy Making in Britain." *Comparative Politics* 25: 275–297.

Orden, D. 2000. "It is Time Domestic Sugar Policy Reform." Testimony presented to the United States Committee on Agriculture, Nutrition, and Forestry, Washington, DC (26 July).

Orden, D., R. Paarlberg, and T. Roe. 1999. *Policy Reform in American Agriculture: Analysis and Prognosis.* Chicago, IL: University of Chicago Press.

Pierson, P. 1994. *Dismantling the Welfare State? Reagan, Thatcher, and the Politics of Retrenchment.* Cambridge, MA: Cambridge University Press.

Runge, C.F., J.A. Schnittker, and T.J. Penny. 1995. "Ending Agricultural Entitlements: How to Fix Farm Policy." Paper prepared for the Progressive Policy Institute, Washington, DC.

Schleich, J. 1999. "Environmental Quality with Endogenous Domestic and Trade Policies." *European Journal of Political Economy* 15: 53–71.

Tweeten, L. and C. Zulauf. 1997. "Public Policy for Agriculture after Commodity Programs." *Review of Agricultural Economics* 19: 263–280.

USDA (U.S. Department of Agriculture). Various years. Internet website: www.usda.gov

Wright, B.D. 1995. "Goals and Realities for Farm Policy," in D.A. Sumner, ed., pp. 9–44 *Agricultural Policy Reform in the United States.* Washington, DC: AEI Press.

Chapter 3

Trade, Uncertainty, and New Farm Programs

Luther Tweeten
The Ohio State University

1. INTRODUCTION

I have contended for some years that the principal economic problem facing commercial agriculture is instability (Tweeten, 1989). It threatens not only farmers' financial viability but also consumers' food security, given consumer desire for stable food availability in the face of intermittently unstable food production.

Some observers expect instability to become a more pressing food-security problem because of increasing liberalization in global farm policy (Johnston and Schertz, 1998; for an alternative view see Collins and Glauber, 1998). Trade policy and farm-commodity policy are inextricably linked. This chapter recognizes that both types of policies are causes and cures for economic instability.

The chapter begins with an examination of the contribution of two key variables (yields and exports) to economic volatility in agriculture that could influence the need for commodity program-stabilization policies. Trends in annual yield volatility give clues as to whether or not changing technology is affecting food-system stability over time. The second section of the chapter quantifies sources of food-system variability in the United States. Emphasis here is on the role of exports as potentially either a stabilizer or a destabilizer of that system.

Finally, the chapter expands in scope to examine instability and public policy considerations within the new paradigm for American agriculture that I have outlined in another paper with Carl Zulauf (Tweeten and Zulauf, 1997). The concluding section raises serious concerns over consequences of political pressure for higher commodity loan rates and for crop and revenue-insurance programs featuring larger public income transfers from taxpayers to farmers.

2. ARE YIELDS BECOMING MORE UNSTABLE?

The major source of annual food production volatility is yields. Nations may be less willing to depend on markets to cope with risk if yield volatility is increasing due to high-yielding varieties, monoculture relying upon a narrow range of genetic stock, extension of crops to marginal lands, global warming, or El Niño and La Niña cycles. The hypothesis that high-yielding varieties display relatively greater yield variability was rejected in an earlier study by Hazell (1985). Later, he concludes that yields have become more variable and more correlated among regions (Hazell, 1993). Correlated yields arose from common genetic stock in high-yielding varieties, from similar wide-spread management practices used to improve yields, and from production narrowed to a few crops.

On the other hand, conventional and bio-engineered breeding has produced drought-, pest-, disease-, and frost-resistant varieties expected to help stabilize crop yields. Whether or not forces for stabilization offset forces for volatility is an empirical question.

Figures 3.1 and 3.2 provide clues. U.S. corn yields have been more volatile since 1980 than they were in the 1950s and 1960s as measured by standard deviations (in percent) from five-year centered average yields (Figure 3.1a). Standard deviations since 1980, however, have been similar to those of decades prior to the 1950s. Thus, the stability of yields in the 1950s and 1960s, rather than the volatility of yields in the 1990s appears to be exceptional.

World coarse-grain yields have been available only since 1961, but show no patterns of increasing variation in recent years (Figure 3.1b). Major deviations occurred in 1983 and 1988 mainly due to drought in the United States (the major global coarse-grain producer).

U.S. wheat yield variation has been less since 1980 than in previous decades with the exception of the 1960s and 1970s (Figure 3.2a). At the global level, wheat yield variation seems to be declining, with the decline most pronounced relative to the 1970s (Figure 3.2b).

FAO data indicate that annual global aggregate food production is remarkably stable, with a coefficient of variation from trend of approximately one percent, a number that has not varied much (FAOSTAT, various years). Thus, opening trade to extend global stability to individual countries can go

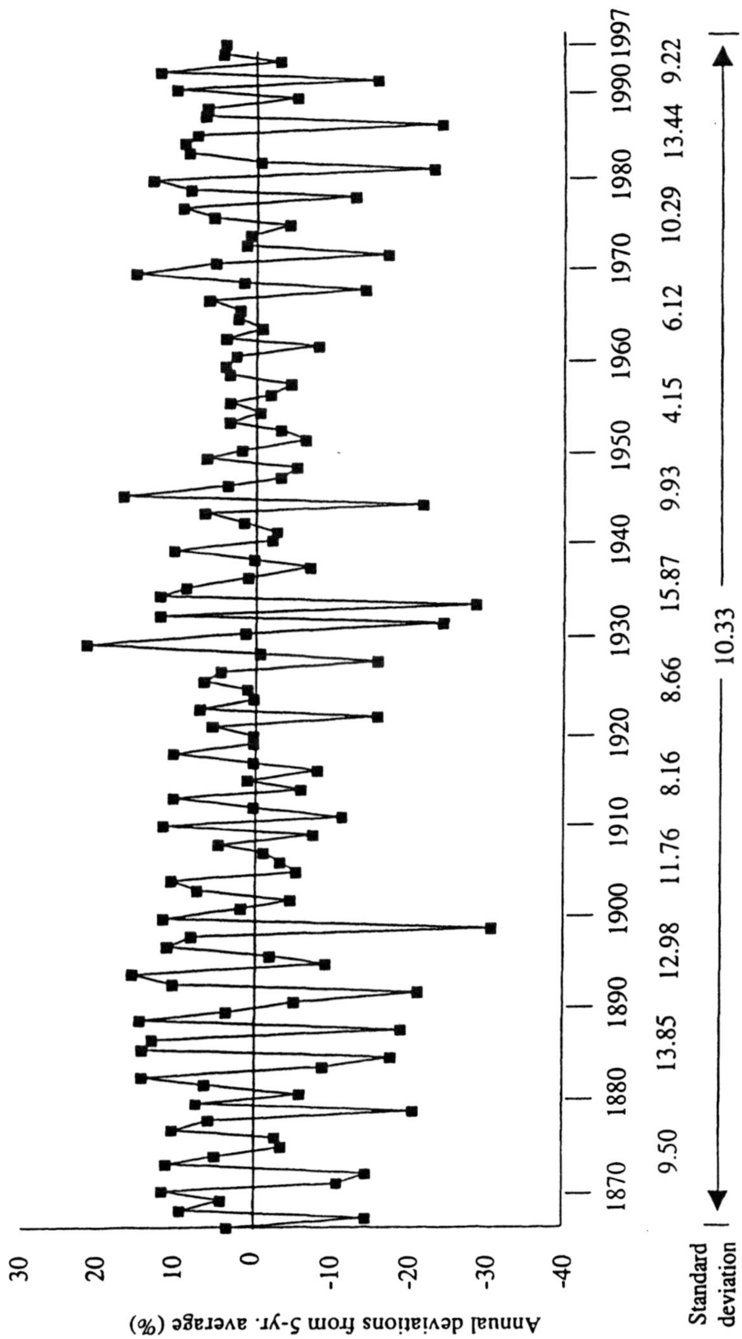

Source: FAOSTAT

Figure 3.1a U.S. Corn Yield Deviations from Five-year Centered Averages from 1868 to 1997

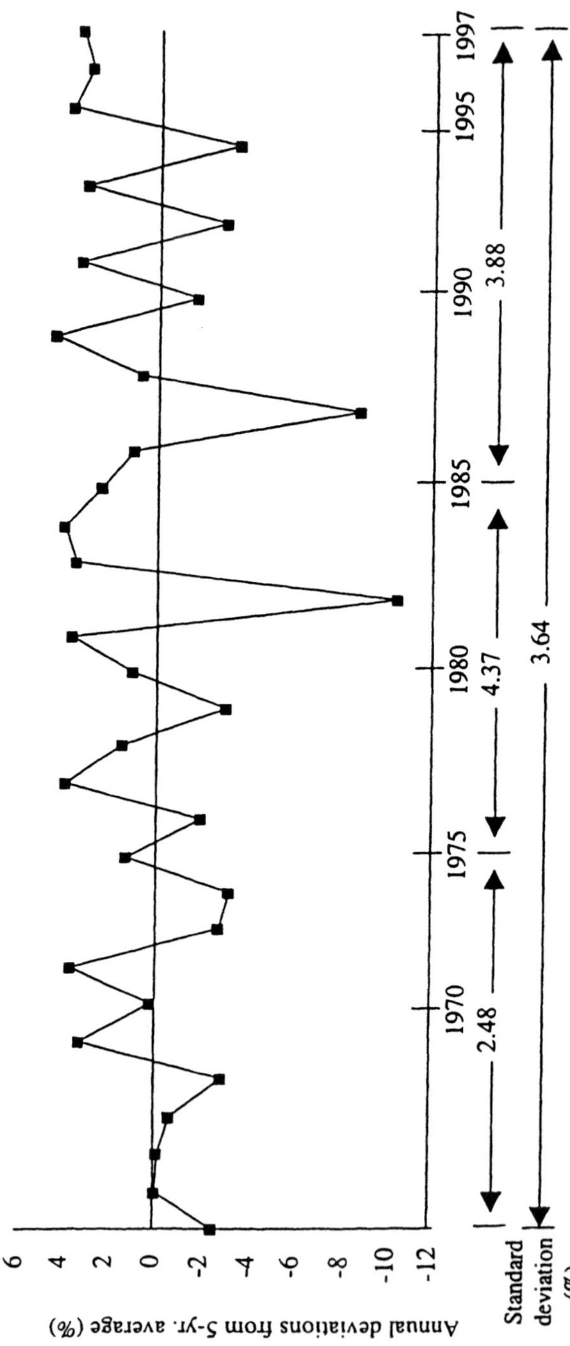

Source: FAOSTAT

Figure 3.1b World Coarse Grain Yield Deviations from Five-year Centered Averages from 1961 to 1997

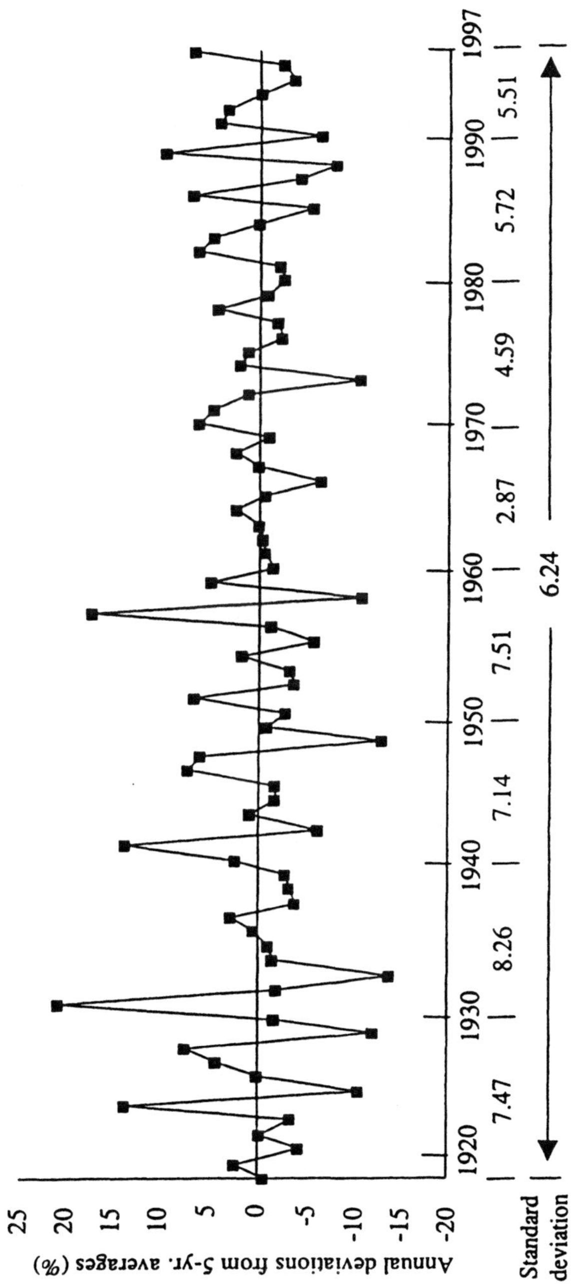

Source: FAOSTAT

Figure 3.2a U.S. Wheat Yield Deviations from Five-year Centered Averages Less from 1919-97

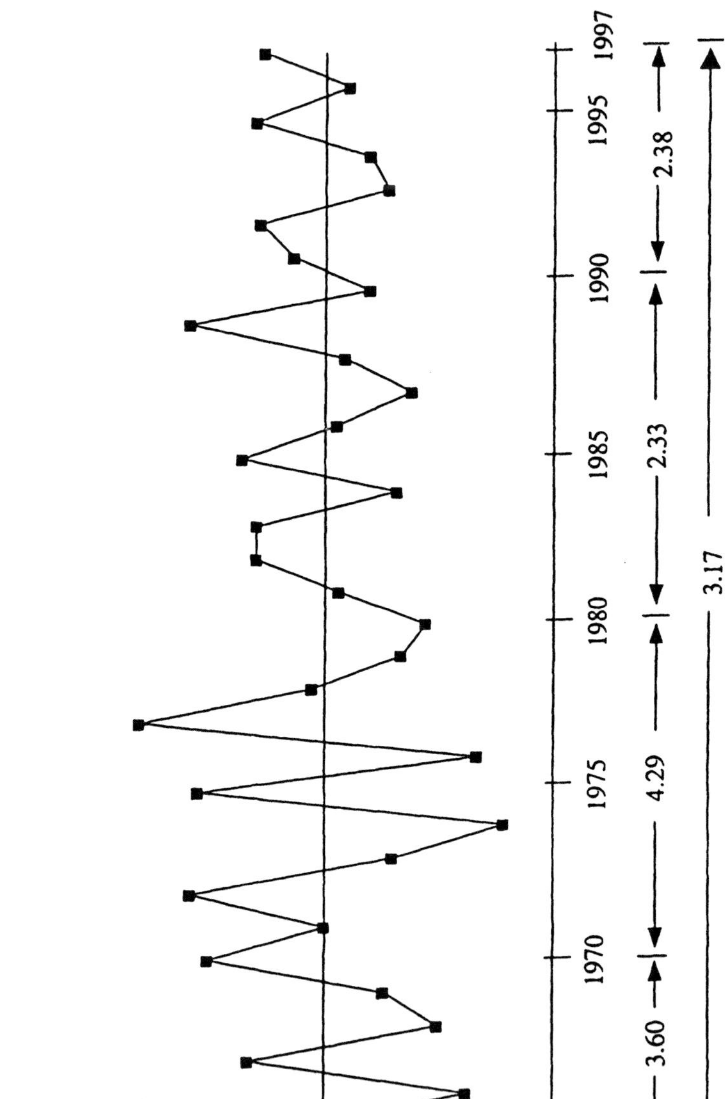

Source: FAOSTAT

Figure 3.2b World Wheat Yield Deviations from Five-year Centered Averages from 1965-97

far to mediate the need for stable food availability in every country in the face of unstable annual food production. More open trade may compensate, in part, for falling global commodity stocks in recent years, especially because of reforms in U.S. and E.U. agricultural policies.[i] Many view open trade as economically destabilizing and a threat to food security. This issue is addressed for the United States in the next section.

3. FARM EXPORTS AND U.S. AGRICULTURAL INSTABILITY

Expected commodity stock carryout may be the best single predictor of commodity price, hence variation in stocks explains variation in commodity prices. Sources of requirements for buffer stocks indicate origins of instability in agricultural markets. Of particular interest is the contribution of exports to market instability.

The contributions of farm production, O, domestic utilization, D, and net U.S. farm exports, X, to instability are quantified for the nation's total agriculture and for feed grain, soybean, and wheat markets (Tables 3.1 and 3.2). Changes in buffer stocks, R, of commodity i are defined as (Tweeten, 1994a):

$$(1)\quad R_i = O_i - D_i - X_i.$$

Estimated variance in buffer stocks $S^2(R_i)$ is thus

$$(2)\quad S^2(R_i) = S^2(O_i) + S^2(D_i) + S^2(X_i) - 2S(O_i{\cdot}D_i) - 2S(O_i{\cdot}X_i) + 2S(D_i{\cdot}X_i).$$

A high $S^2(R_i)$ indicates an unstable system requiring substantial buffer stock (at considerable cost) for stability. The hypothesis that exports destabilize agriculture can be tested from empirical expressions of equation (2). Exports, X, are stabilizing if they increase to offset increased supply (production, O) or reduced domestic utilization, D, all else being equal. On the other hand, exports are destabilizing (require more buffer stocks for stability) if increasing exports are associated with lower production or higher domestic utilization. If exports are unrelated to domestic production and utilization so that $S(O_i{\cdot}X_i)$ and $S(D_i{\cdot}X_i)$ are zero, then the contribution of exports to the need for buffer stocks is simply $S^2(X_i)$, that is, the estimated variance of exports. The role of markets in mediating the impact on stock requirements from changing exports and production cannot be fully defined in one

Table 3.1 Components of Estimated Variance of Stocks of Aggregate Food, Feed, and Fiber and Feed Grains for the United States by Selected Periods, 1950-98

	U.S. aggregate					*Feed grains*				
Variance source[a]	*1950-59*	*1960-69*	*1970-79*	*1980-89*	*1990-96*	*1950-59*	*1960-69*	*1970-79*	*1980-89*	*1990-96*
	billion 1982 dollars					*million metric tons*				
$S^2(O_i)$	15.1	10.9	615.1	229.7	23.2	192.6	180.6	722.8	2097.6	1252.7
(CV, %)[b]	-6.0	-4.3	-35.0	-11.5	-3.9	-12.6	-9.5	-14.2	-18.1	-11.5
$S^2(D_i)$	16.6	23.4	358.0	65.2	27.7	76.6	71.4	159.9	276.9	132.5
(CV, %)	-8.1	-7.8	-31.5	-8.2	-5.7	-8.7	-6.7	-8.8	-8.9	-5.0
$S^2(X_i)$	3.0	1.4	28.5	68.3	5.1	8.3	15.9	223.3	121.9	46.8
(CV, %)	-24.0	-9.5	-58.8	-51.8	-16.5	-41.3	-21.9	-33.9	-18.3	-12.5
$-2S(O_i\ D_i)$	-29.7	-29.4	-932.7	-200.4	-30.3	-218.4	-186.6	-390.4	-809.9	-533.4
$-2S(O_i\ X_i)$	-7.9	-4.2	-251.2	-200.2	-3.9	-78.5	-25.6	-693.3	353.8	-144.8
$2S(D_i\ X_i)$	8.0	4.9	186.7	75.4	-11.2	46.1	19.2	94.7	-94.4	-3.9
$S^2(R_i\ Total)$	5.1	7.1	4.1	38.1	10.6	26.7	74.7	117	1946	750.1
$3S(R_i\ Total)$	6.8	8.0	6.1	18.5	9.8	15.6	25.8	32.4	132.3	82.2
$Mean(D_i+X_i)$	57.2	74.9	69.6	57.3	41.1	107.4	144.4	188.5	124.0	142.7
(3S/Mean, %)	-11.8	-10.7	-8.7	-32.3	-23.7	-14.5	-17.9	-11.1	-106.7	-57.6
$S^2(R_i)$ *share % to:*[c]										
O_i, D_i	39.0	69.0	all	all	all	all	87.0	all	80.0	all
X_i	61.0	31.0	-	-	-	-	13.0	-	20.0	-

[a]*Buffer stock residual* $R_i = O_i$ *(output)* - D_i *(domestic use)* - X_i *(net exports) for commodity i.*

[b]*Coefficients of variation (CV) are standard deviations about the period mean as a percent of the period mean.*

[c]O_i, D_i *share of total variance is* $S^2(O_i) + S^2(D_i) - 2S(O_i\ D_i)$; X_i *share all other.*

Source: Basic data from USDA (1998). See Tweeten (1983) for procedure.

Table 3.2 Components of Estimated Variance of Stocks of Soybeans and Wheat for the United States by Selected Periods, 1950-98

Variance source	Soybeans					Wheat				
	1950-59	1960-69	1970-79	1980-89	1990-98	1950-59	1960-69	1970-79	1980-89	1990-98
					million bushels					
$S^2(O_i)$	12,170	39,627	121,247	37,753	112,098	29,345	24,152	70,193	105,022	49,313
(CV, %)	-25.5	-24	-23.1	-10.3	-14.7	-15.7	-11.7	-14.5	-13.8	-9.3
$S^2(D_i)$	5,541	10,997	79,742	5,616	29,204	1,521	5,474	48,157	18,924	17,272
(CV, %)	-21.6	-18.2	-18.9	-6.4	-11.4	-6.2	-11.2	-20.9	-13.5	-10.8
$S^2(X_i)$	1,080	7,449	21,218	16,256	15,486	11,232	10,565	3,072	69,881	12,345
(CV, %)	-39.5	-36.5	-25.5	-17.3	-16.3	-27.2	-14.4	-7.1	-19.2	-9.9
$-2S(O_i\ D_i)$	-16,232	-39,576	-192,075	-10,178	-94,310	-620	-16,259	3,458	35,273	24,481
$-2S(O_i\ X_i)$	-6,481	-31,116	-98,332	-29,184	-60,543	2,305	13,596	-61,845	-55,426	10,077
$2S(D_i\ X_i)$	4,567	17,867	81,871	1,160	28,907	-1,976	-4,725	-3,269	-32,513	7,014
$S^2(R_i$ Total)	645	5,248	13,670	21,424	30,842	41,987	32,803	59,765	141,162	120,503
$3S(R_i$ Total)	75	216	351	439	527	615	543	733	1,127	1,041
Mean(D_i+X_i)	429	811	2,066	953	1,130	1,017	1,376	1,832	1,189	1,056
(3S/Mean, %)	-17.5	-26.6	-17	-46.1	-46.6	-60.4	-39.5	-40	-94.8	-98.6
$S^2(R_i)$ share (%) to:										
O_i, D_i	all	all	65	all	all	72	41	all	all	76
X_i	–	–	35	–	–	28	59	–	–	24

Source: Basic data from USDA (1998). See Table 3.1 footnotes.

equation, hence equation (2) is only a simplified depiction of the market-stabilization process.

Clues to the role of exports in stabilizing U.S. consumption can be gleaned from the simple correlation between exports and production. A positive correlation coefficient between exports and production indicated by the negative covariance term $-2S(O_i{\cdot}X_i)$ in 16 of the 20 scenarios presented in Tables 3.1 and 3.2 suggests that, on the whole, exports helped to stabilize consumption in the face of unstable production, other things being equal.

Of the three components (O, D, X) of system instability considered in Tables 3.1 and 3.2, the greatest relative instability as measured by the coefficient of variation comes from exports with exceptions only for wheat in 1970-79 and 1990-98. Because exports are small relative to domestic production and utilization, however, direct variation as measured absolute by estimated variance S_i^2 was least for exports for all cases and time periods except for aggregate markets in 1980-96, for feed grains in 1970-79, for soybeans in 1980-89, and for wheat in 1950-69 and 1980-89 in the United States.

Interactions among production, utilization, and exports apparent in covariances offset much of the direct contribution of O, D, and X to instability. Hence, the estimated variances in buffer stocks in all instances in Tables 3.1 and 3.2 are well below the sum of the direct variances S_i^2. Much of the reduction in system variation from interactions resulted from the preponderance of negative covariances between production and domestic utilization $S^2(O_i{\cdot}D_i)$, which probably traces to the tendency for high production to generate low prices and high consumption.

To further gauge the contribution of exports versus domestic production and utilization to system instability, the estimated variance of buffer stocks was divided into a component for exports $[S^2(X_i) - 2S(O_i{\cdot}X_i) + 2S(D_i{\cdot}X_i)]$ and all other sources $[S^2(O_i) + S^2(D_i) - 2S(O_i{\cdot}D_i)]$. The results shown in the last two rows of Tables 3.1 and 3.2 indicate that 61 percent of the system variation measured by $S^2(R_i)$ came from exports for the U.S. aggregate farm economy in the 1950-59 period, and 59 percent of wheat system variability came from exports in the 1960-69 period. For all other periods and crops (18 out of 20 cases), most, or all, of the system net variation as measured by $S^2(R_i)$ arose from domestic production and utilization.[ii]

3.1 Statistically Appropriate Stocks

The calculations in Tables 3.1 and 3.2 also give insights into statistically appropriate buffer-stock requirements over time. Those requirements may increase if exports, or other variables, become more volatile. Volatility as quantified by $S^2(R_i)$ increased in aggregate from U.S. $5 billion in 1950-59 to U.S. $38 billion (1982 dollars) in 1980-89 due mainly to production and utilization variability. For all items shown in Tables 3.1 and 3.2, $S^2(R_i)$ was greatest since 1980.

Two standard deviations $2S(R_i)$ of buffer stockholdings presumably avoid a shortfall of stocks to protect against high utilization, or exports and/or low production, except one out of 50 years, assuming a normal distribution, constant economic structure, and market separability. Another standard deviation is added to cover pipeline stocks necessary for markets to function, aside from seasonal and buffer stocks. This calculation indicates that statistically appropriate U.S. stocks in 1990-98 for this arbitrarily specified level of food security were 82 million metric tons of feed grains, 527 million bushels of soybeans, and 1,041 million bushels of wheat (Tables 3.1 and 3.2). While these stocks were absolutely higher than in most previous periods, they were not necessarily higher as a proportion of the mean utilization. The later relative value was especially high for feed grains in the 1980-89 decade (107 percent) and for wheat (nearly 100 percent) for the 1980-89 and 1990-98 periods. The percentage was only nine percent for aggregate agriculture in the 1970s, despite the sharp rise in exports.

While numbers varied among the individual commodities in Tables 3.1 and 3.2, a consistent pattern of variability in stock requirement related to trade is not apparent. The most notable conclusion is that most system variability traces to production and domestic use rather than to exports.

3.2 Economically Appropriate Stocks

The data in Tables 3.1 and 3.2 suggest statistically appropriate buffer stock levels to avoid shortages for 49 out of 50 years, all else being equal. But of greater interest are economically appropriate stock levels. Studies by Makki et al. (1996) of efficient wheat markets provide several conclusions:

- Under the most reasonable ranges of real interest rates and coefficients of variation in exports and utilization, the economically optimal U.S. carryout of wheat is approximately 400 million bushels—150 million bushels of buffer stocks and 250 million bushels of pipeline stocks. These carry outs are below the numbers presented in Table 3.2 and below actual average carryouts in recent decades that have been inflated by government holdings. Thus, market liberalization in farm policy embodied in the Federal Agriculture Improvement and Reform (FAIR) Act of 1996 can be expected to bring lower carryovers and may bring greater variation in grain prices.
- Optimal U.S. wheat carryout stocks and domestic consumption increase modestly with unilateral and multilateral liberalization of wheat markets,

including a U.S. $.50 per bushel reduction in Export Enhancement Program (EEP) payments. Exports fell because the EEP subsidy ended. A direct payment was estimated to be more cost effective than EEP to raise farm income.

- According to results of the efficient market model employed by Makki et al. (1996), storage and trade offer trade-offs in smoothing domestic prices and consumption in response to domestic and foreign supply variation attending a more market-oriented agriculture. Rational market agents store more and export less when production uncertainty increases, and they store less and export more when production uncertainty decreases.
- An optimal balance of two principal buffers—stocks and trade—in a country depends partly on its share of world markets. The small country optimally relies on trade, holding only enough stocks to last until imports arrive when local production falls short. Large producers, such as the European Union and the United States, optimally rely more on buffer stocks. Models show optimal buffer stocks increase as the share of international production variability of the large country increases (Figure 3 in Makki et al., forthcoming).
- The U.S. and E.U. gain from more open markets even after accounting for the possible need for more buffer stocks with trade liberalization in wheat. Gains to taxpayers and consumers more than offset losses to producers so that annual overall economic gains to the United States were U.S. $394 million and to the European Union were U.S. $357 million with multilateral liberalization versus 1989-93 policies (Makki et al., 1996).
- Each one percentage point increase in the interest (discount) rate reduces optimal buffer stock approximately 11 percent (Table 3 in Makki et al., forthcoming). Thus, the public sector with a real discount rate on capital of approximately three percent has a lower storage cost and price volatility, and a higher optimal buffer stock than the private sector that had a discount rate of seven percent.
- Tyers and Anderson (1998) estimate that trade liberalization would reduce the coefficient of international price variation for dairy products to 7 percent from 16 percent and for wheat to 30 percent from 45 percent. Although trade liberalization made possible by the FAIR Act offers opportunities to reduce volatility of world prices, lower government stocks and set asides may raise domestic price volatility.
- Developing countries do not necessarily fare well with liberalization of U.S. and E.U. policies. The United States and the European Union have provided less-developed countries (LDCs) with very cheap grain. However, market liberalization that provides higher prices to LDC grain producers, coupled with liberal access to global markets for many farm and

non-farm products of developing countries, could compensate LDCs for losses to consumers.

The United States bore an inordinate share of the cost of world food-market stabilization prior to 1996 (Sharples and Martinez, 1993)—a cost it no longer seems willing to bear. Because of the high incremental costs of stabilizing food consumption and prices, complete stabilization of the food economy is not economically justifiable for either the public sector or the private sector. The drop in world-grain stocks in recent years has generated fears of global food crises under more market-oriented agricultural policies. However, it is notable that the most recent world food crises came in 1966-67 and 1972-73 when government interventions were massive. Although reserves and stabilization may be less under the 1996 FAIR Act than under earlier policies, markets may be doing a more efficient job of stabilization—bending but not breaking under the pressure of a potential world food crisis. The situation needs to be monitored for possible improvements, but there is no compelling need to stall trade and farm commodity program liberalization on the grounds that rising yield volatility and freer trade pose food-security risks.

4. THE FOUNDATION FOR NEW FARM PROGRAM DIRECTIONS

This section examines directions for new farm programs in an unstable world. Before turning to program provisions, however, I review ten salient characteristics of the American agricultural economy that will influence trade, stability, and policy. They provide a foundation for the commodity policy conclusions presented later.

4.1 Farm Exports Will Continue to Expand but at a Highly Uneven Rate

Conclusions from foregoing sections suggest that volatile yields and exports need not deter efforts for market liberalization. Indeed, more open trade and technological advances raising crop yields can help to stabilize the economy and food supplies.

With domestic markets for food growing about one percent annually and farm productivity growing 1.5 percent to 2.0 percent annually, the nation's agricultural plant perennially would have to be scaled back with trauma without export market expansion. Despite advantages for trade, America often treats trade illogically. It perceives exports as good and imports as bad. In fact, it is imports that raise living standards; exports are only the means to earn foreign exchange for purchasing imports. The nation's frequent actions against countries that dump their products in the United States at prices below their domestic full production costs are teaching the world a counterproductive lesson that denies low-cost imports to consumers and denies markets to producers. That is, wide adoption and use by foreign countries of U.S. dumping policies will cause them to reject U.S. farm products, which, for sound economic reasons, are often exported by the U.S. below the full cost of production. Ironically, the big losers from trade barriers are the countries that erect them (Makki, Tweeten, and Gleckler, 1994). The United States gains from multilateral and unilateral trade liberalization, but producers gain most from multilateral liberalization.

4.2 The Global Food Supply-Demand Balance for Producers Will Likely be More Favorable

Global crop and livestock yields were increasing considerably faster than aggregate demand for food in the 1950s and 1960s. In the year 2000, trend yields were increasing at about the same rate as population. Carl Zulauf and I (Tweeten and Zulauf, 1997) contend that the old agricultural policy paradigm of a U.S. farming sector beset by technological change in excess of its ability to adjust, has been replaced by a new paradigm recognizing that agriculture can adjust to technological change although annual and cyclical shocks will be painful. Of course, farm productivity gains will continue from science and technology. But few excess resources remain in farming despite the 1998-99 drop in grain and hog prices. The technological treadmill will not stop, but producers can adjust to it.

4.3 The Principal Economic Problem Facing Commercial Farmers Will Continue to be Instability

In response to continuing annual and cyclical instability, farmers will pursue private risk-aversion strategies such as flexibility, diversification, liquidity, insurance, buffer stocks, contracting, and forward pricing. The lines between input supply, farm production, and marketing and processing firms will blur to reduce risk and increase efficiency. For example, when commodity supplies are unusually high, commodity producers often experience low (or negative) profit margins while food processors and retailers experience high profit margins. Use of vertical coordination (for example, contracts and integrated ownership) to share returns reduces risk under such circumstances.

Risk is seemingly not a case requiring government intervention or subsidies to agriculture. The nation does not subsidize risk takers in Las Vegas, the futures market, Wall Street, or small businesses—the latter with higher failure rates than farmers. Private risk-management strategies are effective for farmers. In contrast, government subsidies encourage excessive output and lower farm prices. There is little or no evidence that risk can be better managed by the public sector than by the private sector (Reinsel, 1993; Wright and Hewitt, 1994).

The best argument for public stockholding and other risk-management activities is that private discount rates exceed public discount rates. Hence, some contend that the private sector under-invests in stabilization efforts. This contention holds for virtually all goods and services, not just for risk management. But international experience from the former Soviet Union to New Zealand provides compelling evidence that widespread public mismanagement of risk overshadows other considerations. The weight of experience is on the side of leaving risk management in private hands except for information systems and for other public goods.

4.4 Markets for Farm Commodities Work and are Efficient

Because they are rival, exclusionary, and transparent, farm commodities are market goods, more efficiently allocated by markets than by government (Tweeten, 1989). Markets work for more than just risk management. On average, competent commercial farm operators in the 21st century will likely earn returns as high as elsewhere with or without income support programs for crops and livestock. The best evidence for such a conclusion is the past economic and technological success of commercial farms that have been reasonably well managed.

The farming industry is not a welfare case. To be sure, a few limited-resource farm families need public assistance. But commodity programs are of little help to operators of small farms (the majority of all farms) because they have little to sell, and payments are tied to past production. Commodity programs will not help long-term net farm income because benefits are capitalized into land and lost to renters and to new landowners. Commodity programs help retain family farms in the short run, but not in the long run (Tweeten, 1993).

4.5 Farms Will Become More Industrialized

Farms will be larger, more vertically coordinated, more challenging to manage and finance, and more technological, with precision inputs used to produce designer outputs (Boehlje, 1996). The public and Congress will be less supportive of income transfers to an industrialized, financially robust agriculture over time. This contrasts with 1998, an election year with a budget surplus when partisan politics strongly favored intervention to supplement farm income.

4.6 Most Farms Will be Family Farms that Deal with Fewer, but Larger, Input Supply and Marketing Conglomerates

Some of these conglomerates will be cooperatives. Larger farms will evolve from economies of size, efforts to countervail large agribusinesses, and the need for close coordination of production at all levels to assure quality food tailored and timed to meet discriminating consumer demands.

Agribusiness conglomerates will innovate rapidly and will have substantial bargaining power, but that does not mean low returns for farm resources or low prices for farm products. Even if private agribusiness input supplies were monopolists (they are not, and many are cooperatives), and even if private firms buying commodities from farmers were monopsonists (they also are not, and many are cooperatives), farm resources would still not earn low returns if their resources were mobile (as they certainly are!). Agribusiness conglomerates and contractors must pay producers a return that brings forth needed supplies. Of course, competition must be fostered—an ounce of competition can be worth a pound of regulation.

4.7 Non-commercial Farms Will Decline in Numbers and Share of Output

Commodity programs tied to farm production will not save small farms—only off-farm income will. The mid-size farm is at risk when it is too small to achieve economies of size and too large to allow the operator to spend much time earning off-farm income. Rather than providing costly government payments to all producers, government can cut the cost of saving farms at risk by targeting assistance to them through credit programs.

4.8 One Threat to Family Farms is Cash Flow

Operators who must sink savings into refinancing a farm economic unit (over U.S. $2 million of assets) each generation will continue to live poor and die rich. Cash-flow requirements to service debt can leave operators with little income to pay family living expenses.

Leasing, contracting, and integration, as well as help from parents, are essential for operators to enter and survive on a family farm. Industry-wide government transfers can aggravate the cash-flow problem by raising land values and, hence, the debt burden for entrants.

4.9 Pressure by the Public for Government Regulation of Agriculture Will Increase

Non-point chemicals and large livestock farms will be targets. Farmers will share decisions not only with bankers, input suppliers, and marketing firms, but also with bureaucrats. However, producers can turn some public environmental concerns to their advantage. Carbon sequestration payments to reduce global warming can supplement farm income, build organic matter, save soil, improve water quality and nutrient release, and raise yields (Tweeten, Sohngen, and Hopkins, 1999).

4.10 There are No Decoupled Payments

Government payments cannot be decoupled from production because farmers are short of capital and use payments to purchase fertilizer, machinery, and other inputs that raise output. Government transfer payments to producers are likely to continue for income support and insurance, creating excessive output and low commodity prices.

5. NEW POLICY DIRECTIONS

The foregoing background provides a prism through which to view proposed changes to the 1996 FAIR Act. Critical reform issues include whether or not to raise loan rates, restructure insurance, and continue transfer payments.

5.1 Insurance Reform

Political pressures are mounting to use revenue insurance as a vehicle to channel larger government payments to farmers. Larry Combest, chairman of the U.S. House Agriculture Committee stated in February, 1999: "I think basically we have to scrap the current program, the way we have thought about crop insurance, and completely rebuild it... we have to get away from the idea that crop insurance has to be actuarially sound, ... you've got to have a revenue enhancement feature in it" (Maixner, 1999). Shortcomings of that course of action are evident.

Farmers do not seem to be very risk averse. Their outlook is optimistic; most prefer to self-insure and do not buy crop insurance unless it is heavily subsidized (Gardner, 1994). Farmers and the private sector are clever at devising and using private risk-management tools. One of the most widely used is off-farm income. Producers will buy private fire and hail insurance, but the world offers no evidence that rich or poor farmers will pay the full cost of all-risk insurance. The federal contribution to crop insurance, historically at about one-third of the cost, has doubled in recent years. Yet dissatisfaction with the programs remains high among producers and taxpayers.

A high-income supplement (subsidy) component in insurance is likely to incur disfavor with the World Trade Organization (WTO) because it is a coupled payment. A high income-supplement component generally encourages production, especially in switch regions compared to core regions. Often these switch areas are environmentally fragile as well as economically marginal. Some are best suited for grass or trees. Maintaining production in uneconomic switch regions drives prices down for producers in core areas, and reduces national income, because costs to taxpayers exceed gains to producers and consumers.

Crop and revenue insurance subsidies enable quite uniform premiums to be charged to farmers over all risk situations, creating an attractive moral hazard: Able operators who manage to avoid losses reject programs, while less-careful managers insure and collect payments. The government gets more of what it pays for; that is, unstable yields and revenues.

Another issue is what to insure: yields, prices, revenue, production costs, net farm income, or total net income of farm people from all farm and off-farm sources. Farmers face a bewildering array of insurance choices.

Much can be said for a program such as the Canadian Net Income Stabilization Account (NISA). Begun in 1991, NISA supplements yield insurance. In Canada, a farmer can contribute up to two percent of eligible sales, up to CDN $250,000 per farmer, which is matched by the government, and can make unmatched contributions up to 20 percent of sales (though the latter is little used). Farmers are allowed to accumulate up to 150 percent of the 5-year average of eligible sales. Accounts (in local banks or in a Consolidated Revenue Fund) earn the market interest rate plus three percentage points, the latter from the government. Withdrawals are allowed when net farm income falls below the five-year average or below a low-income threshold. A farmer may withdraw all accumulated funds to drop the program or to retire.

Advantages of NISA are that it covers net farm income and, hence, all crops and livestock. It is cost effective because it is carefully targeted such that net-income variation is reduced at low Treasury per-acre cost. The program in the United States would operate much like an Individual Retirement Account (IRA). It is relatively easy to administer, and it could be operated with low administrative costs by the Internal Revenue Service. A pilot Adjusted Gross Revenue insurance plan currently uses the IRS schedule F information to insure farm gross revenue. Program parameters can be adjusted to keep taxpayer costs as low as deemed appropriate by Congress.

There are, however, disadvantages to NISA. They are that NISA is:

- Regressive. A producer with no income to accumulate an account receives no subsidy.[iii]
- Prone to give out in a run of bad years. With no pooling, a producer's fund may be exhausted quickly.
- Offers no protection for beginning farmers before they have built a fund.
- Offers little help to cash flow. American farmers have grown accustomed to advance deficiency, or to transition payments.
- Ties up financial capital needed to purchase farm production inputs.
- Not cognizant of off-farm income which could offset low farm income.
- Subject, like other government insurance, to abuse through creative accounting if proper administration is lacking.

Many of these shortcomings could be corrected while retaining the advantages of insurance covering income of farm people from farm and off-farm sources.

5.2 Loan Rates

A second major change enacted in FAIR is to raise commodity loan rates. Several considerations suggest that political pressures to raise rates be resisted.

Current loan rates encourage excess production, which drives down commodity prices and creates pressure for public supply management, stock accumulation, and export subsidies. Operators find it advantageous to produce a crop if marginal (as measured by variable costs in the short run) production costs are covered. Loan rates exceeded variable production expenses in 1996 and 1997 for every commodity in Table 3.3. The loan rate was especially high relative to variable expenses for soybeans, and planted acres expanded 13 percent from 1996 to 1998. The high loan rates distort market signals, discourage economic adjustments in resource use, skew the commodity mix, and raise loan-deficiency payments and market-loan costs.

Even loan rates set below free-market equilibrium price or below the cost of production create distortions. Setting the loan rate at the normal market-

Table 3.3 Loan Rates, Variable Expenses, and Total Economic Costs of Production for Major U.S. Crops, 1996 and 1997

Commodity	*Year*	*Loan rate*	*Variable expenses*[1]	*Total economic cost*[2]
Corn	1996	1.89	1.22	2.76
(dollar/bushel)	1997	1.89	1.23	2.69
Soybeans	1996	4.97	2.16	6.30
(dollar/bushel)	1997	5.26	1.88	5.61
Wheat	1996	2.58	2.31	5.94
(dollar/bushel)	1997	2.58	1.96	5.02
Cotton	1996	0.52	0.51	0.85
(dollars/pound)	1997	0.52	0.48	0.80
Rice	1996	0.65	0.611	0.1106
(dollars/hundred weight)	1997	0.65	0.628	0.1170

[1]Variable expenses include seed, fertilizer, chemicals, fuel, repairs, hired labor, purchased irrigation water.
[2]All costs, including overhead and land.
Source: USDA (September 1998: 57–61)

clearing price or average cost of production, p_e, causes the expected market price to rise from p_e to p_l (Figure 3.3). The average, or expected price, p_e, when the entire normal distribution of possible prices is considered in Figure 3.3 is raised to p_l—the average or expected price when the shaded portion of

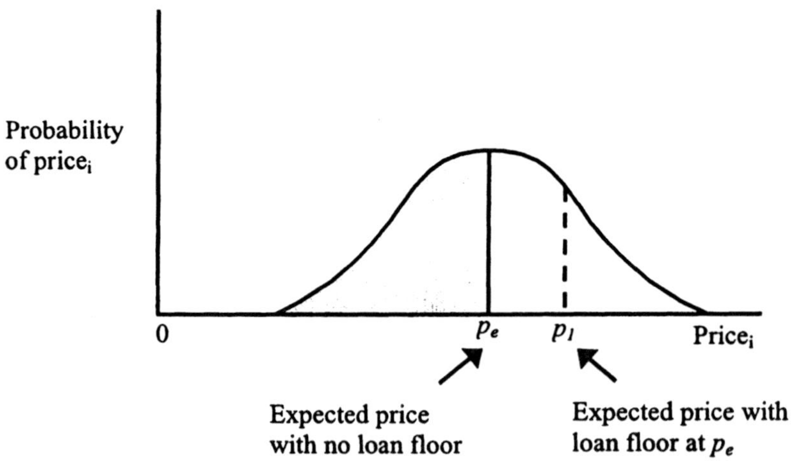

Figure 3.3 Illustration of Impact of the Loan Rate on Price Expectations

the normal distribution is removed by a loan rate set at p_e. Cutting off the lower tail of the market-price distribution with loan rates set at p_e not only raises the expected price from p_e to p_l, but also raises output and depresses market price. Loan rates tied to production costs of average producers also inflate supports because most production occurs below the unweighted average costs of production—larger producers have, on average, lower unit costs.

A high safety net encourages the continuation of excessive production and delays economic recovery in the farming industry. The general rule is that low returns attend low risk as illustrated by bonds versus stocks. Thus, a market orientation is likely to bring more unstable but higher returns, on average.

The economic impact of loan rates set above marginal costs in Table 3.3 is illustrated in Figure 3.4 in which the loan rate set at p_l above the open market clearing price, p_e, gives a new market price, p_m, and a loan deficiency payment of $p_l - p_m$ per unit of output.

Given the short-run supply, *S*, and demand, *D*, for commodity, *q*, welfare gains in monetary value from the loan support rate, p_l, bring in market quantity, q_m, and price, p_m, compared to free market equilibrium price p_e and quantity q_e are:

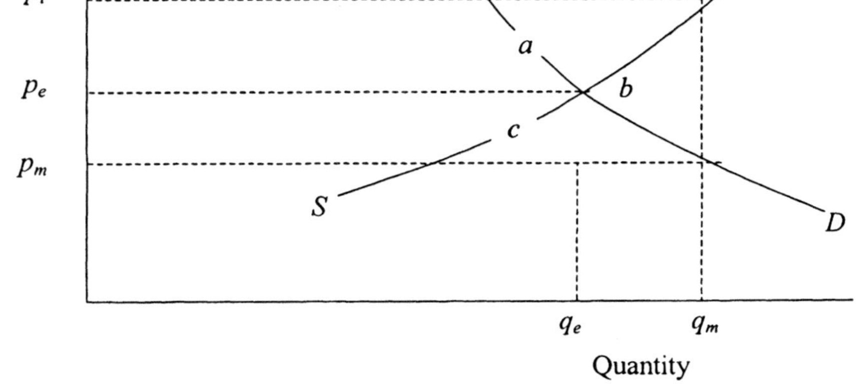

Figure 3.4 Impact of Loan Rates Above the Market Clearing Price

		Area (value)
Gains to:	*Producers*	*a*
	Consumers	*c*
	Taxpayers	*-a-b-c*
	Society	*-b*

5.3 Income Transfers

As indicated earlier, it is difficult to build a case for government-subsidized insurance or any other perennial income transfers to farmers to address economic instability. Output is expanded, commodity prices are reduced, and transfers are from poorer taxpayers to wealthy farmers. Any net economic benefits are bid into land prices. If politics dictates that transfers must occur, careful targeting of specific help, such as credit to small family farms at risk, is more cost effective than are across-the-board payments.

If transfers are a political necessity, then we must consider continuing transition payments that are decoupled from production but coupled to environmental practices. Payments for carbon sequestration could be spread widely among producers because virtually any crop can build soil organic matter, thereby holding carbon in crops and soils rather than in greenhouse gas (CO_2) contributing to global warming. I have elsewhere summarized studies showing that payments for carbon sequestration cannot be justified by the value of reducing global warming alone, but must be justified by complementarity among benefits, such as less soil erosion, better water quality, and improved water-holding capacity and nutrient release (Tweeten, Sohngen, and Hopkins, 1999).

6. CONCLUSION

This chapter makes the case that basic food production instability arising from yield variation has not increased, and that exports can help to stabilize markets. Markets work for farm commodities and for risk management if given the opportunity. Widespread subsidies to producers when farm prices fall can be counter-productive, retarding resource and commodity adjustments to markets that would quickly return profitability to farmers.

I summarize by triaging agricultural commodity policy alternatives for an unstable agricultural economy into three categories: rejected, or blue-box programs; acceptable, or green-box programs; and questionable, or amber-box programs.

6.1 WTO Illegal, Blue Box, and/or Congressionally Rejected Programs

Because of opposition from the WTO, or because Congress is unlikely to turn back the 1996 FAIR Act, the following programs show little promise of success:

- Supply management (set-aside) programs.
- Price supports or guaranteed price above cost-of-production in a coupled deficiency-payment and target-price program.
- Dumping commodities abroad through large export subsidies under EEP.
- Government stock accumulation.

6.2 WTO Legal, or Green Box, Programs

Other programs are more widely acceptable because they have favorable economic pay-offs or because they are not deemed to interfere much with trade. Examples include:

- Science (research, education, extension, and information systems),
- Grades and standards,
- Exports, [promotion (Market Access Program), low interest loans and guarantees, and food aid (PL480)],
- Environmental protection [Conservation Reserve Program (CRP), Environmental Quality Incentive Program (EQIP)], Wetlands Reserve, and technical assistance]
- Infrastructure investment (in ports, rail, and roads),
- Sanitary-phytosanitary (SPS) restrictions based on science,
- FAIR Act: Production flexibility contract (transition) payments, and loan rates (if they are set low enough),
- Disaster assistance, and
- Insurance programs with only nominal subsidy.

Repeated attempts to replace disaster assistance payments with government-subsidized crop or revenue insurance have failed. Hence, a wise course of action may be to keep the subsidy component of insurance programs as low as possible while recognizing that *ad hoc* disaster assistance is likely to continue. Such a policy is probably less resource-use distorting than is reliance on large insurance subsidies—the latter program likely to be either vigorously opposed by competing exporters or (worse yet) adopted by them.

6.3 Amber Programs: Worthy of Caution and Perhaps to be Negotiated in WTO 1999+ Rounds

Several emerging policies were not necessarily anticipated in the Uruguay Round of multilateral trade negotiations or in the FAIR Act but are of growing concern, especially to trading partners or competitors, and need to be addressed by Congress and in trade negotiations:

- Loan or price supports above market or world prices, especially in the absence of production control.
- Yield or revenue insurance, disaster assistance, or export assistance programs with sizable subsidy components.

Risk and environmental problems in agriculture will not go away; addressing these problems begins with use of private markets and of public green-box activities listed above. This chapter makes the case that loan rates (especially for soybeans) could be lowered rather than raised; that risk insurance could be replaced by insurance of net income of farm operators from all sources including crops, livestock, and off-farm income in an IRA-type fund; and that transition payments could be continued (if they are to be continued, they could be based on carbon sequestration and on other environmental goals that are of value to the public).

The lawyer's maxim that tough cases make bad law applies also to economics. The tough economic times in 1998-2000 for farm grain and hog producers were traumatic for some but came as no surprise, given the history of commodity cycles endemic to agriculture. Farmers had completed three of their best years in history, and most were positioned to weather setbacks. The setback could not be blamed on FAIR—hogs were not covered in previous farm bills and grain prices were low partly due to lower U.S. exports to Asia. A useful response would have been temporary credit or cash assistance only to at-risk producers.

Proposals for more permanent economic cures, such as higher loan rates and income transfers through risk insurance, are likely to make the situation worse. American insistence on exemption of agricultural trade from the discipline of the General Agreement on Tariffs and Trade (GATT) opened the gates to a flood of foreign interventions, especially by the European Union, that continue to stifle U.S. farm exports. A federal government venture into heavily subsidized loan rates and revenue insurance could be equally ill-fated. Other nations could contest such ventures through the WTO, and could emulate our actions with serious consequences for U.S. farm exports. Even in the unlikely case that U.S. competitors would not learn from our behavior, the results of higher loan rates and income transfers through revenue insurance are likely to be excessive output; low farm prices and incomes, which lead to government stock accumulation, acreage set asides (or to an expanded CRP to curb output), and large export subsidies. Thus, history would repeat itself with the return to agricultural policies that originated in the 1930s but which are anachronistic for the 21st century.

7. ENDNOTES

[i] Although there is little evidence that public management of commodity stocks or trade can be more efficient than private market stockholding and trade management in meeting needs of people for food and other goods, some may desire an emergency food reserve (such as the 4 million ton U.S. government grain reserve maintained to respond mainly to emergency food needs in developing countries).

[ii] It is not possible to precisely separate variation in X_i from O_i and D_i due to interactions. For example, shifting the terms $-\ 2S(O_i{\cdot}X_i)$ and $+2S(D_i{\cdot}X_i)$ from exports to all other sources would have changed the numbers in the text.

[iii] Creative accounting such as widespread underreporting of income and generous tax deductions allowed to the self-employed can result in much lower taxable income than real economic income (Tweeten 1993: 30, 31). The result could seem inconsistent as operators with no taxable net income are able to add to their stabilization account.

8. REFERENCES

Ash, M., G. Douvelis, J. Castaneda, and N. Morgan. 1995. *Oilseeds: Background for 1995 Farm Legislation*. Agricultural Economic Report No. 715. Washington, DC: USDA/ERS (April).

Boehlje, M. 1996. "Industrialization of Agriculture: What are the Implications?" *Choices* (First Quarter): 30–33.

Collins, K. and J. Glauber. 1998. Will Policy Changes Usher in a New Era of Increased Agricultural Market Variability? *Choices* (Second Quarter): 26–29.

FAOSTAT (Food and Agricultural Organization Statistics Database.) Internet website: http://apps.fao.org/

Gardner, B.L. 1994. "Crop Insurance in U.S. Farm Policy," in D. Hueth and W.H. Furtan, eds., pp. 17–44, *Economics of Agricultural Crop Insurance: Theory and Evidence*. Boston, MA: Kluwer Academic Publishers.

Hazell, P. 1993. "Implications of Grain Trade Liberalization for LDC Food Security," in Robert Reinsel, ed., Chapter 3. *Managing Food Security in Unregulated Markets*. Boulder, CO: Westview Press.

_____. 1985. "Sources of Increased Variability in World Cereal Production since the 1960s." *Journal of Agricultural Economics* 36: 145–159.

Johnston, W. and L. Schertz. 1998. "What Farm Managers Told Us about the 1996 Farm Act." *Choices* (Second Quarter): 21–25.

Lin, W., P. Riley, and S. Evans. 1995. *Feed Grains: Background for 1995 Farm Legislation.* Agricultural Economic Report No. 714. Washington, DC: USDA/ERS.

Makki, S., L. Tweeten, and J. Gleckler. 1994. "Agricultural Trade Negotiations as a Strategic Game." *Agricultural Economics* 10: 71–80.

Makki, S., L. Tweeten, and M. Miranda. 1996. "Wheat Storage and Trade in an Efficient Global Market." *American Journal of Agricultural Economics* 78: 879–890.

_____. Forthcoming. "Storage-Trade Interactions under Domestic and Foreign Production Uncertainty: Implications for Food Security." *Journal of Policy Modeling.*

Maixner, E. 1999. "A Strong Voice and Advocate." *Ohio Farmer* (34–36).

Reinsel, R. 1993. *Managing Food Security in Unregulated Markets.* Boulder, CO: Westview Press.

Sharples, J. and S. Martinez. 1993. *The Role of Stocks in World Grain Market Stability.* Foreign Agricultural Economic Report No. 248. Washington, DC: USDA/ERS.

Tweeten, L. 1998. "Dodging a Malthusian Bullet in the Twenty-First Century." *Agribusiness: An International Journal* 14(1): 15–32.

_____. 1994a. *Agricultural Risk Management in the 21st Century.* Proceedings of "Summit on Risk Management in American Agriculture," Washington, DC. Chicago, IL: Commodity Futures Trading Commission.

_____. 1994b. "Epilogue." in *Countdown to 1995: Perspectives for a New Farm Bill.* Columbus, OH: Department of Agricultural Economics.

_____. 1993. "Government Commodity Program Impacts on Farm Numbers," in A. Hallam, ed., *Structure and the Changing Face of American Agriculture.* Boulder, CO: Westview Press.

_____. 1989. *Farm Policy Analysis.* Boulder, CO: Westview Press.

_____. 1983. "Economic Instability in Agriculture." *American Journal of Agricultural Economics* 65 (December): 922–931.

Tweeten, L., B. Sohngen, and J. Hopkins. 1999. *Assessing the Economics of Carbon Sequestration in Agriculture.* Proceedings of Conference on "Assessment Methods for Soil Carbon Pools." Columbus, OH: Department of Agricultural, Environmental, and Development Economics.

Tweeten, L. and C. Zulauf. 1997. "Public Policy for Agriculture after Commodity Programs." *Review of Agricultural Economics* 19: 263–280.

Tyers, R. and K. Anderson. 1998. "Liberalizing OECD Agricultural Policies in the Uruguay Round: Effects on Trade and Welfare." *Journal of Agricultural Economics* 39: 197–215.

USDA (U.S. Department of Agriculture). 1999. Internet website: http://www.usda.gov/nass/pubs/agr98

_____. 1998. *Agricultural Income and Finance.* Agricultural Economic Report No. AIS-69. Washington, DC: USDA/ERS (September).

Wright, B.D. and J.A. Hewitt. 1994. "All-Risk Crop Insurance: Lessons from Theory and Experience," in D. Hueth and W.H. Furtan, eds., pp. 73–112, *Economics of Agricultural Crop Insurance: Theory and Evidence.* Boston, MA: Kluwer Academic Publishers.

Chapter 4

Has the Importance of Foreign Markets for U.S. Agriculture Been Oversold?

Bruce L. Gardner
University of Maryland

1. INTRODUCTION

From Colonial days until now, trade in cotton, grain, and tobacco have been major items on the economic and political agenda for agriculture. Farmers have expressed no greater political outrage over the last 50 years than they did in response to the grain-export embargoes of the 1970s. Each party's Presidential candidate since 1980 has pledged to abjure embargoes. Secretaries of Agriculture, whatever their party and policy differences, have been relentless promoters of agricultural exports.

With respect to broader trade policy, politicians, as they are wont to do, have spoken on both sides of many trade issues. Recent presidents have been gingerly pro-trade on the whole. Among the more striking overall statements of position is the Reagan-Gore theory that trade restrictions were a prime cause of the Great Depression of the 1930s.[i] With respect to agriculture, recent Secretaries of Agriculture have been uniform in stating the opinion that expanding exports are necessary to farm prosperity.

At the same time, U.S. producers of farm products that have to compete with commodities imported from abroad have sung a different tune. They can call upon less well-accepted economics according to which protection is argued to stimulate economic development. Some historians and economists have given oxygen to the Hamiltonian view that industrial protection crucially stimulated early U.S. manufacturing. Some agricultural economists and historians have argued that price protection promoted productivity growth in U.S. agriculture.[ii] If so, might not less trade in farm products we import, such as sugar and peanuts, be beneficial? We may, nonetheless, argue that agriculture gains from trade overall, because agricultural exports exceed imports, and one has to open up to imports in order to get other countries to accept our exports. But should we acquiesce in this mercantilist view of imports as a necessary evil?

2. WHY SHOULD AGRICULTURAL TRADE BE GOOD FOR FARMERS?

In the realm of economics as it impinges on policy choice, two of the most important results in economic theory were put forth 200 years ago by David Ricardo and Adam Smith in terms that have not been fundamentally improved upon to the present day: the principle of comparative advantage and the theorizing summarized in Smith's statement that "the division of labor is limited by the extent of the market." The latter statement is commonly taken as the basis for the emergence of specialization and economies of scale in an industry, but, as emphasized [for example, Stigler (1951)], its importance as the basis for a theory of economic development may be of even greater significance. The importance for agriculture of a larger market is made even more salient by the recent work on idea-based growth (Jones, 1999). The story is, roughly, that innovative ideas, embodied in technical improvements, are the prime source of productivity gains that underlie the growth of real income in a modern economy. These ideas are pure public goods in the sense that their use in the production of an additional unit of a commodity does not reduce the quantity available for further use. In this context, the physical return to investment in ideas increases with the size of the market (the quantity sold), and the economic return is reduced less with increasing output (the larger the market into which the increased output is sold).

How well do these principles carry over to consideration of agriculture as a sector of the economy? Because economic growth in U.S. agriculture is apparently so strongly linked to technological change, and technological change is linked to new knowledge generated by research, the reasoning has appeal. The billions of dollars spent on research for improvements in grain production have a higher return because the research costs would have been the same whether the new technologies (such as, improved seeds, pesticides, equipment, and timing of operations) were used on 40 million tons of grain or on only half as much. Objections, however, can be raised. What limits the scale of a given set of research findings is not only the quantity of output but the quantity of inputs available, such as soil characteristics and climate, to which the findings are tuned. Lack of homogeneity of farms may be just as limiting as the extent of the market. Suppose, on the other hand, that some research findings have no locational specificity. Then, the problem arises that the ideas may be just as beneficial to foreign as they are to U.S. production, and new ideas give no advantage to the country of origin, no matter what the size of the market. It seems likely that, in reality, innovations originating in one country are more productive in that country than elsewhere, and that while the marginal returns to the expanded use of a given set of innovations declines, the overall return increases with the size of the market. But the size of multifactor productivity gains as the size of the market increases is an empirical question that has not been answered to date.

The more solidly entrenched principle of comparative advantage carries over to the sectoral context less well. Indeed, we have no *a priori* reason to believe any particular sector would gain from increased trade. What does it mean for a sector to gain from trade? It means that the producers of that sector's output gain; that is, that factors of production employed in that sector gain. Since many factors of production receive their opportunity returns, at least in the long run, the gains are increases in returns to specific factors (or in the short run, to quasi-fixed factors). In the case of agriculture, the predominant specific factor is cropland. We have empirical reason to think there would be gains for most producers of agricultural commodities, since these commodities that are aggregated by value are exported, so that increased trade would increase aggregate farm-commodity demand, and, hence, the demand for specific factors in agriculture.

We cannot be sure, because of the heterogeneity of U.S. agriculture. It is possible that sub-sectors of U.S. agriculture have specific factors of their own, for example that particular import-competing commodities, such as those produced in warm climates in the United States, are produced on land having specific location values that would decline under free trade. In such a situation, it is possible that land values for import-competing agricultural areas would be reduced by more than land values in export-commodity areas would be increased by expanded trade, notwithstanding an initially higher aggregate value of land producing exported commodities. (More generally, suppose imported commodities use specific factors with low supply elasticities, while exported commodities use resources with higher supply elasticities more intensively.) The issue is still further blurred by considerations of ongoing domestic-commodity policies, which can reverse the gains from trade from both the national and sectoral viewpoints (Chambers and Paarlberg, 1991).[iii]

3. EVIDENCE ON THE GAINS FROM TRADE TO U.S. AGRICULTURE

Without benefit of econometric studies, farmers who produced exported commodities have long been convinced that they benefit from trade. They could be wrong but, nonetheless, their views on what is good for them economically should be considered with respect. Midwestern grain growers in particular have tended to be ardent advocates of the importance of export markets throughout the 20th century. Southern cotton growers have clung to this priority for an even longer time. Producers of products that compete with imports, including grain producers faced with Canadian competition, have been ardent advocates of trade restrictions. All such viewpoints speak to the perceived importance of trade in the determination of farm income.

Informal investigations of popular- and farm-business journalism have generally accepted the view that participation in world markets makes a big difference for the economic health of U.S. agriculture. In the first decades of the 20th century, a large part of the news reported as relevant to market prices for U.S. crops, for example in the *New York Times*, was about events in European markets, especially in Russia and South America. Discussions of U.S. farm-level wheat prices referred to their relationship with Chicago market quotations but also with Liverpool.

On the academic front, a notable contribution was Schuh's 1974 article arguing that the size of the export market in the 1950s and 1960s was a cause of the farm problem of that time, and that an overvalued dollar depressed the export market. Subsequent work debated the importance of the exchange rate in determining exports, but a maintained hypothesis throughout is that exports matter to farmers. Unfortunately, none of these studies known to me quantified a farm-income effect attributable to export losses due to an overvalued dollar, or to other causes.

Later, a lot of detailed analytical work was done in conjunction with the Uruguay Round of trade negotiations that estimated commodity price effects of increased trade resulting from reduced barriers to agricultural trade. While this work gave varying answers (Gardner, 1989), simulated world-price effects were generally such that producer prices in agricultural exporting countries would increase substantially. But farm income increases were not estimated. The approach taken in such work indicated that such increases would take the form of increased returns to farmland and, hence, in land prices (World Bank, 1986).

Economic models used by U.S. Department of Agriculture (USDA) embody quantitative conclusions about the effects of export demand on farm commodity prices and incomes. In USDA's *Supply and Demand Estimates*, each monthly accumulation of information about export markets is assessed for its potential effect upon farm-level commodity prices in the current and following crop years. The periodically updated *Agricultural Baseline Projections* of the Economic Research Service (ERS) incorporates the outlook for exports in its baseline of future farm prices and incomes, looking forward up to ten years. When estimating the effects on the baseline of external forces, such as world supply/demand balance, the USDA uses multipliers from simulation models, notably the Food and Agricultural Policy Simulator (FAPSIM), to estimate the effects of exogenous changes in exports on farm commodity price and farm income.[iv] For example, some USDA analysis translates an additional million metric tons of wheat exports to a farm-level wheat price increase of five cents per bushel.[v] This is a short-run effect, meaning an increase in exports comes out of domestic use and stocks. Since a million tons is about two percent of U.S. average annual use plus ending stocks, and five cents is slightly less than two percent of the U.S. average farm price, the implied total (that is, use plus stock) demand elasticity is about −1. The longer-run effects of a permanent increase of a million tons in U.S. exports would be less (and it would require a larger increase in export demand to increase exports by a million tons), because of supply response to higher wheat prices in the United States and abroad; but, on the other hand, stock reductions would not be a source of increased supply in the long run.

The decline in export prospects since 1997, attributable to the Asia crisis and to other factors, affords an opportunity to consider concretely the implications of current empirical knowledge of exports as related to farm income.[vi] The USDA baseline published in 1999, based on information available through November 1998, contained substantial changes from the previous baseline largely because of changes in the export situation and outlook. USDA listed the causes of the changed outlook as declining economic growth abroad, exchange-rate developments unfavorable to exports, strong export performance by competitors, and China's trade policies (USDA, 1999a).[vii] The changes in projected exports, and associated changes in net farm income, are shown in Table 4.1. In 1999, exports during 2000-05 were expected to average U.S. $13 billion less than had been expected a year earlier. Net farm income was projected to average U.S. $5.7 billion less over the same period, indicating a revealed parameter estimate that each U.S. $1 decline in exports generates a 44-cent decline in net farm income. This is not a parameter in an explicitly specified equation, but summarizes the results of a lot of parameters that trace through the supply/demand consequences of a particular commodity mix of changes in export prospects. The point is that this is a large impact. It says that if one-half of the U.S. $65 billion in exports projected for 2000-05 were to disappear completely, then net farm income would fall by U.S. $14.3 billion; that is, about one-third of net farm

Table 4.1 Changes in Projected Exports and Associated Changes in Net Farm Income

Year	*Agricultural exports*[a]		*Net farm income*[a]		*Net income gain per dollar increase in exports*
	USDA (1998)	*USDA (1999)*[b]	*USDA (1998)*	*USDA (1999)*[b]	
	billion dollars		*billion dollars*		*dollars*
1997	57.3	57.3	45.3	49.8	–
1998	58.5	53.6	46.2	48.0	–
1999	59.7	50.5	47.2	46.4	–
2000	62.6	50.2	49.0	43.8	0.42
2001	66.2	53.9	49.9	45.3	0.37
2002	69.4	56.7	50.8	46.0	0.38
2003	72.6	59.9	53.3	47.3	0.47
2004	76.0	62.8	56.0	49.0	0.53
2005	79.8	65.2	57.3	50.5	0.47

[a]*Fiscal year for exports, calendar year for income.*
[b]*1999 estimates based on information as of November 1998; 1998 estimates based on information as of November 1997.*
Source: USDA (1999a).

income would disappear (even though one-half of exports amount to only 15 percent of farm product sales).

The multipliers that translate changes in demand for exports into farm-level prices and output are based on econometric supply/demand equations estimated from annual time-series data. The effects on farm income are less well grounded econometrically. Using estimates of variable production costs for purchased farm inputs, the reduced costs generated by output decreases in response to decreased exports and are subtracted from reduced revenues (that is, price times quantity plus government payments). But the cost decreases are estimated from constant cost-per-acre data that do not reflect how input prices respond to changes in input demand. In fact our knowledge is quite weak about input-supply elasticities in either the short or long run. This is important because, if purchased input prices fall, this could significantly moderate the putative decreases in net farm income. On the other hand, if the returns to farmer-owned labor, equipment, and land were more responsive to demand declines, this would make the detrimental effect on farm income even larger—indeed, the returns to these factors is what net farm income consists of. Unfortunately, we have even less solid knowledge about the decrease in demand for these specific factors of production that an export-demand decrease generates, than we have about the factor-supply response to any short-term decrease in factor returns. In short, despite this chorus of favorable opinion, and the pessimistic farm income effects of export declines that USDA projects, analytical support for the importance of agricultural trade for farm income, especially over the long term, is thin.

Moreover, for welfare analysis of trade, it is not sufficient to relate aggregate farm income to exports. We need per-person income, or factor prices. For example, the short-run decline of farm income shown in Table 4.1 for the year 2000 might be mainly a matter of factor returns declining with relatively constant factor quantities (as factor supplies are less elastic in the short run). But the longer run farm income effects shown in 2005, and for later years not shown, might be a result of less labor, land, and farmer-owned capital in agriculture, with prices not much lower (no lower if factor supplies were perfectly elastic). This, after all, is how real wage rates in agriculture relative to non-farm employment have managed to stay constant, or even to rise, over the last 50 years even though the demand for labor in agriculture has fallen by more than one-half. This provides even more skepticism about how worried we have to be to maintain the export market.

What about more direct evidence of association between agricultural exports and farm income? Just look at the data for the 20th century. Figure 4.1 shows the value of agricultural exports is quite steady at about U.S. $15 billion (that is, in 1992 dollars) during most of 1900-30, with about twice that amount in 1915-20. Agricultural imports, however, mostly of crop products such as coffee, rubber, and topical products not directly competitive with U.S. products, more than doubled in real value between 1900 and 1930. Thus, agriculture went from being a contributor to a positive U.S. balance of

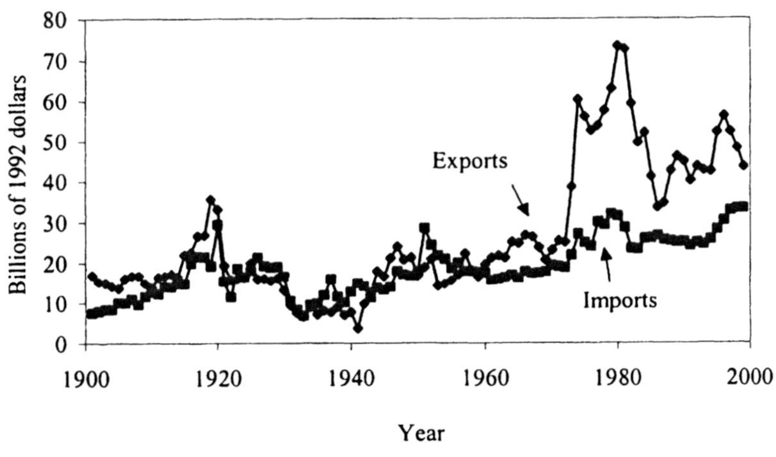

Figure 4.1 Agricultural Exports and Imports, Value in 1992 dollars

merchandize trade in 1900-20 to being a trade-deficit sector in 1925-30. During the Great Depression, both exports and imports plummeted. This has been attributed to growing protectionism, as exampled by the Smoot-Hawley Tariff of 1930 (although tariffs on agricultural imports had been raised substantially well before this in the 1920s). More broadly, the collapse of farm trade in the 1930s is part of the worldwide shrinkage of trade that accompanied and contributed to the Great Depression. Agricultural exports rebounded to about U.S. $20 billion annually after World War II. Farmers came to believe in the benefits of strong world markets vividly at this time—not only market-driven sources of demand, but also the demand for U.S. products created by the Marshall Plan. This experience paved the political way so that when agricultural exports began to weaken in the 1950s, food aid and export-subsidy programs were introduced to maintain and build foreign markets. In the 1960s, agriculture began again to be seen as a significant contributor to a positive U.S. trade balance, and a source of economic benefits to the economy generally, as well as to farmers.

The major story of agricultural exports in our century really did not get under way until the 1970s. It began with a bang, when the Soviet Union massively and unexpectedly bought a quantity of wheat equal to about one-fifth of worldwide exports in 1972. U.S. wheat prices soared, within a matter of weeks turning discussion from grain surpluses to bakers testifying before Congress that something had to be done to protect U.S. consumers from the "grain robbery" fostered by the Soviets. For years, commentators ignored agriculture, and publications, such as Morgan (1979), blamed the large international grain-trading companies for profiteering and acting so as not to permit the farm-level price of grain from reflecting the prices the Soviets were paying. Within two months of the Soviet entry into the markets (July, 1972), the Nixon Administration imposed an embargo upon further shipments of grain to the USSR while government-to-government negotiations were held to determine the level of grain sales that would be permitted. For the next eight years the U.S. government became the regulator of U.S. grain exports, with quantitative restrictions imposed by Presidents Nixon, Ford, and Carter at various times. Nonetheless, the value of agricultural exports reached unprecedented heights for almost a decade.

Despite grumbling about exploitation by the grain-exporting companies, it is hard to argue against the conclusion that farmers benefited greatly from the export boom. The average price of wheat received by farmers increased from U.S. $1.34 per bushel in 1971 to U.S. $3.95 per bushel in 1973, and to U.S. $4.09 in 1974. The farm price of corn rose from U.S. $1.08 in 1971 to U.S. $3.03 in 1974. Grain growers responded to the astonishing profits which resulted by buying new machinery, expanding acreage planted, and bidding up the price of cropland to previously unimaginable levels.

The longer-run farm income effects of increased exports are less clear. Even in the case of the 1970s, it could be argued that the long-run benefits to farmers have been questionable, because supply response to higher prices generated subsequent low grain prices that were at the heart of the farm crisis of the mid-1980s.

An overall indicator of the importance of trade in the U.S. farm economy is the net agricultural trade balance (that is, the value of exports minus the value of imports) as a percentage of the market value of all U.S. farm output. This export-intensity index indicates the fraction of demand for U.S. farm products that is accounted for by foreign markets (Figure 4.2). Fluctuations in this index indicate that trade was important in the commodity booms of World War I, World War II, and the 1970s. But as a cause of long-term market trends in the 20th century, the role of trade is less clear. Until the mid-1970s, trade never became more important as a share of the U.S. market than it was before 1920. Since the 1960s, U.S. agriculture has entered a global market to an extent unprecedented in this century. Net exports during the 1990s were twice as large as the farm-share of commodity sales were from 1910-20, but net exports were still only 10 percent of sales.

We show the time-series relationship between real net exports and net farm income (Figure 4.3). Not only do the two series have trends with opposite signs, but the short-term co-movements of the two series do not look as strongly positive as one might expect. Both net exports and farm income rise

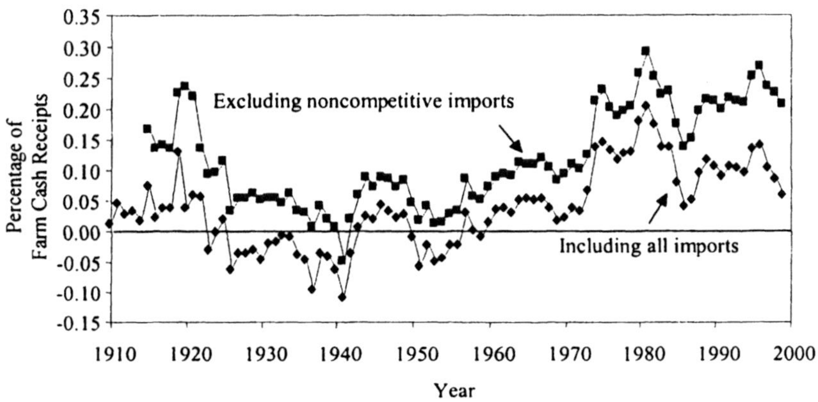

Figure 4.2 Real Agricultural Exports minus Imports as a Percentage of Farm Cash Receipts

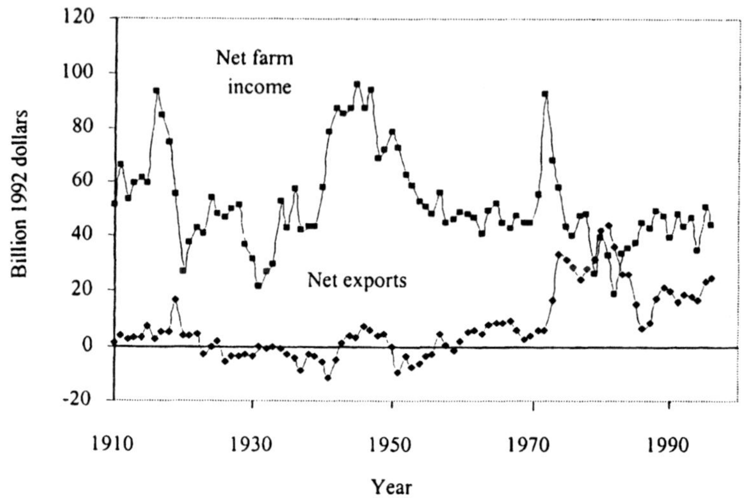

Figure 4.3 Agricultural Net Exports and Net Farm Income Beginning 1910

in the 1940s and in the early 1970s, but in the later 1970s exports rose while farm income fell. In the 1980s and 1990s the connection between the two series was not visually obvious.

4. DISAGGREGATION AND ECONOMETRICS OF EXPORTS

Econometric analysis of agricultural trade and farm income has to deal with a number of problems, some of them intractable. The first is aggregation over commodities. The previous discussion treats agricultural output as one good. In the simplest economic model of trade it is odd to have the same good both exported and imported at the same time. This happens because of locational factors, such as when the U.S. imports wheat from Canada to Minneapolis while it exports wheat from U.S. ocean ports. This happens because of seasonal factors, such as when the United States imports some fruits in the winter and exports them in the summer (generating simultaneous imports and exports within the annual data). This also happens when a product is exported at one stage of growth and is imported at another, as with calves and cattle between the United States and Mexico. In all these cases, we have to define our commodities quite precisely in order to understand what is happening in terms of the standard economics of trade. In these cases, interesting as they are, it is unlikely that aggregation distorts our understanding of the greater picture of the role of trade in the U.S. farm economy. However, more gross aggregation may cause such problems.

An example of this is when the net agricultural-export balance aggregates wheat and coffee. If a world coffee supply increase causes a surge in U.S. coffee imports, it is unlikely to have more of an effect on farm income than if the imports of coffee cups increase. That is a negligible effect when compared to the effect of a foreign wheat-supply increase that had the same dollar value on the agricultural-trade balance. The USDA recognizes this, and, in their statistics, they divide agricultural imports into competitive and noncompetitive categories.[viii] We show both net agricultural exports in Figure 4.2, and perhaps the more meaningful indicator omits noncompetitive imports. This latter competitive-trade balance is negative only in 1941. Over time, the behavior of both indicators of the agricultural-trade balance becomes similar. By either measure, net agricultural exports declined substantially between the pre-World War I period and the Great Depression era of the 1930s. Following the mid-1950s, net-agricultural exports steadily increased. Probably the best overall indicator of how important trade is to the U.S. farm economy is to estimate the total of net competitive exports expressed as a percentage of farm-product sales receipts. (In the late 1990s, this was about 25 percent.)

Figure 4.4 shows the percentage of agricultural imports that are noncompetitive. This percentage has declined since the 1950s from about 60 percent

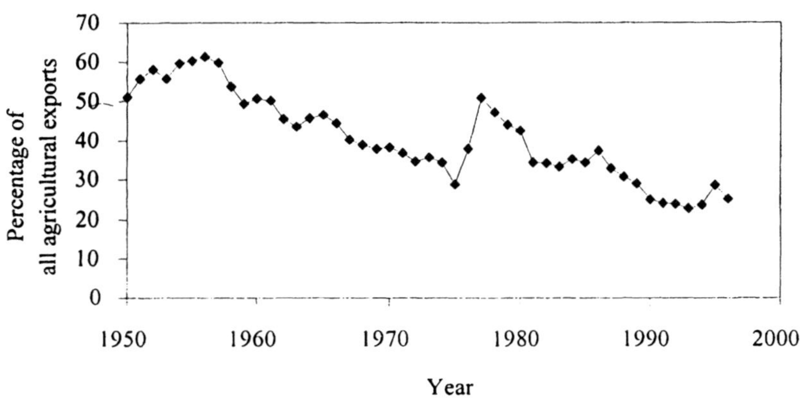

Figure 4.4 Percentage of U.S. Agricultural Imports that are Noncompetitive

to 25 percent. This could be taken as an indication that U.S. producers have learned to grow a wider range of products, such that ones that were formerly noncompetitive have now become competitive. However, the USDA has not changed its list of noncompetitive commodities, so this cannot explain the statistics. Two other possible economic reasons could be that the noncompetitive commodities have either lower income elasticities or have increased in price relative to the U.S.-produced commodities. The former explanation is not plausible. U.S. commodities are more heavily weighted with bulk staple products and imports are more heavily weighted with exotic luxuries. So, if anything, the noncompetitive share should have increased over time. With respect to relative prices, the most important of these commodities, (that is, coffee, cocoa, and tropical fruits), have not increased much in price relative to U.S. farm-commodity prices.[ix]

Given the lack of straightforward market explanations for the reduced share of noncompetitive imports, a possible explanation of the long-term trends in Figure 4.4 could be trade protection. Using several measures, protection increased in the 1920s and 1930s, and decreased in the 1960s and beyond. In addition, protection was not sought for noncompetitive imports. Thus, the noncompetitive-import share of total imports increased in the 1920s and decreased in the 1960s and after.

Within the set of exported commodities itself, aggregation may cause problems. Likely, the most serious is aggregation over different stages of production or processing. USDA trade data are subdivided into two large categories—animal products and vegetable products. A substantial fraction of the output of the latter category is an input into production of the former. In addition, processed food products are included in the data. The problem this could cause for the present issue is that, notwithstanding the current buzz favoring value-added exports, the share of an export dollar going back to farmer-owned resources, particularly land, is highest for grains and other bulk products and lowest for processed products. So the impact of a dollar of export sales on farm income might decline over time if the share of high-value exports is increasing. Fortunately (for purposes of stability over time in regression coefficients) this share does not have a strong trend over time. Table 4.2 shows the commodity breakdown from 1908-10, 1950, and 1995. Applying the USDA definition of bulk products, which defines bulk products as grains, oilseeds, cotton, raw tobacco, and a few other raw products, but not flour, meals, oils, meats, or other processed products, the share of U.S. agricultural exports accounted for by the bulk products listed is 63 percent for 1908-10; 52 percent for 1919; 71 percent for 1950; 60 percent for 1980; and 47 percent for 1996.

Table 4.2 Ordinary Least Squares (OLS) Regression Coefficients

Independent variable	*Farmland price*	*Change in farmland price*
Exports	6.20	4.40
	(3.5)[a]	(6.04)
Exports lagged 1 year	1.06	1.55
	(0.4)	(2.06
Exports lagged 2 years	3.14	1.97
	(1.13)	(2.64)
Exports lagged 3 years	3.82	2.10
	(2.13)	(2.91)
R2	.904	.446

[a]Numbers in parenthesis denote t-statistics.
Source: Estimated coefficients.

Apart from data problems, time-series analysis of exports, farm prices, and farm income is fraught with pitfalls. Prices and exports are mutually determined using ways that standard-time, series-regression methods cannot handle well because there is no stable underlying causal relationship between the two variables. Some changes in exports, like those changes generated by Soviet sales in the 1970s, are obvious shifters of the demand for U.S. farm products. But other changes, such as the sustained growth of U.S. broiler exports in the 1980s and 1990s, are most likely caused by U.S. supply-side changes that have reduced costs.

Attempts have nonetheless been made to estimate the effects of exports on prices using ordinary least-squares (OLS) regressions on time series for short periods in which foreign demand fluctuations were believed to be the main factors in play. One of the few of these estimation efforts to appear in print is (USDA, 1986: 5–22):

$$(1) \qquad \underset{}{(P_t - P_{t-1}) = \underset{(3.16)}{3.704} + \underset{(7.76)}{3.555}(X_t - X_{t-1})}$$

in which P_t is the USDA index of crop prices received by farmers, and X_t is the value of U.S. agricultural exports in 1970 dollars, in year t. The equation was estimated using data from 1971-85. Figures in parentheses are t statistics, and indicate a highly positive effect of export changes on prices. The parameter estimate says that a billion dollar increase (about 3 percent) in real export value generates a 3.55 point increase (about 2.5 percent) in farm-level prices. This estimate suggests that each billion dollars in exports generates even more farm income than the USDA baseline adjustments did in the earlier discussion. But one cannot be confident in the magnitude of the estimated coefficient because of simultaneous equation bias, or in the estimated significance because of likely non-stationarity of the variables. Time series for longer periods, such as those estimated by Gardner (1996) to explain real net-farm income over the 1911-90 period, show the classical symptoms of spurious regression, notably significant t values for explanatory variables whose signs and magnitudes are highly sensitive to the inclusion, or the exclusion, of other variables.

Recent attempts to sort out the causal relationships between export quantities and prices have used vector autoregression to determine Granger causality (defined in terms of statistical significance of one series leading another) without attempting to uncover the underlying economic structure. Babula, Ruppel, and Bessler (1995) undertook such an analysis to consider the long-standing issue of exchange-rate effects on U.S. exports and commodity prices, using monthly data for corn from 1978-92. This topic lends itself to time-series analysis because there is a reasonable prior expectation that the exchange rate between the U.S. dollar and an average of other currencies is exogenous with respect to corn sales and prices. The study focused on the distinction, which is important for establishing the causality of exogenous variables upon exports, between announced sales (reported to the USDA) and export shipments that often are months later. Since sales are the first public information about exports, changes in sales might be expected to lead price changes more consistently than do export shipments. However, Babula, Ruppel, and Bessler (1995) find no significant relationship among these variables, except weak short-term (1 and 2 month) effects of exchange rates influencing prices, and of shipments (counter-intuitively) leading sales. For present purposes the key results are the lack of any long-run relationship between the corn price and either measure of exports, and in the short run it is prices that cause exports rather than exports causing sales.

Given the preceding results, it is hard to see how U.S. data evidence from time series could uncover the effects of exports on farm income in the absence of a fully specified structural model of U.S. agriculture, that is, a model that incorporated all the relevant supply- and demand-side forces that influence farm prices and quantities. Despite valiant attempts, no such model has been achieved. History has not provided enough independent observations of alternative levels of important variables during a time period long enough to provide statistical significance, but not so long that the underlying structural parameters cannot reasonably be assumed to have remained constant. A piecemeal approach, as followed in the USDA equation reported above, may provide useful evidence, but it leaves the question of whether the selection of data it uses really represents the usual, or the current, situation. Moreover, even with a well-specified model there remains ambiguity in the question of how to estimate the effect of export markets on farm income. The thrust of such a model, and of the preceding discussion in this chapter, is that the putative benefits to farmers comes from increased export demand that drives commodity prices up. It is likely, however, that many of the benefits to U.S. agriculture from exports have not been driven by exogenous increases in demand, rather they have been driven by having the export market available as an outlet for U.S. products when their prices fall because of increases in U.S. agricultural productivity. For 2001, the best examples are in livestock products, particularly poultry, but also pork and dairy products.[x]

The approach that seems to me most promising for isolating the long-term effects of export markets on farm income is an indirect one, namely looking at the relationship between exports and farmland prices. Land is the closest approximation to a specific factor in agriculture, so that if exports are important either as an exogenous source of demand or as an endogenous source of response to lower U.S. prices, the beneficial effects should reveal themselves in land prices.[xi] Figure 4.5 shows the time series of real land prices and of exports from 1910 to 2000. There is an obvious co-movement within the series. Nonstationarity is rejected for both series, suggesting an error-correction model (ECM) to test for cointegration between the series. The ECM[xii] equation is:

$$(2) \qquad (P_t - P_{t-1}) = \beta_0 + \beta_1 (P_{t-1} - X_{t-1}) + \beta_2 (X_t - X_{t-1}) + \varepsilon_t,$$

where P is the price of farmland, X is exports (P and X are in U.S. 1992 dollars), t is a time subscript, β is the parameter to be estimated, and ε is an independent and identically distributed (i.i.d.) error term. When estimated on the data shown in Figure 4.5, β_1 is not significantly different from zero (t = 0.2), but β_2 is significant with a coefficient of 4.43 and t = 5.4. The estimate

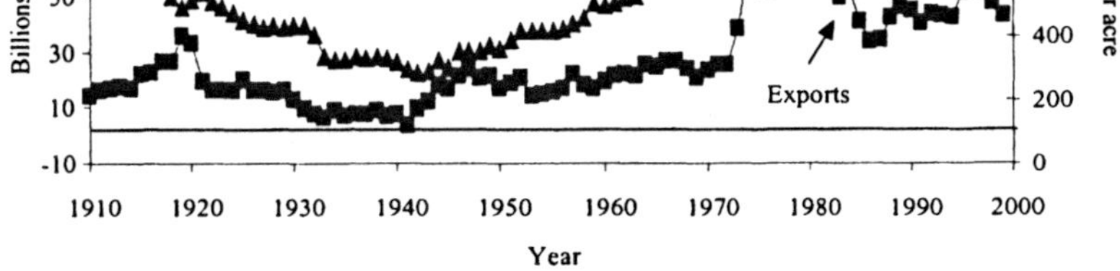

Figure 4.5 Agricultural Exports and Real Farmland Prices

of $\beta_2 = 4.43$ means a billion-dollar increase in exports increases the price of land by U.S. $4.43.

Given the lack of significance of the error-correction term, we considered models in which the level of farmland price is a function of current and lagged values of exports, and alternatively, changes in farmland price are a function of lagged changes in exports. Lags were added until the longest lag generated a statistically insignificant effect. The estimated results for a 3-year lag specification for the period 1913-96 are presented in Table 4.3.

Note that the 4.40 coefficient on exports is almost the identical result of the error-correction model above. (The right-hand side variables in the price-change equation are changes in exports.) Assuming negligible effects after 4 years, meaning a 4-year sustained increase in exports is the most that registers on land price, a permanent increase in exports of U.S. $1 billion generates a land price increase of U.S. $14, that is, the sum of right-hand side coefficients, from the levels equation or U.S. $10 from the price-change equation. We also estimated restricted versions of these two models in which the only right-hand side variable is the 4-year average of exports, or changes, in exports. The coefficients are then 14.5 in the price-level regression and 9.8 in the price-change regression, indicating that summing the coefficients above does not give misleading results. As an alternative, these equations were estimated using Iowa cash rents for the 1923-96 period as the dependent variable, with an estimated effect of a sustained U.S. $1 billion increase in exports being U.S. $1.40 to U.S. $1.70 in per-acre rental price. Given historical rent-value ratios for U.S. farmland, the rental price and land value results are consistent.

The above results indicate that a sustained increase in agricultural exports of U.S. $1 billion generates a one-time U.S. $10 to U.S. $14 per-acre gain to farmers. With 900 million acres in farms, this gain amounts to U.S. $9 to U.S. $13 billion. This suggests that if agricultural exports were cut by one

Table 4.3 Three Lag Time Series Model

Independent variables	*Farmland price*	*Change in farmland price*
Exports	6.2	4.4
	(3.5)[a]	(6.04)
Exports lagged 1 year	1.06	1.55
	(0.4)	(2.06)
Exports lagged 2 years	3.14	1.97
	(1.13)	(2.64)
Exports lagged 3 years	3.82	2.1
	(2.13)	(2.91)
R^2	0.904	0.446

[a]Numbers in parenthesis denote t-statistics.
Source: Estimated coefficients.

half, that is by U.S. $25 billion (13 percent of the annual value of farm output), the loss in U.S. farm real-estate value would be U.S. $225 to U.S. $325 billion, or 25 percent to 35 percent of the value of farm assets. Thus, the estimated relationship between exports and real-estate values indicates that the export market is indeed important for U.S. farmers' income situation.

5. CONCLUSIONS

Evidence, which is both anecdotal and statistical, indicates that the export market has been, and continues to be, important to the economic health of U.S. agriculture. The main exception is that historical data do not reveal a positive relationship between exports and U.S. net farm income. When periods of booming export demand occur, the positive effects on farm prices and incomes become apparent. But there are other periods, during which export demand is not increasing, when the dynamics of U.S. agriculture owes more to technological change, to corresponding productivity growth, and to increasing international competitiveness. With inelastic aggregate demand for U.S. farm products, but with a relatively elastic portion of that demand in the export market, it is plausible that during these periods we would see declining net farm income along with increasing exports. Nonetheless, if it had not been for export expansion to absorb a portion of output increases, net farm income would have fallen even more. Similarly, during some years in the 1950s and 1960s, labor massively exited out of agriculture in response to better earning prospects in non-farm work. This tended to raise labor returns in agriculture, and reduce farm output, *ceteris paribus*. The partial effect (the total effect perhaps still is dominated by technological change) of reduced output is less exports associated with higher farm income. A properly specified econometric model of U.S. agriculture could, in principle, identify the separate effects of technical change and other determinants of supply, so that the consequences of exports could be identified. But no such model exists and this chapter has not remedied that lack. Instead we focused on farmland prices, on the grounds that confounding effects of productivity growth and labor migration would be less of a problem. And, we find quite strong effects of exports on farmland prices (and cash rents) that look apparent in the data and are not plausibly the result of spurious correlation. Thus, it appears that the importance many farm groups attach to the export market is not misplaced.

6. ENDNOTES

[i] President Reagan on numerous occasions expressed his view that the hard times of his Depression-era youth could be tied to trade protectionism. Vice-President Gore made a vivid presentation of the theory in his TV debate with Ross Perot when he showed pictures of Senators Smoot and Hawley as representative of the protectionist idols with which Perot allied himself.

[ii] On the agricultural story c, see Brandow (1977), Cochrane and Ryan (1976), or Clarke (1995). J.K. Galbraith highlighted the controversial element in this position by asserting that price supports "made it possible for farmers to invest in fertilizer, machinery, hybrid seed stock and other technology...Thus they enormously enhanced such investment and nourished an increase in output per farm worker that much exceeded the productivity gain in industry. Nevertheless the support prices had never ceased to be a source of distress to economists, who preferred their free market faith to practical achievement." (Galbraith, 1981: 356). On the broader economic front of support for protection, Paul Krugman has been a recent instigator of Hamiltonian notions (referring to Alexander, not the dynamic optimization tool).

[iii] Another policy-based argument is that national gains from trade are sufficient for sectoral gains because sectoral losers can in principle be compensated from the gains of the gainers, leaving all sectors better off. But we cannot be sure that actual compensation is feasible, and we can be sure from empirical observation that the prospect of compensation does not induce import-competing sectors to support increased trade.

[iv] See Raneses et al. (1999) and references therein for details about the FAPSIM model. The basic reference remains Salathe et al. (1982).

[v] Larry Salathe, Office of Economics, USDA, personal communication.

[vi] The discussion here is elaborated in Gardner (2000).

[vii] Other changes of course occurred during 1998, most notably crop failures and low prices, the resulting U.S. $6 billion agricultural relief legislation, and modest strengthening in the U.S. macroeconomic outlook. But these are mainly short-term (1 or 2 year) factors and probably did not account for a significant part of the change in USDA's projections for 2001-2005. Indeed, USDA's "Baseline Highlights" publication lists no factors causing changes in the baseline other than those influencing exports (USDA, 1999b).

[viii] "Competitive agricultural import products consist of all products similar to agricultural commodities produced commercially in the United States, together with all other agricultural products interchangeable to any significant extent with such commodities" (USDA, 1998: Table 15-9). All other agricultural commodities are noncompetitive. This definition makes palm oil, for example, competitive even though it is not produced commercially in the United States. The definition also excludes complementary demand relationships, for example, more coffee might increase the demand for cream. Some substitution that almost surely exists is reckoned insignificant: raw silk, duty free wool, and some other fibers are noncompetitive even though they are likely to substitute for U.S. wool and cotton to some extent. This is just to illustrate that as always in empirical economics it is quite difficult to implement even fairly straightforward concepts cleanly. Before 1980 USDA labeled the two categories "supplementary" and "complementary" products, but the definitions and product coverage of "supplementary" and "competitive" are the same.

[ix] There is no price index for the noncompetitive products, but taking coffee and cocoa, which make up a large fraction of these products, the unit value of imports decreased about 1.7 percent annually in real terms between 1950-51 and 1995-96, while the real prices-

received index for U.S. farm products decreased 1.9 percent annually over the same period.

[x] This point is related to, but distinct from, the arguments recently made about "value-added" or "high-value" exports being preferable to bulk commodity exports. The livestock products mentioned are high-value in USDA's accounting but the reason their export is profitable is because of the U.S. comparative cost advantage, not just because they add value beyond the farm gate.

[xi] Rental rates for farmland would be an even better indicator, because land prices may be too much influenced by interest rates or non-farm opportunities to reflect effects on current income well. But we do not have as long a consistent U.S. data series for rental rates. And, in fact land prices and rental rates correlate quite well over the time period for which data for both are available, even including the period of land price boom and bust of the 1970s and 1980s.

[xii] See Engle, R.F. and C.W.J. Granger. (1987). "Co-Integration and Error Correction: Representation, Estimation, and Testing." *Econometrica* 55 (251–276).

7. REFERENCES

Babula, R.A., F.J. Ruppel, and D.A. Bessler. 1995. "U.S. Corn Exports: the Role of the Exchange Rate." *Agricultural Economics* 13 (November): 75–88.

Brandow, G.E. 1977. "Policy for Commercial Agriculture: 1945–72," in L. Martin, ed., *A Survey of Agricultural Economics Literature, Vol. 1*, pp. 209–292. Minneapolis, MN: University of Minnesota Press.

Chambers, C.R. 1924. "Relation of Farm Land Income to Farm Land Value." *American Economic Review* 14 (December): 673–698.

Chambers, R. and P. Paarlberg. 1991. "Are More Exports Always Better?" *American Journal of Agricultural Economics* 73: 142–154.

Clarke, S. 1995. *Regulation and the Revolution in U.S. Farm Productivity.* Cambridge, MA: Cambridge University Press.

Cochrane, W. and M. Ryan. 1976. *American Farm Policy.* Minneapolis, MN: University of Minnesota Press.

Engle, R.F. and C.W.J. Granger. 1987. "Co-Integration and Error Correction: Representation, Estimation, and Testing." *Econometrica* 55: 251–276.

Galbraith, J.K. 1981. *A Life in Our Times.* Boston, MA: Houghton-Mifflin.

Gardner, B.L. 2000. "International Trade and the Future of American Agriculture." *Journal of Agribusiness* 18 (March): 27–40.

_____. 1996. "The Federal Government in Commodity Markets." *Agricultural History* 70: 177–195.

_____. 1989. "Recent Studies of Agricultural Trade Liberalization," in A. Maunder and A. Valdes, eds., 361–371. *Agriculture and Governments in an Interdependent World,* Internation Association of Agricultural Economists (IAAE) Proceedings. Brookfield, VT: Gower Publishing.

Jones, C.I. 1999. "Growth: With or Without Scale Effects?" *American Economic Review* 89 (May): 139–144.

Morgan, D. 1979. *Merchants of Grain.* New York, NY: Viking Press.

Raneses, A.R., L.K. Glaser, J.M. Price, and J.A. Duffield. 1999. "Potential Bio-diesel Markets and Their Economic Effects on the Agricultural Sector of the U.S." *Industrial Crops and Products* 9: 151–162.

Salathe, L.E., J.M. Price, and K.E. Gadsen. 1982. "The Food and Agricultural Policy Simulator." *Agricultural Economics Research* 34 (April): 1–15.

Schuh, G.E. 1974. "The Exchange Rate and U.S. Agriculture." *American Journal of Agricultural Economics* 56: 1–14.

Stigler, G.J. 1951. "The Division of Labor is Limited by the Extent of the Market." *Journal of Political Economy* 59: 185–193.

USDA (U.S. Department of Agriculture). Various years. *Agricultural Statistics*. Washington DC: U.S. Government Printing Office.

_____. 1999a. *Agricultural Baseline Projections to 2008*. World Agricultural Outlook Board, Staff Report No. WAOB-99-1. Washington DC: U.S. Government Printing Office (February).

_____. 1999b. *USDA Baseline Highlights*. World Agricultural Outlook Board. Washington DC: U.S. Government Printing Office.

_____. 1998. *Agricultural Baseline Projections to 2007*. World Agricultural Outlook Board, Staff Report WAOB-98-1. Washington DC: U.S. Government Printing Office (February).

_____. 1986. Embargoes, Surplus Disposal, and U.S. Agriculture. Economic Research Service, Agricultural Economic Report No. 564. Washington, DC: U.S. Government Printing Office (December).

World Bank. 1986. *World Development Report*. New York, NY: Oxford University Press.

Chapter 5

Trade Liberalization and Small Economies: The Case of the Caribbean Community

Carlton G. Davis[1], Ballayram[1], Edward A. Evans[1], and Clive Y. Thomas[2]
[1]University of Florida and [2]University of the West Indies

1. INTRODUCTION

The Uruguay Round of the General Agreement on Tariffs and Trade/World Trade Organization (GATT/WTO) confirm the trend toward a liberal international trade regime. The chief features of the agreement are: (1) reciprocal multilateral-trade arrangements based on the prohibition of nontariff barriers (NTBs) and the progressive reduction of tariffs; (2) market-driven economic imperatives and performance-based results; (3) regional or sub-regional co-operation oriented toward open regionalism; (4) linkage of trade to other areas of liberalization, especially trade-related property issues, investment, finance, and the movement of skilled workers and management; (5) linkages of trade to certain non-economic areas, such as environment and labor standards; and, (6) reinforcement of the predictability and certainty of the trade arrangements through appropriate compliance and enforcement mechanisms.

The proposed Free Trade Area of the Americas (FTAA) is intended to establish a regional trading arrangement (RTA) that would advance and deepen the WTO's liberalization among the 34 countries of the Western Hemisphere. The FTAA's aims are: (1) to liberalize trade barriers; (2) to eliminate restraints on investment; (3) to provide free labor movement for specialized workers; (4) to harmonize tax and monetary policies; and (5) to establish supra-regional institutions to administer the arrangement and to handle dispute resolution. These five elements take the FTAA process well beyond the traditional extension of trade reciprocity among trading parties (Bryan, 1995).

The commitment to go beyond traditional reciprocity is critical to the realization of one of its major challenges: the integration of countries of very diverse sizes and levels of development. At the first trade ministers' meeting of FTAA countries in June 1995, 11 working groups were established. One of these was the Working Group on Smaller Economies with a mandate to "identify and examine ways to facilitate the participation of smaller economies in the FTAA process, and to provide recommendations as to how this should be done" (Bernal, 1998: 2).

The structural characteristics of the Caribbean Community and Common Market (CARICOM) group of 14 countries highlights the dilemma of size involved in bringing about an operationally effective FTAA by the year 2005.[i] Since agriculture is an important activity in this region, a considerable proportion of the economic and policy adjustments will have to take place with respect to that sector.[ii] This chapter identifies some of the critical agricultural-adjustment challenges.

2. SOME DIMENSIONS OF CARICOM COUNTRIES AS SMALL ECONOMIES

CARICOM has a total population of approximately 6.0 million, a total gross domestic product (GDP), at 1997 market prices, of U.S. $14.08 billion, and a combined land area of 434,000 square kilometers (sq. km.) We present a relative ranking of the 34 countries that comprise the FTAA with respect to these indicators (Table 5.1). The CARICOM countries (underlined in Table 5.1)—which account for only 0.67 percent of the land area, 0.82 percent of the population, and about 0.2 percent of total GDP—are the lowest in the ranking. The United States is the most populous and is the highest-income country in the FTAA; the least populous and lowest-income countries are St. Kitts and Nevis. Barbados has the smallest land area, and Canada has the largest.

Apart from their small size, CARICOM[iii] countries exhibit other characteristics which further define the complexity of the economic dilemmas that they represent within the FTAA. Included among these characteristics are: (1) relatively high degree of openness as reflected in trade-to-output ratios; (2) relatively non-diversified sectoral and general economic structure; (3) export concentration in a narrow range of products, combined with import reliance for a very wide range of products; (4) generally small firm size with limited opportunities to take advantage of economies of scale, except in export areas such as minerals (oil, bauxite, gold) and plantation agriculture (sugar); (5) a high degree of dependency on nonreciprocal trading arrangements, for historical reasons; (6) a limited resource base in the export agri-

Table 5.1 Ranking of FTAA Countries on Basis of Population, Land Size, and GDP

Rank	*Country*	*Population (millions)*	*Country*	*Land size (1,000 sq. km)*	*Country*	*GDP (1995 US$ million 1990=100)*
34	*St. Kitts and Nevis*	0.04	*Barbados*	0.3	*St. Kitts and Nevis*	150
33	*Antigua and Barbuda*	0.06	*Grenada*	0.3	*Dominica*	178
32	*Dominica*	0.07	*St. Kitts and Nevis*	0.3	*St. Vincent and Grenadines*	218
31	*Grenada*	0.09	*Antigua and Barbuda*	0.4	*Grenada*	229
30	*St. Vincent and Grenadines*	0.10	*St. Vincent and Grenadines*	0.4	*Suriname*	334
29	*St. Lucia*	0.14	*St. Lucia*	0.6	*Antigua and Barbuda*	366
28	*Belize*	0.21	*Dominica*	0.7	*Belize*	491
27	*Barbados*	0.26	*Trinidad and Tobago*	5.0	*St. Lucia*	509
26	*Bahamas*	0.27	*Jamaica*	11.0	*Guyana*	602
25	*Suriname*	0.42	*Bahamas*	14.0	*Haiti* [a]	1,642
24	*Guyana*	0.82	*El Salvador*	21.0	*Barbados*	1,717
23	*Trinidad and Tobago*	1.26	*Belize*	23.0	*Nicaragua*	2,590
22	*Jamaica*	2.53	*Haiti* [a]	28.0	*Bahamas*	3,053
21	*Panama*	2.63	*Dominican Republic*	49.0	*Honduras*	3,378
20	*Uruguay*	3.19	*Costa Rica*	51.0	*Jamaica*	4,171
19	*Costa Rica*	3.07	*Panama*	76.0	*Trinidad and Tobago*	5,707
18	*Nicaragua*	4.54	*Guatemala*	109.0	*Bolivia*	6,496
17	*Paraguay*	4.83	*Honduras*	112.0	*Panama*	6,570
16	*El Salvador*	5.64	*Nicaragua*	130.0	*El Salvador*	6,674
15	*Honduras*	5.95	*Suriname*	163.0	*Costa Rica*	7,027
14	*Haiti* [a]	7.18	*Uruguay*	177.0	*Paraguay*	7,177
13	*Dominican Republic*	7.91	*Guyana*	215.0	*Dominican Republic*	7,341
12	*Bolivia*	8.06	*Ecuador*	284.0	*Guatemala*	9,706
11	*Guatemala*	10.62	*Paraguay*	407.0	*Uruguay*	11,431
10	*Ecuador*	11.46	*Chile*	757.0	*Ecuador*	15,132
9	*Chile*	14.20	*Venezuela*	912.0	*Peru*	47,618
8	*Venezuela*	21.64	*Bolivia*	1,099.0	*Chile*	48,326
7	*Peru*	23.53	*Colombia*	1,139.0	*Colombia*	56,379
6	*Canada*	29.61	*Peru*	1,285.0	*Venezuela*	64,980
5	*Argentina*	34.59	*Mexico*	1,958.0	*Argentina*	196,949
4	*Colombia*	35.10	*Argentina*	2,767.0	*Mexico*	250,936

(continued)

Table 5.1 (Cont'd) Ranking of FTAA Countries on Basis of Population, Land Size, and GDP

Rank	*Country*	*Population (millions)*	*Country*	*Land size (1,000 sq. km)*	*Country*	*GDP (1995 U.S.$ million 1990=100)*
3	*Mexico*	94.78	*Brazil*	8,512.0	*Brazil*	432,433
2	*Brazil*	159.22	*United States*	9,809.0	*Canada*	608,658
1	*United States*	263.06	*Canada*	9,976.0	*United States*	6,173,900

[a]Haiti has satisfied all the conditions for full entry in CARICOM, except an instrument of accession. When this instrument is submitted, Haiti will have become a full member of CARICOM.
Source: Bernal (1998).

culture sub-sectors, utilizing technologies oriented toward labor intensity and associated high-cost production structures (West Indian Commission, 1992); (7) high traditional reliance on concessional capital flows and foreign direct investment (FDI); and (8) severe diseconomies of scale and lumpiness of infra-structural investments. We present these structural characteristics in Tables 5.2 and 5.3.

While all the CARICOM countries are small, there is wide variation among them. First, there is the sub-grouping of smaller states (listed in end-note 1) within the Organization of Eastern Caribbean States (OECS). Cooperation between countries in the OECS exceeds that between countries in the larger group. There is a common Central Bank in the OECS, which makes monetizing the domestic debt, or the pursuit of an independent exchange-rate policy less of a problem than in the larger countries. There is also some amount of policy coordination through a joint OECS Secretariat. Both features have an influence on their adjustment capability within the FTAA. Second, there is considerable variation in economic performance and levels of per-capita income within CARICOM. This performance, however, is not uniquely correlated to size. The OECS has outperformed the remaining group of larger CARICOM countries. For example Antigua and Barbuda have higher real gross domestic products than Guyana, even though they are smaller than Guyana. Antigua and Barbuda have populations of 0.07 and 0.26 million, land areas of 0.4 and 0.3 thousand sq. km., and real per-capita GDPs of U.S $7,380 and U.S $6,837 respectively. Guyana has a population of 0.7 million, land area of 215,000 sq. km., and a per-capita GDP of only U.S. $850.

It should be noted that, in spite of these obvious variations, it is much easier to find similarities among the CARICOM than within the FTAA countries generally. For example, most CARICOM countries registered negative agricultural and merchandise trade balances for 1997 (Table 5.2). In Table

Table 5.2 CARICOM Economic Structure and Performance, Selected Years

Country	Area	1997 Population	1997 Real per capita GDP	Annual growth 1991–97	1997 Inflation	1997 Merchandize trade balance	1997 Current account balance	1996 Total external debt	1996 Ag.[a] exports f.o.b.	1996 Ag. imports c.i.f.	1996 Ag. trade balance
	(1,000 sq km)[b]	(millions)[c]	(U.S.$)[c]	(%)[c]	(%)[c]	(U.S.$m)[d]	(U.S.$m)	(U.S.$m)[c]	(U.S.$m)[e]	(U.S.$m)[e]	(U.S.$m)
Antigua and Barbuda	0.40	0.07	7,380	2.4	1.6	-308.67	-70.15	436	1.00	33.00	-32.00
Barbados	0.30	0.26	6,837	0.8	7.7	-600.60	58.40	581	103.00	144.00	-41.00
Belize	23.00	0.23	2,740	4.3	1.0	-80.70	-42.10	288	125.00	51.00	74.00
Dominica	0.70	0.07	3,120	2.3	2.4	-55.00	-36.93	111	25.00	33.00	-8.00
Grenada	0.30	0.10	3,000	1.8	1.1	-131.34	-79.85	120	12.00	43.00	-31.00
Guyana	215.00	0.73	850	7.0	4.0	-24.00	-68.00	1,631	141.00	50.00	91.00
Jamaica	11.00	2.60	1,560	0.5	9.7	-1,781.00	-403.50	4,041	314.00	388.00	-74.00
St. Kitts and Nevis	0.30	0.04	6,160	3.3	2.5	-79.33	-53.00	58	13.00	18.00	-5.00
St. Lucia	0.60	0.15	3,620	3.0	3.0	-269.76	n.a.	142	60.00	67.00	-7.00
St. Vincent and Grenadines	0.40	0.11	2,500	3.0	0.4	-111.67	-73.26	213	39.00	30.00	9.00
Suriname	163.00	0.4	1,200	1.0	7.1	n.a.	n.a.	n.a.	n.a.	n.a.	n.a.
Trinidad and Tobago	5.00	1.30	4,230	1.9	3.9	-534.00	-142.00	2,242	188.00	287.00	-99.00

[a]Agricultural.

[b]Information obtained from Bernal (1998).

[c]CGCED (1998).

[d]Million dollars.

[e]Information obtained from USDA/ERS (1998).

Table 5.3 Openness and Structure of Production and Exports in CARICOM, 1995

Country	Total trade as percent of GDP[a]	Structure of production as percent of GDP				Structure of exports as percent of total exports (goods and services)			
		Agriculture	Mining	Manf[b]	Services	Agriculture	Mining	Manf[b]	Services
Antigua and Barbuda	171	4	17	3	76	5	16	6	73
Barbados	100	5	8	8	79	13	1	25	61
Belize	119	19	14	13	54	58	1	12	29
Dominica	164	26	11	7	56	23	2	53	22
Grenada	100	14	14	5	67	27	1	10	62
Guyana	175	36	26	11	27	36	19	15	30
Jamaica	151	9	20	18	53	13	6	45	36
St. Kitts and Nevis	122	6	14	12	68	18	0	30	52
St. Lucia	136	11	14	7	68	34	0	22	44
St. Vincent and Grenadines	159	18	19	4	59	47	0	28	25
Trinidad and Tobago	79	3	34	9	54	11	33	47	9

[a]GDP = gross domestic product.
[b]Manufacturing.
Source: Adapted from Finger, Ng, and Soloaga (1998).

5.3, apart from Trinidad and Tobago, in 1997 total CARICOM trade as a percentage of GDP was equal to, or in excess of, 100 percent. Indeed, for five of the CARICOM countries, the ratio exceeded 150 percent, which ranks them among the most open in the world. In all but one of the CARICOM countries listed, services accounted for more than 50 percent of domestic activities, and in five of the countries, they accounted for more than 66 percent. Agriculture and mining, the two primary activities, dominated the rest of the GDP. With respect to exports, again the service sector dominates, followed closely by agriculture and manufacturing. Conventional economic analysis would not normally deal with the issue of potentially different economic consequences among firms, or countries, as a result of differences in economic vulnerability. (This dimension is important for small states because the Briguglio (1995) vulnerability index shows that, of the 20 most vulnerable countries worldwide, ten are from the CARICOM region.[iv]) In Table 5.2, despite their vulnerability, the levels of per-capita GDP, with the exceptions of Guyana and Haiti, place most CARICOM countries in the category of middle-income developing countries. (Indeed, in relation to the human development index (HDI) of the United Nations Development Program (UNDP, 1998), all the countries, with the exception of Haiti, are in the middle- and high-level categories of development. Ten countries rank within the high human-development category.)

While being small does not necessarily put countries at a permanent handicap with respect to economic integration, it does limit the range of policy adjustment options—particularly in the short- to medium-run. This is accentuated for countries that are both small and underdeveloped (Bernal, 1998). However, while the latter characteristic is recognized by the WTO rules as a basis for differential treatment, these rules do not allow for further differential treatment among less-developed countries (LDCs), based on configurations of both smallness and underdevelopment. This is a potentially contentious issue within the integration movement. Many of the potential FTAA countries feel strongly that asymmetries (including size and level of development) must be explicitly accounted for in the operational modalities of such a movement. As of 2001, progress has been slow in dealing with this issue. Resolution of this issue will have significant implications for the FTAA and, indeed, for the evolving post-WTO architecture of global trade.

3. CARICOM AGRICULTURAL SUPPORT AND TRADE POLICIES: AN OVERVIEW

Indicators of the importance of agriculture to the CARICOM countries in 1996 show that: (1) Agriculture accounts for 10 percent or more of GDP in five countries; (2) food exports account for 20 percent or more of total exports in nine of the countries; 3) food imports account for more than 10 percent of all imports in eleven of the countries; and (4) the labor force in agriculture exceeds 20 percent in six countries (Table 5.4). Most of the agriculture is export-based. Food and live animal exports comprise about 84 percent of agricultural exports. The principal products are bananas (34 percent), sugar (33 percent), and cereals (8 percent). Other significant products include coffee, cocoa, spices, citrus, and non-traditional items (such as fruit and cut flowers) (CARICOM Secretariat Statistical Office, n.d.). There is also the unmeasured, but economically important, cultivation of illegal narcotic crops. Agricultural production for domestic consumption is concentrated in root crops, fruits, and vegetables.

Over the years, the relative contribution of agriculture to the GDP has declined. But, because of its contributions to foreign exchange and its status as a major employer of labor, it remains an important activity in several countries (Tables 5.3 and 5.4). Overall, the fastest growing, and now the most important, economic sector in the region is tourism. The link between this activity and domestic agriculture is weak, and the bulk of the agricultural produce consumed by tourists is imported.

The stated policy objectives of the regional governments for the agricultural sector are: (1) improved competitiveness; (2) food security; (3) improved value-added agricultural products and linkages to other sectors; (4) sustainable rural development; and (5) product and market diversification. The identified constraints include:

- Natural factors related to the prevalence of natural disasters (flood, drought, and hurricane), hilly and mountainous topography, small size, endemic biological pests, and plant and animal diseases;
- Policy environment that has a particularly faulty macro-policy, which produces disincentives to agriculture, the lack of institutional support for agriculture, poor marketing and information, inadequate training, weak social policies for rural areas, and the absence of a serious land-reform program;
- Systemic production features, such as low yields/productivity, poor quality of produce, significant post-harvest losses, inappropriate technology, inadequate storage and processing, and defective rural infrastructure;
- Imbalance of the roles of the private and public sectors, because most of the basic services, infrastructure, and institutional oversights are provided by the public sector. Private-sector investment is limited and is frequently dispersed among a large number of small, but independent farmers.

Table 5.4 Importance of Agriculture to CARICOM Countries, 1996

Country	Agriculture in GDP[a]	Agricultural food exports in total exports	Agricultural food imports in total imports	Labor force in agriculture[b]
		percent		
Antigua and Barbuda	3.7	n.a.	n.a.	4.0
Barbados	6.1	40.6	14.6	5.8
Belize	21.1	84.8	16.9	32.8
Dominica	19.8	48.1	18.9	31.1
Grenada	9.7	81.4	23.7	17.1
Guyana	38.9	48.0	14.0	28.4
Jamaica	8.3	20.6	12.5	26.8
St. Kitts and Nevis	5.6	68.3	17.0	13.8
St. Lucia	9.4	69.9	22.0	25.6
St. Vincent and Grenadines	12.6	82.6	23.2	24.8
Suriname	17.8	18.1	21.1	4.3
Trinidad and Tobago	1.9	5.4	11.0	11.7

[a]Gross domestic product.
[b]Data for 1991; n.a.= not available.
Source: CARICOM Secretariat Statistical Office (n.d.).

When seeking to overcome these constraints, the orientations of agricultural support and trade policies were already high-priority in the debate on the region's strategies. Policy reform that was oriented toward liberalization had begun in CARICOM countries even before their present engagement in the FTAA negotiations, or indeed, in the Uruguay Round negotiations that preceded the FTAA.

3.1 Liberalization in CARICOM: Retrospective and Contemporary Dimensions

Conventionally, trade liberalization is defined as any change that makes a country's trade system more open or neutral. Thus, an export promotion (EP) system, in contrast to an import substitution (IS) system, is one that is more open or neutral. In an EP system, incentives for import substitution do not outweigh the incentives for export promotion (Lal and Rajapatirana, 1987; Bhagwati, 1988). The question is: Where does CARICOM stand along this continuum?

Partly, but not exclusively, as a result of externally financed economic stabilization and of structural adjustment programs, considerable progress has been made in CARICOM toward the moderation of NTBs against both intra-regional and extra-regional imports (CARICOM Community Secretariat, 1994). Further, economic reforms have led to market-determined exchange rates, the reduction of fiscal deficits, the substantial privatization of government-owned enterprises, the removal of various price controls and subsidies, and the acceleration of the trade liberalization process (Bernal, 1994; CARICOM Community Secretariat, 1994). The most significant coordinated economic policy aimed at accelerating sub-regional trade competitiveness has been the implementation of a common external tariff (CET). We present the various categories of commodities and tariff rates covered by the CET from 1992-98 (Table 5.5). These 15 policy reforms represent a radical departure from the inward-looking development strategies that the CARICOM countries favored 15 years ago.

Despite these trade reforms, significant export taxes remain in most countries.[v] The main NTB system still in place is the Oils and Fats Agreement (OFA), which is essentially a system of managed trade in vegetable oils, copra, and derivative products (World Bank, 1994). CARICOM countries have also instituted various tariffs. We show the nominal and effective rates of protection for four countries (Table 5.6). Most categories of commodities would have been subject to a tariff ranging from 5 percent to 20 percent in 1998. The exceptions are agricultural products (40 percent rate), agricultural inputs and other sensitive goods (zero percent rate). It has been pointed out that, despite recent reforms, in contrast to the export incentives given to the nonagricultural sectors, traditional agricultural exports are directly and/or indirectly taxed (World Bank, 1994). However, the fragmentary nature of the data presented in this study suggests the need for further research in this area.

3.2 Trading Arrangements

Trade between the United States and CARICOM falls primarily under the Caribbean Basin Initiative (CBI) and the Generalized System of Preferences (GSP); with the European Union (EU), it falls under the Lomé/African-Caribbean-Pacific (ACP) group of countries preferential arrangement; and with Canada, it falls under a preferential trade agreement. The region also has preferential bilateral arrangements with the Dominican Republic, Venezuela, and Colombia. Developments in the last decade, however, raise concerns about the longevity and even the relevance of these nonreciprocal trade relationships to which CARICOM has become accustomed. Both the CBI and CARIBCAN have been under threat from the North American Free Trade Agreement (NAFTA), while the proposed FTAA further complicates this. CARICOM's trade with the European Union is also being threatened for a number of reasons:

Table 5.5 CARICOM CET Rates, 1992-98

Categories	*2/1/91 to 1/1/93*	*1/1/93 to 12/31/94*	*1/1/95 to 12/31/96*	*1/1/97 to 12/31/97*	*From 1/1/98*
			percent		
Noncompeting inputs:					
Primary	0–10[a]	0–5	0–5	0–5	0–5
Intermediate	0–10[a]	0–5	0–5	0–5	0–5
Capital	0	0–5	0–5	0–5	0–5
Competing inputs:					
Capital	30	20	15	10	10
Intermediate	30	25	20	15	15
Selected export	30	20	15	10	10
Noncompeting final goods	30	25	25/30[b]	20/25[b]	20
Agroindustry	45	30/35[b]	25/30[b]	20/25[b]	20[b]
Garments	45	30/35[b]	25/30[b]	20/25[b]	20[b]
General manufactures	45	30/35	25/30[b]	20/25[b]	20[b]
Sensitive Goods:					
List A	——Suspended Rates——				Deleted
List B	——Suspended Rates (LDCs)——				Deleted
List C	——Minimum Rates——				
List D	——Suspended Rates (LDCs)——				Deleted
Safety		0	0		Deleted
Cost of Living		0–20	0–20		Deleted
Socioeconomic and cultural		0–20	0–20		Deleted
Agriculture		40	40	40	40
Agriculture inputs	0	0	0	0	0

[a]*Zero rates for less developed countries (LDCs).*
[b]*The lower of the two rates refers to the countries implementing the trade reform on a fast-track basis, that is, Jamaica, St. Lucia, St. Vincent, Antigua, and Guyana.*
Source: McIntyre (1993).

- NAFTA-type reciprocal free-trade arrangements and other nonreciprocal-unilateral concessions arranged between the European Union and other countries, especially those in Latin America and Central America (Nurse and Sandiford, 1995);
- As the European Union diverts its attention to Eastern Europe, less priority, aid, and investment are accorded to CARICOM (Tincani, 1996);

Table 5.6 Nominal and Effective Rates of Protection, Selected CARICOM Countries (Percentages)

	1981-83	*1984-86*	*1987-89*	*1990-92*	*1993*
			NRP[a]		
Barbados	58.0	54.7	49.0	68.3	n.a.
Guyana	113.3	154.7	118.3	32.0	16.0
Jamaica	85.3	56.0	48.0	49.0	47.0
Trinidad and Tobago	77.0	87.0	83.7	71.7	58.2
			ERP[b]		
Barbados	222.7	210.3	188.0	263.0	n.a.
Guyana	340.0	464.0	354.7	96.0	48.0
Jamaica	255.7	168.3	144.0	147.0	141.0
Trinidad and Tobago	215.7	243.7	211.3	200.3	162.0

[a] *Calculation of the nominal rates of protection (NRP) is based on average tariff and surcharge levels, with a rough adjustment for the effects of nontariff barriers and the black market premium.*

[b] *Calculation of the effective rates of protection (ERP) is based on an assumption that domestically produced import substitutes have an average value-added of 30 percent.*

Source: Gonzales (1993).

- The European Union is being extended as new members join and dilute the original colonial ties that inspired the Lomé Agreement;
- Under Article 174, Section 2a, of the Lomé/ACP Agreement, the inclusion of CARICOM countries in the FTAA could cause these countries to lose their preferential market with the European Union. This could occur because the agreement states that these countries are obligated to provide no less favorable treatment to the European Union than they provide to any developed country (Group of Experts, 1998);
- The effort to maintain WTO compatibility under the Lomé Agreement is leading to reduced support for the nonreciprocity principle upon which present trade agreements are based;
- Concerns over the increased budgetary cost of domestic agricultural subsidies in the European Union to which the price paid for CARICOM's main export to the European Union—sugar—is linked;
- Expressed concerns that preferential access for agricultural commodity exports from CARICOM to the European Union is discouraging its industrialization and the development of services.

From the foregoing we may conclude that, despite reforms already made and those in progress, it seems likely that preferences and the complacency that they engender will not be a sustainable part of CARICOM's future trade. The community, therefore, has to look for and optimize options that are consistent with the creation of a dynamic export sector that is sustainable without trade preferences. The challenge lies in deciding how best to accomplish this within the FTAA framework.

4. WTO AND CARICOM PLATFORM RULES ON AGRICULTURE: SYMMETRIES, ASYMMETRIES AND ECONOMIC ANXIETIES

There are several areas in which WTO and CARICOM Agreement rules are consistent with each other. In particular, both espouse trade liberalization and place emphasis on competitiveness through the reduction and/or rationalization of subsidies and other forms of protection for local agricultural producers. Further, the WTO Agreement clauses on allowable exceptions and on special treatment indicate the WTO's awareness of the special characteristics of developing countries. There are, in addition, areas in the WTO Agreement that are major causes of concern. These areas are discussed here under the headings of (1) market access and preferential arrangements; and (2) domestic support.

4.1 Market Access and Preferential-trade Arrangements for Agriculture

Traditionally, market access refers to the tariff and non-tariff treatment that governments give to products imported from their trading partners. This, however, overlooks three key aspects of market access that are integral to the rights and obligations embodied in trade arrangements: (1) the binding of tariffs; (2) the rules and disciplines on the use of other trade-related government interventions; and (3) the institutional arrangements for monitoring and enforcing compliance. The WTO process recognizes these and has institutionalized rules and regulations governing them. As Blackhurst, Enders and François observe:

> . . . the implications [of market access] for development depend not on external conditions of trade alone, but on the interaction of external conditions with internal conditions. In the end, external conditions are opportunities, but whether or not such opportunities are realized will depend on the internal policies that are pursued, including reductions in the country's own trade barriers, and its acceptance of stronger multilateral rules and disciplines on its own trade-related policies. (1996: 150)

We present some details of the preferential-trade arrangements to which CARICOM countries belong (Table 5.7). As might be expected, the number of beneficiaries is largest in the case of the GSP, but the Lomé/ACP agree-

Table 5.7 CARICOM Preferential Trade Arrangements, 1992

Preference-giving country/region	*Scheme*	*Number of beneficiary countries*	*Number of agricultural commodities*
European Union	GSP[a]	168	530
	ACP[b]/Lomé	69	1,333
United States	GSP	133	467
	CBI[c]	22	957
	U.S. Sugar Quota	40	1
Canada	GSP	182[e]	n.a.
	CARIBCAN[d]	CARICOM[f]	All

[a]*Generalized System of Preferences.*
[b]*African-Caribbean-Pacific Group.*
[c]*Caribbean Basin Initiative.*
[d]*Caribbean-Canada Trade Agreement.*
[e]*Data for 1998.*
[f]*Caribbean Community and Common Market.*
Source: Adapted from Yamazaki (1996) and other data supplied to the authors.

ment provides the widest coverage of agricultural commodities. Under the CARIBCAN Agreement of 1986 (renewed, under a WTO Waiver extension, for a period of 10 years in 1996), Canada offers duty-free access for essentially all CARICOM exports, except sensitive items like textiles. The GSP provides for reduced tariff rates below the most favored nation (MFN)[vi] tariffs or even the duty-free treatment for eligible products exported from developing to developed countries. The significant feature of the nonreciprocal preferential-marketing schemes is that the prices of commodities are usually higher and more stable compared to world prices.

The benefit of a preferential scheme ultimately depends on the relation of the preferential rate to the MFN. Under the WTO, the larger the import volume and/or the greater the reduction in MFN rates, the larger is the erosion of preferences. Benefits could also be eroded because of its new trade rules. Thus, for example, ACP (and hence CARICOM) rice exports into the European Union have recently been subjected to reduced quotas under GATT-legal safeguard measures imposed by the European Commission (*Caribbean Week*, 1997). Also, the dispute over CARICOM banana exports to the European Union has drawn worldwide attention, as it has bordered on the edge of a trade war between the European Union and the United States. Pressure, applied by the WTO to bring an end to nonreciprocal arrangements, and WTO support for generalized access by all developing countries to markets of developed countries have generated a high level of insecurity in CARICOM.

Yamazaki (1996) provides estimates of the extent to which the WTO rules are likely to reduce the overall global benefits accruing from the preferential schemes offered by the European Union, the United States, and Japan. These estimates indicate a total U.S $632 million reduction (from U.S $1,853 million to U.S $1,221 million) of potential benefits, which is 34 percent of the pre-Uruguay Round benefits. A partial breakdown of this reduction is shown in Table 5.8, which indicates that Central America and the Caribbean are expected to lose U.S $58 million, or 19 percent, of their pre-Uruguay Round benefits. These losses represent 41 percent in the E.U. market and 43 percent in the U.S. market.

Yamazaki's (1996) estimates also indicate that the largest reductions in benefits are coffee/tea/spices (95 percent), vegetable oils/fats (66 percent), beverages/spirits (63 percent), and fruits and nuts (44 percent). For coffee/tea/spices and vegetable oils/fats, most of the losses are expected in the E.U. market (96 percent and 70 percent, respectively). The estimated total reduction for sugar is negligible (1.6 percent). It should be noted, however, that 28.6 percent of the pre-Uruguay Round preferential margin for sugar in the U.S. market would be potentially eliminated, compared to only 1.3 percent in the European Union. This would occur because E.U. sugar imports from the ACP countries receive the internal E.U. support price paid to E.U. beet farmers, which is unaffected by the MFN tariff cut. Countries are also allowed to count special trade arrangements as part of their minimum-access commitments and to allocate their minimum access to exporters that have special arrangements. Since the current sugar exports of CARICOM to the United States and to the European Union meet their minimum-access requirements, both continue to allocate their sugar imports as they did formerly (Hathaway and Ingco, 1996).

Although CARICOM is not identified as a group of countries in Yamazaki's (1996) article, a number of relevant inferences can be made from the data presented in Table 5.8 and elsewhere. First, the potential erosion in trade

Table 5.8 Value of Preferential Margins by Region, Selected Periods

Country/region	*Trade in preference-receiving commodities in 1992*	*Value of preferential margins in 1992*	*Value of preferential margins after GATT 1994*
	U.S $ million		*percent* [a]
Africa	4,270	675	509 (24.6)
Far East	3,409	285	105 (63.2)
Near East	1,091	49	23 (53.1)
Central America and the Caribbean	1,997	303	245 (19.1)
South America	4,266	432	246 (43.4)
Europe	245	19	8 (57.9)
Oceania	320	90	84 (6.7)
TOTAL	15,599	1,853	1,221 (34.1)

[a] *The values in parentheses are the percentage reduction of preferential margins for each region.*

Source: Adapted from Yamazaki (1996).

preferences will differ by region and by commodity. While losses in preferential margins for Central America and the Caribbean are relatively small (19.1 percent) compared to the other regions, in absolute terms these losses are by no means insignificant. All three of CARICOM's main exports—sugar, bananas, and rice—enjoy preferential treatment. With respect to sugar, the average world-market price received by the region was U.S $280 per ton in 1995, compared with the E.U. and U.S. preferential price of U.S $637 per ton and U.S $340 per ton, respectively (Durrant and Kendall, 1997). Further, the average cost of production in the region is more than double the existing world prices. The banana industry is unlikely to survive in CARICOM without E.U. preferential prices. In the smaller OECS countries, some members earn more than 50 percent of their total commodity export earnings (70 percent for Dominica) from the banana industry, which employs mainly small farmers and accounts for as much as 30 percent of the labor force (Table 5.4). Additionally, there is the non-quantified benefit from preferential-trade arrangements resulting from their more stable prices as compared to world prices. These preferential schemes, apart from being non-reciprocal, usually combine with a package of economic benefits. Finally, one has to consider that, on the negative side, such preferences contribute both to the discouragement of value-added processing of commodity exports, and to the postponement of efforts to improve crop yields and efficiencies.

A further area of concern for CARICOM relates to the WTO tariff-reduction commitments, which require that developing countries reduce tariffs on agricultural products by 24 percent during a ten-year period, beginning in 1995 with a minimum tariff reduction of 15 percent for each product allowed. However, the CARICOM CET—as presented in Table 5.5—stipulates a constant 40-percent tariff on agriculture. It would appear, therefore, that this CET rate could be in violation of the WTO rules. Our preliminary assessment of this situation, however, suggests that this difficulty may be more of an apparition than it is a reality, a point that is elaborated in a later section.

The final aspect of this discussion relates to the WTO rules, to disciplines on sanitary and phytosanitary (SPS) measures, and to other technical barriers to trade (TBT). Many developing countries complain about the disingenuous use of technical measures as nontransparent means of creating obstacles to the international exchange of agricultural goods (Thornsbury et al., 1997). While the WTO rules stipulate that these measures should not constitute barriers to international trade, or be applied in an arbitrary and discriminatory manner (WTA/GATT, 1994), it is tempting to conclude that the tariffication of NTBs as well as the new GATT/WTO rules on SPS measures and TBT will not eliminate these barriers. As Thornsbury et al. point out, "While it is potentially constructive to have disciplines on technical barriers in place, their formal existence does not guarantee that greater discipline will be imposed on the international use of technical trade barriers" (1997: 1).

Unfortunately, empirical studies of TBT have been hindered by a lack of data. What is available seems to suggest that these barriers have been used as formidable obstacles to trade. Thus, a recent study of the broiler industry in Jamaica notes that exports of its broiler meat to the United States, the United Kingdom, and the European Union faces significant TBT (Witter, 1992). A 1996 USDA survey of NTB to U.S. agricultural exports also identified 326 TBT with a total estimated trade impact of U.S $5,002.86 million (Thornsbury et al., 1997).

4.2 Domestic Support for Agriculture

CARICOM countries have been net importers of agricultural products, particularly food and feed grains, since the early 1970s. Trade liberalization, therefore, has serious implications for its declared food-security objectives, its sustainable rural development, its social stability, and its employment. There is general concern that the WTO rules on domestic agricultural support will prevent these countries from providing agriculture with the kind of domestic support given in the past, or what is needed as the sector strives for efficiency and getting its prices "right."[vii]

Broadly, these WTO rules allow the total support for agriculture to be composed of the following five items: aggregate measured support (AMS); *de minimus* support; special and differential treatment support (SDT); production-limiting support; and WTO-Green Box support. Of these five components, only the first is considered trade distorting and, as such, is subject to reduction commitments. The AMS is the sum of expenditures on non-exempted domestic support, aggregated across all commodities and policies. The rules specify that, in the future, this sum should not exceed that provided in the base AMS, and further, it has to be reduced by at least 13.3 percent by 2005 in the case of the developing countries, and by 20 percent for developed countries by 2000. While reduction commitments refer to total AMS, there are no commodity- or policy-specific reduction commitments. Article 13 of the agreement on due restraint places implicit constraints on switching support from one commodity to another.

Of the other four support items, the *de minimus* support refers to product and non-product-specific measures, each of which amounts to less than 10 percent of the value of farm-gate production in developing countries (less than 5 percent for developed countries). The special and differential treatment support is allowed for developing countries only, and includes allowable support for investment subsidies, agricultural input subsidies to poor producers, and support to producers to encourage diversification away from narcotic crops. The production-limiting support provides direct payments to farmers, such as the set-aside payments of the European Union and the deficiency payments of the United States. The WTO green-box support qualifies for exemption from reduction commitments, with the proviso that the support is provided through a publicly funded government program. This support does not involve transfers from consumers nor does it have the effect of providing price support to producers. It should have no, or at least minimal, trade-distorting effects on production. Assessment of the extent to which countries have used domestic support measures that are WTO legal shows that Green Box policies are clearly more prevalent in the developed countries, while developing countries favor general services in support of their agricultural sector (Konandreas and Greenfield, 1996). Direct payments to producers are rarely pursued. Direct price-support policies may not be available to most developing countries as a consequence of the zero or insignificant levels of base AMS that these countries have included in their WTO schedules.

Konandreas and Greenfield (1996) estimate that the value of the transfers associated with Green-box policies is less than five percent of agricultural GDP, although for some countries such values were more significant. Similarly, support under the SDT is insignificant at less than one percent of agricultural GDP for most countries. Developing countries that have supported their farmers in the past may, therefore, still continue to do so through a combination of tariffs, price intervention, input and investment assistance, and a variety of Green-Box measures. We conclude, therefore, that the real constraints to agricultural support in developing countries will come not from WTO-Round rules but from domestic budgetary restrictions. A similar conclusion was reached by Valdes and McCalla (1996).

Some of the findings of FAO's (1995) assessment of the impact of the Uruguay Round on major agricultural markets are presented in Table 5.9. The total price increase expected for main food commodities by the year 2000 ranges between zero percent and 41 percent. The price changes projected in Table 5.9 were used to project food imports for developing countries, and the results of some of these projections are shown in Table 5.10. For the ten Latin American and Caribbean (LAC) low-income food-deficit (LIFD) countries, the food import bill is projected to be U.S $2.4 billion in the year 2000. For the LAC group as a whole, the overall food import bill is projected to increase by 58.8 percent, from U.S. $8 billion in the base year (1987-89) to U.S. $12.7 billion in 2000, of which U.S. $300 million is attributable to the Uruguay Round.

The CARICOM food import bill in 1984 was U.S. $703.7 million. Cereals and legumes, excluding feed grains, accounted for the largest proportion of

Table 5.9 Projected Percentage Change in Real World Food Prices for Selected Commodities by the Year 2000 (1987-89 = 100)

Commodity	*Baseline*	*GATT/WTO effect*	*Total effect*[a]
		Percent	
Wheat	-3	+7	+4
Rice	+7	+7	+15
Maize	+3	+4	+7
Millet/sorghum	+6	+4	+10
Other grains	-3	+7	+5
Fats and oils	-4	+4	0
Oilmeal proteins	+3	0	+3
Bovine meat	+6	+8	+14
Sheep meat	+3	+10	+13
Pig meat	+13	+10	+24
Poultry	+5	+8	+14
Milk	+32	+7	+41

[a] *The total does not necessarily equal the two effects.*
Source: FAO (1995).

Table 5.10 Food Import Bills of Developing and Low-Income Food-Deficit (LIFD) Countries, Past and Projected[a]

	No. of Countries	*Actual (1987–89)*	*Projected (2000)*	*Size of increase*	*GATT/WTO Round Effect*
			U.S.$ billion		
World					
All developing countries	137	40.0	64.7	24.7	3.6
LIFD countries[b]	72	17.8	27.6	9.8	1.4
Africa					
All developing countries	52	6.0	10.5	4.5	0.5
LIFD countries	43	3.5	6.3	2.8	0.2
Latin America and the Caribbean					
All developing countries	46	8.0	12.7	4.7	0.3
LIFD countries	10	1.6	2.4	0.8	0.1
Near East					
All developing countries	19	11.5	16.8	5.3	0.8
LIFD countries	6	3.7	4.7	1.0	0.1
Far East					
All developing countries	20	14.5	24.7	10.2	2.0
LIFD countries	13	9.0	14.2	5.2	1.0

[a] *Food comprises cereals, oilseeds and products, and meat and dairy products.*
[b] *Net cereal-importing countries with per capita income less than the cut-off point defined by the World Bank for International Development Association (IDA) eligibility (US $1,345 in 1993).*
Source: Adapted from Greenfield, de Negris, and Konandreas (1996).

food imports; however, almost 80 percent of feed grains used in the meat and poultry industry were imported. Meat, meat preparations, and dairy products are the next two important categories of food imports (CARICOM Community Secretariat, 1994). We present data on selected agricultural imports into the CARICOM region from the United States (Table 5.11). Historically, the United States has been a major source of CARICOM food imports, and the data show that, in 1994, imports of feed grain from the United States were valued at U.S. $236.3 million. Imports of poultry meat and other meats were U.S. $135.8 million and U.S. $46.6 million, respectively. Of the total value of the selected agricultural imports from the United States in 1994, feed grains accounted for 43.18 percent; poultry meat, 24.82 percent; and other meats, 8.42 percent. Relatively small fluctuations in the total value of these imports were recorded over the 1991-94 period (USDA/ERS, 1995).

FAO's (1995) price projections point to potentially large increases in prices of meats, dairy products, and grains, consequent upon the WTO effects. In light of their relatively large food- and grain-import bills, these projected price changes could be viewed as opportunities for CARICOM countries to respond in terms of domestic production. The FAO (1995) study also reports that the agricultural trade balance of all developing countries is projected to improve by U.S. $1.9 billion in the year 2000 (Table 5.12). For

Table 5.11 CARICOM Imports of Selected Agricultural Products from the United States, Selected Years

	1991	*1992*	*1993*	*1994*
			U.S. $1,000	
Meats and meat products	38,413	23,451	16,982	17,885 (3.27)[a]
Beef and veal (incl. preserved)	33,743	15,800	12,364	12,385 (2.26)
Lamb, mutton, or goat	425	250	266	314 (0.6)
Pork (incl. preserved)	12,884	8,471	6,736	8,276 (1.51)
Variety meats and edible offals	14,765	11,058	7,399	7,197 (1.32)
Poultry and poultry products	143,876	126,096	135,244	135,832 (24.82)
Eggs	10,200	7,692	7,588	9,519 (1.74)
Dairy products	20,720	19,411	20,181	21,331 (3.90)
Fat, oils, and lard	8,151	9,136	10,358	9,402 (1.72)
Wheat (unmilled)	46,912	55,821	60,562	55,268 (10.10)
Wheat flour and other products	1,914	1,011	2,284	1,477 (0.27)
Corn	38,991	36,583	33,297	32,065 (5.86)
Feed grains and products	278,468	247,016	248,324	236,266 (43.18)
TOTAL	649,462	561,796	561,585	547,217

[a] *The values in parentheses are percentages of the total for 1994.*
Source: USDA/ERS (1995).

Table 5.12 Trade Balances of Agricultural Commodities, Past and Projected (f.o.b. U.S.$ billion)[a]

	Actual (1987-89)	*Baseline (2000)*	*Uruguay Round (2000)*
		f.o.b. U.S. $ billion	
World			
Imports	275.5	334.6	362.0
Exports	280.4	340.4	366.2
Developed countries			
Imports	208.7	236.4	256.3
Exports	198.5	223.1	240.0
Developing countries			
Imports	66.8	98.2	105.7
Exports	81.8	114.3	122.6
Africa			
Imports	8.4	13.9	14.9
Exports	9.4	12.4	13.5
Latin America and the Caribbean			
Imports	10.5	15.9	16.8
Exports	30.9	42.3	45.6
Near East			
Imports	17.8	25.5	27.1
Exports	6.5	7.7	8.2
Far East			
Imports	30.1	42.9	46.9
Exports	35.0	51.9	55.3

[a]Both exports and imports valued at world export-unit prices (f.o.b. basis).
Source: Adapted from FAO (1995).

the LAC group, the trade balance in agricultural commodities is projected to increase from U.S. $20.4 billion for 1987-89 to U.S. $28.8 billion in the year 2000, of which U.S. $2.4 billion is due to the Uruguay Round effect. According to the FAO (1995) study, several countries in the LAC are low-cost, efficient producers and, therefore, stand to gain from the Uruguay Round, both through higher market prices and gains in market shares. This improvement is unlikely to benefit small economies, which have limited capacity to respond to increased opportunities, to the same extent that it would benefit such countries as Brazil and Argentina.

5. SOME CARICOM ECONOMIC ADJUSTMENT CHALLENGES

In addition to benefits under the GSP, CARICOM's external trade in agricultural products benefits from a large number of special nonreciprocal preferential arrangements: the Lomé Convention; the CBI; the U.S. Sugar Quota; CARIBCAN; and the special trade relations with Venezuela, the Dominican Republic, and Colombia. However, since the FTAA is being negotiated under the umbrella trade-liberalization regime of the WTO, these special preferential arrangements are at considerable risk. While the WTO has acknowledged the existence of these ongoing preferential arrangements and, therefore, has introduced an element of security and predictability to them, its long-term focus is toward GSP-type generalized benefits being made available to all developing countries uniformly. This holds true even though it recognizes the existence of different levels of development as a moderating factor in arriving at a liberal trade regime. The *de facto* existence of special nonreciprocal arrangements does, however, put pressure on theFTAA to go beyond the concept of simple reciprocity among trading partners. The Working Group on Small Economies is one response to this pressure, suggesting that CARICOM's participation in the FTAA does not have to mean unrestricted openness and unbridled free trade.

The nexus between agricultural trade and domestic policy issues is important for the adjustment process in the FTAA. At the global level, this link is captured in the WTO rules pertaining to market access, domestic support, and export competition. Additionally, the nexus between agricultural trade and non-trade domestic areas of economic activity is also important. The evidence shows that domestic agricultural and macroeconomic policies are frequently the root cause in the prevalence of agricultural trade-distorting policies in most countries (Hathaway and Ingco, 1996). This requires that the rules/behavior governing agricultural trade should not be viewed exclusively through a trade lens.

Many of the agriculture-related adjustment challenges directly associated with the FTAA will be faced by all economies and while, in theory, CARICOM countries will not necessarily be permanently economically disadvantaged because of their small size, smallness will limit the range of policy options and will increase their vulnerability to external shocks. The need to go beyond reciprocity must, therefore, be accorded high priority and must achieve parallel progress in negotiations. An example of how this might be done is through some form of sectoral exemption (whether it is temporary or permanent) as was the case for cultural industries under NAFTA (Group of Experts, 1998).

5.1 Consistency of Policies Within a WTO Framework

As signatories to the WTO accord, CARICOM's conformity to its rules is likely to be dictated largely by its interpretation of how these rules impact their developmental objectives. As a general rule, all countries manage their trade through various forms of interventions. Even though all WTO member countries have committed themselves to apply trade restrictions in a nondiscriminatory manner and to participate in negotiations to reduce these restrictions, we discern a thrust in the opposite direction, namely, permitting trade restrictions of various kinds, based on different interpretations of the rules. An example of this is that, while the liberalization process ostensibly seeks to achieve export promotion (neutrality) in trade, liberalized trade is simultaneously deemed to be consistent with government intervention, thereby permitting a variety of export promotion intervention policies in specific products and activities. With such ambiguity, an important challenge for CARICOM countries is to identify distinct areas for their own policy interventions that are WTO legal and that, in the short run, can be utilized in a manner consistent with its long-term developmental objectives. In the long run, however, interventions must be designed to increase international competitiveness, rather than to prolong special treatment. This calls for high-quality strategic/selective research and policy formulation.

5.2 Erosion of Preferences

The two commodities that benefit most under existing preference schemes are sugar and bananas. In the case of sugar, the region receives preferences under both the E.U./ACP Sugar protocol and under the U.S. Sugar Quota. Apart from guaranteed quotas, the prices paid are based on the subsidized price paid to domestic beet farmers in the European Union and to domestic cane and beet farmers in the United States, all of which are well above the world free-market prices. In the case of bananas, the region benefits through a system of quotas placed on non-ACP dollar bananas by the European Union, and receives a price that compensates their relatively higher-cost producers. This arrangement has, so far, been successfully challenged by the United States and supported by Central American banana producers in the WTO. The threat to the erosion of preferences is that WTO rules—which are to convert NTBs into tariffs and to progressively reduce these tariffs over time—as well as the reduction in the MFN rates, will eventually attract new entrants into the markets in which CARICOM countries have historically enjoyed preferential treatment. In the long run, international competitiveness is the only sustainable basis, therefore, for CARICOM producers to secure their presence in these markets.

The immediate challenge is to secure policy interventions that are timely, focused, and less distortionary than before. These can influence the direction and speed of liberalization as well as the environment within which micro-level resource allocation decisions are made. Thus, in the case of bananas, the particular challenge is how best, on the one hand, to restructure, diversify, and introduce efficiency-enhancing measures into the supply function and, on the other, how to market the product based on quality differentiation strategies, which create a unique Caribbean banana that can withstand competition from dollar bananas. With respect to sugar, the challenge is to increase productivity, to stimulate divestiture (*Caribbean Update*, 1997), and to diversify, where possible, into sugar-based industries, while trimming the industries to leave only the commercially viable producers. Like bananas, there will be the need, also, to identify niche markets where quality and special properties matter more than price does.

5.3 Revenue Implication

A recent Organization of American States (OAS, 1997) study has found that, of 11 CARICOM countries studied, eight rely on trade taxes for more than 20 percent of government revenue. A major challenge, which will result from a reciprocal-trading arrangement, is the need to find alternative sources of government finance.

5.4 Stabilization of Foreign Exchange

CARICOM countries' agricultural exports are not necessarily confined to one or two products, but these products face significant price fluctuations for the portions sold on world markets. Because the existence of preferences has served to somewhat mitigate these price fluctuations, their erosion is likely to expose these countries to greater fluctuations in price and foreign exchange earnings. The challenge here will be finding ways to stabilize earnings.

5.5 Maintaining Macroeconomic Balance and Completing Trade Liberalization

Although, as we have seen, several CARICOM countries have taken steps—unilaterally, regionally, and multilaterally—to introduce economic reforms and to restructure their economies, this effort will need to be maintained and, perhaps, accelerated in some situations. As indicated earlier, substantial trade restrictions still remain, and tariff rates are still relatively high compared to some of the other developing countries in Latin America.

5.6 The Challenge of Inclusion in the Process of Globalization

As in other developing countries, the global trade liberalization process has been the source of many concerns in CARICOM. One concern is that the time period of ten years, given to developing countries for adjustment to the WTO rules, is considered too short and is made shorter by the fact that the Agreement will come up for renegotiation. Another is that the Agreement on Agriculture is seen largely as the product of agreement among the developed countries, particularly the European Union and the United States, with developing countries' concerns largely ignored. Further, because CARICOM is about to enter territory quite different than the sheltered preferential trading systems to which they have become accustomed, there is concern about the social and political consequences of external shocks. Finally, the region produces agricultural exports that are comparatively high-cost and noncompetitive at prevailing world-market prices.

Despite these concerns, it is important for CARICOM countries to exercise whatever leverage they may have at their disposal and to be proactive rather than reactive in their efforts to strategically reposition themselves in the global economy. The fundamental difficulty that it faces lies not only in the high cost, noncompetitive structure of its agriculture, but in the fact that, at this juncture, it has also embraced the open regionalism model of economic integration. This model caters to the formation of sub-regional trading blocs within the group (for example, the OECS) as well as supra-regional blocs (for example, the Association of Caribbean States (ACS) and the ACP), which facilitate linkages with still other blocs, ultimately synthesizing into a global free-trade area. The various groupings are expected to function as building blocks, and not stumbling blocks. The belief is that, in the absence of such concentric relationships, small economies might not be able to secure the necessary assistance for themselves to survive as regional entities (Dookeran, 1996). The model permits preferential elements in these trading relations, but generally, import barriers are expected to be low. While open regionalism is an appealing model for CARICOM, it should be recognized that, from an operational point of view, it could create a serious dilemma in that too much openness, and too many outside relationships, could reduce the inner discipline required for tight decision-making and bargaining. The challenge remains: how best to negotiate economic arrangements that will go beyond trade and trade policies, and that will also foster development of the Region.

5.7 Development of Special Products

A recent Latin American Economic System (SELA, 1998) report indicates that, during the past 20 years, there has been a shift away from trade in bulk processed and unprocessed agricultural commodities—such as grains, raw sugar, oilseeds, flour, and oils—to consumer-ready processed and unprocessed commodities—such as fresh fruits, vegetables, breakfast cereals, pastas, refined sugar, beverages, and processed fruits. The study also notes that LAC lags behind the rest of the world in the movement to upgrade its processing sector in order to capture this value-added. The export challenge facing the region's agriculture is to get the process of identifying niche markets and such special products started early, and to press the need for the liberalization of these more than the liberalization of agricultural raw materials. This calls for an aggressive program, as it is unlikely that small countries will be able to compete head-on with their larger counterparts in the same primary-product markets.

6. CONCLUDING OBSERVATIONS

A number of important FTAA-CARICOM-related issues were not examined in this chapter. First, whether or not the smaller sub-grouping of the micro-states that comprise the OECS qualifies for a truly special status, along the lines of the group of LDCs within the WTO. Second, the FTAA negotiations are occurring almost concurrently with the renegotiation of the Lomé IV E.U./ACP Agreement. The outcome of each of these processes will inevitably influence the other. The Europeans have already made it clear that CARICOM cannot expect to protect its preferential status with the European Union while, at the same time, giving reciprocal status to the United States. Third, a broader Atlantic trade arrangement, involving Europe or European-Latin American integration (particularly with MERCOSUR), might well overtake the FTAA process, particularly as the Clinton administration has been denied fast-track authority. Fourth, the forthcoming WTO round, while still uncertain at this stage, may well open up new areas of liberalization that are not clear at the moment. The efforts of WTO seem aimed at bringing agriculture into line with other commodities in world trade, offering greater security, transparency, and predictability in the hope that this would encourage countries to abandon extreme domestic producer-friendly measures. Already, the next round of negotiations seems set to deal with further tariffication, further reduction in export subsidies, and containment of governments' role in agriculture as well as NTB. We might well find that a speedy movement to global free trade, combined with delays in getting the FTAA off the ground, could result in the latter being overtaken by events.

This chapter has looked at the FTAA as a trade integration framework for the deepening of the outward-oriented trade policy rules and disciplines of the WTO. The FTAA is seen as being intended to take the regional integration process beyond the trade reciprocity dimensions of partnership among countries. Consequently, many of the agricultural adjustment challenges that have resulted from more liberal regional and global trading regimes will be faced by all countries in the Americas. Our conclusion is that small economies will likely experience special adjustment challenges by virtue of their limited range of policy options and resource bases as they seek to react to external economic shocks. Given the relative importance of agriculture in the Americas and within CARICOM, the adjustment challenges associated with the WTO—plus liberalized rules on market access, export competition, and domestic support in agriculture—are areas in which CARICOM is likely to experience its most profound adjustment challenges and dilemmas.

On the one hand, the rules of the WTO seem to indicate potential losses of preferential margins and market access as well as reductions in protective tariffs and domestic support. These losses have instilled justifiable fear and apprehension about these rules. On the other hand, critical analysis of the rules also suggests that substantial elbowroom exists for countries to maneuver and to make adjustments as a basis for short-term survival. To make this work, CARICOM countries need to make substantial intellectual investments if they are to be able to exploit the legitimate conditionalities in the rules and use these conditionalities toward the growth and development of their economies. These short-term WTO-legal intervention mechanisms, however, must be seen not as ends, but as transitory coping policies as CARICOM proactively invests in knowledge-based, longer-term sustainable strategies consistent with the global imperatives of reciprocity and market-driven and performance-based results led by private-sector initiatives.

Both in the short term and in the longer term, however, the adjustments that CARICOM countries make within the WTO-legal framework must involve trade-related and domestic policies. This is a direct result of the way in which the WTO rules have been framed to impact not only trade, but also non-trade economic activities. Adjustment policies must, therefore, focus on: (1) the productivity and international competitiveness nexus, and (2) incentive-neutral trade intervention. These are key mechanisms with which CARICOM countries can strategically assist in the negation of the more binding structural impediments, such as size and insularity that it is confronting.

If CARICOM and other small economies of the Americas are to achieve full integration into the FTAA, the beyond-reciprocity principles of this RTA process must be given equal and parallel status in negotiations. Open regionalism as a CARICOM integration/development model could facilitate the mutually supportive roles of sub-regional adjustments and regional integration. With this type of model, there is increased scope for the synchronization of transitional arrangements via negotiations among trading blocs. According to the Group of Experts (1998), such intra-bloc negotiations should encompass, among other things, (1) the definition of reciprocity, (2) the coverage of agreements, and (3) the adjustment period.

Although CARICOM constitutes a fairly homogenous group of small developing countries, there is significant variation in key economic and resource indicators as well as economic performance levels of well being. The danger is that this could lead to conflicts of interest within the FTAA process. The survival of CARICOM as an integrating entity is key to the containment of such risks and to the pursuit of harmonious policies that benefit, not only CARICOM, but the wider FTAA of which it hopes to be a vibrant part.

Finally, we address two issues that are extremely contentious within the framework of our FTAA-WTO-CARICOM analyses. The first relates to the long-term benefits likely to be derived from any type of special and differential trading arrangements accorded to CARICOM countries under FTAA-WTO legal regimes. This issue becomes a focal point since some analysts argue that the historical preferential-trading umbrella enjoyed by CARICOM has been part of the root cause of its present day economic dilemma. The second has to do with the potential adverse effects of the so-called free-rider problems, with respect to CARICOM's external negotiating propensity, which some analysts argue have prevented it from forming an effective coordinating and cooperating external negotiating machinery.

With respect to the first issue, this chapter advances the proposition that economic size is not neutral with respect to the benefits and costs of trade liberalization and economic integration under the FTAA-CARICOM situation. The discussion in this chapter argues that size poses limits to the range of policy adjustment options available to small countries, such as those of CARICOM, as compared to larger entities. As such, a logical extension of this argument could be one that the benefits of unbridled trade liberalization may: (1) not be as robust for small economies such as CARICOM; (2) carry adjustment costs (economic and social) that are disproportionately larger; and (3) engender a combination of (1) and (2). On this particular issue there is paucity of both theoretical and empirical material in the economics literature. Conventional trade theory (Ricardian and Heckscher-Olin) based exclusively on the core concept of comparative advantage concludes that small economies benefit more from the process of trade liberalization than larger ones do. However, as demonstrated by the new international trade theories of the last decade, increasing returns to scale is as fundamental a cause of international trade as is comparative advantage. What are the welfare implications of the new international trade theories for the liberalization phenomena as it relates to small economies? There is a lacuna in the literature on these issues; this notwithstanding, we have attempted to build a convincing case from the limited evidence available, regarding the relationship between economic size and the intensity and inter-temporal dimensions of economic and social disequilibria associated with trade liberalization initiatives. Reference was made to the fact that the WTO regime explicitly recognizes less-developed status as a basis for special and differential treatment as it relates to trade reciprocity requirements. In this regard we proposed here that smallness *cum* underdevelopment is an equally compelling argument for granting of preferential treatment to CARICOM countries within the FTAA-WTO liberal regimes. Such action is certainly not contradictory with the beyond-reciprocity principles of the FTAA.

How does our argument for selected preferential treatment for CARICOM compare to the counter argument advanced by some analysts that this would reduce the potential benefits of trade liberalization to these states? To add perspective to this issue we point to the fact that conventional trade theories assert that the so-called benefits from liberalization take the form of a joint efficiency/welfare gains product. Further, these theories assert that it is through the synergistic attributes of this joint product, or benefits, that countries are propelled along a higher growth and development trajectory. We would like to point out that these theoretical arguments often overlook the fact that, according to accepted economic theory, the classical gains from trade must be related to some specified welfare functions before any pronouncements about welfare benefits can be made. Technically speaking, in the absence of optimizing income redistribution policy, the social indifference map need not indicate changes in social welfare. Within the context of these considerations, we take the position that liberalized trading regimes are not ends in themselves. Rather, they should be seen as one set of means associated with raising the standard of living of people and countries. The FTAA's beyond-reciprocity principle explicitly acknowledges this element of the integration movement. More recently, the European Union made a bold move to bring this principle to the center stage of the WTO regime by proposing an everything-but-arms trade concession to the world's poorest countries. We therefore argue that within the context of a more robust definition of trade benefits, selected preferential treatment to small developing economies, such as CARICOM, is consistent with their long-term growth and development objectives.

There is little doubt that the multiple-track trade negotiations in which CARICOM is engaged could be taxed to the limit, and possibly could exhaust the resources of almost any negotiating body. Historically, CARICOM has a long tradition of coordinating its external trade policies *vis-à-vis* trading partners offering non-reciprocal arrangements (for example, E.U./LOMÉ, U.S./CBI, CARIBCAN). However, under such arrangements they were effectively insulated from having to deal with the wider multilateral negotiation issues outside of these bi-lateral arrangements. In other words, they accrued trade-related benefits from free riding on the negotiating skills of their major trading partners. A major outcome of this free-rider problem is the incapacity of the region to mount an effective negotiating stance consistent with the requirements of the new international forum. This could pose a major challenge to CARICOM's ability to strategically reposition itself in the new global environment. The potential negative impact of such a problem should not be underestimated. CARICOM should move proactively to improve its readiness to deal with the negotiating issues. It would appear that there is some recognition given to the urgency of this issue. CARICOM has created a regional negotiating machinery for international economic negotiations. The purpose of this unit is to formulate and implement joint negotiating strategies for talks with WTO, FTAA, the European Union, the United States and other trading entities. A high-level negotiator has been retained to head the regional negotiating machinery and a technical advisory group as well as special technical unit has been put in place to support the work of an intergovernmental negotiating working group. The negotiating capacity of the regional negotiating machinery is being assisted by technical and financial resources from national and multilateral agencies. To be sure the regional negotiating machinery has a daunting task ahead when charting CARICOM's path in the global-liberal environment.

7. ENDNOTES

[1]CARICOM countries include: Antigua and Barbuda, Bahamas (a member of the Community, but not of the Common Market), Barbados, Belize, Dominica, Grenada, Guyana, Montserrat, Jamaica, St. Kitts and Nevis, St. Lucia, St. Vincent and the Grenadines, Trinidad and Tobago, and Suriname. Haiti has satisfied all the terms and condition for full membership in CARICOM, except an instrument of accession; when the latter formality has been completed, Haiti will become a full member of CARICOM. Associate members and Observer states have special relationships with CARICOM. Associate Members are: British Virgin Islands, and Turks and Caicos Islands. Observer States are: Anguilla, Dominican Republic, Netherlands Antilles, Puerto Rico, and Venezuela. Within CARICOM, the smaller states are recognized as a subgroup, the Organization of Eastern Caribbean States (OECS). It includes: Antigua and Barbados, Dominica, Grenada, Montserrat, St. Kitts and Nevis, St.

Lucia, and St. Vincent and the Grenadines. Anguilla and the British Virgin Islands are associate members.

[ii]Any intra-FTAA trade regime must be in compliance with Article 24 of the WTO Agriculture Agreement to be WTO-legal. This article requires comprehensive reduction in trade barriers among the members of RTAs, but requires no raising of trade barriers to nonmembers.

[iii]CARICOM countries (including Haiti), along with the Dominican Republic, constitute the Caribbean Forum (CARIFORUM) component of the African, Caribbean, and Pacific (ACP) group of countries. This group, for historical reasons, has enjoyed preferential and nonreciprocal trading arrangements with the European Economic Community/European Union (EEC/EU) under successive Lome conventions dating back to 1975. The current convention (Lome IV) expired in the year 2000 and is currently being renegotiated.

[iv] For a detailed discussion of the definition and construction of the Briguglio economic vulnerability index, see Briguglio (1995).

[v] For an overview of the tax regime of CARICOM countries, see Davis et al. (1999).

[vi] The Most Favored Nation (MFN) clause of a commercial trade treaty is a binding contract by the signatories to confer upon each other all of the most favored trade concessions that any of the signatories may grant to any other nation subsequent to the signing of the agreement.

[vii] As alluded earlier, there might be some question as to whether or not CARICOM agriculture is a net beneficiary of protective subsidy in light of some evidence that the sector is overtaxed. This is an empirical issue not addressed in this paper.

8. REFERENCES

Bernal, R.L. 1998. *The Integration of Small Economies in the Free Trade Area of the Americas, Volume 9*. Washington, DC: Center for Strategic and International Studies.

_____. 1994. "From NAFTA to Hemispheric Free Trade." *Colombia Journal of World Business* 29(3): 23–31.

Bhagwati, J. 1988. "Export-Promoting Trade Strategy: Issues and Evidence." *World Research Observer* 3: 27–57.

Blackhurst, R., A. Enders, and J.F. François. 1996. "The Uruguay Round and Market Access: Opportunities and Challenges for Developing Countries," in W. Martin and L.A. Winters, eds., *The Uruguay Round and the Developing Countries*. Cambridge, MA: Cambridge University Press.

Briguglio, L. 1995. "Small Island Developing States and Their Economic Vulnerabilities." *World Development* 23(9).

Bryan, A.T. 1995. "Copying with the New Dynamics," in A.T. Bryan, ed., *The Caribbean: New Dynamics in Trade and Political Economy*. Miami, FL: North-South Center, University of Miami.

CARICOM Community Secretariat. 1994. *Review and Analysis of the Performance of the Agricultural Sector in CARICOM Member States (1988–1992)*. Georgetown, Guyana.

CARICOM Secretariat Statistical Office (n.d.). Georgetown, Guyana.

Caribbean Update. 1997. "Banana Trade Row Ruling." 3(4).

Caribbean Week. 1997. "E.U. files appeal." 21 June to 4 July.

CGCED (Caribbean Group for Cooperation in Economic Development). 1998. *Caribbean Economic Overview*. Washington, DC: CGCED.

Davis, C.G., Ballayram, E.A. Evans, and C.Y. Thomas. 1999. "Trade Liberalization and Small Economies in the Americas: The Case of the Caribbean Community." International Working Paper IW99-1, Food and Resource Economics Department, University of Florida, Gainesville, FL (February).

Dookeran, W.C. 1996. *Choices and Change: Reflections on the Caribbean*. Baltimore, MD: Johns Hopkins Press.

Durrant, N. and H. Kendall. 1997. "Structural Effects of Non-Reciprocal Preferential Trading Arrangement on the Agricultural Sector of Guyana in the Context of Structural Adjustment." Paper presented at the Twenty-Second West Indies Agricultural Economics Conference, Barbados, West Indies (27–30 August).

FAO (Food and Agriculture Organization of the United Nations). 1995. *Impact of the Uruguay Round on Agriculture CCP:95/13*. Rome, Italy.

Finger, M., F. Ng, and I. Soloaga. 1998. "Trade Policies in the Caribbean Countries: A Look at the Positive Agenda." Paper prepared for discussion at the meetings of the Caribbean Group for Cooperation on Economic Development (CGCED), Washington, DC (June).

Gonzales, A.P. 1993. "Trade Liberalization, Growth, and Employment in CARICOM." Paper presented at the Conference on Trade Liberalization, Growth, and Employment in the Caribbean Basin, Washington, DC (8–9 September).

Greenfield, J., M. De Negris, and P. Konandreas. 1996. "The Uruguay Round Agreement on Agriculture: Food Security Implications for Developing Countries." *Food Policy* 21(4): 365–375.

Group of Experts. 1998. *Overcoming Obstacles and Maximizing Opportunities: Smaller Economies and Western Hemisphere Integration*. Kingston, Jamaica (March).

Hathaway, D. and M. Ingco. 1996. "Agricultural Liberalization and the Uruguay Round," in W. Martin and L. Winters, eds., *The Uruguay Round and the Developing Countries*. Washington, DC: The World Bank.

Konandreas, P. and J. Greenfield. 1996. "Uruguay Round Commitments on Domestic Support: Their Implications for Developing Countries." *Food Policy* 21(4/5): 433–446.

Lal, D. and S. Rajapatirana. 1987. *World Bank Research Observer* 2(2 July): 189–217.

McIntyre, A.M. 1993. *A Paper on Regional Trade Policies*. OECS Secretariat, Castries, St. Lucia, West Indies.

Nurse, K., and W. Sandiford. 1995. *Windward Islands Bananas: Challenges and Options Under the Single Economic Market*. Kingston, Jamaica: Friedrich Ebert Stiftung.

OAS (Organization of American States). 1997. *Small and Relatively Less Developed Economies and Western Hemispheric Integration*. Washington, DC: Trade Unit.

SELA (Latin American Economic System). 1998. "Implementation of the Uruguay Round Agreement on Agriculture: Issues for Latin America and the Caribbean." SP/DRE/Di No. 22-98, Permanent Secretariat of SELA, Caracas, Venezuela.

Thornsbury, S., D. Roberts, K. DeRemer, and D. Orden. 1997. "A First Step in Understanding Technical Barriers to Agricultural Trade." Draft copy of contributed paper presented at the XXIII International Conference of Agricultural Economics, Sacramento, California (10–16 August).

Tincani, A. 1996. "The European Union and the Caribbean: Challenges Ahead," in W.C. Dookeran, ed., *Choices and Change: Reflections on the Caribbean*. Baltimore, MD: Johns Hopkins Press.

UNDP (United Nations Development Program). 1998. *Human Development Report*. New York, NY: UNDP.

USDA/ERS (U.S. Department of Agriculture/Economic Research Service). 1998. *Free Trade in the Americas: Situation and Outlook Series.* WRS-98-1, Washington, DC (November).

_____. 1995. "Supplementary Tables." *Foreign Agricultural Trade of the United States.* Washington, DC: USDA/ERS.

Valdes, A. and A. McCalla. 1996. "The Uruguay Round and Agricultural Policies in Developing Countries and Economies in Transition." *Food Policy* 21(4/5): 419–431.

West Indian Commission. 1992. *Time for Action: Report of the West Indian Commission. 2nd ed.* Kingston, Jamaica: University of the West Indies Press.

Witter, M. 1992. *The Poultry Industry in Jamaica: The Impact of the Reform of the Common External Tariff (CET).* Consortium Graduate School, University of the West Indies, Jamaica, West Indies.

World Bank. 1994. *Coping with Changes in the External Environment.* Report No. LAC 12821, Caribbean Division. Washington, DC: The World Bank.

WTA/GATT (World Trade Agreement/General Agreement on Tariffs and Trade). 1994. "GATT, Agriculture, (MTN/FA II-A1A-3)." Internet website: http:itl.irv.uit.no/trade_law/documents/freetrade/wta-94/art/iia 1 a3.html.

Yamazaki, F. 1996. "Potential Erosion of Trade Preferences in Agricultural Products." *Food Policy* 21(4/5): 409–417.

Chapter 6

Agricultural and Trade Policy under Administrative Water Regimes

Charles B. Moss and Chris de Bodisco
University of Florida and Vanderbilt University

1. INTRODUCTION

Urban growth in Florida over the past twenty-five years has been nothing short of spectacular. In 1975, Florida was the ninth largest state in the nation with a population of 2 million people. By 1996, the population in Florida had risen to 12 million people—the fourth most populous state in the nation. This growth has led to resource conflicts between urban demands and Florida's traditional agricultural economy, particularly where land and water are concerned. Between 1970 and 1995, freshwater withdrawals for public supply increased by 134 percent, while self-supplied withdrawals for agriculture increased 54 percent (Marella, 1995). The conflict between agriculture and urbanization in the state is complicated by recent changes in agricultural and trade policy. Specifically, agriculture's water use tends to be dominated by the state's high-valued crops that compete with imports. In addition, dominant crops in north Florida are heavily reliant on traditional commodity programs. Hence, the changes implicit in the North American Free Trade Agreement (NAFTA) and the 1996 Federal Agriculture Improvement and Reform Act (FAIR) may have significant ramifications for both cropping patterns and water use in Florida. Given the history of water-management in Florida, the emerging conflicts between agricultural and urban-water use will likely be resolved less through markets or courts and more through a unique regulatory environment under administrative law.

This chapter examines the impact of NAFTA and FAIR on water allocation in the state of Florida within the state's administrative water policy. The next section presents an overview of Florida's water-management districts that form the institutional framework for the allocation of water between competing uses in the state. Following the discussion of the water-management district structure, we present the results of an econometric analysis of the value of water in northwest Florida and potential changes that may result from NAFTA and FAIR. We present the implications of these changes for the allocation of water within the state.

2. INSTITUTIONAL FRAMEWORK

The Water Resources Act of 1972 changed the standard rules of water ownership in Florida by bringing all water resources under state control. Previously, the common law system of riparian rights settled disputes in court on the basis of reasonable use of resources. The Water Resources Act changed common law to administrative law by establishing five water-management districts each charged with the orderly allocation of water to the maximum benefit of the state (Carriker, 1988). To that end, water-management districts monitor such facets as water quality, the issuance of use permits, and the maintenance of flood control. We present a map of the water-management districts (Figure 6.1). To fund water-management activities, each district has the authority to levy *ad valorem* taxes. Most funding, however, is legislated directly by the state for specific purposes, such as land acquisition.

Under the water-management scheme in northwest Florida, the establishment of the water-management structure increased the fixed cost of irrigation. Specifically, a farmer now applies for a permit specifying the crops to be grown and an estimate of the volume of water that will be required. The water-management district then approves that use subject to sufficient local resources. In a few cases, such as the Telogia Creek area in Gadsden County, permits may be denied because the existing water resources are already completely allocated. In these cases, the right to water may resemble a right of first claim, albeit with greater uncertainty due to the district's power to deny future permits. Recent concerns over specific water resources, such as those in Gadsden and Jackson counties, have led to higher producer cost due to a requirement for more detailed monitoring of water use.

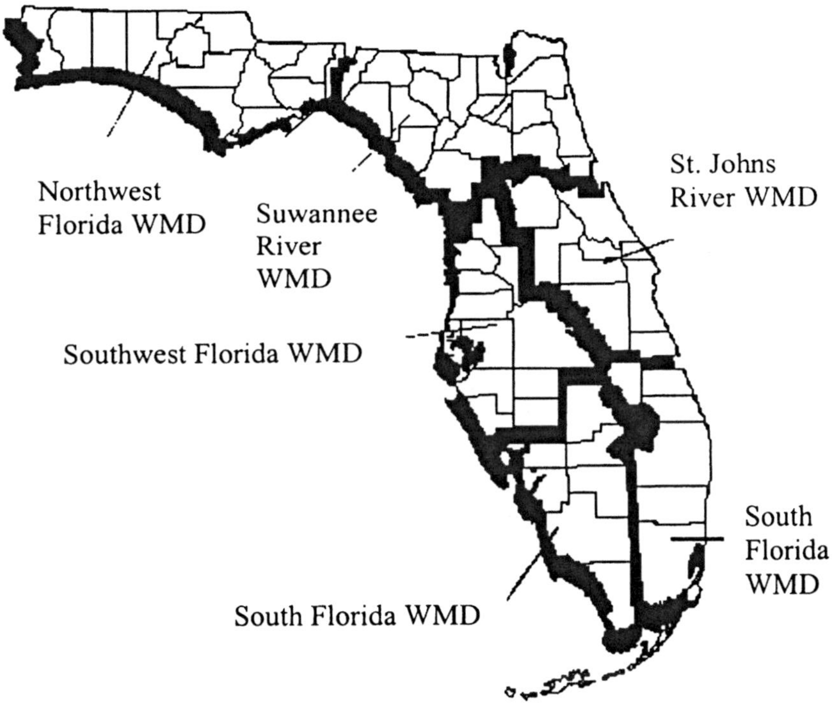

Figure 6.1 Five Water Management Districts in Florida

3. VALUING AGRICULTURAL WATER UNDER ADMINISTRATIVE LAW

As a part of the regulatory process, the Florida legislature requires the districts to project the demand for, and reserves of water in, the state through the year 2020. This section presents the results from the empirical model estimated by Moss and de Bodisco (1998) to project the agricultural demand for water in the Northwest Florida Water Management District (NWFWMD) under this directive. In addition, we extend these results to suggest the potential effects of NAFTA and FAIR.

In general, the counties of the NWFWMD are more dependent on agricultural income than are other counties in the state. On average, agriculture accounts for 0.9 percent of private income in the entire state of Florida, whereas agriculture accounts for 1.4 percent of private income for the counties in the NWFWMD. Furthermore, agriculture in these counties is primarily composed of traditional crops (such as corn, cotton, and peanuts), although Gadsden County has experienced an increase in tomatoes. As a result of this reliance on program crops, agricultural-water demand in the NWFWMD region is expected to be more sensitive to changes in agricultural price supports under FAIR than the state is as a whole.

3.1 Empirical Model

As indicated in the development of the institutional framework, the decision to irrigate tends to be an investment decision for farmers in Florida. In addition to the direct and indirect permitting cost required by the water-management districts, farmers invest in irrigation pumps and delivery systems.[i] Once these investments have been made, the irrigation potential is largely fixed.[ii] The relative inflexibility of irrigation decisions suggests treating irrigation as a quasi-fixed input within the traditional dual profit-function framework. The profit function assumes that farmers react optimally to changes in input and output prices by reallocating across variable inputs to produce a vector of outputs. However, this optimizing behavior is constrained by the levels of quasi-fixed inputs that cannot be instantaneously varied in response to changes in price. Economically, the value of the quasi-fixed level of irrigation is referred to as the shadow value of acres under irrigation, and is quantified by the derivative of the profit function with respect to the level of the quasi-fixed input. The level of the quasi-fixed input is assumed to change over time through investment, or disinvestments, in response to changes in its shadow values.

Moss and de Bodisco (1998) used a Translog specification for the profit function based upon five output prices (corn, wheat and soybeans, cotton, peanuts, and tomatoes) with five outputs. Input categories are seeds, chemicals, fuel, labor, and other. Both input and output prices are Divisia price indices based upon United States Department of Agriculture (USDA) historical data for the United States. In addition, we appended two quasi-fixed inputs: fraction of cropland irrigated and intermediate per-acre investment. Water-use data is obtained from Census Figures and Permit numbers from the NWFWMD for metered and non-metered wells. The estimated shadow value of irrigation and of intermediate investment along with the per-acre profit from the left hand side of the profit function were then used to estimate investment equations for irrigation and for intermediate investment along with a relationship for land in crops.[iii]

3.2 Baseline Results

We estimate the per-acre profit and the shadow value of irrigation for the years 1970 to 1995 (Table 6.1). During the last ten years, the average shadow value of irrigation has been \$35.88 with a standard deviation of \$5.18 in 1996 U.S. dollars. We present the predicted acres in agriculture, acres under irrigation, intermediate per-acre assets, per-acre profit, value of irrigation, and value of intermediate assets using the output price forecast from the Food and Agricultural Policy Research Institute (FAPRI) through 2006 (Table 6.2). These results indicate that the acreage under irrigation increases from 1989 until 1996. In 1996, the number of acres under irrigation is the highest level since 1981. The shadow value of irrigation reaches a peak of U.S. \$50.46 in 1990, and declines to a ten-year minimum of U.S. \$29.31 in 1997 (Figure 6.2). The value of irrigation then increases fairly steadily through year 2006.

The empirical results indicate that the shadow value of water is approximately equal to the amortized cost of investment in an irrigation system, plus the operating cost within the sample period. Using the average cost of a center pivot irrigation system (Smajstrla, Zazueta, and Haman, 1988) including maintenance, repairs, and depreciation (Thompson, Spiess, and Krider, 1980) and discounting at a real rate of 4.65 percent (Moss, Featherstone, and Baker, 1987), the per-acre cost of irrigation is U.S. \$38.68. This equivalence is consistent with the sustained level of irrigation in the district. In addition, projecting the shadow price of water using the FAPRI baseline implies an

Table 6.1 Value of Irrigation

Year	*Estimated profit*	*Shadow price of irrigation*	*Shadow price of int.*[a] *assets*	*Year*	*Estimated profit*	*Shadow price of irrigation*	*Shadow price of int.*[a] *assets*
1970	1.756	28.278	0.026	*1983*	1.391	27.187	0.016
1971	1.809	31.830	0.028	*1984*	1.130	24.406	0.015
1972	1.790	28.128	0.026	*1985*	0.822	21.720	0.012
1973	3.329	45.861	0.032	*1986*	1.251	28.917	0.018
1974	3.010	33.125	0.029	*1987*	1.568	31.859	0.022
1975	2.125	42.390	0.026	*1988*	1.974	31.589	0.023
1976	2.956	44.156	0.031	*1989*	1.742	34.424	0.023
1977	1.480	30.931	0.017	*1990*	2.198	46.898	0.028
1978	1.837	37.298	0.019	*1991*	2.053	38.418	0.026
1979	1.512	37.655	0.015	*1992*	2.095	39.987	0.028
1980	1.460	35.667	0.015	*1993*	2.051	33.797	0.024
1981	1.218	28.441	0.015	*1994*	1.920	38.321	0.025
1982	0.790	24.318	0.010	*1995*	2.115	34.563	0.028

[a] *intermediate assets.*

Source: Moss and de Bodisco (1998).

Table 6.2 Projected Acres, Irrigated Acres, and Shadow Values of Irrigation and Assets

Year	*Acres in agriculture*	*Acres under irrigation*	*Intermediate investment per acre*	*Per-acre profit*	*Shadow value of irrigation*	*Shadow value of intermediate assets*
1996	227,492	25,472	139.35	2.47	35.11	0.04
1997	232,843	24,127	137.98	1.97	29.61	0.03
1998	232,115	24,504	138.63	2.03	30.50	0.03
1999	238,198	23,519	141.75	2.02	32.81	0.03
2000	244,774	23,139	144.06	2.09	34.78	0.03
2001	253,237	22,299	142.32	2.00	34.50	0.03
2002	260,957	22,477	143.69	2.12	36.72	0.03
2003	272,721	22,814	144.56	2.17	37.84	0.03
2004	285,093	23,222	145.39	2.23	38.93	0.03
2005	299,101	23,281	145.84	2.22	39.68	0.03
2006	314,236	23,740	147.43	2.31	41.05	0.03

Source: Projections based on Moss and de Bodisco (1998).

increasing value to irrigation through 2006. The investment model based upon this trend depicts an increasing number of acres under irrigation from 2001 to 2006. The short-run decline in irrigation is due to low shadow values in 1997, 1998, and 1999. Thus, the results predict a gradual increase in the value of irrigation and in the level of irrigation in the near future.

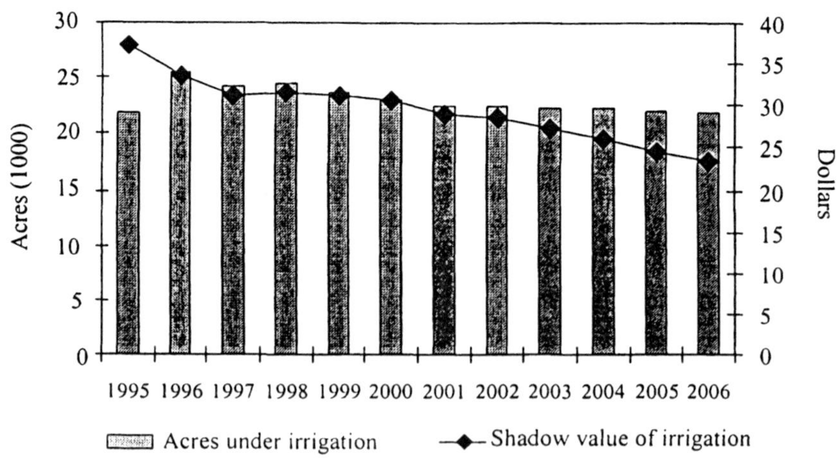

Figure 6.2 Projected Irrigated Acres and Shadow Values

Perhaps counter-intuitively, the total amount of water applied to row crops actually declines in the next decade, even as irrigated acreage increases. As relative input and output prices change, the share of vegetables and cotton in the farmers' optimal crop mix declines, while the share of soybeans, wheat, and peanuts increases. Since soybeans, wheat, and peanuts use relatively less water per acre, total water use falls. However, water use will eventually increase as the share of irrigated acres continues to increase.

A key implication of these results is that water use in row-crop agriculture will not contribute significantly to increases in the demand for water in the near future. This is consistent with the understanding of the managers of the NWFWMD. Future increases in water demand are expected to arise from urban and suburban growth, the development associated with suburbs, and environmental necessities. One example of development associated with suburban growth is nurseries. A simple trend-line growth estimate of nurseries in the district predicts a significant increase in water use due to nurseries. Unfortunately, the limited data on nursery cost and returns over time make a systematic inclusion of these crops into the farmer's decision process impossible. This emphasizes the point that future water demand depends heavily on farmers' optimal crop mix. As price changes alter cropping patterns, both the intensity of irrigation and the return to investment in irrigation equipment will change, thereby changing the demand for water. This suggests that the implications of changes in farm costs and returns should not be neglected in policy.

3.3 The Price of Water Resources and Trade Distortions

Trade distortions can occur when a resource is not priced according to its social marginal product. As Schmitz, Schmitz, and Dumas (1997) show, the cost of subsidizing water inputs may outweigh the gains to trade in cotton. In our case, Florida's implicit water subsidy would shift the supply of each crop outward, reducing the aggregate price and possibly affecting trade flows. If water were priced at its social optimum, the supply curves for each crop would shift back, leading to decreased domestic production. Since northwest Florida is a minor producer of commodities for export, any changes in its output should have little effect on aggregate prices. Nevertheless, if its water resources are given non-zero prices, it stands to reason that the level of irrigation should fall, as producers shift into less water-intensive crops over time.

Let λ_t be the shadow value of irrigation at time t. Imposing a positive price for water, ρ_t^w, leads to an adjusted shadow price of λ_t^*:

$$(1) \quad \lambda_t^* = \lambda_t - \rho_t^w$$

As long as λ_t^* is positive, the producer will apply water in period t. Over time, however, the level of irrigation will decline as the payoff to investing in irrigation equipment falls. Working backwards from the value of water, if we assume that the value of water is equal to its average shadow price of U.S. \$35.88, then levying a ten percent surcharge for water would make ρ_t^w = U.S. \$3.50 per-acre foot of water per year. (Rather than make ρ_t^* an annual ten percent charge of an implicit price, we fixed the surcharge at U.S. \$3.50, or ten percent of the average shadow price, as being more representative of an administrative policy.) As expected, the increased water price eventually decreases the acreage under irrigation by five percent. Overall, profits in the region decline by up to nine percent, while agricultural acres decline slightly. Interestingly, by the year 2006 the shadow value of water falls more than 20 percent from baseline estimates (Table 6.3). All of these changes arise as producers reduce investment in new irrigation and shift resources into the less water-intensive field crops. As a result, the share of irrigated acres in soybeans and wheat increases, while the share in all other crops, especially cotton and vegetables, declines. Since, in our projections, the price of field crops declines more rapidly than do prices for the other commodities, this multiplies through to the shadow value of water. That is, the marginal-value product of water falls, even if its marginal-product increases as the lower output price swamps any other effects.

Table 6.3 Percent Change in Projected Acres, Irrigated Acres, and Shadow Values of Irrigation and Assets when a Surcharge is Added

Year	*Acres in agriculture*	*Acres under irrigation*	*Per-acre profit*	*Shadow value of irrigation*
1996	0.0	0.0	0.0	-11.2
1997	0.0	-0.7	-1.2	-12.6
1998	-0.1	-0.8	-1.1	-12.5
1999	-0.1	-3.2	-5.4	-14.5
2000	-0.5	-3.4	-5.2	-14.7
2001	-0.5	-3.9	-6.2	-15.9
2002	-0.6	-3.9	-6.2	-16.2
2003	-0.6	-4.5	-8.1	-18.2
2004	-0.7	-4.6	-8.0	-18.8
2005	-0.7	-5.0	-9.0	-20.3
2006	-0.8	-5.1	-9.2	-21.2

Source: Projections based on Moss and de Bodisco (1998).

3.4 Implications of NAFTA and FAIR

In order to examine the potential impact of FAIR and NAFTA, we examine changes in agricultural-water use from the FAPRI baseline due to a 10 percent reduction in traditional program crops, in the case of FAIR, and of vegetables, in the case of NAFTA. While some academic debate may persist as to whether or not the recent changes in output prices are indicative of these policy changes, empirical observations of farm level prices for 1997 and 1998 are consistent with a general decline in agricultural prices due to both policy changes.

We depict the effects of FAIR (Table 6.4). After a five-year lag, irrigated acreage falls roughly 4 percent from baseline projections, while total water increases 1.8 percent by the year 2006. Irrigation primarily moves out of peanuts (a decline of approximately 10 percent) and into vegetables (eventually increasing as much as 20 percent to 30 percent). The per-acre shadow value of irrigation declines from the baseline by 12 percent to 15 percent. The per-acre profit is 9 percent below the baseline in the first year, and continues to fall to roughly 14 percent lower after eight years.

The NAFTA policy change has much less impact in this region due to the comparatively small amount of vegetables produced. The number of acres irrigated in Table 6.5 remains essentially unchanged from baseline estimates. The per-acre profit falls 5 percent immediately, but essentially returns to baseline estimates by 2006. The shadow value of irrigation initially falls 6 percent below baseline, but ends up only 2 percent below baseline in 2006. However, irrigation applied to vegetables falls from baseline projections by larger and larger increments, so that, in 2006, it is 23 percent below the baseline. Irrigation in cotton eventually increases by a maximum of 9 percent.

In summary, the results of FAIR and NAFTA depict changes in the optimizing behavior that mitigate partial equilibrium effects of output prices on water use. Declines in output prices, traditionally supported under agricultural policies, lead to increases in total water use because vegetables are substituted for corn and peanuts. On the other hand, reductions in water use, implied by declining vegetable prices under NAFTA, are partially offset by increases in cotton acreage. Undoubtedly, the effects of both FAIR and NAFTA will bring about a reduction in overall water use as agriculture in the district contracts.[iv]

Table 6.4 Percent Change in Total Acreages and Water Use by Crop due to FAIR

	1996	*1997*	*1998*	*1999*	*2000*	*2001*	*2002*	*2003*	*2004*	*2005*	*2006*
						percent					
Acres in agriculture	0.0	-0.9	-0.7	-0.8	-0.8	-1.4	-1.2	-1.2	-1.2	-1.4	-1.3
Acres under irrigation	0.0	0.0	-0.8	-0.8	-3.5	-3.4	-3.7	-3.7	-4.4	-4.2	-4.4
Intermediate investment	0.0	0.0	0.3	0.0	-0.9	-0.7	-0.8	-0.9	-1.3	-1.2	-1.2
Fraction irrigated	0.0	0.9	-0.1	0.0	-2.7	-2.0	-2.5	-2.5	-3.2	-2.9	-3.1
Per-acre profit	-8.9	-7.3	-8.5	-8.8	-14.2	-12.7	-13.3	-13.3	-14.8	-14.0	-14.1
Shadow value of irrigation	-11.9	-11.2	-11.4	-11.7	-14.7	-13.9	-14.2	-14.3	-15.2	-14.8	-14.8
Shadow value of intermediate assets	-10.4	-8.8	-10.2	-10.3	-14.8	-13.4	-14.0	-14.0	-15.2	-14.5	-14.6
Shadow value of irrigation per acre foot	-13.1	-12.2	-12.8	-12.8	-16.1	-16.9	-17.6	-18.2	-19.5	-19.6	-20.0
Total Water Use	1.3	1.2	0.8	0.5	-1.9	0.1	0.4	0.9	0.7	1.5	1.8
Corn	3.7	3.6	3.2	2.6	-1.2	1.5	1.4	2.2	1.7	2.6	2.8
Field	–	-18.5	–	9.5	28.9	3.9	3.6	2.2	3.1	0.7	0.4
Vegetables	2.7	2.2	1.0	1.3	-3.0	3.5	6.4	10.6	13.1	22.4	33.9
Cotton	7.4	8.7	6.7	6.6	-2.5	2.7	2.3	3.5	0.6	3.7	3.7
Peanuts	-10.8	-11.1	-10.6	-10.5	-15.3	-12.3	-12.1	-11.2	-12.1	-10.9	-10.6

Source: Projections based on Moss and de Bodisco (1998).

Table 6.5 Percent Change in Total Acreages and Water Use by Crop due to NAFTA

	1996	1997	1998	1999	2000	2001	2002	2003	2004	2005	2006
					percent						
Acres in agriculture	0.0	0.0	0.0	0.0	0.0	0.0	0.1	0.2	0.2	0.3	0.3
Acres under irrigation	0.0	0.2	0.1	0.1	0.0	0.0	0.0	0.2	0.3	0.5	0.7
Intermediate investment	0.0	0.1	0.1	0.1	0.0	0.0	0.0	0.0	0.1	0.1	0.2
Fraction irrigated	0.0	0.2	0.2	0.1	0.0	0.0	-0.1	0.0	0.1	0.3	0.3
Per-acre profit	-4.9	-4.7	-4.9	-4.5	-4.1	-3.3	-2.8	-2.1	-1.5	-0.8	-0.2
Shadow value of irrigation	-6.3	-6.3	-6.5	-6.0	-5.5	-4.7	-4.3	-3.7	-3.3	-2.7	-2.3
Shadow value of intermediate assets	-5.6	-5.6	-5.7	-5.2	-4.8	-4.0	-3.6	-3.0	-2.4	-1.7	-1.2
Total water use	-0.1	-1.3	0.1	-1.4	-1.3	-0.8	-0.9	-0.8	-0.8	-0.7	-0.7
Corn	0.2	-1.7	0.8	-1.3	-1.1	-0.4	-0.4	-0.2	-0.1	0.2	0.2
Field	–	27.1	–	29.9	6.7	-0.1	-0.5	-1.5	-2.0	-2.6	-2.7
Vegetables	-1.8	-3.6	-1.9	-4.9	-5.3	-6.0	-7.8	-9.3	-11.6	-15.7	-22.8
Cotton	-1.9	-3.7	-2.3	-4.1	-3.2	-0.9	-0.1	1.6	3.2	6.4	9.0
Peanuts	2.8	0.8	1.7	-1.5	-1.0	0.2	0.3	0.8	1.1	1.8	2.1

Source: Projections based on Moss and de Bodisco (1998).

4. INTERREGIONAL IMPLICATIONS OF NAFTA, FAIR, AND POPULATION GROWTH

Focusing simply upon relative changes within one water-management district in the state conceals some of the broader water-management issues facing Florida under changing agricultural and trade policy. Specifically, neither population growth nor agricultural intensity is uniformly distributed across the state. In fact, both tend to be concentrated in the southern part of the state. This concentration will affect the emerging debate over the allocation of water to meet urban versus agricultural needs in two ways. First, the concentration of urban-water use and the intensity of agricultural-water use will accelerate the debate over the allocation of water. Second, a current moratorium on water transfers between water-management districts will come under increasing scrutiny as water resources in south Florida come under increasing pressure.

In order to frame the interregional debate, we propose the use of the informational measures of entropy and distance used by Theil (1967). Theil relies on Shannon's (1949) mathematical definition of entropy:

$$(2) \quad J = -\sum_{i=1}^{N} p_i \ln(p_i)$$

where p_i is the probability of event i. If all the events are equally likely, the entropy measure assumes its maximum value of $\ln(N)$. At the other extreme, as one of the events approaches certainty, the value of entropy approaches zero. The difference between the theoretical minimum and the observed value of entropy can then be used as a measure of concentration.

The entropy results indicate that the observed level of entropy is on average 64 percent of the maximum level of entropy (Table 6.6). This statistic supports the assertion that Florida's water use tends to be concentrated in certain water-management districts. To fully develop the implications of this concentration, we turn to another informational measure developed by Theil (1967). Theil presents an informational index that measures the value of a second signal, given the prior probabilities before that signal is observed. Mathematically, this measure is defined as:

$$(3) \quad I = \sum_{i=1}^{N} p_i \ln\left(\frac{p_i}{q_i}\right),$$

Table 6.6 Index of Agricultural-water Share vs. Urban Water Share High-value Water Share by State and Region

Year		*1975*	*1980*	*1985*
		Measure of entropy and information gain by district		
Entropy	J	0.972	1.013	1.144
Information gain		0.637	0.596	0.465
		Water share versus urban-water share by state and region		
All districts	I	0.087	0.069	0.053
Average	I_g	0.047	0.057	0.040
Region	I_r	0.040	0.012	0.014
Regional percent		46	17	25
		Water share versus population by state and region		
All districts	I	0.142	0.129	0.067
Average	I_g	0.091	0.097	0.039
Region	I_r	0.051	0.032	0.028
Regional Percent		36	25	42
		Water share versus value of agriculture by state and region		
All districts	I	–	–	0.026
Average	I_g	–	–	0.013
Region	I_r	–	–	0.013

Source: Authors' empirical estimates.

where p_i is the value of the original probability of event *i* and q_i is the probability of event *i* after observing the second signal. Intuitively, if the probability of event *i* after observing the second signal is approximately the same as the prior probability, then the second event was not informative. Mathematically, if $p_i = q_i$ for all *i*, then *I* in equation (3) is zero. Alternatively, as the two probabilities become dissimilar, the value of *I* increases. In the current context, the share of agricultural-water use is p_i and the share of other factors is q_i. If the other factor tends to be highly correlated with agricultural-water use, then the informational index will fall.[v]

One advantage of this methodology is the decomposibility of the index. Specifically, when analyzing the concentration of water use in Florida, we are particularly interested in the north-south division of the state. Similarities between the northern water-management districts and southern water-management districts may have implications for interregional transfers of water. It can be demonstrated that:

$$I = I_R + \sum_g P_g I_g$$

$$P_g = \sum_{i \in g} p_i, \quad Q_g = \sum_{i \in g} q_i$$

$$(4) \qquad I_R = \sum_g P_g \ln\left(\frac{P_g}{Q_g}\right)$$

$$I_g = \sum_{i \in g} \frac{p_i}{P_g} \ln\left(\frac{p_i / P_g}{q_i / Q_g}\right)$$

Thus, the overall correlation in concentration can be decomposed into an average correlation within a region and a correlation across regions.

Given these informational measures, we show the difference in information between the share of agricultural-water use, the share of urban-water use, the share of the state's population, and the share of highly valued agriculture across water-management districts and across the north-south axis of the state. Numerically, agricultural-water use tends to be highly correlated with each of these factors. However, the results indicate that agricultural-water use tends to be more highly correlated with urban-water use than it is with population. This result suggests that regions with increased levels of agricultural-water use also have higher water usage per urban resident. In addition, 46 percent of the information in urban-water use is attributable to north-south differences, compared to 36 percent of the information in population share (Table 6.6).

Information in highly valued crops, indicates that agricultural-water use in the state of Florida tends to be highly related to the level of highly valued crops (Table 6.6). These results are consistent with the empirical results from NWFWMD with respect to tomatoes. One may expect that the effect of highly valued crops on agricultural-water use would be magnified in the south due to the dominance of sugar, citrus, and winter vegetables. Results indicate that the north-south division accounts for 51 percent of the information due to highly valued crops.

Given these results, we conclude that agricultural-water use in the state of Florida is fairly concentrated with higher relative usage in the southern part of the state. Extending the informational approach we see that agricultural-water use is more highly correlated with urban-water use than with general population. Hence, resource conflicts arising from increased urbanization will occur more rapidly in the southern part of Florida. Finally, the share of agricultural-water use tends to be highly correlated with highly valued agricultural crops, which will exacerbate the north-south dimension of the administrative water debate.

5. CONCLUSIONS

Florida's rapid growth rate during the past quarter century has led to the emergence of significant conflict over several natural resources in the state. The conflict may be most dramatic for water. The resolution of these disputes occurs within a unique regulatory environment whereby water is allocated across competing uses by the administrative authority of Florida's five water-management districts.

In general, both FAIR and NAFTA will have significant consequences for agricultural-water demand in the state. FAIR will have its greatest impact in northern Florida, where agriculture is more dependent upon traditional program crops. The price declines expected under FAIR are projected to reduce the overall level of program crops in northern Florida. However, this decline may actually result in increased water use as farmers shift into vegetable production.

A similar phenomenon may be observed for NAFTA. Specifically, reduction in vegetable and fruit prices are anticipated to shift farmers out of these enterprises, but the reduction in water use will be partially mitigated by an increased planting of cotton and peanuts.

The most dramatic consequence of these policies, however, may involve regional shifts in agricultural production in the state. The informational results indicate that agricultural-water use is concentrated in south Florida. Further, agricultural-water use appears to be highly correlated with urban-water use. This correlation is likely to intensify the conflict over water in south Florida. The north-south dimension of the problem may be affected by NAFTA. Specifically, agricultural-water use in south Florida is a function of the highly valued crops that will be directly affected by NAFTA. However, any relief due to shifts away from these crops will be partially mitigated by increased levels of other crops.

6. ENDNOTES

[i] Standard irrigation delivery systems in the region include center pivot and drip irrigation under black plastic. Center pivot systems represent a more fixed investment, but can be used for a variety of crops. Drip irrigation under black plastic is typically only used for vegetable crops and must be replaced every year.

[ii] Once in place, the salvage value of a typical irrigation well and center pivot system is only a fraction of its original cost. Further, the cost of drilling the well and the permitting process is, shall we say, completely sunk.

[iii] In general, the quasi-fixed input formulation has been used to test for agricultural asset fixity (Chambers and Vasavada, 1983; Taylor, 1990). Under this formulation the researcher tests for whether the shadow value of agricultural assets equals an observed market price. If the shadow value cannot be distinguished from the market price then no evidence of asset fixity exists. Another formulation is to estimate the investment equations simultaneously with the dual profit function. This approach typically requires some additional simplifying assumptions for tractability. In our approach we estimate the investment function in ancillary regressions. In doing this we opt for flexibility over statistical efficiency.

[iv] Implicitly, imposing FAIR and NAFTA simultaneously on the baseline simply represents a relative increase in real input prices.

[v] We use the term correlation with great reservation. The common statistical use of correlation is typically associated with the normalized covariances from the multivariate normal. In our application, correlation denotes a measure of joint likelihood that is not dependent on the multivariate normal. Cover and Thomas (1991) use the phrase mutual information to denote this concept.

7. REFERENCES

Carriker, R.R. 1988. "State Water Management Policy: The Florida Experience," in G.M. Johnston, D. Freshwater, and P. Favero, eds., *Natural Resource and Environmental Policy Analysis: Cases in Applied Economics.* Boulder, CO: Westview Press.

Chambers, R.G. and U. Vasavada. 1983. "Testing Asset Fixity For United States Agriculture." *American Journal of Agricultural Economics* 65: 761–769.

Cover, T.M. and J.A. Thomas. 1991. *Elements of Information Theory.* New York, NY: John Wiley and Sons.

Marella, R.L. 1995. *Water Withdrawals, Use and Trends in Florida, 1995. Water Resources Investigations Report 99-4002.* United States Department of the Interior, U.S. Geological Survey, Tallahassee, Florida.

Moss, C.B. and C. de Bodisco. 1998. *Historical Agricultural Water Use in the Northwest Florida Water Management District, 1975–1995.* Grant Report to the Northwest Florida Water Management District, Havana, Florida.

Moss, C.B., A.M. Featherstone, and T.G. Baker. 1987. "Agricultural Assets in an Efficient Multi-period Investment Portfolio." *Agricultural Finance Review* 47: 82–94.

Schmitz, T.G., A. Schmitz, and C. Dumas. 1997. "Gains from Trade, Inefficiency of Government Programs, and the Net Economic Effects of Trading." *Journal of Political Economy* (April): 637–648.

Shannon, C.E. 1949. *The Mathematical Theory of Communication.* Urbana, IL: The University of Illinois Press.

Smajstrla, A.G., F.S. Zazueta, and D.Z. Haman. 1988. "Center Pivot Irrigation Systems and Applications in Florida." Florida Cooperative Extension Service, Circular 804, Institute of Food and Agricultural Sciences, University of Florida, Gainesville, Florida (June).

Taylor, T. 1990. "A Test of Asset Fixity in Southeastern U.S. Agriculture." *Southern Journal of Agricultural Economics* 22: 105–111.

Theil, H. *Economics and Information Theory*. 1967. Amsterdam, Netherlands: North-Holland Publishing Co.

Thompson, G.T., L.B. Spiess, and J.N. Krider. 1980. "Farm Resources and System Selection," in M.E. Jensen, ed., pp. 45–58. *Design and Operation of Farm Irrigation Systems*. ASAE Monograph 3, St. Joseph, Michigan.

USDA. (United States Department of Agriculture). 1999. Internet website: mannlib.cornell.edu70

USDA/ERS (U.S. Department of Agriculture, Economic Research Service). 1998. *Crops County Data*. Washington, DC: USDA/ERS.

_____. 1998. *Prices Received by Farmers*. Washington, DC: USDA/ERS.

Chapter 7

Liberalization with Protection: Import Management in Korea (with Emphasis on Rice)

Daniel A. Sumner and Jung-Sup Choi
University of California at Davis and Korea Rural Economic Institute

1. INTRODUCTION

Classic discussions of trade liberalization often deal with the implications of reducing tariffs or expanding import quotas. Of course, as soon as we take one step away from the textbook or chalkboard, we face a world of ingenious instruments that governments have devised to protect local capitalists from foreign capitalism. This chapter examines some of those ingenious instruments.

The Uruguay Round Agreement on Agriculture (URAA) provided a framework to facilitate the liberalization of agricultural trade. The degree to which liberalization was actually achieved varies widely from country to country and from commodity to commodity. The four key areas we discuss here are market access; export subsidies; internal subsidies that affect trade; and illegitimate technical barriers that appear in the guise of sanctions designed to protect human, animal, or plant health. Each of these issues is complex and contains many sub-topics.

In the area of market access, the URAA made progress on several fronts. First, it reduced bound-tariff rates on most agricultural commodities. (The degree of reduction varied widely across countries and across commodities but was significant in many cases.) Second, it established the principle of current access (such that changes made to import regimes would not be used to reduce quantitative access to any market). Third, the URAA established a minimum-access quantity (generally three percent of the base-period consumption of the commodity) to open up previously closed markets. Finally, the URAA required that non-tariff barriers, in all their various forms, be converted to tariffs, and that these tariffs be subject to reduction and to other provisions of the agreement.

The most common instrument used to convert non-tariff barriers into tariffs was the tariff rate quota (TRQ). While the TRQ often performs economically as a quota, it is, in fact, considered a tariff instrument by the General Agreement on Tariffs and Trade (GATT) lawyers because it is comprised of two tariffs. A TRQ typically includes a low tariff on imports up to some set quantity, and then a higher, perhaps prohibitive, tariff on import volumes above that quantity. This tariffication process often creates huge controversy, but relatively little market opening. It has, arguably, set the stage for substantial trade liberalization in the future (Josling, Tangermann, and Warley, 1996; Sumner and Tangermann, forthcoming).

The spread of TRQs and the administration of the quota quantities under TRQs have raised a number of interesting questions about how much liberalization can occur when protectionist interests, or their agents within government, tend to be the ones who administer the tools of market opening. This issue has become particularly interesting in the context of State Trading Enterprises (STEs), which operate the agricultural import regimes for many major commodities in a number of major agricultural trading nations. Asia, China, Japan, Indonesia, and Korea all use STEs to manage imports for some of their important commodities, such as wheat and rice.

TRQ administration methods include first-come, first-served; use-it or lose-it; lottery; historical allocation; auction; license on demand; and more arbitrary systems operated by STEs or by producer groups and associations. (All these systems have detailed administrative rules that affect outcomes.)

The questions related to TRQ administration are twofold. First, internationally, is it administered with commercial considerations on a most-favored nation (MFN) basis? And second, what are the welfare implications domestically? Even more specifically, the domestic issue must address who gets the quota rent and revenue, and how that allocation will affect producer and consumer surpluses under alternative methods of TRQ administration.

In this chapter, we examine the effects of alternative quota administrative schemes, focusing mainly on the Korean government's management of rice imports since the implementation of the URAA in 1995. We also provide some background on import regimes for a few other commodities, but rice is the major issue for Korea, and remains a troublesome point in trade relations.

Rice assumed special status in the URAA when, in the last stages of negotiations, a special class of import-access provisions was created expressly to accommodate the political sensitivities of Japan and Korea. Both countries banned rice imports in the decade prior to the URAA, with the exception of

1993 when Japan imported rice after a severe domestic-yield collapse. In the 1994 agreement, Japan was allowed to maintain an absolute quota on rice imports in exchange for a larger quantity of minimum-access requirements than would have been required under a TRQ schedule. Over the first four years of the implementation of the URAA, Japan imported 44 percent more rice than it would if it had accepted a TRQ and had used prohibitive second-tier tariff rates (as most other countries have done). To forestall even more excess imports Japan switched, in 1999, to a TRQ with a tariff rate of about 1,000 percent.

Japan has purchased rice from a variety of suppliers. It has purchased both long-grain indica rice and the short- and medium-grain japonica rices that are strongly preferred by Japanese consumers. About 50 percent of its imports have come from the United States—mainly from California. Most of the remaining imports have come from Australia, Thailand, and China. Much of this imported rice has probably been of a quality that would be acceptable to many Japanese consumers. Although it has met the letter of its World Trade Organization (WTO) commitments, Japan, through its state import process, government storage, and food aid, has allowed little (if any) of the foreign rice to reach domestic consumers in the form of table rice.

Korea is the second most important japonica rice market, next to Japan. There is also much international speculation on Korea's possible import demand for table rice. Korea is a major importer of other agricultural products, with a low self-sufficiency rate in grain in total.

Korea negotiated a particularly limited market-access commitment for rice. It maintains an absolute quota that was set at one percent of the domestic market in the base year, which has grown gradually to four percent of that quantity over a 10-year phase-in period. In the following sections of the chapter we provide some context and then a model of quota administration that captures the essence of the Korean policy. We focus especially on the welfare consequences of alternative quota administrative regimes for various groups in Korea. We also consider how quota policy relates to domestic farm-subsidy policy, along with the implications of significant expansions of imports under a variety of trade scenarios. This chapter illustrates, with a simple analytical framework and simulation, how the results depend upon how the quota is administered. The issues raised, the approach, and the broad results have implications for the new round of WTO trade negotiations for agriculture that were launched in the spring of 2000.

The next section describes TRQ management in Korea. It also provides some overview on Korean agriculture and agricultural trade policy under the URAA. Subsequently, we provide a simple conceptual framework for the analysis of quota management. Some illustrative calculations offer an idea of the importance of the issues for Korean rice-import management.

2. IMPORT MANAGEMENT AND AGRICULTURE IN KOREA

2.1 Imports and Import Policy

Korea started a large-scale elimination of non-tariff import restrictions in 1989. This process was accelerated with the URAA implementation in Korea. For example, the number of agricultural products subject to the non-tariff restrictions fell from 220 in 1994 to 22 in 1997. Following the tariffication of beef imports on January 1, 2001, that number was reduced to 14 out of 1,312 agricultural products. All remaining items belong to the rice group.

The Korean agricultural market has become more open as a result of these import reforms. Recent data reveal that the annual growth rate of agricultural imports showed a significant increase in 1995. Imports of highly income-elastic products, including fruits and dairy products among others, increased substantially during 1995-96 (Table 7.1).

The import pattern varies widely across commodities. For beef, the TRQ is to be expanded from 187,000 tons in 1998 to 225,000 tons in 2000. (All tons referred to in this chapter are metric tons.) The private trading portion of the TRQ has been increasing 10 percent each year, and reached 70 percent in 1999 and 2000. The additional levy on top of the import tariff fell from the current 20 percent to zero percent in 2000. Thus, all non-tariff import barriers on beef were removed in January 2001. The tariff, which was set at 44 percent in 1995, will be reduced to 40 percent in 2004. For pork and chicken, all quantitative import restrictions were eliminated on July 1, 1997, and tariffs will be reduced annually toward the year 2004. For oranges, the TRQ will be increased from 28,125 tons in 1998 to 57,017 tons in 2004. The in-quota tariff for oranges is 50 percent and the current out-quota tariff of

Table 7.1 Korean Imports of Major Agricultural Products, 1992-97

Product	*1992*	*1993*	*1994*	*1995*	*1996*	*1997*
			tons[a]			
Grain	911	1,398	1,513	1,898	2,597	1,917
Pulses	78	66	38	58	144	118
Soybeans	349	316	385	423	501	535
Fruits	214	205	250	315	323	347
Vegetables	71	66	152	140	175	180
Oilseeds	164	157	208	244	252	280
Meat	575	428	621	797	799	849
Dairy products	29	63	89	150	172	158

[a]*Metric tons.*

Source: MAF (1998).

79.4 percent will be reduced to 50 percent in 2004. Orange juice importing was completely liberalized with the implementation of a tariff rate of 60 percent on July 1, 1997.

The URAA created TRQs for 190 items. Among those 190 items, 104 were designated minimum-market access, and the remaining were designated current-market access items. Of the total, 97 items were put under the control of the country's STEs. Other items are subject to quota auction or import license (Table 7.2).

The administration of TRQs may be classified in terms of who gets the rent and revenue:

- Open tender under state trading: Government or STEs acquire rent and revenue. (Examples include rice, some beef, and oranges);
- Quota auction: Government gets the rent and the importing firm that won the importing right gets the revenue. (Skim milk and sesame oil are two examples);
- Private trading: Some state-trading products and import-license products belong to this category. Revenue goes to the importing firm. (Feed grain is an example); and
- Private trading with mark-up: In this case, rent and the mark-ups go to the government and revenue goes to the importing firm. (Some beef imports, for example).

Table 7.2 Classification of Agricultural Products Group under Tariff-rate Quota (TRQ)

Classification	*Agricultural products*
State trading	Rice, barley,[1] beef,[1] natural honey,[1] potatoes, onions,[1] garlic,[1] buckwheat, pepper,[1] oranges,[1] beans, sesame,[2] ginger,[2] soybeans (edible),[1] peanuts,[1] pine nuts, ginseng
Quota auction	Skim milk powder,[1] whole milk powder, evaporated milk, sesame oil[1]
Import license	Live bovine animals, swine, fowl, whey, butter, eggs, bone powder, silkworm eggs, fruit trees, mulberry trees, seed potatoes, manioc, manioc pellets, sweet potatoes, roots and tubers, chestnuts, pine nuts, jujube, green tea, rye (seed), malting barley, oats, soybean (manufactured), corn (seed), corn, grain sorghum (seed), millet (seed), other cereal flours, potato flour, malt, wheat starch, potato starch, manioc starch, sweet potato starch, forage products, lactose, artificial honey, ethyl alcohol, meat flours, oil cake, other mixed feeds, supplementary feeds, cocoons, silk

[1]*Part of tariff- rate quota TRQ is administered by private trading companies with import licenses.*

[2]*Part of TRQ is sold to the private sector through quota auction.*

Note: Korea notified 67 groups of agricultural products as TRQ commodities in 1994. Pork, chicken, and orange juice were removed from the list as of July 1, 1997.

Source: Republic of Korea (1994).

For any of the TRQ products except rice, over-quota importation is possible—or at least it is legal. In reality, imports are highly probable only for products with lower over-quota tariffs such as oranges, whey, and butter. In some cases, for domestic policy purposes, TRQs are expanded for raw materials used for feed or for other manufacturing purposes. (For example, the TRQs for about 20 product groups were expanded every year in 1995, 1996, and 1997. Those product groups included corn, soybeans, barley, and sesame.)

2.2 Rice Imports and the Role of Rice in Korean Agriculture

As noted in the introduction, the URAA mandated that Korea open its rice market. The import ban of the last decade or two was lifted, and a minimum-market quota was established for 51,307 tons (one percent of the domestic market) in 1995, increasing to 205,228 tons (four percent of the domestic market) in 2004 (Table 7.3).

The Supply Administration of Republic of Korea (SAROK) operates an open-tender with a sealed-bids system. The general rule has been to buy from the lowest bidder who meets an acceptable minimum quality. For example, the rice tender notification of October 30, 1998, specifies rice as "Non-glutinous brown rice, medium- or short-grain (japonica type), in 40 kg jute bag or P.P. bag; Crop year: 1997 or 1998; and Grade: U.S. No. 3 or better for the classes of brown rice unit quantity of 20,000 M/T net" (SAROK, 1998). In 1996, out of nine importing companies that applied for bidding, five companies passed the sample examination. Those that passed the sample examination included Daewoo (Chinese rice: U.S. $437.50), Daeboon (Australian rice: U.S. $544), GCA International (California rice offered by Commodity Specialists: U.S. $549), and GCA International (California rice offered by Comet Rice: U.S. $549). Daewoo won the right of supply with the lowest price.

Rice still dominates Korean farming and the Korean diet. Korea is now 98 percent self-sufficient in rice and the government stresses self-sufficiency for rice in every agricultural policy. Vegetables are consumed as side dishes, and most are produced domestically. Wheat and feed grains are almost all imported and a significant proportion of beef is now imported also. Korea is 30 percent self-sufficient in grain.

Table 7.3 Tariff-rate Quota (TRQ) and Tariff Rates for Major Product Groups, 2004

Product	*TRQ*	*In-quota tariff*	*Tariff equivalent*	*TRQ allocation classification*
	1000 tons	*percent*		
Rice	205.0	5	n.a.	STE
Barley	243.0	20	229.70	License
Corn	6,102.0	3	328.00	License
Soybeans	1,302.0	5	487.00	STE, License
Potatoes	19.0	30	304.00	STE, License
Onions	21.0	50	135.00	STE, License
Garlic	15.0	50	360.00	STE, License
Pepper	7.0	50	270.00	STE
Oranges	57.0	50	50.00	STE, License
Beef	225.0*	20	41.25	STE, License
Skim milk powder	1.0	20	176.00	Auction
Whole milk powder	0.6	40	176.00	Auction
Whey	54.0	20	49.50	License
Butter	0.4	40	89.00	License

**Tariff-rate quota (TRQ) of 2000.*
Source: Republic of Korea (1994).

In 1997, rice represented 42.6 percent and horticultural crops represented 34.5 percent of rural household agricultural revenues (Table 7.4). Moreover, rice is grown everywhere, while horticultural crops have high regional concentration due to weather and to other farming conditions. Livestock farming (which includes beef cattle, dairy cattle, hogs, and chickens) is also important for cash income for rural households in Korea.

Table 7.4 Composition of Farm Household Agricultural Revenue in Korea, 1990 and 1997

Crop	*1990*	*1997*
	percent	
Rice	48.2	42.6
Barley	1.9	0.7
Horticulture	28.1	34.5
Miscellaneous crops	4.1	3.7
Livestock	17.7	18.0
*Total**	100.0	100.0

**The total does not add up to 100 percent because of rounding.*
Source: MAF (1998).

2.3 Modeling the Impacts of the Rice Import Quota

Product characteristics (such as type, quality, and perhaps even nation of origin) affect the buyer's willingness to pay for agricultural commodities. So, even if we model rice or wheat as single homogeneous commodities, we know that commercial transactions specify a whole series of characteristics that affect market prices. Furthermore, in some cases, two types within a commodity aggregate may not even be particularly close substitutes. As an example, for premium pasta makers, other wheats are not close substitutes for durum, and the markets for durum wheat and for hard-red-winter wheat may move quite independently at times.

The issue of differentiation is important for the analysis of quotas and TRQs because, usually, many distinct types and qualities of a commodity fit within a single-quota category. Thus, for example, Japanese quota administrators imported both low-quality indica rice and high-quality japonica rice under the same quota, and chose the share of each type using criteria that were not made public and were widely assumed to be primarily political.

Consider now the simplest possible approach to this situation. We assume the relevant commodity has only two types and two uses and that the two types are independent in demand. In Korea, high-quality japonica rice is served as table rice each day at all three meals. Over any expected range of prices, and within the set of incomes observed in recent decades, even after the financial crisis of 1997-98, indica rice, or very low-quality japonica rice, is used only for the small food-processing market. These are quite separate uses and it seems a reasonable first approximation to treat these two markets as independent.

The market of low-grade manufacturing rice is not well defined in Korea: it is a residual market. Also, the rice-processing industry is not well developed. There was a ban, which lasted until the early 1980s, on the use of rice for processing. That ban accompanied a program to require production of low-quality, high-yielding hybrid rice, and reflected a push toward self-sufficiency. Since the early 1980s, low-quality rice, or old rice, has been used for manufacturing products such as rice wine or rice cakes.

Since 1995, imported rice has been used for processing or it has simply been stored. The Korean government allocates imported rice to the registered rice processors at cost. In 1995, long-grain indica rice was imported. During 1996-98, however, some low-quality, medium-grain rice was also imported.

Of course, even within the japonica-rice type there is a range of quality. Thus, one could easily complicate the analysis presented below by considering many subtypes, or even by considering a continuum of qualities. If one were to take that route, it would be important to consider substitution across types or qualities. We expect that the basic results found here would continue to hold.

The second major simplification we make is to treat Korea as a small country in the world market for either type of rice. That seems to be a relatively harmless assumption over the range of the quota projected to 2004. Even in a relatively thin world market for japonica rice, the 200,000 tons of rice that Korea is scheduled to buy is a small share of the total. However, if Korea's purchases expand substantially, as some expect, it would be appropriate to generalize our results to the large-country case and, then, to consider the impact on export suppliers.

In Figure 7.1, panel 1 denotes the market for low-quality processing rice while panel 2 represents the market for high-quality table rice. D and S (not in figure) represent autarky domestic demand curve and supply curve, respectively. S_i' (for i = 1 or 2) is the supply curve with the TRQ imports. Q_1' in panel 1 and Q_2' in panel 2 are both TRQ quantities that are allocated to Market 1 and Market 2, respectively, where $Q_1' + Q_2'$ equals Q^T, the total quantity of quota. P^w denotes world price.

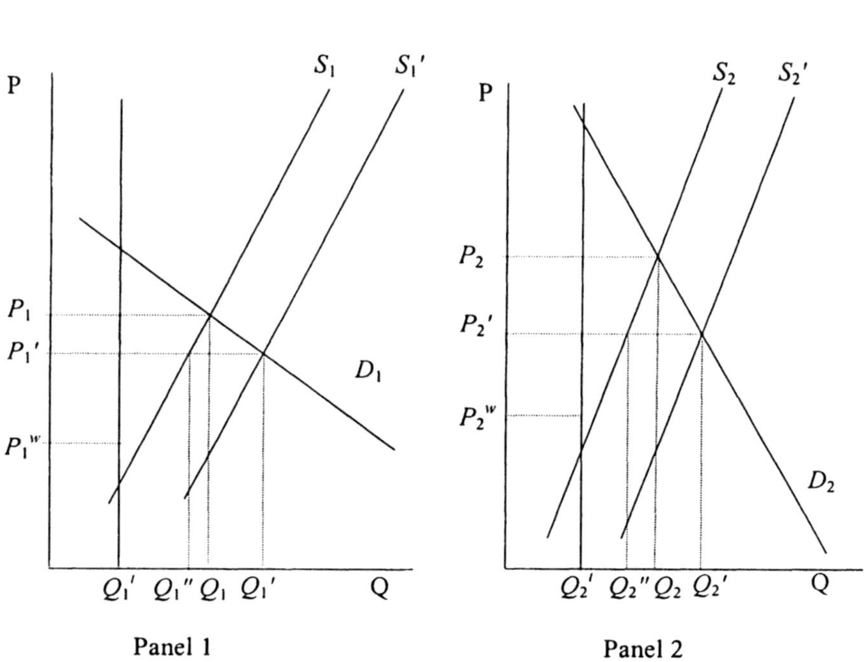

Figure 7.1 Differentiation of Market by Quality

The quota rent, producer surplus, and consumer surplus are given by the areas in each panel: $(P_i' - P_i^w)Q_i'$, $(P_i - P_i')Q_i'' + ½(P_i - P_i')(Q_i - Q_i'')$, and $(P_i - P_i')Q_i + ½(P_i - P_i')(Q_i' - Q_i)$. The figure is drawn in a stylized way, but could be used to represent some key points of the quantitative analysis in the next section. The price of low-quality rice is much lower than the price of high-quality rice both in the world market and in the Korean market. Further, the price differential between the internal price and the world price is much smaller in the low-quality market. The demand is more inelastic in the high-quality market and we have no sense that the supply elasticities differ. Figure 7.1 does not indicate that the low-quality market is only about two percent as large as the high-quality market.

From this figure and the simple set of equations that it represents, one can analyze the impact of allocating rice import quotas to one market or another. We do this by using alternative objective functions that the government might entertain.

The following list of alternative objectives for the administration of the Korean rice quota may be considered. For convenience, we substitute unobserved points — P_1, Q_1, P_2', and Q_2'— with observed points — P_1', Q_1', P_2, and Q_2 — using information from a further specification of the markets.[i] In each case, optimum allocation from the first order condition is expressed in terms of observable points and elasticities:

- To maximize quota rent for a given quota import is equivalent to choosing the highest price gap between world and domestic markets: Max $W = QR_1 + QR_2 = (P_1' - P_1^w)Q_1' + \{P_2 - P_2^w - F(Q^T - Q_1')\}(Q^T - Q_1')$;
- To minimize losses in producer surplus: Min $Y = \Delta PS_1 + \Delta PS_2 = BQ_1'Q_1' - \{B(2d+b)/2b+2d)\}(Q_1')^2 + FQ_2(Q^T - Q_1') - (F^2/2h)(Q^T - Q_1')^2$;
- To maximize the sum of changes in quota rent and producer surplus may be interesting if quota rents were to be transferred to producers: Max $U = QR_1 + QR_2 - \Delta PS_1 - \Delta PS_2$; or
- Maximizing the social welfare maximizes the sum of changes in quota rent, producer surplus, and consumer surplus, and perhaps also accounts for the deadweight cost of taxation:

$$\text{Max } V = QR_1 + QR_2 - \Delta PS_1 - \Delta PS_2 + \Delta CS_1 + \Delta CS_2 = (P_1' - P_1^w)Q_1' + (P_2' - P_2^w)(Q^T - Q_1') + ½B(Q_1')^2 - ½F(Q^T - Q_1')^2.$$

2.4 Internal Subsidy Policy for Rice

Before we turn to the simulations, we consider the internal subsidy for rice so that the border barrier may be placed in the appropriate context. Broad policy measures for Korean agriculture have recently included the designation of the rice-growing area by land zones, government investments for irrigation and land consolidation, and research and development. In addition, the government subsidizes the construction of large-scale milling facilities. This set of policies has an annual budget of approximately 15 percent of farm revenue. Despite the obvious supply-inducing nature of these policies, they are accepted as green-box, non-distorting policies by the WTO.

We now consider the rice-market policy instruments that Korea uses to subsidize rice-farmer income. There are three key elements of the policy. First, the government buys rice at harvest time [directly and through the National Agricultural Cooperatives Federation (NACF)] and pays above-market price for the rice it buys. Second, the government sells most of the rice it has bought back into the market later in the year through regular market channels using NACF as an agent. The government also uses some rice it has bought at inflated prices for military and other official purposes. Third, because it offers a pre-specified, above-market price for rice, the government must limit its total purchase and allocate that to individual suppliers. This is done by a kind of farm-level quota scheme.

With this outline in mind, we consider in somewhat more detail how the program operates. Each year around harvest time, the Korean Ministry of Agriculture and Forestry proposes next year's government purchase price to the national assembly. After the negotiation and approval by the national assembly, the President of Korea sets the final purchase price (about 5 percent above the harvest-period market price in the late 1990s). The aggregate quantity purchased has been approximately constant for a decade and does not vary directly with yield or with other factors. The total has usually been in the range of 20 percent of the crop, but we stress it is a specific quantity, not simply a share or expected share of national output. Each year before planting time, rice farmers make contracts with local agricultural cooperatives for certain quantities and receive a portion of the total contract amount, roughly 40 percent. After harvest, a farmer has the option to receive the remaining amount and deliver the rice, or to pay back the advanced amount with interest and sell the rice through the market.

Since 1995, the total budget used in this scheme has been restricted by Korea's WTO Aggregate Measurement of Support (AMS) commitment. The AMS is not calculated as the budget cost of this scheme. Rather, it is the government-purchased quantity times the price gap between the Korean government purchase price and a pre-determined, base-period, world-market price. To meet its AMS reduction commitment under this scheme, Korea must either bring down the purchase quantity or lower the purchase price gradually over the next several years.

The aggregate government quantity is allocated first to the nine provinces (do) based on some rough historical rice-production share. It is then allocated to counties (kun), sub-counties (eup, myon, or dong), and villages (ri). Finally, each rice farm in each village is assigned an annual ceiling on the right to sell to the government. We will model this assignment as a fixed-quantity quota. In reality, there is likely to be some muted connection between the right to sell to the government and the current year output, but that is the subject for another paper. We next provide a simple representation of the welfare impacts of the domestic rice policy just outlined (Figure 7.2).

The initial supply curve is shown as S and the demand curve is shown as D. Aggregate demand could be affected by the purchase and resale scheme outlined above for two reasons. First, the income shift from urban to rural could affect the aggregate consumption of rice if the income elasticity of rice was higher among rice farmers than it was among the others. Second, flattening the seasonal path of prices to consumers could also affect aggregate consumption marginally. We assume these two effects are too small to notice; therefore, the demand function is shown as unaffected by the policy.

Given the fixed quantity purchased by the government, the aggregate quantity supplied is also invariant to the government purchase price or to the government purchase quantity. In Figure 7.2, the government purchases Q_g tons of rice at price P_g with expenditures of $P_g abP_1$. Farmers treat this as a

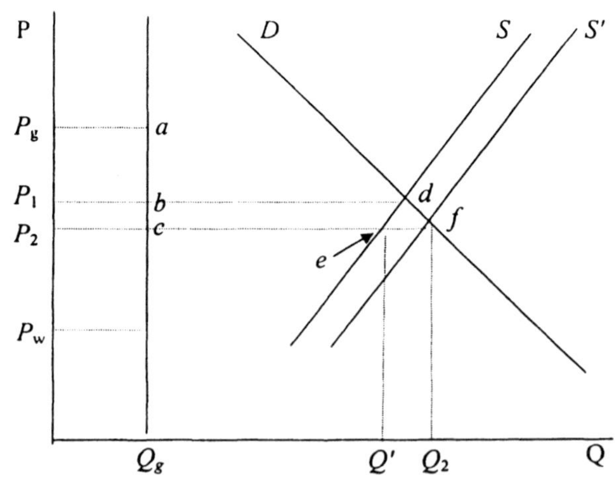

Figure 7.2 Welfare Impacts of the Korean Rice Program and the Effect of Imports

direct payment that is decoupled from their individual output choices. Note that, in our analysis, the domestic policy has little, or no, production subsidy element, and any such effect is almost surely second order. Nonetheless, these operations are notified to the WTO in the trade-distorting AMS. Indeed, given the importance of this single commodity in its agriculture, Korea has had one of the few binding AMS commitments for a major agricultural trader.

Now consider the impact of the import quota Q_i in the context of this domestic policy (Figure 7.2). An import quota shifts the market supply curve from S to S'. Consequently, market price falls from P_1 to P_2, and the new equilibrium quantity consumed is Q_2. The change in producer surplus is P_1deP_2. Notice that the government expenditure is increased by P_1bcP_2, which reduces the loss in producer surplus from the quota by the same amount. The influence of the domestic subsidy on the impacts of the import quota is simply to shift some of the change in producer surplus to the government in the form of higher outlays. If we consider consumer surplus, the gains in consumer surplus make up for the change in government expenditure and net loss in producer surplus, *bdec*, and incur a net social gain of the triangle *dfe*.

3. MEASUREMENT AND SIMULATION

In order to assess the impact of alternative quota-import rules, we investigate how imported rice enters into the two different markets. We use data from 1995-97, inclusive, as our starting point. Two observations are useful at the outset. (1) As noted above, the market for processing rice in Korea accounts for only about two percent of total rice consumption. Also, the domestic price in this market is only about 144 percent or U.S. $433 per ton above the import-offer price, compared to about 265 percent or U.S. $1,460 per ton above the import-offer price in the table-rice market (Tables 7.5 and 7.6).

These facts mean that a minimum import-price rule will always select rice for the processing market. Also, to maximize quota rent, all imports will be allocated to the high-quality market. Given the large size of the table-rice market relative to the size of the quota, minimizing producer-surplus losses drives all imports to the low-quality processing rice market. Indeed, given the relative prices and sizes of the two markets, it is optimal to allocate the entire quota to one market or the other under each of our four alternative objective functions.

Table 7.5 Prices and Quantities Before and After Import into the Low-quality Processing Rice Market

	Autarky	*With imports*	
Import price	—	U.S. $300/ton	
Domestic price	U.S. $1,466/ton	U.S. $733/ton	(-50 percent change)
Domestic production	100,000 tons	33,000 tons	(-67 percent change)
Total supply	100,000 tons	133,000 tons	(+33 percent change)

Note: $\eta_{low}=1.0$; $\eta_{high} = 0.25$; $\varepsilon_{low} = \varepsilon_{high} = 1.0$. ($\eta$: demand elasticity; ε: supply elasticity; low: low-quality market; and high: high-quality market).
Source: Authors' estimations.

We show the initial prices and quantities and the changes implied by imports into each market under the baseline set of supply and demand elasticities (Tables 7.5 and 7.6). To aid comparison and to give a feel for the magnitude of the impacts, all changes are expressed as a percentage of the total revenue in the total Korean rice market (about U.S. $9.6 billion). We provide the welfare effects of allocating the import quota to the processing rice market by purchasing from the low-price bidder, or by allocating the import quota for the table-rice market by opening the import decision to competition (Table 7.5).

Under these baseline assumptions, if the import quota shifts from the processing-rice market to the table-rice market, then the quantity of domestic supply and the market price for table rice in Korea would each fall by about 1.6 percent (Table 7.6). We examined the sensitivity of the results to alternative parameters, but none of our basic conclusions changed even over a rela-

Table 7.6 Prices and Quantities Before and After Import into the High-quality Table Rice Market

	Autarky	*With imports*	
Import price	—	U.S. $550/ton	
Domestic price	U.S. $1,910/ton	U.S. $1,879/ton	(-1.6 percent change)
Domestic production	4,900,000 tons	4,821,000 tons	(-1.6 percent change)
Total supply	4,900,000 tons	4,921,000 tons	(+0.4 percent change)

Note: $\eta_{low}=1.0$; $\eta_{high} = 0.25$; $\varepsilon_{low} = \varepsilon_{high} = 1.0$. ($\eta$: demand elasticity; ε: supply elasticity; low: low-quality market; and high: high-quality market).
Source: Authors' estimations.

tively wide range of elasticities. The main point that stands out from Table 7.5 and Table 7.6 is that the size of the two markets dominates the percentage results.

From looking at the price differentials, the quota rent earned by Korea is much larger if the quota is allocated to the table-rice market. But, as shown in Table 7.6, the producer surplus loss is also much greater if the quota is allocated to the table-rice market rather than being segregated into the processing market (– 1.6 compared to – 0.2). Indeed, the sum of quota-rent and producer-surplus loss indicates that quota rent is large enough to compensate farmers if the quota is allocated to the low-quality processing market. If the quota is reallocated, then the sum of these two effects is negative, (0.3 compared to – 0.2). Thus, ignoring consumers (which seems a fair representation of the key features of Korean rice policy) the government may be quite sensible in allocating the quota as it does.

When we consider consumer-surplus gains and the net-welfare effect, allocation to the high-quality, table-rice market is optimal. (This net welfare effect is larger if we allow for a deadweight cost of raising government revenue.) By allocating its tiny import quota of 100,000 tons to the processing-rice market, Korea is giving up welfare in the range of about one percent of total rice revenue, or about U.S. $100 million, which is a social waste of about U.S. $1,000 for every ton of rice imported, a value approximately equal to the loss in quota rent per ton.

Next we reconsider the reallocation of quota from the processing market to the table-rice market, while allowing for the interaction with the domestic-rice policy described above (Table 7.7). We assume that the government program for rice applies to the table-rice market, and we initially assume that the government leaves the purchase quantity and price unchanged. If the quota is allocated to the table-rice market, then the market price decline of 1.6 percent leads to an increase in budget cost for the rice purchase-sales scheme. Since the government purchase price is about 1.1 times the original market price, the price drop, which affects the government resale price, raises the cost of the government program by roughly 15 percent, from about 2.0 percent of the total rice-market revenue to about 2.3 percent. The net affect on the government budget of this quota reallocation is the outlay change plus the change in quota rent, or a value of about 0.6 percent of total rice revenue [(1.4 – 0.5) – 0.3].

Under the Korean rice program, producers do not lose as much as they could stand to lose without the option to sell 20 percent of their crop at a fixed, inflated, price. Now, producer surplus falls by 1.3 percent of total revenue (–1.6 + 0.3), or 1.1 percent more than when the quota was allocated to the processing market. And finally, given the character of the internal rice-subsidy policy, it would be relatively easy to leave producers unaffected

Table 7.7 Changes in Market and Welfare Measures with Rice Imports

Changes	*Into the low-quality processing rice market*	*Into the high-quality table rice market*
	(change as percentage of rice total revenue)	
Total revenue	-1.5	-3.2
Quota rents	0.5	1.4
Producer surplus	-0.2	-1.6
Quota rents plus producer surplus	0.3	-0.2
Consumer surplus	0.3	1.6
Net welfare	0.6	1.4
Net welfare with deadweight cost of taxation (=0.2)	0.7	1.7

Note: $\eta_{low}=1.0$; $\eta_{high}=0.25$; $\varepsilon_{low}=\varepsilon_{high}=1.0$. *($\eta$: demand elasticity; ε: supply elasticity; low: low-quality market; and high: high-quality market.*
Source: Authors' calculations.

by the shift in the import quota by simply increasing the transfer from the purchase/sales program. Indeed, this approach could be used to raise the quantity of the import quota as well.

4. CONCLUSION

At first glance, the Korean system of allocating the minimum-access import quota among potential suppliers seems to be an open, competitive system. Since there had been no imports prior to liberalization, no historical market shares could be applied and an open-price bid system allowed all suppliers to compete. Indeed, if the internal prices of goods from alternative exporters were equal, then the low-price bid would be selected in a purely competitive market. That is, maximizing quota rent would have the same result as minimizing the supply price.

In this case, the Korean government did not specify a particular set of quality characteristics for imports that minimize the affect of imports on the domestic market. Instead, knowing the characteristics of the domestic market, they were able to open the quota to competitive bidding while keeping the relevant domestic market closed. In doing so, they minimized the loss in domestic-producer surplus while foregoing substantial consumer gains and import quota-rent income.

This chapter analyzed the Korean rice case quantitatively and showed the magnitude of the various welfare effects relative to alternative quota administration policies. We show that the current system of minimizing the import price, by shifting imports to the low-quality market, saves producers approximately 1.1 percent of total rice revenue in surplus losses. It also lowers quota rent by about 0.9 percent of total rice revenue, and reduces consumer surplus gains by approximately 1.3 percent of total rice revenue. Further, the loss of producer surplus from shifting to imports of industrial rice versus table rice is larger than the gain in quota rents. Thus, it would be impossible to compensate farmers for the shift using quota rents alone.

This situation changes somewhat when we consider the import policy in the context of the pre-existing domestic policy for subsidizing rice-farmer income. Allowing imports to enter the table-rice market lowers the market price and raises the cost of the government transfer to rice farmers. This offset also reduces the loss to farmers.

Indeed, recognizing the nature of the domestic rice policy suggests that one way to improve social welfare dramatically, while leaving farmers unharmed, would be to increase payments through the domestic program enough to offset losses from importing table rice. However, using the current program, an increase in payments entails raising the government purchase price, or expanding the amount of rice covered by the purchase scheme. Unfortunately, both of these policy moves would violate Korea's AMS commitment under the URAA. Thus, the implementation of the internal subsidy agreement of the URAA does nothing to induce trade liberalization and, in fact, makes it more difficult for Korea to liberalize.

5. ENDNOTES

[i] Define the curves in market 1 such as inverse demand curve: $P_1^d = a - bQ_1^d$; autarky inverse supply curve: $P_1^s = c + dQ_1^s$; and inverse supply curve with imports: $P_1^s = c + d(Q_1^s - Q_1^I)$, where a, b, c, and $d > 0$. Then $P_1 - P_1' = bd/(b+d)Q_1^I$. For convenience, define $B = bd/(b+d) > 0$. Also we get $Q' = Q_1 + d/(b+d)Q_1^I$ and $Q_1' = Q_1' - b/(b+d)Q_1^I$. In market 2, we get similar relationship by defining the curves as $P_2^d = e - fQ_2^d$; $P_2^s = g + hQ_2^s$; and $P_2^s = g + h(Q_2^s - Q_2^I)$, where e, f, g, and $h > 0$. Then $P_2 - P_2' = fh/(f+h)Q_2^I$ and defining $F = fh/(f+h) > 0$, we have $Q_2' = Q_2 + h/(f+h)Q_2^I$ and $Q_2' = Q_2' - f/(f+h)Q_2^I$.

6. REFERENCES

Choi, J., D.A. Sumner, and J. Song. 1998. "Importing STEs in Korea and Japan: Evolution, Operation, and Implications." Paper presented at the Role of State and Agricultural Trade Workshop, Co-organized by the North American Forum, Stanford University and Agricultural Issues Center, University of California (20–22 November).

Josling, T., S. Tangermann, and T. Warley. 1996. *Agriculture in GATT*. New York, NY: St. Martin's Press Inc.

Lee, J., D. Kim, Y. Sakong, J. Suh, and J. Im. 1994. *Direction of Import Management after the Liberalization*. Korea Rural Economic Institute. Seoul, Korea (April).

Lee, J. and Y.K. Choi. 1997. *Evaluation of Agricultural Market Liberalization and Direction of Import Policy*. Seoul, Korea: Korea Rural Economic Institute (December).

MAF (Ministry of Agriculture and Forestry). 1998. *Annual Report on the Situation of Agriculture*. Seoul, Korea: Ministry of Agriculture and Forestry (October).

Moschini, G. 1991. "Economic Issues in Tariffication: An Overview." *Agricultural Economics* 5: 101–120.

Republic of Korea. 1994. *Country Schedule*. Seoul, Korea: Republic of Korea.

Sumner, D.A. and S. Tangermann. Forthcoming. "International Trade Policy and Negotiations," in B. Gardner and G. Rausser, eds., *Handbook of Agricultural Economics*. Amsterdam, Netherlands: North Holland Press.

SAROK (Supply Administration of the Republic of Korea). 1998. Government Report, *Invitation for Bids*. Seoul, Korea: Administration of the Republic of Korea (19 September).

USDA/FAS (U.S. Department of Agriculture, Foreign Agricultural Service). "FAS Outline-Pacific Rim," Internet website: http://www.fas.usda.gov/itp/policy/pacific%Frim.html.

WTO (World Trade Organization). 1995. *Notification Concerning the Administration of Tariff Quotas of Korea*. WTO Report No. G/AG/N/KOR/1, Geneva, Switzerland (31 May).

_____. 1996. *Review of Korea*, WTO Report No. PRESS/TPRB/42, Geneva, Switzerland (1 October).

PART II

TARIFFS QUOTAS AND RENT-SEEKING BEHAVIOR

Chapter 8

Market Conduct and the Economic Impacts of a Tariff-Rate Quota Policy: The European Banana Case

Roland Herrmann and Richard J. Sexton[i]

University of Giessen, and University of California at Davis and Giannini Foundation of Agricultural Economics

1. INTRODUCTION

The European Common Market Organization for bananas (CMOB) is currently the most controversial tariff-rate quota (TRQ) policy among the increasing number of TRQs in international agricultural trade. The CMOB was introduced under the pressure of the Single European Act in July 1993. It has affected various countries of the European Union (E.U.) differently, as the previous national banana policies of the member countries implied differential levels of protection. The new policy regime, with a tariff quota as its major feature, has lowered the level of agricultural protection in formerly more protected countries (such as France). But a new trade barrier was introduced for countries that had more liberalized banana-trade policies before 1993. In particular, Germany has experienced higher prices due to the new policy. It is mainly these new barriers which have been discussed in public and criticized for being another policy-induced distortion in the international banana trade (Borrell, 1997).

Within the debate on recent protectionism in the European banana market, selected economic impacts of the new policy, including price, consumption and welfare effects, have been presented in some recent studies (Matthews, 1992; Behr/Ellinger, 1993; Kersten, 1995; Guyomard, Laroche, Le Mouël, 1999; Herrmann, 1999). However, these studies have all assumed a competitive market structure, which Guyomard, Laroche, and Le Mouël (1999) have identified as a key limitation of work conducted to date on the

CMOB. Although most economists view the CMOB as highly distortive, its economic effects, in particular the distributive effects (including the quota rents), are not clear. Guyomard, Laroche, and Le Mouël (1999) have identified the analysis of quota rents as an additional key research need in order to fully assess the impacts of the CMOB.

The goals of this chapter are to address these remaining key research issues concerning the CMOB, and to make some general points about the importance of considering and evaluating alternative plausible models of market conduct when undertaking applied policy analysis. We concentrate on the German banana market: the largest segment of the E.U. banana market and a fully liberalized market prior to the implementation of the CMOB. We show that the welfare effects of the CMOB depend crucially on assumptions regarding the existing market structure, particularly whether or not Germany is a large country in respect to its banana purchases and whether or not oligopoly power was exercised by banana importers prior to the imposition of the CMOB. Specifically, we set forth the conceptual framework to evaluate the welfare effects of the CMOB under each of three hypotheses regarding the market:

- *Hypothesis 1*: Germany was a small country in a competitive import market for bananas prior to the CMOB (perfect-competition/small-country hypothesis).
- *Hypothesis 2*: Germany was a large country in a competitive import market for bananas prior to the CMOB (perfect-competition/large-country hypothesis).
- *Hypothesis 3*: The German import price was raised by imperfect competition prior to the CMOB (imperfect-competition/small- or large-country hypothesis).

We then estimate econometrically each of the models and conduct statistical tests of the hypotheses concerning (a) the large versus the small importer designation, and (b) the competitiveness of the German banana market. Finally, we analyze how the different hypotheses regarding market conduct in the banana economy affect the welfare impacts of the TRQ policy.

We show that a researcher with an *a priori* notion about behavior in this market would probably have been able to find econometric evidence in support of it, and we expect that this situation will also often be true in other applications. The welfare analysis would then typically be based on the maintained model, ignoring the possibility that other behavioral hypotheses, yielding different welfare conclusions, might find even more support in the data. Thus, in addition to providing specific results regarding the structure of the German banana market and the welfare effects of the CMOB, this chapter offers more general conclusions regarding welfare analyses of trade or other market policies. Such analyses should be made only after a careful consideration of alternative plausible specifications of market behavior.

2. WELFARE IMPLICATIONS OF THE CMOB UNDER ALTERNATIVE MARKET MODELS

Only the elements of the CMOB that are relevant to modeling policy impacts are summarized in this section. Thagesen and Matthews (1997) provide a detailed survey of the Banana Framework Agreement (BFA), which succeeded the initial framework in 1995. The WTO Panel Report of 1999 (WTO, 1999) surveys later changes in the CMOB as well as the WTO-E.U. trade dispute on bananas, and assesses, in great detail, which elements of the CMOB have been inconsistent with WTO rules. We refer to the initial CMOB, as introduced in 1993.

On the internal E.U. market, a deficiency-payment system for E.U. banana producers was introduced. This system was intended to cover earning shortfalls due to the new policy. Payments to producers covered the whole difference between historical reference earnings and the actual earnings from banana production. They were limited to a maximum quantity of 854,000 metric tons of bananas in the traditional banana-producing E.U. regions (including Guadeloupe and Martinique).

With regard to external trade, tariff quota and import-licensing systems were put into force. The quota was 2 million metric tons of bananas either imported from third-world countries, or imported as non-traditional African, Pacific, and Caribbean (ACP) exports to the E.U. Non-traditional ACP exports to the European Union were defined as quantities above a traditional pre-CMOB export level of 857,700 metric tons. A tariff of 100 European currency units (ECU) per metric ton was imposed on banana imports from third-world countries within this quota; non-traditional ACP banana exports to the E.U. market within the quota were duty-free (Article 18). The quota could be raised or lowered in principle, depending on forecasts of banana needs in the European Union. It was increased to 2.1 million metric tons in 1994. For exports exceeding the quota, a prohibitive tariff of 750 ECU per metric ton for non-traditional ACP banana exports to the European Union and of 850 ECU per metric ton for imports from third-world countries was set. In addition to this quota, which was mainly filled by exporters of dollar bananas (bananas originating in Central and South America), traditional imports from ACP countries were allowed to enter the E.U. market duty free.

A further distinctive feature of the TRQ policy was the initial distribution of import licences (Article 19). Only 66.5 percent of the whole quota was allocated to participants who had marketed dollar bananas and/or non-traditional ACP bananas (so-called A-licences). 30 percent of the import licences were given to market participants who had marketed E.U. and/or traditional ACP bananas (so called B-licences). The residual 3.5 percent of the quota was reserved for beginners who had started to sell non-traditional ACP bananas in the European Union. The allocation of licences within the two major groups was further regulated according to functions of the firms in the distribution channel: 57 percent went to primary importers, 15 percent to secondary importers, and 28 percent to ripeners (Thagesen and Matthews, 1997: 617). The relevant benchmark quantities were computed by using a moving average of the applicable exports in the last three years. The licences were generally tradable.

2.1 Hypothesis 1: Small Country and Perfect Competition

Four stages of price formation in the German banana market can be distinguished: (1) the export level, at which foreign suppliers in the producing countries meet multinational banana firms on the demand side; (2) the import level, at which multinational firms typically offer bananas to ripeners and to wholesalers; (3) the wholesale level, at which ripeners and wholesalers sell bananas to retailers; (4) the retail level, at which the consumers buy the final product from retailers. This analysis focuses on the import level because the TRQ policy is operational at the import level, and because high levels of concentration among importers suggest the possible presence of oligopoly power.

The German share of the world banana market is about ten percent. Under the small-country hypothesis, Germany faces a perfectly elastic supply curve for its banana imports. Price, trade, and the welfare impacts of a TRQ policy, such as the CMOB policy under the small-country hypothesis, are set forth by Abbott and Paarlberg (1998). The relevant analysis is presented in Figure 8.1. The export supply curve of dollar-banana exporters on the German banana market, $S(q)$, is perfectly elastic at the world-price level. (We concentrate on the supply of dollar bananas, as more than 95 percent of all bananas consumed on the German market are dollar bananas in the situations with and without the CMOB.) A sizeable substitution between dollar and ACP or E.U. bananas has not occurred. The German import demand curve is $D(q)$, and free trade, the pre-policy situation, results in an equilibrium import price p_0 and German banana imports of q_0.

With the TRQ introduced under the CMOB a quota is fixed at the European level, of which Germany receives q_1. Licences for the quotas are

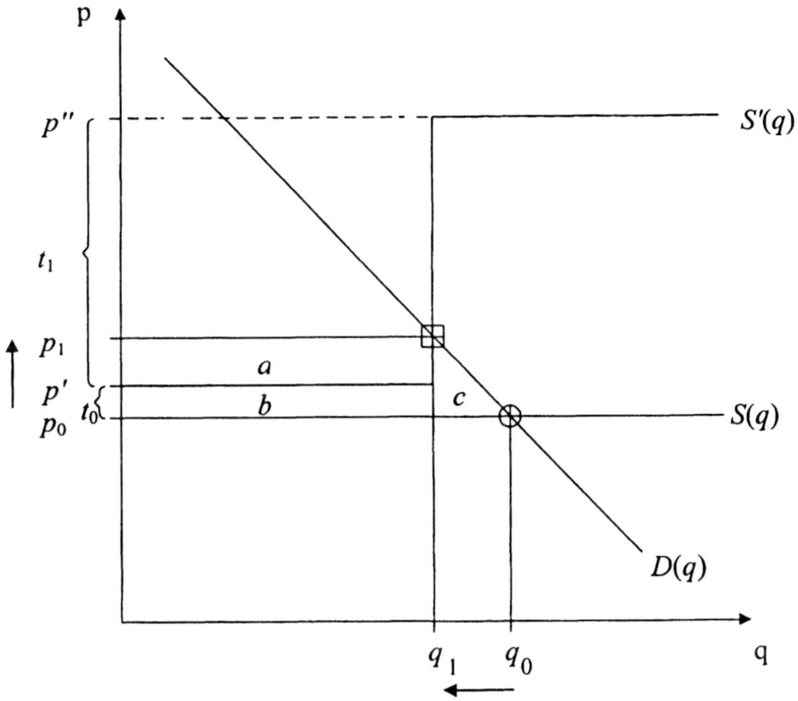

Figure 8.1 Economic Implications of the TRQ Policy on the German Banana Market (Small-country Assumption)

allocated to importers in the original CMOB, thus leading to quota rents for importers. A two-part tariff is introduced: t_0 on all imports up to the quota and (t_0 + t_1) as a prohibitive tariff rate above the quota. This leads to the kinked export supply curve $S'(q)$ with a vertical part between p' and p'' at the quota quantity, q_1. $S'(q)$ now intersects $D(q)$ at p_1. Thus, the TRQ raises the import price from p_0 to p_1 and restricts imports from q_0 to q_1.

Welfare analysis from the European point of view has to incorporate welfare effects on E.U. consumers, E.U. traders, and the E.U. budget. Consumers in the German market for banana imports lose the area (a+b+c) in consumer surplus as a consequence of rising prices. A part of the price increase is due to the introduction of the tariff t_0. A budgetary gain of area b arises at the E.U. level. Due to the TRQ, the equilibrium price rises to p_1 rather than p' as in the pure tariff case. Quota rents occur in the magnitude of area a on the German market. Because licenses for import quantity, q_1, are held by importers under the original rules, the quota rents accrue in part to multinational firms. It is thus unclear to what degree the rents are captured by Europeans as opposed to agents outside of Europe. We define γ ($0 < \gamma < 1$) as the portion of the quota rents captured by Europeans. From the European point of view, the CMOB policy on the German market causes losses to E.U. consumers that exceed the sum of the E.U. tax revenues and the gains by E.U. traders (even if the quota rents are captured entirely by Europeans). The net welfare loss is the area $c + (1 - \gamma)a$.[ii]

2.2 Hypothesis 2: Large Country and Perfect Competition

Given the ten percent German share of the world banana market from 1960-92, the small-country assumption could be questioned. Changes in German import demand might affect the German import price. This possibility is captured in Hypothesis 2, in which we posit a competitive market and the large-country hypothesis (Moschini, 1991). In this case, the residual supply of bananas to the German market is upward-sloping (Figure 8.2). In the large-country case, the upward-sloping residual supply curve to Germany represents the excess supply curve of all dollar-banana exporters minus the total import demand for dollar bananas from all importing countries excluding Germany. The equilibrium import price in the free-trade situation is p_0 and German banana imports are q_0.

Under the CMOB a quota is fixed at the European level, of which Germany receives q_1. As in Figure 8.1, the two-part tariff leads to an export supply curve, $S'(q)$, with a vertical portion between p' and p'' at the quota level. $S'(q)$ now intersects $D(q)$ at p_1. Thus, the TRQ raises the import price from p_0 to p_1 and restricts imports from q_0 to q_1.

The price and trade impacts are thus similar to hypothesis 1 with the exception that the export supply curve under the policy influence remains upward-sloping, with a vertical segment at the quota level. This causes the exporting countries to be affected differently by the CMOB under Hypotheses 1 and 2. Whereas the exporting countries' price was p_0 in Figure 8.1, it is now p_{EXP} in Figure 8.2. Therefore, exporting countries are worse off on the German banana market, since prices decline under the new policy but they do not decline under the small-country case. Importers capture the quota rent as in hypothesis 1, but the amount of the quota rent increases when the export price declines to p_{EXP}, compared with the small-country case.

Due to the CMOB, consumers on the German market for banana imports lose the area $(a+b+c)$ in consumer surplus under hypothesis 2. A part of the underlying price rise from p_0 to p_1 is a consequence of the introduction of the

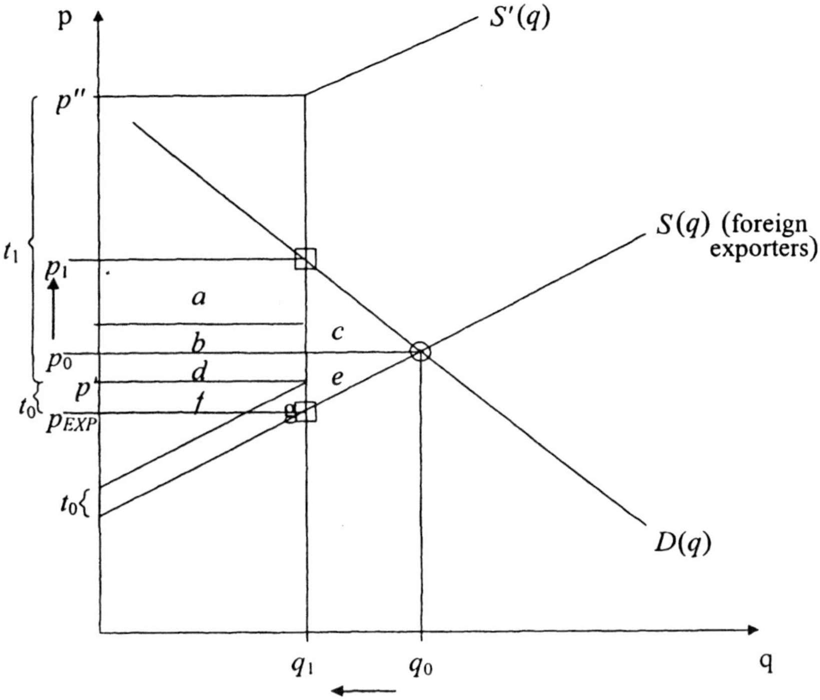

Figure 8.2 Economic Impacts of the TRQ Policy on the German Banana Market (Large-country Assumption)

tariff t_0. A budgetary gain of the area $(f+g)$, which is equal to area b from Figure 8.1, arises at the E.U. level. The quota raises the import price more than in the pure tariff case. Traders may use their import licences to sell dollar bananas to consumers at p_1. Price received by the exporting countries falls to $p_{EXP.}$ Traders earn a quota rent on the German market equal to $(a+b+d)$ of which $\gamma(a+b+d)$ is captured by Europeans. Thus, the import price p_1 includes the exporting countries' price, p_{EXP}, the tariff, $p'-p_{EXP}$, and a quota rent for importers per imported unit of p_1-p'.

The net welfare impact of the tariff-quota policy under Hypothesis 2 from the European perspective is $\gamma(b+d) - c - a(1-\gamma)$. The consumers' welfare loss may, or may not, exceed the sum of welfare gains by E.U. traders and the E.U. budget. Thus, the net welfare impact from an E.U. perspective is uncertain and is dependent upon the relative magnitudes of the deadweight loss, quota rents, and the extent to which the quota rents are captured by European (as opposed to international) traders. The size of the quota rent is also strongly affected by the large-country hypothesis. The quota rent is higher under Hypothesis 2 than under Hypothesis 1, and the additional quota rent under Hypothesis 2 is paid by suppliers from exporting countries. Non-E.U. exporters lose the area, $d+e+f+g$, in producer surplus on the German banana market as a consequence of the new policy.

2.3 Hypothesis 3: Imperfect Competition and Small Country

The recent analyses of the impacts of the European banana policy use perfect-competition models. Other studies, however, cast doubt on the veracity of the perfect-competition hypothesis (Deodhar and Sheldon, 1995; McCorriston and Sheldon, 1996). Information on market shares tends to support these doubts. McCorriston and Sheldon (1996) report that three multinational firms—United Brands (Chiquita), Standard Fruit (Dole) and Del Monte—captured 70 percent of the world market in bananas and 66 percent of the European market. Three firms—United Brands, Standard Fruit and Noboa—accounted for 72 percent of the German banana market. This concentrated market structure is also associated with political influence. Pedler (1994) shows that the fruit companies were very active players in lobbying prior to the introduction of the CMOB.

Hypothesis 3 posits the case of importer oligopoly power on the German banana market and small-country status for Germany as an importer.[iii] We interpret prices and quantities observed prior to the implementation of the CMOB as an oligopoly equilibrium rather than perfect competition (Figure 8.3). The curve, $S(q)$, is the export supply curve under perfect competition, and represents the aggregated marginal cost curve of importers under imperfect competition. Consistent with the small-country assumption, firms' marginal costs are assumed to be constant and aggregate easily to the industry level.

In the absence of the TRQ policy, output is q under perfect competition and q_0^M under pure monopoly. As Holloway (1991) and others have shown, an oligopoly solution can be represented by defining a conjectural variations parameter, $\xi \in [0,1]$, with 0 denoting perfect competition, 1 denoting monopoly, and intermediate values representing various oligopoly outcomes. Larger values of ξ in the interval $(0,1)$ represent higher levels of oligopoly power. In this setting, the equilibrium output occurs where the curve is defined as the linear combination, $\xi MR(q)+(1-\xi)D(q)$, of the marginal revenue, and demand curves intersects the marginal cost curve. This output, and the associated market price, are denoted as (q^o,p^o) in Figure 8.3, where the O superscript denotes an oligopoly equilibrium.

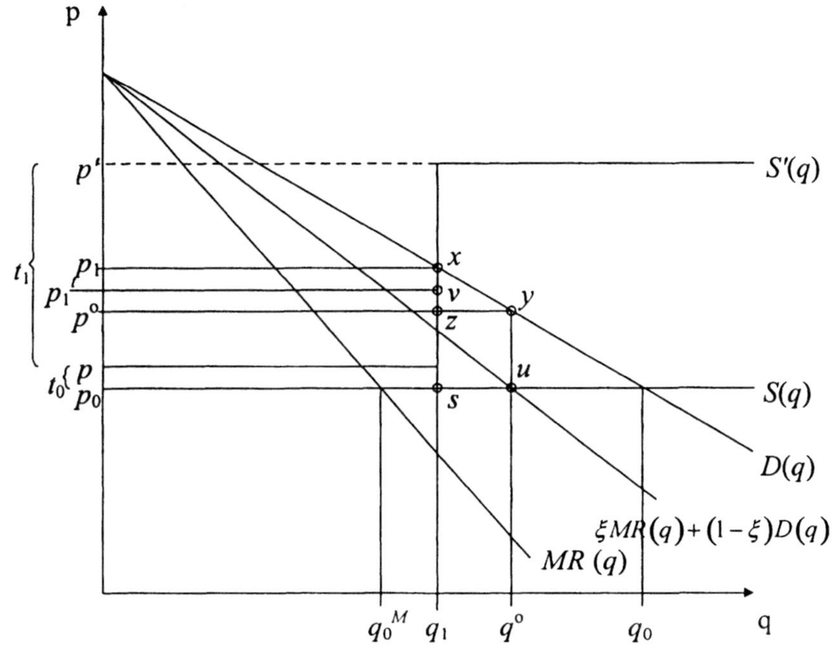

Figure 8.3 Economic Impacts of the TRQ Policy on the German Banana Market (Small-country Assumption; Market Power)

The tariffs t_0 and t_1 are represented as parallel shifts upward in the marginal cost curve, with a vertical segment occurring at the quota level, q_1. Because the imposition of the TRQ on the German market clearly results in reduced banana sales and higher prices, Figure 8.3 depicts $q_1 < q^o$, that is, whatever type of conduct prevailed on the German banana market, the TRQ reduced sales to it.

In addition, Figure 8.3 assumes that the quota, q_1, is binding, that is, the $\xi MR(q)+(1-\xi)D(q)$ curve intersects the $S'(q)$ curve in the region of its vertical segment. In this case, which certainly applies to the German banana market, the impacts of the CMOB on price and imports are identical under imperfect and perfect competition. Quantity is set at the quota level, q_1, and price is set off the demand curve at p_1. The only difference is that we interpret the hypothetical equilibrium (q^o, p^o) that would have prevailed in the absence of the policy as the outcome of imperfect competition. This point is important because, beginning with the seminal work of Brander and Spencer (1985), various authors have shown that trade policies can perform a strategic role by altering the extent of active rivalry in the market. However, if the policy instrument is a quota, and the quota is observed to bind, then output and price in the post-quota regime are determined simply by the quota, and not by changes in strategic behavior among exporting firms.[iv]

What are the welfare implications of the TRQ on the German banana market if the pre-policy equilibrium is based on importer oligopoly power? Due to the price increase, consumers lose surplus defined by the area $p^o p_1 xy$, equivalent to the area $(a+b+c)$ in Figure 8.1. A part of the price increase is due to the tariff t, and this leads to a budgetary gain at the E.U. level defined by the rectangle $p^o p_1^t vz$, equivalent to the area b in Figure 8.1. E.U. traders, however, are affected differently if the equilibrium without policy is imperfectly competitive. They gain the rectangle $\gamma(p_1^t p_1 xv)$ as a quota rent, but they lose the rectangle $\gamma(szyu)$ of their profits from the non-policy situation. This latter area represents the profits to E.U. traders that are foregone by restricting sales from the pre-quota level, q^0. The impacts on the groups result in an aggregate welfare loss from the European point of view equal to the area $zxy + (1-\gamma)(p_1^t p_1 xv) + \gamma(szyu)$. The first area is the deadweight loss on the consumption side from restricting output to q_1. The second area is the quota rents that are not captured by E.U. traders, and the third area is the European traders' share of the oligopoly profits lost due to restricting output below the oligopoly solution. The aggregate welfare loss from the European point of view is larger in Figure 8.3 (imperfect competition) than in either hypotheses 1 or 2 (perfect competition), because E.U. traders gain less from the introduction of the TRQ scheme when the situation without policy involves their exercise of oligopoly power. These traders lose some of the oligopoly profits when sales are restricted below the imperfect-competition equilibrium.

In summary, the welfare and redistributive implications of the TRQs differ substantially according to the model of market conduct which is utilized to explain observed prices and quantities in the pre-CMOB situation. We next proceed to estimate each of the market models set forth here, and we evaluate which, if any, of them find support in the data.

3. ECONOMETRIC TESTS OF THE ALTERNATIVE HYPOTHESES ON MARKET STRUCTURE AND CONDUCT

The general practice in quantitative policy analysis is to start from one market-conduct hypothesis, to rationalize that hypothesis, and to base policy analysis on assumed or estimated coefficients of an econometric model which is consistent with the selected hypothesis. Our econometric analysis starts from each of the three theoretical models set forth in the previous sections. We specify econometric models that conform to the respective theories. In each case, the question is posed as to whether or not econometric evidence can be found that is consistent with the theoretical model.

In analyses of the CMOB, the perfect competition hypothesis has been assumed. Some studies have used the small-country and others have used the large-country assumption for either Germany or the European Union. Deodhar and Sheldon (1995) tested the hypothesis of imperfect competition in the German banana market, but did not analyze the impact of the CMOB under imperfect competition.

Under any of the market scenarios, the import price for Germany from 1993 onward in the hypothetical situation without the CMOB must be forecast to provide a baseline for comparison to the alternative CMOB scenarios. We used a simple price transmission equation to forecast the German import price, p, measured in Deutch Marks (DM), as a function of the world price p_W, defined as the U.S. dollar price for South and Middle American bananas of class I in New York, and the exchange rate, e, expressed as the price of one U.S. dollar in DM. The following price transmission equation was found to fit best:

$$\begin{aligned} (1) \quad & \ln(p) = \underset{(1.78)}{0.1212} + \underset{(12.53)}{0.9212}\ln(p_w) + \underset{(6.82)}{0.8746}\ln(e), \quad \bar{R}^2 = 0.89, \\ & DW = 1.32 \end{aligned}$$

where t-statistics are indicated in parentheses below each coefficient.[v]

Equation (1) allows for a differential impact of the world price and the exchange rate on the German import price for bananas. The world price and exchange rate transmission elasticities are close to unity and are highly significant. This indicates that alterations in world prices and exchange rates were nearly completely transmitted to the German market within the individual years from 1970-92. About 90 percent of the total variation in logarithms of German import prices can be explained by the price-transmission equation. This result, however, does not imply that the import market was competitive (for example, McCorriston (1997) has shown that the price transmission may also be very high in oligopolistic markets).

The analysis of price transmission as well as import demand included a comprehensive investigation of cointegration with the estimation of error-correction models (Herrmann, 1999). The relevant time series do seem to be cointegrated. Hence, we can interpret the price-transmission equation (1), as well as all presented import-demand functions, as cointegrating regressions in the sense of Engle and Granger (1987), revealing the long-run relationship between independent and dependent variables.

Equation (1) is used to forecast what the German import price would have been in the absence of the CMOB policy under each of the market scenarios.

3.1 Econometric Results for Hypothesis 1 (Small Country and Perfect Competition)

Hypothetical imports have to be modelled on the basis of estimated import demand functions for bananas. Under Hypothesis 1, foreign exporters supply bananas at the German border for a given international price. The import price can be treated as exogenous in the econometric estimates of import demand, justifying ordinary least-squares (OLS) estimation.

Economic theory suggests that per-capita banana import demand, q^c, depends on the import price of bananas p, the import price of major substitutes in consumption, and the per-capita income level, Y^c. Previous studies of the fruit market suggest that the major substitute for bananas is apples (Weiß, 1995). If the world apple price, p^A, is introduced, and we also consider a dummy variable, D, for the effects of German reunification, demand can be expressed as:

$$(2) \quad q^c = g\left(p, p^A, Y^c, D\right).$$

All monetary variables are expressed in DM and denominated in real terms, denoted subsequently by the subscript R, by deflating with the German consumer-price index (1985 = 100). Further details on the demand analysis are available in Herrmann and Sexton (1999).

Although the results for all demand models were broadly consistent, the following linear functional form produced better statistical properties than a double-log alternative:

$$(3) \quad \begin{array}{l} q^c = -5.6561 - 5.8171 p_R + 0.7029 p_R^A + 0.00073 Y_R^c + 4.1880 D \\ \qquad\quad (1.04) \quad (5.46) \qquad (0.46) \qquad (5.72) \qquad (8.87) \\ \left(\bar{R}^2 = 0.96, DW = 1.94, \text{Ramsey's RESET } 2 = 0.02\right) \end{array}$$

Absolute t-statistics are in parentheses. The model explains more than 95 percent of demand variations, with highly significant explanatory variables, and no indication of autocorrelation or misspecification. German import demand for bananas has a significant negative correlation with real per capita income, but the real import price for apples did not have a significant effect. Apparently apples are no longer regarded as good substitutes for bananas.

The German unification also increased per-capita imports of bananas. Evaluated at the data means, the estimated own-price elasticity of import demand for bananas in Germany is –0.535, and the income elasticity is 1.906. The magnitude of our price elasticity compares closely with values estimated, or assumed, in prior studies; our income elasticity is somewhat higher than that obtained in previously completed work.[vii]

In general, the import demand function formulated on the basis of the small-country hypothesis performs very well. A Hausman test supported treatment of the import price as an exogenous variable (Herrmann and Sexton, 1999), thereby also supporting the small-country hypothesis.

3.2 Econometric Results for Hypothesis 2 (Large Country and Perfect Competition)

Given that Germany has a market share of about ten percent of the world banana market, one might argue that Germany is a large, rather than a small, country. A direct test of this argument is to investigate whether or not export supply to the German market is less than infinitely price-elastic. This is feasible within a simultaneous model for the German banana import market containing an import demand function, an export supply function, and an equilibrium condition.

Import demand was specified in accordance with the estimates under hypothesis 1, dropping the insignificant apple price variable. The export supply curve of bananas to the German import market is a residual supply curve. Hence, the demand and supply shifters in all other regions of the world should enter the export supply curve on the German market. A specification of this type generates obvious econometric problems. Many demand shifters, like income variables in the major banana-consuming countries and import prices of major substitutes, are highly correlated. Econometric models that include many of these variables in the export supply function suffer from severe multicolinearity problems. We selected, therefore, a very parsimonious specification of the export supply curve, which is oriented at the purchasing power of exports to the German market. The latter can be decomposed into the German import price for bananas and the real exchange rate, measured by the U.S.-dollar price of one DM multiplied by the ratio between the consumer price levels in Germany, P^G, and in the United States, P^{US}. In an equilibrium model under hypothesis 2, which explains the market price in DM, the real exchange rate has to be included in the export supply function, because foreign exporters plan in the international currency, namely the import price in DM(p) times the real exchange rate of the DM in U.S. dollars $\left(e_R^{\$,DM}\right)$. Theory suggests that foreign suppliers increase banana exports to the German market if the import price, denominated in DM, increases and/or if the real exchange rate of the German Mark increases.

The import demand and export-supply functions of the implicit four-equation model were estimated by two-stage least squares.[viii] Instruments used in the first stage to obtain the predicted prices, $\hat{p}$, and $\hat{p}_R$, included Y_R^c, e_R and D. The t-values are those of the structural model and not of the second stage, which would be obtained by doing two-stage least squares literally in two stages (Hall and Cummins, 1997: 32).

The estimated import demand function is:

$$q_D^c = -3.8048 - 5.7448\hat{p}_R + 0.00069Y_R + 4.2396D \tag{4}$$
$$(1.34) \quad (5.19) \quad (7.99) \quad (8.94)$$
$$\left(\bar{R}^2 = 0.96; DW = 1.89\right)$$

and the estimated export supply function is:

$$q_S^C = -53.8581 + 31.0675\hat{p} + 53.7034e_R^{\$,DM} \tag{5}$$
$$(2.11) \quad (2.33) \quad (2.33)$$
$$\left(\bar{R}^2 = 0.11; DW = 1.56\right)$$

Results indicate some support for the hypothesis of a simultaneous market model of the German import market for bananas despite Hausman's specification test not indicating any misspecification due to simultaneity under the small-country assumption implicit in equation (3). Although the statistical test criteria are clearly better for the import demand than the export supply function, the signs of all coefficients of the export supply functions are plausible and the statistical significance level of each coefficient is 95 percent or higher. The results reveal that export supply from foreign exporters to the German banana market increased with a rising import price for bananas on the German market, and a rising real exchange rate of the German Mark in U.S. dollars. Elasticities computed at the data means are plausible both as to sign and magnitude. The price elasticity of import demand is –0.528, and the income elasticity is 1.798. Thus, both elasticities are of similar size as in the earlier econometric estimations featuring the small-country assumption. The mean price elasticity of export supply on the German banana market is 2.795 and the exchange rate elasticity of export supply is 3.036. This implies that the price elasticity of export supply on the German market for banana imports is high, but not infinitely elastic as implied by the perfect-competition hypothesis.

For a researcher who believed *a priori* in hypothesis 2, the estimation results appear to justify the conclusion that importers on the German banana market face an upward sloping export supply curve. Shifts in import demand would then affect the German import price for bananas, and, thus, Germany could not be regarded as a price-taker on the world banana market.

3.3 Econometric Results for Hypothesis 3 (Imperfect Competition)

Deodhar and Sheldon (1995) provide the only formal tests for the existence of market power in the German banana economy. They estimate a structural model of the German market for banana imports and derive a conjectural variation parameter to measure the degree of market power. Deodhar and Sheldon (1995) conclude that the German market for banana imports was not perfectly competitive from 1970-92. Rather, their results supported a hypothesis that firms demonstrated Cournot-Nash behavior. Because we had concerns regarding the demand model used by Deodhar and Sheldon (1995), we re-estimated their structural model using the demand framework that emerged from our own investigation.

Another departure from Deodhar and Sheldon (1995) is that we also tested the hypothesis of imperfect competition at the import level rather than the retail level. The import level is the level at which multinational firms supply bananas to wholesalers and ripeners and the level at which the first stage of price formation for bananas occurs. Market power could be suspected here given the high levels of seller concentration noted previously.

We use a demand equation similar to the specifications given in equations (3) and (4). We regard this specification as superior to the Deodhar and Sheldon (1995) formulation, in which market demand, not per-capita demand, was specified as a function of price, magnitude of the German population age 65 and above, a time trend, and a squared time trend.[ix]

As discussed previously, under the small-country hypothesis, the industry equilibrium condition in each period under importer oligopoly power can be expressed as:

$$(6) \quad (1-\xi)D(q^c)+\xi MR(q^c)=MC \ ,$$

where $p = D(q^c)$ is the inverse import demand at time t (expressed as a function of per-capita consumption), $MR(q^c)$ is the associated marginal revenue function, and MC is the constant marginal cost, assumed to be identical across importers. Subscripts to denote time dependence of the expressions are omitted to avoid notational clutter. $\xi \in [0,1]$ is the market power coefficient which measures the degree of oligopoly power in the industry. Measured values of ξ that significantly exceed zero represent statistical evidence of importer market power in the industry. Equation (6) can be rewritten slightly into the following form that is useful for estimation:

$$(6') \quad p + \xi q^c \left({dp}/{dq^c} \right) = MC$$

To test for importer market power in the German banana economy, we begin with the per-capita demand function in (2), but omit the real apple price as a substitute variable, since it has been demonstrated to be insignificant:

$$(2') \quad q^c = \alpha_0 + \alpha_1 p_R + \alpha_2 Y_R^C + \alpha_3 D + \varepsilon_1$$

where ε_1 is an error term. The marginal cost function at the import stage of the German banana market is formulated as:

$$(7) \quad MC = \beta_0 + \beta_1 p_R^{EX} + \beta_2 e + \beta_3 TC_R$$

Marginal costs are assumed to depend on the costs of the raw product at the previous marketing stage, that is, the export price of bananas expressed in DM, p_R^{EX}, the real DM per U.S. dollar exchange rate, e, and a transportation cost index, TC, expressed in DM, which was derived from average spot transportation rates for grains between the U.S./Canada and Europe.[x] Prices, income and transport costs were expressed in real terms, subscript R, by dividing by the German CPI (1985 = 100). The theoretical hypotheses on the coefficients are: $\beta_1 > 0$, $\beta_2 > 0$ and $\beta_3 > 0$.

When equation (7) is introduced into (6'), and an error term, ε_2, is appended, the following import supply relationship is obtained:

$$(8) \quad p_R = \beta_0 + \beta_1 p_R^{EX} + \beta_2 e + \beta_3 TC_R + \beta_4 q^c + \varepsilon_2$$

where $\beta_4 = (-\xi/\alpha_1)$, since $dp_R/dq^c = 1/\alpha_1$. The market power coefficient ξ is subsumed within the regression coefficient $\gamma_4 : \xi = -\beta_4 \alpha_1$. The simultaneous equation system (2') and (8) was estimated with two-stage least squares. The results are as follows:

$$q^C = -4.0482 - 5.5474\,\hat{p}_R + 0.00069Y_R^c + 4.2868D, \qquad \bar{R}^2 = 0.96;$$
$$(1.39) \quad (4.60) \quad (7.96) \quad (8.75) \qquad DW = 1.84,$$

$$p_R = -0.1325 + 0.00167\hat{q}_c + 0.1675\,p_R^{EX} + 0.2874\,e + 0.00033\,TC_R,$$
$$(0.31) \quad (0.09) \quad (0.40) \quad (1.64) \quad (0.83)$$
$$\bar{R}^2 = 0.64; \; DW = 1.31$$

$$\hat{\xi} = -(-5.5474 \times 0.00167) = 0.01, SE(\hat{\xi}) = 0.1053,$$

where ^ indicates estimated values from the first stage. The demand specification reaffirms the highly significant influence of the import price, income, and German unification on German imports of bananas. However, the estimated market power coefficient is very low: $\hat{\xi} = 0.01$. As the product of two normal variables, $\hat{\xi}$ may itself not be distributed normally. However, it is clear, given the estimated standard error of $\hat{\xi}$ that any confidence interval for $\hat{\xi}$ will include zero and, thus, the hypothesis that the market has been competitive cannot be rejected.[xi]

Although three firms supply the majority of the bananas to ripeners and wholesalers in Germany, the import stage of the market seems to be very competitive. This result differs from the Deodhar and Sheldon (1995) estimate of $\xi = 0.29$, which they interpreted as consistent with Cournot-Nash behavior.[xii] The apparent competitiveness of the market despite its oligopolistic structure could be due to several factors. The leading firms may simply compete aggressively with one another. Alternatively, they may be disciplined by the competitive fringe of sellers, or by countervailing power on the part of wholesalers and ripeners.[xiii]

4. WELFARE ANALYSIS OF THE CMOB UNDER DIFFERENT ASSUMPTIONS ON MARKET CONDUCT

We have formulated three alternative hypotheses on market conduct. Based on the available literature and the econometric evidence shown in Section 3 of this chapter, a researcher could rely on any of these hypotheses and rationalize the choice with empirical results. We found strong empirical evidence in support of the small-country and perfect-competition model but some evidence was also consistent with the large-country and perfect-competition model. The small-country and imperfect-competition hypothesis could also be utilized in a synthetic approach by appealing to the Deodhar and Sheldon (1995) results. In contrast to Deodhar and Sheldon (1995), we did not find evidence to support a hypothesis of importer market power. However, the Deodhar and Sheldon (1995) analysis suffices to demonstrate that a model of this market can be formulated and estimated which does demonstrate some support for the imperfect competition hypothesis. Thus, as there is some empirical support for all three hypotheses, we will show welfare impacts of the CMOB in this section based on all three hypotheses for a comparative view.

We show how the alternative hypotheses on market conduct in the pre-CMOB situation affect the computation of economic impacts induced by the CMOB (Tables 8.1 and 8.2). The price, trade and welfare effects are calculated as illustrated in Figures 8.1, 8.2, and 8.3. The price-transmission equation estimated in (1) is free of any assumptions regarding market conduct.[xiv] It is used to model price formation in the hypothetical free-trade situation for all market scenarios, and the estimated price effects of the CMOB are identical for all hypotheses. The estimated impacts on import quantity and consumer welfare are also identical across the three market scenarios because the estimated price elasticities of demand were very similar in all estimates and, hence, we utilized the price coefficient in equation (2) for all computations. We present these results for prices, quantities, and expenditures for 1993 and 1994 (Table 8.1). In general, the effects of the CMOB are much larger in 1994 when the CMOB affected the market for the whole year as opposed to only six months in 1993, as the CMOB entered into operation in July of 1993.[xv]

Table 8.1 Economic Effects of the CMOB on Imports, Import Prices and Expenditures

Economic variables	*1993*	*1994*
Actual import price for bananas (DM/kilogram)	1.256	1.569
Hypothetical import price for bananas (DM/kilogram	0.856	0.834
Policy-induced impact on the import price for bananas (percent)	+46.8	+88.1
Actual export price of bananas for third-country exporters (DM/kg)	0.855	0.834
Hypothetical export price (DM/kilogram)	0.813	0.764
Policy-induced impact on the export price for bananas (percent)	-4.9	-8.4
Actual imports (and exports) of bananas (metric ton)	1,186,391	1,128,320
Hypothetical imports of bananas (metric ton)	1,375,512	1,476,249
Policy-induced impact on imports for bananas (percent)	-13.7	-23.6
Actual import expenditures for bananas (million DM)	1490.11	1770.33
Hypothetical expenditures for bananas (million DM)	1176.82	1231.64
Policy-induced impact on import expenditures for bananas (percent)	+26.6	+43.7

Source: Authors' calculations.

Table 8.2 Welfare Effects of the CMOB on the German Import Markets: How Markets Conduct Matters (million DM)

Year/ assumptions on market conduct[a]	*Consumers of imported bananas*	*European traders (quota rent)*	*E.U. budget*	*Aggregate E.U. welfare change*	*Exporters on the German market*
1993					
Hypothesis 1	-513.0	+(γ358.8)	+116.3	-396.7+(γ358.8)	0
Hypothesis 2	-513.0	+(γ408.8)	+116.3	-396.7+(γ408.8)	-53.9
Hypothesis 3	-513.0	+(γ269.1)	+116.3	-396.7+(γ269.1)	0
1994					
Hypothesis 1	-956.8	+(γ609.0)	+220.0	-736.8+(γ609.0)	0
Hypothesis 2	-956.8	+(γ688.3)	+220.0	-736.8+(γ688.3)	-91.6
Hypothesis 3	-956.8	+(γ448.0)	+220.0	-736.8+(γ448.0)	0

[a]Hypothesis 1: small country/perfect competition; Hypothesis 2: large country/perfect competition; Hypothesis 3: small country/imperfect competition.
Source: Authors' calculations.

The CMOB raised the German import price for bananas substantially. In 1994, the import price was 1.57 DM per kilogram (kg) under the new policy compared to a predicted 0.83 DM/kg without the policy change; in other words, the import price for bananas nearly doubled. German banana imports were reduced by 24 percent in 1994. Import expenditures increased by 44 percent.

From the European perspective, changes in the German banana economy affect the German consumers, the traders from E.U. countries on the German banana market, and the E.U. budget. We summarize the estimated welfare effects under the alternative market hypotheses (Table 8.2). Under hypothesis 1 the estimated loss to German consumers (957 million DM) is offset partially by the 220 million DM in tax revenue that is generated and the share, γ, of the total quota rents (609 million DM) that accrue to European traders. Even under the unlikely scenario that Europeans capture all the quota rents, the policy reduces net welfare from the E.U. perspective.[xvi]

Under hypothesis 2, the welfare implications are substantially different from the small-country hypothesis. As the export price declines for foreign exporters when import licences are allocated to importers, higher quota rents occur for banana traders on the German market. The magnitude of these quota rents is estimated to be 409 million DM for 1993 and 688 million DM for 1994. In fact, the estimated total welfare gains achieved by European traders and by the E.U. budget (116.3 million DM) in 1993 exceeded the welfare loss of consumers (513 million DM) for $\gamma > 0.97$, making it conceivable under Hypothesis 2 that the policy resulted in an aggregate welfare gain from the European point of view. However, the policy is welfare diminishing under Hypothesis 2 for any value of γ in 1994.

Because estimated importer oligopoly power was both low and insignificant based on our estimates, we report welfare results under Hypothesis 3 using Deodhar and Sheldon's (1995) estimate of $\xi = 0.29$ to gain perspective on the sensitivity of the welfare analysis to the possible presence of market power. Under the market power scenario estimated by Deodhar and Sheldon (1995), importers gained, in total, 269 million DM in quota rents in 1993 and 448 million DM in 1994: 25.0 percent and 26.4 percent less, respectively, than estimated under the perfect-competition and small-country scenario. Again, the reason is that profitable sales are sacrificed under the quota, and these losses offset the rents created by the quota. Among the scenarios considered, the estimated aggregate welfare loss from the European point of view is always highest in the imperfect-competition scenario, with the precise estimate of the welfare differential depending upon the share of the quota rents and of the oligopoly profits captured by European traders.

5. CONCLUSIONS

As of 2000, the European CMOB contains the most controversial TRQ policy among the increasing number of TRQs in international agricultural trade. The objectives of this chapter were to evaluate the welfare affects of the TRQ policy on the German banana market from the E.U. perspective, to analyze how different assumptions on market conduct affect the welfare impacts of the TRQ policy, and to determine which of the market-conduct hypotheses is supported empirically.

We showed that the redistributive and welfare effects of a TRQ depend on whether perfect or imperfect competition existed prior to the introduction of the new policy and whether the residual supply of bananas to Germany is perfectly elastic (the small-country hypothesis) or upward sloping (the large-country case). The size of the traders' quota rent and the total welfare impacts are particularly sensitive to the choice of market model.

When measuring the economic impacts of the European banana policy on the German market, the modeling of hypothetical prices in the benchmark, no-policy situation was crucial. We assumed that the same price transmission between the export market and the German import market as in the pre-policy period would have taken place had the policy not been enacted. This price transmission might have been associated with different forms of market conduct. Given that the no-policy price forecasts were generated by the same price transmission equation for all market conduct scenarios, and that the CMOB quota was observed to bind in the German market throughout the study period, the policy's impacts on price and, due to uniform import price elasticity of demand, also on trade and import expenditures, were identical across the alternative market structures. The analysis revealed that the impacts of the CMOB policy have been substantial. In 1994, for example, estimates were that the German import price rose by 88 percent, imports declined by 24 percent, import expenditures increased by 44 percent, and German consumers lost 957 million DM in surplus.

The estimates of welfare impacts on traders and the aggregate welfare impacts were affected by the choice of market model. However, traders were shown to be the major beneficiaries of the new policy under any scenario. Their gains were the highest under hypothesis 2 (large-country case and competition) and lowest under hypothesis 3 (imperfect competition and small-country). Based on hypothesis 2, it is possible that Europe achieved a net welfare gain from the TRQ policy in 1993 if the European share of the quota rents was very high. However, we would conclude that Europe lost on net under hypotheses 1 and 3, and in 1994 also under hypothesis 2, even in the unlikely scenario that the quota rents were captured entirely by E.U. traders. Thus, although the qualitative impacts of the CMOB seem relatively clear regardless of the preferred model of market conduct, evaluation of the quantitative impacts would clearly benefit from a detailed study of the distribution of the quota rents.

On balance, we believe the evidence concerning the German banana market is most supportive of a small-country, perfect-competition model. The imperfect-competition hypothesis, despite the high concentration in the banana economy, was not supported by the econometric analysis. The argument set forth in some policy analyses on the CMOB, that it is important to move from perfect-competition to imperfect-competition models, is not confirmed here. Based on our analysis, to impose an oligopoly model on this market would result in underestimation of the quota rents earned by traders on the German banana market.

In terms of research methodology, we conclude that it is crucial to consider and test for plausible alternative forms of market conduct in agricultural trade research. Looking at one plausible hypothesis on market conduct in isolation may well lead to acceptable econometric results. However, an alternative model of market conduct may perform even better and yield rather different policy conclusions.

6. ENDNOTES

[i] Senior authorship is not assigned. Ideas for this paper were developed when Herrmann visited the Department of Agricultural and Resource Economics, University of California, Davis, in February, March, and June of 1998. A grant from Deutscher Akademischer

Auslandsdienst (DAAD) is gratefully acknowledged. Thanks are due to Christina Moennich for very helpful comments on an earlier version.

[ii]The CMOB is a Common European Policy and, therefore, re-distributional implications of the common financial system have to be taken into account (Koester, 1977). The national point of view is different from the European point of view. The welfare analysis from the German point of view would consider German consumers, German traders and the national government's budget. The distinction between the European and the German points of view has been explained in detail elsewhere (Herrmann, 1999) and was applied to an analysis of the CMOB under different market conducts in Herrmann and Sexton (1999).

[iii]The combination of imperfect competition and the large-country case is discussed in Herrmann and Sexton (1999).

[iv]The possible impacts of tariffs or quotas on strategic behavior depend crucially on whether firms are competing in prices or quantities prior to the imposition of the policy. For further discussion on this point, see Hwang and Mai (1988), Krishna (1989), Helpman and Krugman (1989), and McCorriston and Sheldon (1996). McCorriston and Sheldon argue that imposition of a quota on the German banana market will cause the industry to behave in a Cournot manner, that is, rivals cannot change output in response to a given firm's change in output. They suggest, based on a calibration model, that behavior prior to the imposition of the CMOB policy was more competitive than Cournot. Thus, McCorriston and Sheldon argue that the policy would have the effect of diminishing competition. However, this logic does not apply if the quota binds in aggregate because an increase in any firm's output (for example, through purchasing import licenses) must be matched exactly by a decrease in other firms' output. In other words, output is fixed by the quota regardless of the market power exercised by the importing firms prior to imposition of the TRQ.

[v]Data for the econometric analyses reported in this section are from Weiß (1995), Deutsche Bundesbank , FAO, and Statistisches Bundesamt.

[vi]Ramsey's RESET 2 is a Lagrange multiplier test of functional form, where the residuals are regressed on the independent variables and the square of the fitted dependent variable (Hall and Cummins, 1999: 31). This test has a chi-squared distribution with one degree of freedom.

[vii]The price elasticity of import demand is somewhat higher in absolute terms than the estimate of –0.34 by Weiß (1995: 162), whose analysis differed in terms of model specification and the period analysed, and the estimate of -0.4 by Islam and Subramanian (1989). The latter study concentrated on aggregate banana imports from developing countries within a structural market model. Our income elasticity of demand is higher than the 1.42 estimate in Weiß (1995) and much higher than the value of 0.58 in Islam and Subramanian (1989). The price elasticities of import demand posited in Guyomard, Laroche, and Le Mouël (1999), and Kersten (1995) were –0.4 and -0.5 respectively.

[viii]It can be easily shown that the import demand function and the export supply function of a simultaneous model consisting of the equations (4) and (5), the equilibrium condition, and a definition $p_R = (p \cdot 100)/P^G$ fulfil the order criterion for a 2SLS estimation. ^ indicates estimates from the first stage of the two stage least-squares regressions.

[ix]With regard to the age variable, Deodhar and Sheldon (1995: 343) argue that bananas are a health food, and are considered an important part of the diet of the sick and very old. However, data from the German Nationale Verzehrsstudie show that per-capita consumption of bananas does not rise with age. For example, it was higher in 1987 at age 19-24 (43.3 grams per day) than at age 64 and above (39.0 grams per day). Thus, although Deodhar and Sheldon (1995) obtain a positive and marginally significant coefficient for

their age variable, it must be acting as a proxy for other, unmeasured effects. One possibility is that Deodhar and Sheldon captured the impact of German reunification. Because they estimated total, not per-capita demand, reunification (which raised both total population and the population age 65 or older) caused a positive but spurious correlation between the two in their results. In results not reported here but are available from the authors, we show that the proportion of the population 65 or older has no significant impact on per-capita demand.

[x]Unfortunately, no transportation cost data for bananas were available. We used the transportation rates for grain between the U.S./Canada and Europe as a proxy, since (i) the difference between p_R and p_R^{EX} is affected by transport costs on the same shipping route (U.S. to Europe); and (ii) transportation rates for bananas and grains are affected by some joint determinants such as the price of oil. Hence, both transportation rates should be positively correlated.

[xi]See Mood, Graybill, and Boes (1974) regarding the formula for the variance of a random variable that is the product of two random variables.

[xii]With a homogeneous product oligopoly and with identical firms, $\xi = 1/n$, where n is the number of sellers. With the market dominated by three sellers, Deodhar and Sheldon (1995) considered that $n = 3$, in which case $\xi = 0.333$. The DS estimate of $\xi = 0.29$ is, thus, qualitatively close and not statistically different from the Cournot-Nash value.

[xiii]In the expanded version of our paper (Herrmann and Sexton, 1999), we discuss generalization of the Deodhar and Sheldon (1995) approach to incorporate the large-country hypothesis of upward sloping import supply to Germany based on Bresnahan's (1982) approach.

[xiv]In other words, although we may not know specifically the form of market conduct that prevailed prior to implementation of the CMOB, we assume that the pattern of price transmission to the German market that prevailed prior to the CMOB would have continued if the CMOB had not been implemented.

[xv]The price effect of the CMOB based on the monthly price transmission equation (1) was estimated for the period July-December 1993 and a weighted impact on the annual price in 1993 was computed.

[xvi]The licensing procedure under the original 1993-94 CMOB is important to gain perspective on the possible size of γ. As indicated in Section 2, licenses to import from third countries or from non-traditional ECP exports to the E.U. were distributed at the import stage within the E.U.: 57 percent to primary importers, 15 percent to secondary importers, 28 percent to ripeners. Primary importers are mainly multinational firms which are vertically integrated and have sites in the E.U., whereas secondary importers are independent E.U.-based importers. Thus, a first principal question is whether E.U. branches of multinational firms should be treated (i) as E.U. firms or (ii) as foreign firms according to the country of the main branch. If (i) is chosen, $\gamma = 1$ as far as the "rights to rent" are distributed within the E.U. like under the original CMOB. If (ii) is chosen, $\gamma = 0.43$ if the assumption is made that the distribution of rents follows the distribution of licenses and that the share of primary importers in licenses equals the share of multinational firms in overall quota rents. The treatment of multinational firms in national welfare analyses as well as implications for measuring γ under the CMOB deserve a separate treatment in future research.

7. REFERENCES

Abbott, P. and P. Paarlberg. 1998. "Tariff Rate Quotas: Structural and Stability Impacts in Growing Markets." *Agricultural Economics* 19: 257–267.

Behr, H.C. and W. Ellinger. 1993. Die Bananenmarktordnung und ihre Folgen. (Materialien zur Marktberichterstattung, Band 1), Zentrale Markt- und Preisberichtstelle, Bonn.

Borrell, B. 1997. "Policymaking in the E.U.: The Bananarama Story, the WTO and Policy Transparency." *The Australian Journal of Agricultural and Resource Economics* 41(2): 263–276.

Brander, J.A. and B.J. Spencer. 1985. "Export Subsidies and International Market Share Rivalry." *Journal of International Economics* 18: 83–100.

Bresnahan, T.F. 1989. "Empirical Studies of Industries with Market Power," in R. Schmalensee and R. Willig, eds., *Handbook of Industrial Organization* 2: 1,011–1,057. Amsterdam, Netherlands: North Holland.

_____. 1982. "The Oligopoly Solution Concept Is Identified." *Economics Letters* 10: 87–92.

Deodhar, S.Y. and I.M. Sheldon. 1995. "Is Foreign Trade (Im)perfectly Competitive?: An Analysis of the German Market for Banana Imports." *Journal of Agricultural Economics* 46(3): 336–348.

Deutsche Bundesbank. Various years. *Devisenkursstatistik. Statistische Beihefte zu den Monatsberichten*. Frankfurt, Germany: Main.

Dole. 1997. Unpublished statistics on the banana market. Hamburg, Germany.

Engle, R.E. and C.W.J. Granger. 1987. "Cointegration and Error-Correction: Representation, Estimation, and Testing." *Econometrica* 55: 251–276.

FAO (Food and Agriculture Organization). Various years. *Quarterly Bulletin of Statistics*. Rome, Italy.

Guyomard, H., C. Laroche, and C. Le Mouël. 1999. "An Economic Assessment of the Common Market Organization for Bananas in the European Union." *Agricultural Economics* 20: 105–120.

Hall, B.H. and C. Cummins. 1999. *Time Series Processor Version 4.5 User's Guide Including an Introductory Guide*. Palo Alto, CA: TSP International.

Helpman, E. and Krugman, P. 1989. *Trade Policy and Market Structure*. Cambridge, MA: MIT.

Herrmann, R. 1999. "Economic Impacts of the New European Banana Market Regime: The Case of Germany." *Jahrbücher für Nationalökonomie und Statistik* 218(1 and 2): 63–84.

Herrmann, R. and R. Sexton. 1999. "Redistributive Implications of a Tariff-rate Quota Policy: How Market Structure and Conduct Matter." *Discussion Papers in Agricultural Economics*. Giessen, Germany.

Holloway, G. 1991. "The Farm-Retail Price Spread in an Imperfectly Competitive Industry." *American Journal of Agricultural Economics* 73: 979–989.

Hwang, H. and C.C. Mai. 1988. "On the Equivalence of Tariffs and Quotas Under Duopoly." *Journal of International Economics* 24: 373–380.

Islam, N. and A. Subramanian. 1989. "Agricultural Exports of Developing Countries: Estimates of Income and Price Elasticities of Demand and Supply." *Journal of Agricultural Economics* 40(1): 221–231.

Kersten, L. 1995. "Impacts of the E.U. Banana Market Regulation on International Competition, Trade, and Welfare." *European Review of Agricultural Economics* 22(3): 321–335.

Koester, U. 1977. "The Redistributional Effects of the Common Agricultural Financial System." *European Review of Agricultural Economics* 4(4): 321–345.

Krishna, K. 1989. "Trade Restrictions as Facilitating Practices." *Journal of International Economics* 26: 251–270.

Matthews, A. 1992. "The European Community's Banana Policy after 1992." Discussion Papers in Agricultural Economics, No.13, University of Giessen, Giessen, Germany.

McCorriston, S. 1997. "Price Transmission in Vertically Related Markets under Imperfect Competition," in S. Bauer, R. Herrmann and F. Kuhlmann, eds., pp. 59–68, *Märkte der Agrar- und Ernährungswirtschaft. Analyse, einzelwirtschaftliche Strategien, staatliche Einflussnahme*. Münster-Hiltrup, Germany: Landwirtschaftsverlag.

McCorriston, S. and I.M. Sheldon. 1996. "Incorporating Industrial Organization into Agricultural Trade Modeling," in D. Martimort, ed., pp. 313–330, *Agricultural Markets: Mechanisms, Failures and Regulations*. Amsterdam, Netherlands: Elsevier.

Moschini, G. 1991. "Economic Issues in Tariffication: An Overview." *Agricultural Economics* 5: 101–120.

Mood, A.M., F.A. Graybill, and D.C. Boes. 1974. *Introduction to the Theory of Statistics, 3rd ed.* New York, NY: McGraw Hill.

Pedler, R.H. 1994. "The Fruit Companies and the Banana Trade Regime (BTR)," in R.H. Pedler and M.P.C.M. van Schendelen, eds., pp.67–91, *Lobbying the European Union: Companies, Trade Associations and Issue Groups*. Aldershot, Hants: Dartmouth.

Statistisches Bundesamt. Various years. Fachserie 17: Preise. Reihe 11: Preise und Preisindizes im Ausland Wiesbaden.

Thagesen, R. and A. Matthews. 1997. "The E.U.'s Common Banana Regime. An Initial Evaluation." *Journal of Common Market Studies* 35(4): 315–327.

Weiß, D. 1995. *Preis- und Wechselkurstransmission auf Märkten der Agrar- und Ernährungswirtschaft—Theoretische Analyse und empirische Zusammenhänge am Beispiel des Bananenmarktes*. Gießen, Germany: Wissenschaftlicher Fachverlag.

WTO (World Trade Organization). 1999. *European Communities-Regime for the Importation, Sale and Distribution of Bananas—Recourse to Article 21.5 by Ecuador*. Report of the Panel, WT/DS27/RW/ECU. Geneva, Switzerland (12 April).

_____. 1997. *European Communities Regime for the Importation, Sale, and Distribution of Bananas*. Report of the Panel. WTO-Document WT/DS27/RW/ECU, Geneva, Switzerland (22 May).

Chapter 9

Rent Seeking and International Trade in Agriculture

Harry de Gorter, Gordon C. Rausser, and Andrew Schmitz
Cornell University, University of California at Berkeley, and University of Florida

1. INTRODUCTION

Rent seeking is an important activity in international agricultural trade. Special-interest groups initiate many protectionist trade measures. For example, the United States was recently in a trade war with the European Union over bananas and beef, and with Canada over wheat. The source of each of these particular disputes provides a good example of the various policy instruments generating contestable rents and hence trade frictions: trade rents from import quotas (bananas); domestic rents altered by own and other countries' domestic policy (wheat); and the non-trade concerns rents (beef).[i] "Non-trade concerns" rents refers to rents sought under the hubris of the new trade agenda arising from trade protectionist policies seeking to control food quality, achieve food security, internalize negative and positive externalities (that is, environment and the landscape, respectively), and enhance viability of rural areas.

Rent seeking in the United States and around the world in agriculture takes many forms, all of which directly or indirectly affect international trade. Several key features of rent seeking can be highlighted. For example, the 1996 Federal Agriculture Improvement and Reform (FAIR) Act in the United States included both *ex ante* and *ex post* rent seeking. Farmers were able to lobby successfully *ex ante* for subsidies greater than were otherwise proposed in the new Farm Bill [Orden, Paarlberg and Roe (1999); Dumas and Schmitz (this volume)]. Also farmers have received, from *ex post* rent seeking, new policy interventions on their behalf. Rent-seeking activities in agriculture also include institutional rent seeking whereby state trading enterprizes (STEs) may be vulnerable to cost padding, excessive capital expenditures, cross-subsidization and other self-preserving activities by members of the organization and dependent private sector agencies (Economics, Policy and Strategy Consultants, 1996). In a rent-seeking framework, there may be substantial resource costs beyond the normal efficiency measures of the pricing policies emanating from such institutions. Clearly, supply management, such as practiced in Canadian agriculture, involves significant rent seeking (Schmitz, de Gorter and Schmitz, 1996). Groups within the supply management sector, such as broiler producers, are able to maintain their supply-control status even in the face of the Canada-U.S. Trade Agreement (CUSTA) and the North American Free Trade Agreement (NAFTA), and the Uruguay Round.

Rent seeking also involves coalitions between domestic and foreign groups in collusive or cooperative strategies that often do not involve formal trade policies. The 1998 compromise in the fresh-tomato trade dispute between Mexico and the United States is a classic example of overt rent seeking where minimum import prices were established, thereby resulting in an import-export, cartel-type arrangement for vegetable producers which mimicked the effects of a voluntary export restraint (Bredahl, Schmitz and Hillman, 1987; VanSickle, 1997). Traditional rent-seeking analysis looks at consumers, producers and taxpayers, but often ignores the role of processors or upstream and downstream industries in the agri-food sector (Alston, Carter and Wohlgenant, 1989).

Many perceive the Agricultural Agreement (hereafter referred to as the Agreement) of the Uruguay Round as a success in reducing the distortions of agricultural policies on international trade by bringing agriculture under the General Agreement on Tariffs and Trade (GATT) discipline, thereby eliminating or greatly reducing rent-seeking activities (Rausser, 1995). However, Table 9.1 shows that total rents to producers in agriculture for a subset of countries, policies, and commodity sectors averaged U.S. $203 billion in 1995-97. Indeed, recent Organization for Economic Cooperation and Development (OECD) data show that producer subsidy equivalents have increased in 1998 to over U.S. $350 billion, despite support reduction commitments in the Agreement.[ii] Column (2) in Table 9.1 provides estimates of total rents to farmers from trade barriers, with a breakdown of import quota rents and tariff revenues.[iii] Annual average quota rents and tariff revenues are in the order of U.S. $44 billion. Column (3) gives an estimate of rents derived from domestic agricultural policy while the final column shows the proportion of rents that are self-defeating since world prices decline due to own- and other-country policies. A significant proportion of rents generated is therefore phantom in that there is mutually assured destruction amongst agricul-

Table 9.1 Empirical Measures of Rents to Agricultural Producers in World Agriculture (1995-97)

	(1) Total rents	(2) Rents from trade barriers: (a) Total	(2) Rents from trade barriers: (b) Tariff revenues and import quota rents	(3) Non-trade rents	(4) Percent of rents created by world price declines due to policies
			billion U.S. $		percent
U.S.	70	32	4.0	38.0	63
E.U.	320	170	20.0	150.0	41
Canada	12.2	4.5	0.2	7.7	47
Norway	8.6	4	0.2	4.6	11
Switzerland	23.3	12.3	2.4	11.0	8
Japan	124	104	69.2	19.5	13
Mexico	3.8	–0.3	–.03	4.1	n.a.
Korea	50	47.7	40.7	2.3	10
Total	611.8	374.6	131.0	237.2	

Source: Estimates from OECD data files (from AGSTAT Databank). World price changes taken from Roningen and Dixit (1987).

tural policies worldwide. Hence, much of these rents are dissipated due to rent seeking, and as such do not represent net gains to any group.

This chapter focuses specifically on the inefficiencies resulting from rent seeking with the new 2-tier, tariff rate import quotas (TRQs) for agriculture implemented in the Agreement. The degree of rent seeking is affected by the allocation of quotas to exporting countries and licenses to importing and exporting firms. Receiving quota rights does not necessarily imply that the rights to rents are assigned to a firm or country. We show that the distribution of rents and the degree of rent seeking can be affected by the various alternative administration methods in use allocating these rights to rents. Market power enjoyed by firms in imperfectly competitive markets does not necessarily mean they have the bargaining power to obtain the rents. Nontradability of quotas and licenses can also affect the economic inefficiency of rent seeking.

2. THE THEORY OF RENT SEEKING IN PERSPECTIVE

Rent seeking refers to the waste of resources to society at large[iv] to capture (or to create) the private benefits and/or to avoid (or to prevent) the private costs resulting from actual (or potential) government policy or institutional settings. One can use two different reference points to measure the level of resource waste due to rent seeking. First, one can deem that all rent seeking results in reduced social welfare. This assumes that all property rights are complete and enforced without cost, and that all market failures are corrected with no transaction costs incurred by a government that has full information. Hence, in this hypothetical world there is no trade-off between welfare-increasing and welfare-decreasing type government policies (Rausser, 1992). Rational agents in the pursuit of private gain waste resources to rent seek, relative to the situation of socially optimal levels of government policy or institutional settings in which no resources are expended. Using this reference point means all rent seeking reduces social welfare, even if agents are lobbying for more welfare-increasing public goods, for a reduction in welfare-decreasing price supports or lobbying provides information that possibly increases social welfare (Rausser and Zusman, 2001).

Second, one can use actual policy as a reference point (instead of socially optimal policy with no transaction costs as assumed above). This means that the net outcome of rent seeking on social welfare is indeterminate.[v]

Rent seeking involves public policy, even if one is analyzing rent seeking between private individuals or institutions. This is because private-property rights and legal-dispute mechanisms involve government coercion of one type or another (for example, private institutions like land tenancy or profit-sharing arrangements, or inheritance disputes with no inheritance tax, nevertheless require implicit public policies in that private property rights, for example, need to be enforced). Therefore, rents are artificial in that they are created by public decree and do not refer to competition (through innovation, for example) for Ricardian rents that are part of the social welfare-increasing aspect of market forces.

The general and theoretical modern literature on rent seeking began with Tullock (1967) [see the early book of readings by Buchanan, Tollison, and Tullock (1980) and surveys by Tollison (1987), Bhagwati (1982), and Brooks and Heijdra (1989)].[vi] The first formal modern study of rent seeking due to agricultural policy is by Dennen (1977). He shows how the Homestead Act forced families to line up for announcements, required land rushes, and forced farmers to farm the land immediately rather than waiting for the optimal time to do so. A more recent study, by de Gorter and Fisher (1993), shows how voluntary participation in U.S. crop programs with endogenous base acreage (upon which subsidies are based) had farmers foregoing current subsidies, and incurring extra production costs to increase current production, in order to increase subsidies for both producing and not producing more in the future. Joseph Heller, in *Catch 22*, described this phenomenon:

> He was a long limbed farmer, a God fearing, freedom-loving, law-abiding rugged individualist who held that federal aid to anyone was creeping socialism. . . His specialty was alfalfa, and he made a good

thing of not growing any. The government paid him well for every bushel of alfalfa he did not grow. The more alfalfa he did not grow, the more money the government gave him, and he spent every penny he didn't earn on new land to increase the amount of alfalfa he did not produce. . . He invested in land wisely and soon was not growing more alfalfa than any other man in the country. (1970: 82–83)

Similarly, in the former Soviet Union, prices received by farmers doubled for production above and beyond individual farm production targets. As in the United States, production targets could be manipulated by strategic farmer behavior. Clearly, in such systems, farmers have an incentive to reduce production targets in the short run in order to produce larger increments above targets in the future: the so-called ratchet effect (Johnson and Brooks, 1983). A key characteristic in many rent-seeking examples is the dynamic or inter-temporal perspective whereby investments are made to capture increased benefits (or to create benefits in the first place), or to avoid (or to prevent in the first place) the costs of the rents (Rausser, 1992).

Individuals in interest groups (and their lobbies) rent seek, both over the level of transfer and the type of policy instrument. Politicians are central to the political equilibrium. Constituents will prefer those instruments that obfuscate the level of transfers and their effects in order to maximize the expected value of positive transfers (and losers will lobby for more transparent measures that minimize costs). Once the policy is in place, different instruments will generate different incentive structures to rent seek and hence will have differential impacts on social costs. We therefore delineate two categories of, or stages in, rent seeking, defined as an investment (or transaction cost) to:

1. increase (or avoid a decrease in) income as a result of securing (blocking) a change in public policy (or institutional arrangement); or
2. maximize the benefits (or minimize the costs) once these policy changes have created non-exclusive rights

The first category treats policy as endogenous if a change in policy results, and the second treats policy as exogenous. Rent seeking therefore occurs when agents try to change the structure of entitlements and/or occurs in response to the involuntary change in entitlements by trying to bring about a transfer (with either case explicitly or implicitly involving government policy).

From a mechanism-design perspective, the issue of what political and institutional conditions affect the amount of resources wasted in vying for transfer levels and the particular policy instruments used to determine and implement those transfers is critically important. In other words, how do alternative institutional settings or governance rules (including economic conditions) induce a variety of incentive distortions to be chosen in the form of alternative (inefficient) policy instruments? Once the institutional and economic characteristics are identified in the outcome of a policy choice, the degree of social waste induced by agents maximizing benefits or minimizing losses from the policy may have feedback effects on the incentives for lobbies, voters, and politicians to augment the policy instrument choice later. Under these circumstances, the separable categories of rent seeking become simultaneous or contemporaneous events.

The conditions determining the degree to which all rents are dissipated with rent seeking include the competitiveness of the activity to obtain the rents in the first place, the elasticity of the supply of inputs used in rent seeking, the ability of rent seeking groups to organize and control free riding (and likewise the ability of competitor groups or those opposing the government policy to do likewise), discount rates, risk aversion and uncertainty as to whether or not the policy will remain in place (Rausser and Zusman, 2001; Posner, 1975; Fisher, 1985; and Varian, 1987). It is possible that total resources wasted in rent seeking exceed the rents available if more than one group competes for the rents (Baysinger and Tollison, 1980).

Hillman and Riley (1989) analyze the interaction of lobbyists trying to change policy where the prize itself is determined endogenously. They evaluate the strategic nature of bargaining games and political or institutional rules that determine the outcome of lobbying by different groups. Conditions under which rents are fully dissipated by lobbying are determined by, and depend critically on, asymmetric valuations of potential benefits. Hillman and Riley determine that the equilibrium number of participants (and non-participants) is critically dependent on the rival's evaluation of the potential rents from policy (for both a pre-existing policy-contrived rent and for contestable income-transfer policies).

Coggins (1995) extends the work of Hillman and Riley (1989) and others by specifying a model of government behavior that is represented by a relative price-setting function that takes lobbying contributions as arguments. Opposing lobbies spend resources to influence the government choice of a price policy determined from the Nash equilibrium of a non-cooperative game. If equally matched opposing lobbies are involved, it is possible that in the end lobbies will have no affect on the policy they are trying to change. Both groups would be better off if they ceased and desisted. In Nash's model, which assumes symmetric or equally matched opponents, there are many situations where both groups lobby energetically, but nothing changes. Over-dissipation of rents occurs when total lobbying costs exceed the size of the prize, where the latter is measured by the compensating variation for both agents due to the change in relative prices resulting from the lobbying game. If two groups or lobbyists (such as agricultural producers and food consumers) are unequally matched (as producers and consumers are, since lobbies on behalf of food consumers for lower prices are relatively weak), then under-dissipation is likely.

Although the values of contributions by lobbyists are small relative to rents created by agricultural policy in the United States (Table 9.2), the productivity of the lobby expenditures may be large. The literature by Hillman and Riley (1989) and Coggins (1995) can also be relevant to analysis of the conditions under which politicians choose inefficient policy instruments, implemented as a result of lobbying; the instruments can provide incentives that create huge economic losses. Politicians do not respond passively to either interest- or pressure groups, leaving a role for entrepreneurial politicians to not only extract rents for themselves, but also to determine the optimal policy type to maximize political support (de Gorter and Swinnen, 1994, 1995). Politicians threaten to take away through taxation, and promise to give through subsidization (Bates, 1990), with asymmetric information and differential responses by groups because of differences in the concentration/dispersion of per-capita benefits/costs of the policy. The political economy of rent seeking focuses on category 1 of rent seeking defined earlier in this chapter: why do constituents support governments and why does the latter choose these many inefficient and often peculiar policy instrument types? From a prescriptive standpoint, legislation on how to regulate lobbying to minimize category 1 rent seeking is incomplete. Emphasis must be placed on designing institutional arrangements that lead to policy instrument types that incur less economic waste due to category 2 rent seeking.[vii]

Table 9.2 Total Contributions and Contributions from Agricultural Sector to Federal Candidates, 1989-96 (Contributions from Individuals Giving U.S. $200 or More and from Political Action Committees)

	Total contributions	*Contributions from agriculture sector*	*Contributions from the rest of the sectors*
1989-90	$230,268,586	$17,836,291 [7.8 percent]	$212,432,295 [92.3 percent]
1991-92	$368,001,139 (159.8 percent)	$24,892,124 (139.6 percent) [6.8 percent]	$343,109,015 (161.5 percent) [93.2 percent]
1993-94	$366,862,216 (99.7 percent)	$24,210,866 (97.3 percent) [6.56 percent]	$342,651,350 (99.9 percent) [93.4 percent]
1995-96	$433,582,938 (118.2 percent)	$29,540,396 (122 percent) [6.8 percent]	$404,042,542 (117.9 percent) [93.2 percent]

Source: Makinson (1992), Makinson and Goldstein (1994, 1996), Center for Responsive Politics (1997). The percent change in contributions compared with the previous time period is shown in parentheses, and the percent of total contributions are shown in brackets.

3. TRADE RENTS AND THE AGRICULTURAL AGREEMENT OF THE URUGUAY ROUND

The Agreement converted all non-tariff barriers into tariffs with the first-tier tariff applied to in-quota imports while the higher second-tier tariff is applied to over-quota imports only. However, import quotas were an obligation to improve market access beyond the anticipated effects of scheduled tariff reductions. Rent seeking occurred immediately in the form of dirty tariffication (Ingco, 1996) whereby tariff rates were purposely inflated, and countries reduced low tariffs even further to meet the minimum 36 percent, unweighted-average, tariff-reduction requirement. All countries were expected to allow imports of at least 3 percent of domestic consumption and rise to 5 percent of domestic consumption. These provisions of minimum access were in addition to the preservation of pre-existing market access termed current access.

High tariffs and safeguard provisions required countries to implement TRQs to meet current access obligations. Rent seeking also occurred in the form of dirty quotification with the purposeful manipulation of base consumption levels by using net (instead of gross) imports, and through the choice of year or commodity aggregation level. In most cases, TRQs maintain traditional trade flows, and few new markets have been created by quotas, with the exception of Japan and South Korea (Choi and Sumner, 1998). A total of 35 countries have scheduled 1,370 TRQs for agricultural commodities in the Agreement. There is no uniformity across countries or commodities either in the setting of the absolute or relative levels of the first- and second-tier tariffs or in their reduction, resulting in unequal quota rents across countries and commodities. Quota rents are generated even if the second-tier tariff, plus the world price, equals the domestic price, because the first-tier tariff is lower for in-quota imports.

The degree of inefficiency due to rent seeking with TRQs depends critically on if and how import licenses are assigned for the right to import, and export quotas (and/or export licenses) are assigned for the right to export. There are many possible combinations for allocating the rights to rents between importing and exporting firms, and between importing and exporting countries. Assigning rights implicitly allocates the rights to rents, but it may be possible that those owning the rights will not always obtain the rents (Boughner and de Gorter, 1998; Hranaiova and de Gorter, 1999). When no rights are assigned, we will argue that rents are then most vulnerable to rent-seeking practices, which generates further economic waste.

3.1 Export Quotas and/or Licenses, Rent Dissipation, and Shares

Importing countries often allocate non-tradable export quotas to specific countries, sometimes requiring export licenses for exporting firms. The exporting firm has the potential to obtain the rents in this case, depending on whether or not import licenses are granted to importing firms, and depending on the degree of imperfect competition between trading firms. The resulting bargaining power depends on the allocation of the rights to rents, the degree of imperfect competition and the method of quota administration.

To isolate the inefficiencies of non-tradable, export-country quota allocations by importing countries, consider the case in which licenses are allocated to importing firms but not to exporting firms. If we assume perfect competition among exporting firms, then the importing country captures the quota rents. However, the level of rents will be lower than otherwise if high cost exporters have received the export quota rights. Part (and maybe even all) of the rents are dissipated.

If licenses are instead issued to exporting firms (or imperfect competition prevails in the exporter market) and not to importing firms, then the rights to rents (and the rents themselves in this case) are obtained by the exporting firms. Exporting firms have bargaining power now relative to that of the perfectly competitive importing firms. As a result, governments in exporting countries have less incentive to lobby for changes in the TRQ scheme through the WTO. The outcome is analogous to the incentives of an exporter under a Voluntary Export Restraint (VER) whereby rents accrue to the exporting country. Exporters obtain all of the rents (again, a part or a whole thereof may be dissipated if quotas are given to high-cost exporters only) because we assume competition among importing firms (who are without licenses) such that exporters have all of the bargaining power.

Another possible scenario is when both the rights to export to quota-restricted markets and to import have been allocated (or when imperfect competition exists among both importing and exporting firms). This results in an equal rights to rents, and relative bargaining power determines the distribution of rents (the level of which, again, depends on the amount distributed to high-cost exporters receiving the quotas).

The New Zealand Dairy Board (NZDB) owns a substantial proportion of outstanding TRQs in world dairy markets. Table 9.3 indicates that approximately U.S. $200 million are captured by the NZDB in quota rents alone. It has obtained these quota rights through direct allocation, through the reallocation of surrendered rights, or through exporting under the "any country"

Table 9.3 TRQ Rents for New Zealand Dairy Product Exports, 1997

Country/product	Rents[a]	Imports	Unit value	Wholesale price	Rent	Tariff revenue
	U.S.$	kg	U.S.$/kg[b]	U.S.$/kg	U.S.$/kg	U.S.$
European Union 1997						
Butter	90,883,804	61,644,000	2,414	3.888	1.474	53,556,307
Cheese for processing, cheddar	16,099,333	9,500,000	2,590	4.285	1.695	1,620,700
European Union 1997 total	106,983,137					55,177,007
United States 1997						
Dried skim milk	1,210,914	1,563,701	1.967	2.742	0.774	49,257
Dried buttermilk/whey	51,150	63,000	1.542	2.354	0.812	2,424
Dried whole milk	708,466	911,375	2.025	2.802	0.777	43,281
Butter	2,075,475	2,869,016	1.733	2.456	0.723	355,299
Butter substitutes	5,139,131	5,314,510	1.833	2.800	0.967	936,469
Other cheese NSPF	39,987,857	11,476,984	2.500	5.200	2.700	4,318,789
Cheddar	6,768,411	5,509,044	2.470	3.699	1.229	1,673,097
American other than cheddar	1,816,431	1,991,439	2.530	3.442	0.912	39,829
Swiss cheese/emmentaler	701,409	217,020	2.510	5.742	3.232	47,744
United States 1997 total	49,459,245					7,466,189
Japan 1997						
Milk powder	50,982,998	20,561,000	2.063	4.543	2.480	12,725,203
Butter	1,625,986	259,000	1.919	8.197	6.278	173,950
Japan 1997 total	52,608,984					12,899,153
Mexico 1996[c]						
Milk powder	31,089,990	26,219,000	2.295	3.481	1.186	—
Cheese	45,095,591	21,628,000	0.809	2.894	2.085	8,751,770
Mexico 1996 total	76,185,581					8,751,770
Venezuela 1996[c]						
Milk powder	29,228,200	21,439,000	2.437	3.800	1.363	—
Venezuela 1996 total	29,228,200					—
Total rents 1997	314,465,147					
Tariff revenue to governments	84,294,119					
Net TRQ rents to quota holders	230,171,028					

[a](Wholesale price-unit value) imports; [b]Unit value is equal to the c.i.f. price; [c]1997 Import data for Mexico and Venezuela was unavailable; *Sources*: U.S. Census Bureau, USDA/FAS (various years), USDA/AMS (various years), UNCTAD (TRAINS database), Centro de Estadistica Agropecuaria (Mexico), Agroplan (Venezuela).

category. For example, the NZDB dominates the export market for U.S. Dairy products (Table 9.4). In the Uruguay Round, the right to export was allocated to New Zealand for many dairy products and in most cases, the NZDB exported more than the TRQ share initially allocated to them. The NZDB purchased subsidiary companies that own historical import licenses in the U.S. (NZDB, Annual Report, 1998). The NZDB overcomes the transaction costs of the bargaining problem by buying out the importer and thereby possibly reducing the deadweight costs of rent seeking (but then they introduce the potential problem of imperfect market power). Western Dairy Products, Inc. (owned by the NZDB) are the sole owners of New Zealand's import license for dried whole milk, while Dorman (also owned by the NZDB) is the owner of an import license under the "any country" category for dried milk. In 1997, New Zealand imported 59 percent of the TRQ for dried whole milk. For dried buttermilk/whey, Western Dairy Products, Inc. are again the sole owners of the import license for New Zealand. In 1997, New Zealand owned 28.4 percent of the import quota into the United States with Canada owning the remainder. However, Canada surrendered its right to the quota to New Zealand. Therefore, New Zealand imported 100 percent of the buttermilk/whey TRQ for the United States (Boughner and de Gorter, 1998).

An important consideration is how the NZDB redistributes quota rents back to farmers. In addition to quota rents, the NZDB pools revenues from downstream profits and from domestic price premia, all of which total about 30 percent of the total value of milk production (Table 9.5). Another product for which New Zealand has control over import-quota rents as an exporter is lamb exports to the European Union. The regime was a VER previously but now New Zealand administers prices and trade on both sides of the Atlantic. For some products, the specifications on the TRQs are defined so narrowly that the exporter automatically obtains the rents. A good example is the import license for mozzarella imports in the European Union. that is specified for Pizza Hut products, thereby resulting in Pizza Hut obtaining the rents regardless of who owns the license. Another example of a bargaining game between importer and exporter, and the distribution of rents, is the E.U.'s import regime for fruits and vegetables (Martin and de Gorter, 1998). The entry price system (EPS) for imports of fruits and vegetables will guarantee a minimum entry price (MEP) (plus a conventional tariff, or CVT) using variable levies (tariff equivalents) and a fixed per-unit tax (the maximum tariff equivalent, or MTE) for world prices below 92 percent of the MEP. The importer declares an entry price for each consignment upon which three tariffs are added (the CVT, tariff equivalent and MTE). The importer can choose between three different methods of declaring the entry price.

Table 9.4 TRQ Rents for U.S. Dairy Product Imports (1997)

Product	Rents	New Zealand rents	Imports	Total imports[a]	Imports	U.S. wholesale price	Rent[b]
	U.S.$	percent of total	kg	percent	% of TRQ	U.S.$/kg	U.S.$/kg
Dried skim milk	1,684,803	71.88	2,171,826	98.86	75.91	2.742	0.776
Dried buttermilk/whey	51,144	100.00	63,000	100.00	28.00	2.354	0.812
Dried whole milk	1,331,090	53.04	1,542,875	99.94	99.46	2.802	0.863
Butter	3,781,917	54.88	4,873,239	97.25	94.13	2.456	0.776
Butter substitutes	5,134,403	100.00	5,314,510	99.99	117.56	2.800	0.966
Other cheese NSPF	61,338,928	50.54	28,063,971	83.16	62.73	5.200	2.110
Blue mold	4,050,996	—	1,991,572	78.29	73.19	6.358	2.034
Cheddar	8,427,133	80.51	6,939,948	93.48	80.34	3.699	1.214
American other than cheddar	2,631,967	69.05	2,842,399	94.49	81.85	3.442	0.926
Edam and gouda	10,307,037	—	4,782,652	98.35	75.72	5.951	2.155
Italian type	23,962,609	—	8,511,151	68.18	64.08	7.098	2.815
Gruyere processed	13,921,350	—	5,424,911	99.45	70.78	5.740	2.566
Other cheeses	9,662,422	—	3,043,086	99.99	58.58	6.446	3.175
Swiss cheese/emmentaler	52,359,354	1.34	24,576,219	99.76	73.33	5.742	2.130
Total	198,645,152		101,141,359				
Tariff revenue to government	34,078,711						
Rents to quota holders	164,566,441						

[a]Does not add up to 100 percent because it excludes imports of high-cost dairy products.
[b](Average U.S. wholesale price – c.i.f. prices) imports.
Table prepared by Isabelle Schluep.
Source: U.S. Census Bureau, USDA/AMS (Various years), USDA/FAS (Various years).

Table 9.5 Total Rents of New Zealand's Dairy Policy[1]

Year	Downstream profits[2]		Tariff-rate-quota (TRQ) rents[3]		Domestic price premia[4]		Total rents	Total production value	Percent of production value from:			
	Actual in M US$[5]	Percent of total rents	Actual in M US$	Percent of total rents	Actual in M US$[5]	Percent of total rents	Actual in M US$[5]	Actual in M US$[5]	Down-stream profits	TRQ	Domestic price premia	Total rents
1990	410.25	73.75	114.87	20.65	31.179	5.60	556.30	1,641	25.00	7.00	1.90	33.90
1991	550.75	73.75	154.21	20.65	41.857	5.60	746.82	2,203	25.00	7.00	1.90	33.90
1992	620.25	73.75	173.67	20.65	47.139	5.60	841.06	2,481	25.00	7.00	1.90	33.90
1993	641.50	73.75	179.62	20.65	48.754	5.60	869.87	2,566	25.00	7.00	1.90	33.90
1994	644.75	73.75	180.53	20.65	49.001	5.60	874.28	2,579	25.00	7.00	1.90	33.90
1995	801.50	73.75	224.42	20.65	60.914	5.60	1086.83	3,206	25.00	7.00	1.90	33.90
1996	831.25	73.75	232.75	20.65	63.175	5.60	1127.18	3,325	25.00	7.00	1.90	33.90
1997	812.50	71.38	230.17	20.22	61.719	5.42	1138.21	3,250	25.00	7.00	1.90	33.98

[1] *Table prepared by Isabelle Schluep, Cornell University.*
[2] *Downstream profits have been calculated to be approximately 25 percent of total production value (New Zealand Business Roundtable, 1996).*
[3] *TRQ rents have been calculated for 1997 and are projected back; they are assumed to be 7 percent of total production value.*
[4] *Domestic price premia have been calculated to be approximately 1.9 percent of total production value (New Zealand Business Roundtable, 1996).*
[5] *Million U.S.$.*
Source: OECD (2000), New Zealand Business Roundtables (1996 and 1998).

Once the importer has chosen a method, that entry price is compared to the MEP to determine the import duty the importer will have to pay. The import duties are determined according to the following:

- If the entry price is greater than or equal to the MEP, then the importer pays the CVT.
- If the entry price is greater than 92 percent of the MEP but less than the MEP, then the importer pays the CVT and the tariff equivalent, in which the tariff equivalent is equal to the difference between the entry price and the MEP.
- If the entry price is less than 92 percent of the MEP, then the importer pays the CVT and the MTE.

The entry-price system generates rents equal to the difference between the world or cost price and the entry price. The rights to these rents are not allocated by the E.U. government in any specific manner. Thus, the distribution of rents between importers and exporters depends on the degree of bargaining power between the two groups. Exporting countries derive bargaining power from negotiating preferential agreements with the European Union and from creating imperfect competition. An exporting country that negotiates an agreement with the European Union that lowers the import duties applicable to its produce will have a distinct advantage in drawing importing firms to itself; it will be able to offer the importer the possibility of selling on the E.U. market at a price lower than that established by the MEP. If an exporting country concentrates its supply using an STE or marketing board, then it can better prevent its standard import value from falling below the MEP, and so it can more closely control the price which it offers importing firms. Importing firms also derive bargaining power from imperfect competition, but, more importantly, the importing firm chooses the method by which the entry price is declared. The importing firm can thus choose to negotiate a favorable invoice price, or it can use a special deductive method allowed by the European Union when the exporting country's standard import value is deemed too low.

In addition to allowing the importer and exporter to distribute the rents among themselves, the entry-price system provides opportunities for rent seeking. The use of invoices obviously lends itself to rent seeking through misrepresentation of the entry price of a consignment or part of a consignment. Traders can also choose, as the point of entry for their consignments, a member state which implements the methods of entry-price declaration, particularly the deductive method that involves determining a final selling price with less stringent conditions. Finally, importers can control the standard import value by selling part of a consignment at a high price on the representative wholesale markets and then selling the bulk of their produce directly to retailers or through other marketing channels at much lower prices (Grethe and Tangerman, 1998).

3.2 Methods of Allocating Import Licenses

There are no specific provisions in the 1994 GATT for specific methods of allocating import licenses by importing countries. The WTO lists various methods of allocation and the number of TRQs in each category (Table 9.6). Approximately 50 percent of the 1370 TRQs posted with the WTO are administered with applied (in-quota, first tier) tariffs, which means that neither the license nor the quota (export nor import) determines imports. Except for a first-come, first-served (FCFS) system, all other allocation procedures listed in Table 9.6 can involve import licenses. Export quotas and/or licenses can be allocated for any category in Table 9.6. Note that an export country quota allocation can be made by the importing country but any firm from a third country may be able to export the product (peanuts from Argentina to the United States by a Canadian firm is one example, despite protests by the Argentine government).

There are inconsistencies among countries over the period of validity for import licenses, the size of the licenses, eligibility requirements for reallocation of unused licenses, and requirements for the use of the license. The Agreement on Import Licensing requires that the application process for obtaining and renewing a license be simple, and that all rules and information concerning the procedures be published (Boughner and de Gorter, 1998).

The E.U. banana regime is very inefficient in transferring rents to African, Caribbean and Pacific State (ACP) countries with the latter receiving less than 25 percent of the U.S. $1.6 billion worth of rents created (Borrell, 1997). Tangermann (1998) shows how the change in the European Union licensing procedures for bananas is to eliminate the 30 percent allocation to ACP countries and to give them direct aid instead. But importers can receive these licenses based on historical imports and so find it attractive to waste resources in markets they otherwise would not be in just to capture some of these rents (what Tangermann calls "taxpayers pay, traders cash in"). Chiquita, for example, is expanding output in Africa to capture rents as an importer and an exporter.

Unlike the one-time allocation of the right to export by importing countries in the Agreement, the right to import can and has been allocated more frequently. Most import licenses are allocated for use during a specific time period (a year or a season, for example). Often the licenses are allocated with no consideration given to the economic efficiency of the importing firms.

Table 9.6 Principal Methods of Administering TRQs

Method of administration		*Number of TRQs*	
	1995	*1996*	*1997*
Applied tariffs	650	650	643
License on demand	325	333	342
First-come, first-served	102	104	102
Historical allocation	49	60	67
Auction	32	30	32
State trading	22	22	20
Producer groups	8	8	8
Mixed allocation	42	43	42
Other	22	21	15
Not specified	9	7	99
Sum of above	1261	1278	1370
Total number of TRQs in schedules	1370	1370	1370

Additional conditions		*Number of TRQs*	
	1995	*1996*	*1997*
Domestic purchase requirement [1]	46	46	46
Limits on TRQ shares [2]	102	111	118
Export certificates [3]	25	25	25
Past trading performance [4]	71	71	70
Past trading performance and limits on TRQ shares	3	3	3

[1]A condition requiring the purchase of domestic production of the product in order to be eligible.
[2]Limits the maximum share or quantity of the quota allowed.
[3]Requires an export certificate administered by the exporting country.
[4]Limits eligibility to established importers of the product concerned.
Source: WTO (1997).

Moreover, the sizes of the licenses are usually determined on the basis of availability and not on the cost structure of the importing firms. Equalizing rents across members of a group leads to inefficiency and to premiums for particular licenses (Anderson, 1988). A competitive auction would ensure that the most efficient, low-cost importing firms would gain access to the rights to import if the rights are tradable after the initial auction (Bergsten et al., 1987). However, the granting of a valuable property right in itself creates a constituency that will rent seek and lobby governments to resist trade liberalization, possibly rendering auctions as an inefficient method of allocation.

The non-tradability feature of an import-licensing scheme exacerbates the economic waste due to inefficient allocation across product categories at the outset. The inefficiency of an importing country's arbitrary allocation of licenses across products within a TRQ category will depend on the elasticity of substitution in demand between the products (Anderson, 1985). The higher the elasticity of substitution, the less harm an arbitrary allocation of non-tradable import quotas across products will be because consumers will substitute away from the more expensive products. On the other hand, a low elasticity of substitution results in consumption closer to the optimal quantity, despite the arbitrary import quota set on each product category within a group. The net result is then an empirical question.

Nevertheless, Anderson (1985) concludes that sub-categorization of non-tradable import licenses results in significant inefficiency. Krishna and Tan (1998) question this result by showing various situations in which sub-categorization is desirable. The outcome depends on the economic environment, including whether or not the country exerts market power on world markets and on the objective function of the government (Spencer, 1996).

3.3 Rent Contestability: Rights to Rents versus Distribution of Rents

Because large rents exist under the TRQ system in the WTO for agriculture, there is a huge potential for waste of resources through rent seeking above and beyond the traditional deadweight loss triangles. The extent of the rent seeking depends on whether rights to rents are assigned or not, the method of assignment and/or the degree of imperfect competition between trading firms. Rights rent seeking occurs when rights to import or export are assigned and interested agents rent seek in an attempt to attain those rights, and rents rent seeking occurs when no rights are assigned and interested agents rent seek to attain the rents directly. These correspond to the two theoretical categories of rent seeking defined earlier.

Rights rent seeking takes place when rights are assigned, and so firms waste resources to obtain the rights to export or import. For example, exporting countries spend resources lobbying importing countries to attain a country-specific portion of their TRQ, or exporting firms spend resources attempting to obtain the export license. In fact, some rights to export were achieved during the Uruguay Round by conceding to some of the demands of the importing countries (that is, in return for lower tariff rates on some products, exporters would obtain the right to export other products).

If an importing country requires an import license, rights rent seeking will occur when importing firms spend resources in competition to obtain the rights to import, whether or not the right to export was allocated. Because the rights to import are allocated more frequently (yearly or seasonally), rent seeking is induced on a more continuous basis (unlike with the fixed allocation of the right to export). Initially, securing the rights to import can vary from bribing civil servants prior to the allocation process, or through allocating resources to influence the expected size of the import licenses (Kreuger, 1974). In some cases, importing firms may attempt to circumvent the rules of the license-allocation process by applying for amounts greater than they need, or by splitting their company into many subsidiaries to be eligible for greater amounts (IATRC, 1997).

The relevant literature to model rights rent seeking should include game-theoretic models on interest groups, lobbying, and politician-voter interaction with campaign contributions. Agents (single or non-strategic) do the rent seeking or lobby for a share of the rents. Transfers have deadweight costs associated with them in this context, and not all dissipation of rents is pure waste (Varian, 1987). To model rents rent seeking in our framework, the literature on rent-seeking contests becomes relevant with Tullock's (1980) strategic winner-take-all contests as the starting point (Nitzan, 1994). The outcome is determined by the nature of the contest (an N-player strategic game where N>2) and the characteristics of the contested rents (the winner takes all) with players spending resources to increase the probability of winning the rent. This literature is concerned with the existence and characterization of Nash equilibria and the relationship between rent-seeking outlays and the value of the contested rent. This becomes particularly relevant for import-quota rents (Hillman, 1989). The analysis of the rent-seeking game depends on the assumptions regarding the number of players, the source of the rent, the nature of the rent, and the nature of the rent setter. The outcome will depend on transaction costs, property right assignment to quota rents, the method of quota allocation, and the bargaining game.

The central difference between the rights versus rents rent seeking models delineated above is that losers in the latter only lose rent-seeking expenditures, while in the former, the rent is not cash-on-the-table but involves a transfer from one party to another as well.

If quota/license rights are assigned for both the rights to export and/or to import, then the rights to rents have been allocated to the maximum possible extent. If rights to rents are not assigned, then all-out rent seeking is possible and the dissipation of all or more rents can occur. If rights are allocated, a difference can emerge between who receives the rights and who actually receives the rents. The distribution of rents depends not only on the assignment of rights, but also on the degree of bargaining power held by importers relative to that held by exporters. It is very possible that imperfect competition results in market power for one side of the trading market and hence bargaining power. Because of the many outcomes possible in allocating the rights to rents, it is possible that importing or exporting firms have bargaining power with perfect competition in the economic marketplace. Rent seeking is more likely if a difference emerges between who receives the rights and who actually obtains the rents because of the allocation of rights and market power.

Table 9.7 summarizes the expected degree of wasteful rent seeking for alternative combinations of right assignments and the resulting distribution of rents. Clear property rights are defined if rents are allocated to one side only (importer or exporter), thereby resulting in the prediction that few resources will be wasted in rent seeking for rents directly (last column of Table 9.7). On the other hand, assigning rights to both importing and exporting firms results in ill-defined property rights, so the distribution of rents will depend on the relative degree of bargaining power, itself a function of rent-seeking activities. Both sides are now vulnerable to waste resources while attempting to extract the largest share of rents. As a result, we argue that the WTO needs stricter regulations on the assignment of property rights that are more clearly defined and are tradable. Having equal rights to rents may seem fair, but can result in rent-seeking practices that reduce social welfare. At the other extreme, no rights allocated whatsoever may also seem fair, but can lead to an open-access, rent-seeking environment. In this case, the TRQ is administered on a first-come, first-served basis (FCFS); that is, no quotas or licenses are allocated. Even though the intent of the FCFS allocation method may be to induce competition among exporting countries, inefficiency may occur because of competition over rents, resulting in the dissipation of rents. Exporting firms waste resources by altering shipment dates (or by shipping higher quality products if in competition with other countries) to ensure their product is the first to enter the importing country's port, thereby maximizing their share of the rents. Resources can also be wasted when trading firms collude or lobby governments in order to gain immediate access to markets before others.

We argue that imperfect competition among importing and exporting firms may result in less rent seeking. When equal rights to rents are assigned under perfect competition, firms waste resources when bargaining over and trying to obtain rents. Firms with market power can overcome this by having direct access to the rents. However, two caveats are in order: (1) resources may still be wasted in rights rent seeking; and (2) imperfect competition may directly reduce imports below the import quota, thereby reducing social welfare even more. Table 9.8 summarizes the combinations of import and export license requirements for the 1,370 TRQs logged in at the WTO (shaded areas give the total number in each quadrant while examples are given in each quadrant for the European Union, the United States, and Canada). A striking outcome is that 1,044 TRQs have no licenses or quotas required whatsoever, resulting in the majority of rents generated by TRQs being vulnerable to open-access rent seeking.

Table 9.7 The Matrix of Rent Assignment Versus Rent Distribution

Right assignment			*Rent distribution*		*Resources wasted in rent seeking*	
X quota (country)	*M license (firm)*	*X license (firm)*	*Who most likely obtains rents*	*How rents are obtained*	*For "rights" to licenses and quotas*	*For "rents" directly*
√	√	√	Importer and/or exporter	*Licensing affords bargaining power to both importers and exporters*	High	Medium high
√	√		Importer	Bargaining power in importer's favor	Medium	Medium low
√		√	Exporter	Bargaining power in exporter's favor	Medium	Medium low
	√		Importer and/or exporter	Relative bargaining power of importing versus exporting firms	Lower	Very low
			Importer	Bargaining power in importer's favor	Lower	Very low
			Importer and/or exporter	"Open access" rent seeking		Very high

Source: Authors' compilation.

Table 9.8 Import and/or Export License Allocation Matrix

X License Required	M license required							
	YES				NO			
	TRQ[a]	Canada	European Union	United States	TRQ[a]	Canada	European Union	United States
YES	19	Beef and veal	Live sheep and goats Sheep or goat meat Butter Cheese for processing Cheddar cheese Manioc Bananas Cane or beet sugar Boneless beef Rice		5		Meats Fruits and vegetables Fructose Other cereals	Meats Dairy Olives Peanuts Other sugars/syrups Cocoa/chocolate Ice cream Cotton/cotton fibers
NO	302	Cream Whey Other dairy Margarine	Sweet potatoes Raw cane sugar Maize Grain sorghum Broken rice Millet Manioc starch Bran Pig meat Poultry and eggs Skim milk powder Butter Cheeses Wheat Oats	Butter Butter substitutes Cheese Dried milk Whey	1044	Meats Dairy Eggs Wheat Barley	Live sheep Mushrooms Orange juice	Cane sugar Tobacco

[a]*Tariff-rate quotas logged in at the WTO.*
Source: Calculated from WTO, 1998.

Bargaining power may also be inherent to the industry in question, independent of the allocation of licenses. For example, exporting firms have bargaining power based on characteristics such as the quality of their product or reliability in delivering the product. Imperfect information also provides bargaining power, either through economic or political markets. A firm with bargaining power may capture the rents available, but may not necessarily engage in imperfectly competitive practices. Imperfect market structure depends critically on the number of firms, the structure of the market, and the like, while bargaining power is critically affected by the allocation mechanism of the rights to rents.

Rent seeking can also occur if future quotas allocated are a function of current or past import levels. Firms may reduce profit margins in the short run in order to obtain more quota rights (and profits) in the long run (Krishna and Tan, 1993; Krueger, 1974; de Gorter and Fisher, 1993). Such a strategy will also occur if a stipulation has been placed on an importing firm either to use the license or to lose it (or a portion of it) the following year. Countries can outline provisions that require a certain percentage fill rate, or they can place penalties on importing firms that do not use the entire portion allocated, or they can require a security deposit that will be refunded after the license is used. Provisions such as these can actually encourage higher fill rates, but can induce economic inefficiency (as in the case of water rights held by farmers in California).

3.4 Rent Seeking under Alternative Import Quota Administration Methods

The Agreement does not insist that each TRQ be filled. A low quota fill rate does not necessarily imply inefficiency (there may be unavailable supply, or insufficient demand, or the first tier tariff may be effective) nor does a fill rate of 100 percent or more imply efficiency. Trade may have occurred with high-cost importing firms or export countries/firms, or STEs may have fulfilled WTO commitments by importing low-quality product or by destroying imports upon arrival (Choi and Sumner, 1998). Independent of export quotas or non-tradability of licenses, the method of allocation of the import license itself can have a direct impact on the quota fill rate, rent seeking, and hence on economic efficiency.

Fill rates for each administration method are given in Table 9.9 for a subset of the total TRQs for agriculture listed in the WTO (number of countries, commodity groups, and observations are also provided for each method). The average fill rates can be misleading because some are equal to zero and others are equal to 100 percent. They are also not weighted by trade volume

Table 9.9 Fill Rate Breakdown, 1996

Administration method	*Countries*	*Product groups*[a]	*Total observations*	*Average fill rate*	*Distribution of fill rates <95 percent*
		Number		*Percent*	*Percent*
Applied tariffs	10	12	154	72	39
Auctioning	4	4	27	38	70
First come, first served	6	9	81	63	56
Historical importers	10	10	67	82	45
Licenses on demand	12	12	242	52	72
Mixed administration	7	8	42	82	55
Not specified	2	2	6	34	83
Other	3	3	8	54	50
Producer groups	2	6	7	53	57
State trading enterprizes	5	7	21	82	24
Total	22	12	655	63	57

[a]*Product groups are: cereals; oilseeds; sugar; dairy; meat; eggs; beverages; fruit and vegetables; tobacco; agricultural fibers; coffee; tea, spices and processed agricultural products from multiple groups; and other.*
Source: WTO (1997 and 1998).

or by value. Nevertheless, STEs and historical importers have the highest fill rate and auctions have the lowest. The last column of Table 9.9 gives the distributions of fill rates falling below 95 percent. We now analyze the incentives for rent seeking which occur under each administration type.

3.4.1 Auctioning

The auctioning of import quota licenses is generally deemed the most efficient in allocating the right to import (Bergsten et al., 1987; Skully, 1997). If there is no uncertainty or no information costs, then an import quota generates the same level of revenues (and hence imports) to the government as an equivalent import tariff (de Gorter, 1985). However, data in Table 9.9 indicate that this method has the lowest fill rate. It is possible for one group to purchase the entire portion of the right to import (domestic or foreign), and then withhold part of the licenses to maximize revenues. Bergsten et al. (1987) argue that procedures can be designed to guard against the monopolization of licenses. Krishna (1988, 1990, 1991) argues that rent seeking can occur under auction systems when foreign suppliers change their price in response to the quota. Schmitz, de Gorter, and Schmitz (1996) show how domestic monopolies can change domestic prices in response to trade liberalization with a quota in place, regardless of who owns the quota. McCorriston (1996) shows how import quotas create oligopsony power in either the domestic- or foreign-product market, or in the market for licenses (including the processing sector).

3.4.2 First Come, First Served (FCFS)

The FCFS method prevails when no rights to importing or exporting firms have been allocated.[viii] Average quota fill rates are 63 percent, perhaps reflecting the uncertainty associated with this method. Exporters may not risk the costs of shipping the product to the port only to find that quota has been filled. The costs of storage until the following TRQ season, of paying the higher second-tier tariff, or of shipping the product elsewhere may be very high. The cost for traders to establish a business relationship over time with importers is also a factor contributing to under-fill. Exporters do not have information on who holds the import license. The FCFS is prone to rent seeking by concentrating imports at the beginning of the season (ships may appear early and wait in line), increasing costs for importers who have to store the product, and discriminating against exporters who have different seasons and higher costs at the beginning of the TRQ year. It opens the door for open-access rent seeking in that there are no property rights allocated to the import quota rents at all.

3.4.3 Historical Importers

Allocating import licenses on the basis of historical import shares can lead to rent seeking by trading firms who import more than what is optimal. Table 9.10 shows that approximately 30 percent of all TRQs have fill rates of over 100 percent. This is not necessarily due to rent seekers trying to pad their import licenses (Table 9.6 showed that almost half of all TRQs are administered by applied tariffs, many of which can result in quota overfill), but firms may behave strategically in order to obtain more import licenses. Chiquita is purported to have expanded imports in the European Union in 1992 in anticipation of the new Common Market Organization for bananas under which import licenses (and export quotas) were to be allocated as a proportion of historical imports. Historical import allocation also enables high-cost firms to operate, which leads to the partial dissipation of quota rents.

Table 9.10 Distribution of Tariff-quota Fill Rates

	Proportion in fill-rate category	
	1995	*1996*
Fill-rate percent category	*percent*	
0-20	24	27
20-40	7	7
40-60	7	7
60-80	7	7
80-100	22	25
>100	33	27

Source: Estimated from WTO (1998).

3.4.4 Licenses on Demand

Import licenses are usually issued before the season begins and are allocated to individual firms *pro rata* on the difference between the TRQ level and the total amount of licenses requested. This again enables higher-cost importers to obtain the rights to the rents, thereby dissipating the rents somewhat. A good example of such a scheme is the E.U. regime for eggs. Any importer can apply for the license but is required to pay 20 percent of the value of the product in advance. This gives an incentive for the importer not to overbid for the license and hence reduces rent seeking. It also provides an incentive for the firm to import the product, thereby reducing the chances of quota under fill. However, the costs of the down payment may deter some firms from participating, and may lead to quota underfill because of uncertainty. Once a firm requests an import license, the firm loses the down payment.

3.4.5 Producer Groups

The control over imports by producer groups leads them to trade-off the benefits from owning the quota rents and the losses in producer surplus with the imports. Failing to fill the quota is advantageous only if the quota rents are smaller than the loss in producer surplus due to increased imports. The outcome depends on the relative domestic supply/demand elasticities for the product, the level of the domestic price with imports (which depends on the world price and excess supply conditions in the world market) and the level of the import quota. There remains a problem of distributing the rents to farmers, which if blended with revenues from sales from domestic production, would cause an increase in domestic production costs and would reduce producer welfare. Another option is to destroy the imported product, but farmers would need to incur the purchase costs at world market prices (unless the domestic government finances it on behalf of the producer group). Producer groups could instead import a product that is of inferior quality, thereby fulfilling their TRQ and simultaneously maintaining income from domestic production, provided the cross elasticity of demand is low (Choi and Sumner, 1998).

3.4.6 State Trading Enterprizes

The quota fill rate is very high for STEs, perhaps reflecting the ability of exporters to overcome transaction and information costs when dealing with many importing firms owning import licenses under other schemes. An STE controls all imports, is visible, and perhaps even scrutinized more by foreign governments. The ability to seek rents is still possible, however. Alston,

Carter and Jarvis (1990) argue that the Japanese Food Agency deliberately allocated export quotas to the United States and excluded Canada in the 1970s. This meant that some rents were dissipated because the United States is a higher-cost exporter and that rents were allocated to the United States like a VER. STEs also have the ability to discriminate with respect to price in their purchases and sell the product at lower prices on the domestic market. Import STEs can restrict imports to help domestic farmers. Choi, Sumner, and Song (1998) show how the Korean STE imports rice through an open bidding system. The lowest bidder (supplier) that meets minimum quality standards wins the right to export, leading to lower quality rice imports that are not a close substitute for domestic rice, thereby protecting farm incomes.

3.4.7 Lottery

An option not discussed in the WTO's documents involves issuing import licenses by lottery. This will reduce rent-seeking outlays in that a firm cannot affect the likelihood of obtaining the license prior to the lottery. Nevertheless, each firm would have to comply with application procedures, and, assuming that each firm is only allowed one draw, there are incentives to break the firm down into many small firms to increase the probability of receiving a license. Because of the random nature of such a procedure, higher cost firms may win the lottery and thus the rents are dissipated somewhat.

3.4.8 Additional Conditions

Table 9.6 lists the additional conditions associated with the administration methods discussed above. Each of these lead to rent seeking as well. Domestic purchase requirement increases the cost for some importing firms that otherwise would not be involved in domestic production. Thus, part of the quota rent is dissipated and fill rates are lower as domestic consumption declines and production increases. Limits on TRQ shares do not allow for economies of size and coordination, again resulting in the dissipation of quota rents. Export certificate requirements tend to allocate the rights to rents to higher cost exporters as does the condition of past trading performance, both of which lead to partial dissipation of quota rents.

4. CONCLUDING REMARKS

Our analysis demonstrates that the Agricultural Agreement in the Uruguay Round of multilateral trade negotiations and the resulting GATT disciplines imposed on agriculture still allow for economic agents to develop strategies that result in a waste of resources due to rent seeking and so are counter to the WTO's goal of trade liberalization. The amount of quota rents and the number of stakeholders may even be higher than they were before the Uruguay Round, resulting in the potential for more rent seeking and economic waste. Political constituencies are formed when rents are created, thereby reducing the chances of policy reform and increasing the level of strategic behavior among stakeholders. Table 9.1 summarizes only some of the enormous levels of rents created by agricultural policies worldwide. Although the Agreement on Agriculture was meant to be a transition from distortion to free trade, the mechanism of TRQs themselves has created incentives to wastefully pursue rents. Those successful in retaining the rents potentially become obstructionists to policy reform.

The incentives along the path of transition have only begun to be documented and this chapter is only a beginning in analyzing the true social costs of rent seeking in world agriculture. TRQs have created both well-defined and ill-defined property rights to rents. Both induce economic inefficiency but the latter can generate more wasteful rent seeking in competition for these rents. Several mechanisms are being used to resolve these problems, but not one is strategy proof in design, such that rent seeking is minimized or eliminated. Any solution, like auctions, can legitimize imperfect competition or can identify vested interests that will then resist policy reform. Rules in the WTO are required to minimize rent seeking and maximize the probability of policy reform to a market-oriented trading regime for agriculture. These should include measures to increase openness and transparency such that participants' awareness of the distribution of income and economic waste generated by TRQs would make such TRQs less sustainable in the political marketplace.

We argue that the WTO needs stricter regulations on the assignment of rights with equal rights to rents and no rights to rents being inferior to clearly defined property rights on rents. Strategic behavior by STEs, national governments and producer groups allows countries to circumvent GATT import commitments by changing domestic price supports or production quotas (Schmitz, de Gorter, and Schmitz, 1996). One problem in designing agricultural policy reform has been the mindset of some looking to maximize their own welfare through strategic trade policy instead of maximizing world welfare. Many commentators, like Krugman (1992), promote the new strategic trade-policy literature in terms of governments maximizing national welfare with "gains from deviating from free trade. . . questioning the academic orthodoxy of free trade. . . potential role for predatory trade policies." We argue, on the other hand, that the goal of the WTO should be to minimize trade distortions (de Gorter, 1992) and to work toward changing the policy instruments employed to achieve an increase in world welfare alone.

5. ENDNOTES

[i] An indicator of the importance of agriculture in trade disputes and hence rent seeking is the fact that over 50 percent of the 163 trade disputes since the inception of the World Trade Organization (WTO) in 1995 have been in agriculture (43 percent in the General Agreement on Tariffs and Trade (GATT) from 1948-90).

[ii] The rents in Table 1 (like OECD data) exclude many countries, especially developing countries, and many commodity sectors such as fruit and vegetables, and the effects of many policies; for example, research expenditures and the bans of genetically altered products.

[iii] Column (1) is total producer subsidy equivalents while column (2a) is production times the domestic-world price gap and column (2b) is imports times the domestic-world price gap.

[iv] Resource waste above and beyond the standard allocation costs represented by deadweight loss triangles in the new welfare economics analyzed in, for example, Just, Hueth and Schmitz (1982).

[v] Bhagwati (1992) chooses sub-optimal levels of government policy, with or without resources expended, as a comparison point—his second best analysis, therefore, allows that the outcome of rent seeking on net social welfare be indeterminate.

[vi] Brooks and Heijdra (1989) review Varian's (1987) path-breaking study and several other studies that show under what conditions all rents (or more) can be wasted in rent seeking by interest groups. See Krueger (1996) for a re-statement of the issues.

[vii] We are ignoring the traditional issues of deadweight loss triangles and equity issues in this discussion, but ultimately it would be ideal to integrate these issues with those discussed in this chapter.

[viii] Table 9.9 shows this makes up most of the cases, but the data do not identify whether specific export quotas are allocated or not. If they are allocated, it is still possible for the FCFS to have the same effect if there are many firms exporting in each country and they behave competitively.

6. REFERENCES

Alston, J., C. Carter, and M. Wohlgenant. 1989. "Who Determines Farm Programs? Agribusiness and the Making of Farm Policy," International Agricultural Trade Research Consortium Working Paper 89-1, St Paul, Minnesota.

Alston, J., C. Carter, and L. Jarvis. 1990. "Discriminatory Trade: The Case of Japanese Beef and Wheat Imports." *Canadian Journal of Agricultural Economics* 38: 197–214.

Anderson, J.E. 1988. *The Relative Inefficiency of Quotas*. Cambridge, MA: MIT Press.

_____. 1985. "The Relative Inefficiency of Quotas: The Cheese Case." *The American Economic Review* 75(1): 178–190.

Anderson, K. 1997. "Environmental and Labor Standards: What Role for the World Trade Organizations," Working paper, Center for International Economic Studies, University of Adelaide, Australia.

_____. 1998. "Domestic Agricultural Policy Objectives and Trade Liberalization: Synergies and Trade-Offs," Paper prepared for OECD Workshop on Emerging Trade Issues in Agriculture, Oct. 26-27, University of Adelaide, Australia.

Bates, R.H. 1990. "The Political Framework for Agricultural Policy Decisions," in C. Eicher and J. Staatz, eds., pp. 154–159. *Agricultural Development in the Third World, 2nd Ed.*, Baltimore, MD and London, UK: Johns Hopkins University Press.

Baysinger, B. and R.D. Tollison. 1980. "Evaluating the Social Costs of Monopoly and Regulation." *Atlantic Economic Journal* 8(Dec.): 22–26.

Bergsten, C.F., K.A. Elliott, J.J. Schott, and W.E Takacs. 1987. *Auction Quotas and United States Trade Policy*. Policy Analyses in International Economics No. 19, Institute of International Economics, Washington, DC.

Bhagwati, J.N. 1982. "Directly Unproductive, Profit-seeking (DUP) Activities." *Journal of Political Economy* 90: 988–1002.

Borrell, B. 1997. "Policymaking in the E.U.: the Bananarama Story, the WTO and Policy Transparency." *The Australian Journal of Agricultural and Resource Economics* 41: 263–276.

Boughner, D.S. and H. de Gorter. 1998. "The Economics of 2-tier Tariff-Rate Import Quotas: The Agreement on Agriculture in the WTO and U.S. Dairy Policy." Paper presented at the International Agricultural Trade Research Consortium Annual Meeting, St. Petersburg, Florida (13-15 December).

Bredahl, M., A. Schmitz, and J. Hillman. 1987. "Rent Seeking in International Trade: The Great Tomato War." *American Journal of Agricultural Economics* 69(1):1–10.

Brooks, M. and B. Heijdra. 1989. "An Exploration of Rent Seeking." *The Economic Record* 65(188): 32–50.

Buchanan, R., D. Tollison and G. Tullock eds. 1980. *Toward a Theory of Rent Seeking Society*. College Station, TX: Texas A&M University Press.

Center for Responsive Politics. 1997. *The Big Picture*. Washington, DC: Center for Responsive Politics. Online. Internet website: http://www.crp.org/crpdocs/bigpicture/defatult.htm

Choi, J. and D.A. Sumner. 1998. "Welfare Implications of Tariff-Rate Quota Management in Korea." Paper presented for discussion at the International Agricultural Trade Research Consortium Annual Meeting, St. Petersburg, Florida (13-15 December).

Choi, J., D.A. Sumner, and J. Song. 1998. "Importing STEs in Korea and Japan: Evolution, Operation, and Implications." Paper presented for the Role of State and Agricultural Trade Workshop co-organized by the North American Forum, Stanford University and Agricultural Issues Center, University of California, Davis (20-22 November).

Coggins, J.S. 1995. "Rent Dissipation and the Social Cost of Price Policy." *Economics and Politics* 7: 147–166.

Coleman, D. "Ethics and Externalities: Agricultural Stewardship and Other Behavior." *Journal of Agricultural Economics*, Vol. 45(3): 299–311.

de Gorter, H. 1992. "Agricultural Policies and the GATT: A Rawlsian Perspective," in T. Becker, R. Gray and A. Schmitz (eds.), *Mechanisms to Improve Agricultural Trade Performance Under the GATT*. Kiel Germany, Wissenschaftsverlag Vauk.

_____. 1985. "An Economic Analysis of Agricultural Import Policy Instruments: The Theory," Preliminary Technical Paper. Agriculture Canada. Ottawa, ON.

de Gorter, H. and E. Fisher. 1993. "The Dynamic Effects of Agricultural Subsidies in the United States." *Journal of Agricultural and Resource Economics* 18(2): 147–159.

de Gorter H. and J. Swinnen. 1995. "The Economic Polity of Farm Policy: Reply." *Journal of Agricultural Economics* 46(3): 403–414.

____. 1994. "The Economic Polity of Farm Policy." *Journal of Agricultural Economics* 45(3): 312–326.

Dennen, R.T. 1977. "Some Efficiency Effects of Nineteenth Century Federal Land Policy: A Dynamic Analysis." *Agricultural History* 51: 718–736.

Dixit, P.M and T.E. Josling. 1997. "State Trading in Agriculture: An Analytical Framework." International Agricultural Trade Research Consortium Working Paper No. 97-4, St Paul, Minnesota.

Fisher, F.M. 1985. "The Social Costs of Monopoly Power and Regulation: Posner Reconsidered." *Journal of Political Economy* 93: 410–416.

Grethe, H. and S. Tangermann. 1998. "The E.U. Import Regime for Fresh Fruit and Vegetables after Implementation of the Results of the Uruguay Round." Paper presented for the Commodities and Trade Division, FAO Economic and Social Department, Gottingen, Germany (October).

Heller, J. 1970. *Catch 22.* New York, NY: Dell.

Hillman, A.L. 1989. "The Resource Cost of Political Contestability." *The Political Economy of Protection.* Geneva, Switzerland: Harwood Academic Press.

Hillman, A.L. and J. Riley. 1989. "Politically Contestable Rents and Transfers." *Economics and Politics* 1: 17–39.

Hornig, E., R.N. Boisvert, and D. Blandford. 1990. "Explaining the Distribution of Quota Rents from U.S. Cheese Imports." *The Australian Journal of Agricultural Economics* 34(1): 1–20.

Hranaiova, J. and H. de Gorter. 1999 "Rent Seeking with Politically Contestable Rights to Import Quotas." Working paper, Department of Applied Economics and Management, Cornell University, Ithaca, NY.

Hudec, R. 1997. "Does the Agreement on Agriculture Work? Agricultural Disputes After the Uruguay Round." Paper presented at the International Agricultural Trade Research Consortium Annual Meeting, San Diego, CA. (14-16 December).

IATRC (International Agricultural Trade Research Consortium). 1997. "Bringing Agriculture into the GATT: Implementation of the Uruguay Round Agreement on Agriculture and Issues for the Next Round of Agricultural Negotiations." Commissioned Paper No. 12, University of Minnesota, St. Paul, MN.

Ingco, M.D. 1996. "Tariffication in the Uruguay Round: How Much Liberalization?" *World Economy* 19: 425–446.

Johnson, D.G. and K. Brooks. 1983. *Prospects for Soviet Agriculture in the 1980s.* Bloomington, Indiana: Indiana University Press.

Josling, T. 1998. "The Uruguay Round Agreement on Agriculture: A Forward Looking Assessment." Paper prepared for OECD Workshop on Emerging Trade Issues in Agriculture, Institute for International Studies, Stanford University (26–27 October).

Josling, T.E., S. Tangermann, and T.K. Warley. 1996. *Agriculture in the GATT.* Houndsmill and London: Macmillan Press.

Just, R., D. Hueth, and A. Schmitz. 1982. *Applied Welfare Economics and Public Policy.* Englewood Cliffs, New Jersey: Prentice-Hall.

Krishna, K. 1991. "Making Altruism Pay in Auction Quotas," in E. Helpman and A. Razin, eds, *International Trade and Trade Policy.* Cambridge, MA: MIT Press.

_____. 1990. "The Case of the Vanishing Revenues: Auction Quotas with Monopoly." *American Economic Review* 80(4): 828–836

_____. 1988. "The Case of the Vanishing Revenues: Auction Quotas with Oligopoly." NBER Working Paper No. 2723.

Krishna, K. and L.H. Tan. 1998. "The Dynamic Behavior of Quota License Prices: Theory and Evidence from Hong Kong Apparel Quotas." Working Paper. International Economics Department, The World Bank, Washington, DC.

________ . 1998. *Rags and Riches.* Ann Arbor, MI: The University of Michigan Press.

Krueger, A.O. 1974. "The Political Economy of the Rent-Seeking Society." *The American Economic Review* 64(3): 291–303.

Krugman, P. 1992. "Does the New Trade Theory Require a New Trade Policy?" *The World Economy* 15(4): 423–441.

Lindland, J. 1998. "Non Trade Concerns in a Multifunctional Agriculture." Paper prepared for OECD Workshop on Emerging Trade Issues in Agriculture, Ministry of Agriculture, Norway (26–27 October).

Makinson, L. 1992. *Open secrets: The Encyclopedia of Congressional Money and Politics, 2nd Edition*. Washington, DC: Center for Responsive Politics.

Makinson, L. and J. Goldstein. 1996. *Open secrets: The Encyclopedia of Congressional Money and Politics, 4th Edition*. Washington, DC: Center for Responsive Politics. Internet website: http://www.crp.org/low/secrets.html

_____. 1994. *Open secrets: The Encyclopedia of Congressional Money and Politics, 3rd Edition.* Washington, DC: Center for Responsive Politics.

Martin, E. and H. de Gorter, 1998. "The Agreement on Agriculture and the CAP: The Reform of the Fruit and Vegetable Common Market Organization," paper presented at the International Agricultural Trade Research Consortium Annual Meeting, St. Petersburg, FL (13–15 December).

McCorriston, S. 1996. "Import Quota Licenses and Market Power." *American Journal of Agricultural Economics* 78: 367–372.

McCorriston, S. and I.M. Sheldon. 1994. "Selling Import Quota Licenses: The U.S. Cheese Case." *American Journal of Agricultural Economics* 76: 818–827.

Meilke, K.D., D. McClatchy and H. de Gorter. 1996. "Challenges in Quantitative Economic Analysis in Support of Multilateral Trade Negotiations." *Agricultural Economics* 14: 185–200.

Moschini, G. 1991. "Economic Issues in Tariffication: An Overview." *Agricultural Economics* 5: 101–120.

Moschini, G. and K.D. Meilke. 1991. "Tariffication with Supply Management: The Case of the U.S.-Canada Chicken Trade." *Canadian Journal of Agricultural Economics* 39: 55–68.

Nitzan, S. 1994. "Modelling Rent-seeking Contests." *European Journal of Political Economy* 10: 41–60.

New Zealand Business Roundtables. Various years. *Annual Reports*. Wellington, NZ.

NZDB (New Zealand Dairy Board). Various years. *Annual Reports*. Wellington, NZ: New Zealand Dairy Board.

_____. 1998. *Annual Report*.

O'Connor and Company. 1997. "Tariff Rate Quotas in EC and GATT Law." Pre-printer's draft of monograph published in March. Brussels, Belgium: O'Connor and Company.

OECD (Organization for Economic Cooperation and Development). 2000. *Agricultural Policies in OECD Countries: Monitoring and Evaluation 2001*. Paris, France: OECD.

Orden, D., R. Paarlberg, and T. Roe. 1999. *Policy Reform in American Agriculture: Analysis and Prognosis*. Chicago, IL: University of Chicago Press.

Posner, R.A., 1975. "The Social Costs of Monopoly and Regulation." reprinted in J.M. Buchanan, R.D. Tollison, and G. Tullock eds., 1980 (71–94). *Toward a Theory of the Rent Seeking Society*. College Station, TX: Texas A&M University Press.

Rausser, G.C. 1995. "GATT Negotiations and the Political Economy of Policy Reform." Berlin, Heidelberg, New York: Springer-Verlag. 357.

Rausser, G.C. and P. Zusman. 2001. "Political Power and Endogenous Policy Formation." Amsterdam, The Netherlands: Elsevier Science, Elsevier North-Holland Pergamon Press.

Rausser, G.C. 1992. "Predatory Versus Productive Governments: The Case of U.S. Agricultural Policy." *Journal of Economic Perspectives* 6(3): 133–157.

Rom, M. 1979. *The Role of Tariff Quotas in Commercial Policy*. New York, NY: Holmes and Meier Publishers, Inc.

Roningen, V.O. and P. Dixit. 1987. "Impact of Removal of Support to Agriculture in Developed Countries," Unpublished paper. Washington, DC: USDA/ERS.

Schluep, I. and H. de Gorter. 1998. "The Law and Economic of Consumer Only Financed Export Subsidies: A Context for the WTO's Dispute Settlement Panel on Canadian Dairy Policy." Paper presented at the International Agricultural Trade Research Consortium Annual Meeting meetings, St. Petersburg, FL, 13–15 December.

Schmitz, A., H. de Gorter, and T.G. Schmitz. 1996. "Consequences of Tariffication." in A. Schmitz, G. Coffin and K.A. Rosaasen, eds., *Regulation and Protection under GATT: Case Studies in North American Agriculture*. Boulder, CO: Westview Press.

Skully, D.W. 1998. "Auctioning Tariff Quotas for U.S. Sugar Exports." Special Article, Market and Trade Economics Division, Economic Research Service, Washington, DC.

_____. 1997. "Economics of TRQ Administration." International Agricultural Trade Research Consortium Annual Meeting. San Diego, California.

Spencer, B.J. 1996. "Quota Licenses for Imported Capital Equipment: Could Bureaucrats Ever Do Better Than the Market?" *Journal of International Economics* 43: 1–27.

Tangermann, S. 1998. "E.U. Banana Reform: An Improvement." *Agra Europe*, 6 February: A/1-A/3.

Tasman Asia Pacific Ltd. and Acil Economics and Policy Ltd. 1996, 1998. "The Economy-wide Effects of Bundling Milk and Non-milk Returns." Paper prepared for the New Zealand Business Roundtable. Wellington, NZ (July).

Tollison, R.D. 1989. "Is the Theory of the Rent-seeking Here to Stay? in C.K. Rowley, ed., pp. 143–157, *Democracy and Public Choice: Essays in Honor of Gordon Tullock*. Basil Blackwell, Oxford.

Tullock, G. 1980. "Efficient Rent-seeking" in J.M. Buchanan, R.D. Tollison, and G. Tullock, eds. pp. 97–112. *Toward a Theory of the Rent-Seeking Society*. College Station, TX: Texas A&M University Press.

_____. 1967. "The Welfare Costs of Tariffs, Monopolies and Theft," reprinted in J.M. Buchanan, R.D. Tollison and G. Tullock eds., 39–50, 1980. *Toward a Theory of the Rent-Seeking Society*. College Station, TX: Texas A&M University Press.

U.S. Census Bureau, Foreign Trade Division, Data Dissemination Branch. Various issues. Washington DC: United States Government. Internet website: http://www.census.gov/foreign-trade/sitc1/2001/c3070.html

USDA/AMS (United States Department of Agriculture, Agricultural Marketing Services). Various years. *Dairy Market Statistics, Annual Report*. Washington DC: USDA/AMS.

USDA/FAS (United States Department of Agriculture, Foreign Agricultural Service). Various years. *Attache Reports*. Washington, DC: U.S. Department of Agriculture. Internet website: http://www.fas.usda.gov/scriptsw/attacherep/default.asp

_____. 1998. "Dairy Monthly Imports: U.S. Licensed Cheese Imports January-December 1996-1997." *Dairy Monthly Imports*. Circular Series FD MI 10-97 (January). Washington, DC: USDA/FAS.

VanSickle, J. 1997. "A Compromise in the Fresh Tomato Trade Dispute." *Florida Journal of International Law* 11: 399–408.

Varian, H. 1987. "Measuring the Deadweight Costs of DUP and Rent Seeking." *Economics and Politics* I(1): 81–95

Vercammen, J.A. and A. Schmitz. 1996. "Imports into Canada: Why Have They Remained Low?" in A. Schmitz, G. Coffin and K.A. Rosaasen, eds., *Regulation and Protection Under GATT: Case Studies in North American Agriculture*. Boulder, CO: Westview Press.

_____. 1992 "Supply Management and Import Concessions." *Canadian Journal of Economics* 4: 957–971.

WTO (World Trade Organization). 1998. *Tariff and Other Quotas*. Report No. AIE/S1/Rev1 1998. Geneva, Switzerland: WTO.

_____. 1997. *Tariff Quota Administration Methods and Tariff-fill Rates*, Report No. AIE/S4/Rev1. Geneva, Switzerland: WTO.

_____. 1997a. *European Communities-Regime for the Importation, Sale and Distribution of Bananas*, Report No. WT/DS27/AB/R. The Appellate Body. Geneva, Switzerland: WTO (9 September).

_____. 1997b. Committee on Agriculture, Summary of the Report of the Meeting Held on 26-27 June, Note by the Secretariat (G/AG/R/11). Geneva, Switzerland: WTO.

_____. 1997c. European Communities-Regime for the Importation, Sale and Distribution of Bananas, Complaint by Ecuador, Report No. WT/DS27/R/ECU by the Panel, 22 May. Geneva, Switzerland: WTO.

_____. 1995. *Agreement on Agriculture*. Geneva, Switzerland: WTO.

_____. 1994a. *GATT 1994*. Geneva, Switzerland: WTO.

_____. 1994b. *Agreement on Import Licensing Procedures*. Geneva, Switzerland: WTO.

_____. 1993. "Modalities for the Establishment of Specific Binding Commitments Under the Reform Programme." Note by the Chairman of the Market Access Group (MTN.GNG/MA/W/24) 20 December. Geneva, Switzerland: WTO.

Chapter 10

Ex Ante Assessment of the FAIR Act

Chris Dumas and Troy G. Schmitz
University of North Carolina at Wilmington and Arizona State University

1. INTRODUCTION

The U.S. Federal Agriculture Improvement and Reform Act of 1996 (FAIR) was designed to reduce government interference in farmers' crop choice and acreage decisions, and to reduce the burden of agricultural price support payments on the U.S. Treasury (USDA, 1996a; USDA, 1998b). Prior to FAIR, agricultural policy was governed by the Food, Agriculture, Conservation and Trade Act of 1990 (FACT) (USDA, 1989; USDA, 1995). Under FACT, the U.S. government intervened in agricultural commodity markets by limiting production through commodity-specific acreage restrictions and by supporting target prices through deficiency and loan deficiency payments. With the passage of FAIR in 1996, target prices and associated deficiency payments were eliminated, planting restrictions were lifted, and lump-sum subsidy payments (called Production Flexibility Contract Payments, or PFC payments) were introduced (USDA, 1996a; USDA, 1997). In contrast to the deficiency payments made under FACT, PFC payments made under FAIR were decoupled from planting decisions. In theory, decoupling subsidy payments from planting decisions would maintain farm income while improving flexibility and efficiency in production, thereby improving U.S. competitiveness in world markets, and reducing the future farm-subsidy obligations of the U.S. Treasury.

Although FAIR may be criticized in hindsight, we believe that any meaningful assessment of the decision to adopt FAIR must be made on an *ex ante* basis (that is, in the context of the time when FAIR was enacted). Some observers questioned whether or not PFC payments to growers were too lavish, because prices for some commodities rose to record heights soon after FAIR was passed. However, recent declines in crop prices have caused others to argue that just the opposite is the case. Some interest groups may find that adopting FAIR was the best *ex ante* decision, even though *ex post* outcomes may have fallen well short of expectations.

The purpose of this chapter is to assess the *ex ante* economic impact of FAIR—that is, the impact expected to result from the program at the time of its implementation—for a major agricultural program crop: upland cotton. We measured the anticipated economic impact of FAIR on three key interest groups: U.S. cotton growers, U.S. cotton consumers and U.S. taxpayers. We developed a model to compare the anticipated (circa 1995) economic welfare of each interest group under FAIR with the anticipated economic welfare of each group under FACT. Further, we conducted simulations to measure anticipated welfare differences across policies for plausible values of model parameters. Because future decisions regarding modifications of the FAIR Act must be made on a similar, *ex ante* basis, improvements in our ability to understand the anticipated economic desirability of the 1995 decision will aid future policy development efforts.

2. THEORETICAL MODEL

Schmitz, Schmitz, and Dumas (1997) developed an economic model of the U.S. cotton sector to assess the net domestic economic effects of U.S. cotton production and export trade under the 1990 FACT Act. They found that in some years the net domestic economic effects of trading cotton were negative for the United States due to market distortions caused by U.S. farm programs. Although Schmitz, Schmitz, and Dumas (1997) considered the welfare impacts of removing the cotton target price, they did not consider the combined effects of removing the target price and adding PFC payments under FAIR, because FAIR had yet to be implemented at the time the article was written. In addition, the model developed by Schmitz, Schmitz and Dumas (1997) did not account for rational producer price expectations or for cotton storage decisions made by either the domestic storage industry or by foreign governments. The present study outlines a two-sector (United States and foreign), partial equilibrium, cotton trade model that includes both rational producer price expectations and storage. To conduct an *ex ante* analysis, model parameters are chosen to reflect market conditions in 1995, and the model is run forward in time to 2002.

2.1 Cotton Consumption Demand

For simplicity, we assume that the annual, aggregate inverse demand function for price-taking U.S. consumers of upland cotton is linear with respect to domestic quantity demanded.[i] The linear inverse U.S. demand function is:

$$(1)\quad P_t^d = a_{d0} - a_{d1} q_t^d ,$$

where t is an annual time index, P_t^d is farm price per 480-pound bale of U.S. upland cotton in real (Producer Price Index-deflated) 1995 dollars, q_t^d is millions of 480-pound bales of U.S. upland cotton consumed domestically, and a_{d0} and a_{d1} are parameters to be estimated. Once the base-year real domestic price, P_0^d, domestic quantity, q_0^d, and price elasticity of demand for U.S. upland cotton, ε_d, are chosen, the demand parameters are calculated as:

$$(2)\quad a_{d0} = P_0^d \left(\frac{\varepsilon_d - 1}{\varepsilon_d} \right), \quad a_{d1} = \frac{-P_0^d}{\varepsilon_d q_0^d} .$$

Similar to our U.S. cotton demand specification, we assume that the annual, aggregate inverse demand function for price-taking foreign consumers of upland cotton is linear with respect to total foreign quantity demanded. The inverse linear foreign demand function is:

$$(3)\quad P_t^f = a_{f0} - a_{f1} q_t^f ,$$

where t is an annual time index, P_t^f is world price per 480-pound bale of cotton in real (Producer Price Index-deflated) 1995 dollars, q_t^f is millions of 480-pound bales foreign cotton consumption, and a_{f0}, a_{f1} are parameters to be estimated. Once the base year average real foreign price, P_0^f, aggregate foreign consumption, q_0^f, and price elasticity of aggregate foreign demand, ε_f, are chosen, the foreign demand parameters can be derived in a manner analogous to that shown in equation (2).

2.2 Cotton Storage Demand

Cotton is a storable commodity; the quality characteristics of baled cotton are relatively stable over reasonable periods of time. Cotton markets exploit storage to smooth the impact of cotton yield variations over time. Cotton storage activity in the United States is significant and variable, ranging from 20 percent to 90 percent of U.S. annual cotton production (Table 10.1).

Table 10.1 United States Cotton Production, Use and Price Data 1984-95

Crop year	Planted acres[1]	Harvested acres[1]	Yield[1]	Carry-in storage[1]	Production[1]	Imports[1]	Mill use[1]	Exports[1]	Statistical discrepancy[1]	Nominal farm price[1]	Nominal farm price[1]	Producer price index[2]
	1,000 acres		lbs/ac.[a]	million 480-lb. bales						(cents/lb.)	($/bale)	
1984	11,145	10,379	600	2.775	12.982	.024	5.538	6.215	.074	57.8	277.44	103.7
1985	10,685	10,229	630	4.102	13.432	.033	6.413	1.960	.154	56.3	270.24	103.2
1986	10,045	8,468	552	9.348	9.731	.003	7.452	6.684	.080	52.4	251.52	100.2
1987	10,397	10,030	706	5.026	14.760	.002	7.617	6.582	.182	64.3	308.64	102.8
1988	12,515	11,948	619	5.771	15.411	.005	7.782	6.148	(.165)	56.6	271.68	106.9
1989	10,587	9,538	614	7.092	12.196	.002	8.759	7.694	.163	66.2	317.76	112.2
1990	12,348	11,732	634	3.000	15.505	.004	8.657	7.793	.285	68.2	327.36	116.3
1991	14,052	12,960	652	2.344	17.614	.013	9.613	6.646	(.008)	58.1	278.88	116.5
1992	13,240	11,143	700	3.704	16.218	.001	10.250	5.201	.190	54.9	263.52	117.2
1993	13,438	12,783	606	4.662	16.133	.006	10.418	6.862	.008	58.4	280.32	118.9
1994	13,720	13,322	708	3.530	19.662	.020	11.198	9.402	.038	72.0	345.60	120.4
1995	16,931	16,007	537	2.650	17.900	.408	10.647	7.675	(.027)	76.5	367.20	124.7

[a]Acre.

[1]*Source: USDA/ASCS, Appendix Table 1. (Various years).*

[2]*Source: USBLS (1999). Producer Price Index Data, Commodities Index, Base Year 1982=100.*

Foreign cotton storage is also significant, amounting to 30 to 50 percent of foreign annual cotton production (Table 10.2). Storage activity affects current and future cotton prices, consumption, and trade. It also influences cotton price expectations, current cotton planting decisions, and expected production and storage. The dynamic links among cotton storage, current and expected prices, and current planting decisions imply that the impacts of changes in cotton policy regimes may be protracted and complex. The magnitude and complexity of cotton storage activity suggest that storage must be considered explicitly in any welfare analysis of the cotton market.

Storage activity is analogous to an extremely simple investment. The term of the investment is essentially one period, because the investment decision (the decision to store) is always reversible in the next period. If we assume a competitive[ii], risk-neutral[iii] storage industry within the United States, domestic storage firm γ treats the current domestic farm price, P_t^d, and the common (industry-wide) expectation in period t of anticipated farm price in the next period $t+1$, $E_t\left[P_{t+1}^d\right]$, as independent of the quantity actually stored by firm γ in period t, $s_{\gamma t}$. Aggregate storage by the domestic storage industry in period t is given by $S_t^d = \sum_\gamma s_{\gamma t}$. At the industry level, the (implicit) equilibrium storage behavior of the domestic storage industry must obey the following rules to eliminate opportunities for intertemporal arbitrage:

$$(4) \quad P_t^d\left(S_t^d\right) = \frac{E_t\left[P_{t+1}^d\left(S_t^d\right)\right]}{(1+r)} - k\,,\; S^d_t > 0,$$

$$(5) \quad P_t^d\left(S_t^d\right) \geq \frac{E_t\left[P_{t+1}^d\left(S_t^d\right)\right]}{(1+r)} - k\,,\; S^d_t = 0,$$

where r is the real interest rate, and k is the (constant) marginal and average cost of storage. The present value of the expected cotton price for the next period (less storage cost) must equal the current cotton price under positive storage. S_t^d is constrained to be non-negative because storage physically cannot fall below zero for the industry as a whole.

While it seems plausible to argue that the U.S. cotton storage industry is competitive, and although some foreign producers (such as Australia) may have similar competitive cotton-storage sectors, the majority of foreign production occurs in countries (such as China, India, and Pakistan) whose governments appear to intervene in the cotton-storage sector to maintain stocks-

Table 10.2 Foreign Cotton Production, Use, and Price Data 1984-95

Crop year	Harvested Acres[a]	Yield[1]	Beginning stocks[1]	Production[1]	Consumption[1]	Imports from U.S.[2]	Exports to U.S.[2]	Nominal world price[3]	Nominal world price[3]
	mil. ac.[b]	lbs/ac.	millions 480-lb bales					cents/lb	$/bale
1984	65.0	497.84	21.0	75.7	65.2	6.215	.024	–	–
1985	60.4	472.86	37.6	66.9	68.9	1.960	.033	–	–
1986	57.1	455.91	38.0	60.8	74.7	6.684	.003	–	–
1987	59.1	479.99	30.3	66.3	76.5	6.582	.002	72.26	346.85
1988	63.7	463.04	26.4	69.0	77.4	6.148	.005	66.42	318.82
1989	61.1	473.75	23.7	67.5	78.1	7.694	.002	82.34	395.23
1990	62.6	488.92	21.8	71.6	76.9	7.793	.004	82.87	397.78
1991	65.3	513.01	24.6	78.1	76.4	6.646	.013	62.90	301.92
1992	61.9	457.69	33.3	66.3	75.0	5.201	.001	56.87	272.98
1993	56.2	461.26	29.7	60.6	74.9	6.862	.006	70.75	339.60
1994	59.1	478.21	22.7	65.9	74.4	9.402	.020	92.66	444.77
1995	65.0	495.16	25.7	75.1	76.3	7.675	.408	85.55	410.64

[a] *Crop year beginning 1 August.*

[b] *Acre.*

[1] *Source: USDA/ASCS (Various years) Appendix Table 16.*

[2] *Source: USDA/ASCS (Various years) Appendix Table 1.*

[3] *U.S. Cotton, c.i.f. Northern Europe, "A Index." Source: USDA/ERS/CAD (1996) Appendix Table 19.*

to-use ratio targets. Hence, we assume that aggregate foreign stocks are expected to adjust so as to maintain a constant, aggregate foreign stocks-to-use ratio target over the duration of FAIR (1996-2002). Under this specification, the level of foreign stocks withheld from the market in period t, S_t^f, is given by:

$$(6) \quad S_t^f = \theta q_t^f,$$

where q_t^f is foreign consumption in period t, and θ is the assumed aggregate foreign-stocks to foreign-consumption target ratio (Table 10.3).

2.3 Stochastic Crop Yield

U.S. and foreign crop yields (lb/acre) are random variables driven by several exogenous factors, including weather. We assume that producers' *ex-ante* decisions regarding whether to support the FAIR Act or to continue the FACT Act in 1995 were based on knowledge of prior crop yields, market prices, target prices, and deficiency payments from 1984-95. In particular, we assume that the expected crop yield distributions for years beyond 1995 are based on past yield data. Historic data on foreign and U.S. yields from 1960-95 are provided in Figure 10.1. Notice that yields experienced a rapid increase in the decade prior to 1984, but appear to plateau from 1984 to 1995. Data on average foreign and U.S. yields for 1984-95 are provided in Tables 10.1 and 10.2. The mean yield in the United States from 1984-95 was 629.83 lb/acre with a standard deviation of 55.47 lb/acre. The mean foreign yield from 1984-95 was 478.14 lb/acre with a standard deviation of 17.80 lb/acre. We assume that domestic and foreign annual cotton yields follow distinct, independent log-normal distributions with means and standard deviations as reported above. The probability mass under each log-normal distribution is partitioned into an eleven-point discrete distribution, with the distance between points equal to one-third standard deviation. One-half of the probability mass on either side of each discrete point is allocated to the point, with the exception of the end points, which carry the additional probability mass of the matching tail of the respective distribution. The additional probability mass in any tail is less than four percent of the probability under either distribution. The value of each discrete point is expressed as a percent deviation from mean yield and denoted v^{dj}, $j = 1...11$, for the domestic distribution, and v^{fw}, $w = 1...11$, for the foreign distribution. The results of this

Table 10.1 Model Parameters

Parameter symbol	and name	Parameter value	Source
ε_d	*U.S. demand elasticity, upland cotton*	-0.3	Duffy and Wohlgenant (1991)
η_d	*U.S. supply elasticity, upland cotton*	0.3	Duffy and Wohlgenant (1991)
ε_f	*foreign demand elasticity*	-0.25	Babula (1987); Monke and Taylor (1985)
η_f	*foreign supply elasticity*	0.25	[1]
θ	*foreign stocks/use target ratio*	0.39	1984-95 average of actual stocks/use, USDA/ASCS (various years)
R	*real interest rate used for discounting*	0.03	Assuming a nominal rate of 6 percent and expected rate of inflation of 3 percent
Z	*shipping cost*	U.S. \$23.51/bale, U.S. 4.90 cents/lb	1995 dollars, USDA/ERS/CAD (1996)
q_0^d	*U.S. consumption, 1995*	10.538 million bales	USDA/ASCS (various years)
P_0^d	*U.S. farm price, 1995*	U.S. \$361.92/bale	(75.40 cents/lb), 1995 dollars, USDA/FSA (1998)
P_N^{dr}	*U.S. target price, 1992*	U.S. \$372.31/bale, U.S. 77.56 cents/lb	1995 dollars, USDA/FSA (1998)
AVC_v^d	*U.S. average variable cost, 1992*	U.S. \$235.59/bale, U.S. 49.08 cents/lb	1995 dollars, USDA/ERS (1999)
$\overline{h}_N^d$	*U.S. planned production, 1993*	17.14 million bales	Derived from USDA/FSA (1998)[2]
q_0^f	*foreign consumption, 1995*	76.3 million bales	USDA/ASCS (various years)
b_1^f	*foreign shut-down price*	U.S. \$220.80/bale	1995 dollars, based on ICAC (1998)[3]
P_0^f	*foreign farm price, 1995*	U.S. \$395.43/bale, U.S. 82.38 cents/lb	1995 dollars, the sum of P^d_0 and z
P_N^{fr}	*foreign supply-inducing price, 1992*	U.S. \$340.56/bale, U.S. 70.95 cents/lb	1995 dollars[4]
$\overline{h}_N^f$	*foreign planned production, 1993*	62.75 million bales	USDA/FAS (1998)[5]
K	*U.S. cotton storage cost*	U.S. \$23.11/bale/yr, U.S. 4.81 cents/lb/yr	1995 dollars, USDA/ERS/CAD (1996)

[1]*Representative value, based on a range of estimates of the elasticity of planted acres with respect to price in several major cotton producing countries. Assumes planted acres equal harvested acres, and assumes production equals planted acres multiplied by average foreign yield per acre. Sources: Mues and Simmons (1988), Gurkan and Kasnakoglu (1991), Coleman and Thigpen (1991), and personal communication with Hudson (1999).*

[2]*Equal to 12.98 million U.S. acres planted in 1992 multiplied by 1984-92 mean U.S. yield of 634.11 lbs/acre.*

[3]*Pakistan was lowest cost foreign producer in 1997-98, with costs equal to approximately \$1.00 per kg. (based on 1997-98 yields). The 1993 foreign shut-down price is derived from this number by adjusting for 1993 yield and deflating.*

[4]*Equal to the sum of U.S. futures price in April 1992 and average shipping cost in 1992. U.S. futures price in April 1992 is the average of the December 1992 U.S. cotton futures price for the last 5 days in April 1992, or \$317.05 per bale (1995 dollars) (Dow Jones & Co., Inc. 1993). Average shipping cost is \$23.51 per bale (1995 dollars) (USDA/CAD 1996).*

[5]*Equal to 62.75 million foreign acres harvested in 1993 multiplied by 1984-92 mean foreign yield of 478.11 lbs/acre.*

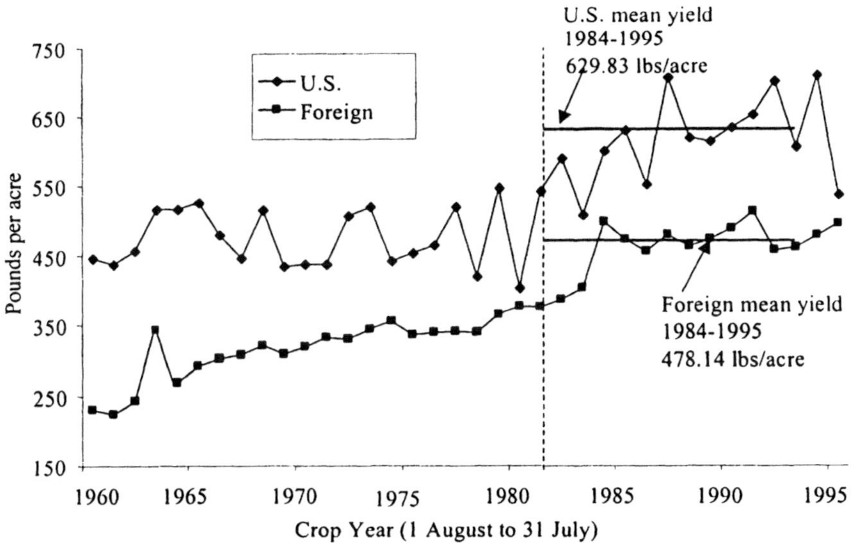

Figure 10.1 Historical Crop Yields (lbs/acre) 1960-95[iv]

log-normal transformation are presented in Figure 10.2. The probabilities associated with the v^{dj} and v^{fw} are denoted $\pi(v^{dj})$ and $\pi(v^{fw})$, respectively.

2.4 Domestic Supply

We assume that realized output of the U.S. domestic cotton industry in period t, h_t^d (measured in 480-pound bales), is the sum of expected output, $\overline{h}_t^d$ (based on expected yield), and a multiplicative error term, $v_{i,t}^d$, which is equal to the realized percent deviation of yield from its expected value in period t:

$$(7) \quad h_t^d = \overline{h}_t^d + \overline{h}_t^d \cdot v_{i,t}^d = \overline{h}_t^d \cdot \left(1 + v_{i,t}^d\right).$$

Cotton producers are assumed to choose expected output by varying acreage subject to a fixed yield distribution. The variable cost incurred by domestic cotton producers, VC^d, to produce expected cotton output $\overline{h}_t^d$ is assumed to following the simple functional form:

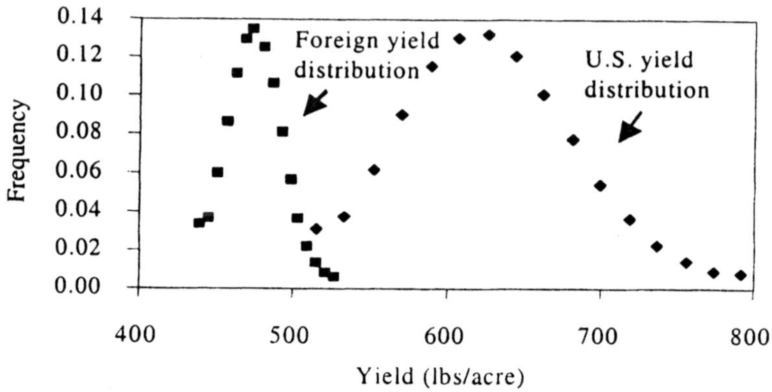

Figure 10.2 United States and Foreign Yield Distributions, 1984-95[v]

$$(8) \quad VC^d = b_1^d \cdot \overline{h}_t^d + b_2^d \cdot \left(\overline{h}_t^d\right)^{b_3^d},$$

where b_1^d, b_2^d and b_3^d are parameters to be estimated. This functional form implies that variable cost is directly related to expected quantity produced plus an adjustment for decreasing returns to scale at high levels of production. The variable cost relationship is assumed to remain constant over the period of time under consideration. Given the functional form for VC^d, average variable cost, AVC^d becomes:

$$(9) \quad AVC^d = \frac{VC^d}{\overline{h}_t^d} = b_1^d + b_2^d \cdot \left(\overline{h}_t^d\right)^{b_3^d - 1}.$$

Equation (9) implies that the shut-down price for the industry (minimum average variable cost) is simply b_1^d. Marginal cost for the industry is:

$$(10) \quad MC^d = \frac{dVC^d}{d\overline{h}_t^d} = b_1^d + b_3^d \cdot b_2^d \cdot \left(\overline{h}_t^d\right)^{b_3^d - 1}.$$

Under this specification for VC^d, MC^d equals AVC^d when $\overline{h}_t^d = 0$, and MC^d rises faster than AVC^d (assuming $b_3^d > 0$) as $\overline{h}_t^d$ increases. Since industry supply is defined as marginal cost above the point of intersection with AVC^d, industry supply under this specification is simply equal to MC^d for all $\overline{h}_t^d \geq 0$.

It is necessary to obtain a value for MC^d in order to determine the parameters in equation (10). However, marginal cost data is not readily available. We resolve this problem by assuming that domestic production expands until MC^d equals domestic producers' incentive price, $P_{t+1}^{d\ r}$, defined by:

$$P_{t+1}^{d\ r} \equiv MAX\left(P_{t+1}^{TP}, \frac{E_t\left(h_{t+1}^d \cdot P_{t+1}^d\right)}{\bar{h}_{t+1}^d} \right)$$

$$\text{(11) where } \frac{E_t\left(h_{t+1}^d \cdot P_{t+1}^d\right)}{\bar{h}_{t+1}^d} = \frac{E_t\left(\bar{h}_{t+1}^d \cdot \left(1 + v_{i,t+1}^d\right) \cdot P_{t+1}^d\right)}{\bar{h}_{t+1}^d}$$

$$= E_t\left(\left(1 + v_{i,t+1}^d\right) \cdot P_{t+1}^d\right) = \sum_i \pi(v_{i,t+1}^d) \cdot (1 + v_{i,t+1}^d) \cdot P_{t+1}^d$$

P_{t+1}^{TP} is the (pre-announced) target price in period $t+1$ (note that P_{t+1}^{TP} is identically equal to zero under FAIR), and $\frac{E_t\left(h_{t+1}^d \cdot P_{t+1}^d\right)}{\bar{h}_{t+1}^d}$ is domestic producers' expected average revenue in period t, received in period t+1 (Williams and Wright, 1991). Although each producer is a price-taker, each assumes that an average yield disturbance at the industry level will have the same proportional effect on individual output as on industry output, and, hence, expected average revenue is the same for all producers.

Equating MC^d from equation (10) with $P_{t+1}^{d\ r}$ from equation (11) and solving for $\bar{h}_t^d$ yields the following expression for expected industry supply:

$$\text{(12) } \bar{h}_t^d = \left(\frac{(P_{t+1}^{d\ r} - b_1^d)}{(b_3^d \cdot b_2^d)} \right)^{\frac{1}{b_3^d - 1}}.$$

Substituting equation (12) into the formula for elasticity of domestic supply,

$$\text{(13) } \eta_d = \frac{d\bar{h}_t^d}{dP_{t+1}^{d\ r}} \frac{P_{t+1}^{d\ r}}{\bar{h}_t^d},$$

yields the following:

$$(14)\quad \eta_d = \frac{b_3^d \cdot b_2^d + b_1^d \cdot (\bar{h}_t^d)^{1-b_3^d}}{b_3^d \cdot b_2^d \cdot (b_3^d - 1)}.$$

Using equations (9), (10), and (14) and assuming that producers adjust planned output such that MC_0^d equals $P_0^{d^r}$, all supply parameters, b_1^d, b_2^d and b_3^d, can be derived in the following sequence:

$$(15)\quad b_3^d = \frac{P_0^{d^r}}{\left(P_0^{d^r} - AVC_0^d\right)} \cdot \frac{1}{\eta_d},$$

$$(16)\quad b_2^d = \frac{\left(P_0^{d^r} - AVC_0^d\right)}{\left(b_3^d - 1\right)} \cdot \left(\bar{h}_0^d\right)^{(1-b_3^d)}.$$

$$(17)\quad b_1^d = AVC_0^d - b_2^d \cdot \left(\bar{h}_0^d\right)^{(b_3^d - 1)}$$

2.5 Foreign Supply

The development of expressions and notation for realized foreign output in period t (measured in 480-pound bales), h_t^f, expected foreign output, $\bar{h}_t^f$, and foreign yield error term, $v_{i,t}^f$ is analogous to that for domestic supply in equation (7). The expressions for foreign marginal cost, MC^f, and foreign price elasticity of supply, η_f, are the same as equations (10) and (14) except that superscript *d* is replaced with superscript *f*. However, because data are not available on the average variable cost of cotton for foreign producers, AVC^f, we must find an alternative method of determining the parameters contained in the marginal cost relationship. Given an estimate of the foreign shut-down price parameter b_1^f, and assuming foreign producers adjust planned production until MC^f equals $P_N^{f^r}$, expressions for parameters b_2^f and b_3^f may be obtained by combining and manipulating the equations for MC^f and η_f:

$$(18)\quad b_3^f = \frac{P_N^{f^r} + \eta_f\left(P_N^{f^r} - b_1^f\right)}{\eta_f\left(P_N^{f^r} - b_1^f\right)},$$

$$(19)\quad b_2^f = \frac{\left(P_N^{f^r} - b_1^f\right) \cdot \bar{h}_N^{f\left(1-b_3^f\right)}}{b_3^f}$$

2.6 Trade

Gross exports of U.S. Upland Cotton have generally exceeded gross imports by more than an order of magnitude since 1933 (Table 10.1). Hence, although the model allows for the case of domestic net imports, this discussion assumes that domestic net exports are positive. We assume that the marketing loan, loan deficiency payment, and marketing certificate provisions of the 1990 and 1996 Farm Bills ensure that net exports in period t, Z_t, adjust until an equilibrium is achieved whereby foreign consumption price per bale, P_t^f, equals domestic farm price, P_t^d, plus the average shipping cost per bale, P_z:

$$(20)\quad P_t^f\left(Z_t\right) = P_t^d\left(Z_t\right) + P_z.$$

2.7 Model Solution

Given that our model features uncertainty, forward-looking rational expectations and non-negative storage, it is difficult to deduce the competitive equilibrium solution analytically using dynamic programming, except in a limited number of special cases. Hence, we follow Williams and Wright (1991) and derive a numerical solution to the problem via the method of polynomial approximation. Williams and Wright's solution algorithm relies on the stationarity of the relationship between domestic storage in period t and expected farm price in period t+1 under an infinite planning horizon. We implement a modified version of Williams and Wright's storage algorithm with an annual time step that allows for two trading regions with idiosyncratic yield distributions and elastic supply, storage in each region, and trade between the regions. The numerical solution algorithm essentially consists of multiple nested numerical approximations that reach convergence simultaneously when the relationship between domestic storage and expected farm price achieves stationarity. The solution algorithm is described fully in Dumas and Schmitz (1999). We implement the solution algorithm via Microsoft EXCEL 97 SR-1 software (Microsoft, 1999). Given the stationary relationship between storage and expected farm price, we then run the model forward in time (given a sequence of annual realized yield disturbances) under alternative policy regimes representing FACT and FAIR.

2.8 Welfare Measures for the U.S. Cotton Market

The ultimate objective of this study is to perform a welfare analysis on the ex-ante anticipated outcomes associated with retaining the target price beyond 1995 under FACT versus eliminating the target price and moving to FAIR. After calibrating the model to market conditions in 1995, we estimate the economic welfare of three key interest groups: U.S. cotton producers, U.S. cotton consumers, and U.S. taxpayers, under both the target price policy of the 1990 FACT Act and the PFC policy of the 1996 FAIR Act, for each year from 1996 to 2002. The *ex ante* impact of FAIR on the economic welfare of each interest group is calculated as the present value (using a real interest rate of three percent) of the stream of differences in each group's anticipated economic welfare under each of the two policy regimes.

The change in the economic welfare of domestic consumers due to change in farm policy regime is measured by the change in domestic consumer surplus.[vi] The anticipated change in domestic consumer surplus in period t resulting from a change in farm policy regime is given by:

(21)
$$Change\ in\ CS_t = -\int_{P_t^{d1}}^{P_t^{d2}} q_t^d \, dP_t^d = -\left[\left(\frac{a_{d0}\cdot P_t^{d2}}{a_{d1}} - \frac{(P_t^{d2})^2}{2a_{d1}}\right) - \left(\frac{a_{d0}\cdot P_t^{d1}}{a_{d1}} - \frac{(P_t^{d1})^2}{2a_{d1}}\right)\right],$$

where CS_t denotes domestic consumer surplus in period t, and P_t^{d1} and P_t^{d2} denote farm price in period t under FACT and FAIR, respectively.

The change in the economic welfare of domestic producers due to change in farm policy is measured by the sum of the change in domestic producer surplus and changes in deficiency, loan deficiency and PFC payments received from the government. The anticipated change in domestic producer surplus in period t resulting from a change in farm policy regime is given by:

(22)
$$Change\ in\ PS_t = \left(\int_0^{h_t^{d2}(P_t^{dr2})} P_t^{d2} - MC^d \quad dh\right) - \left(\int_0^{h_t^{d1}(P_t^{dr1})} P_t^{d1} - MC^d \quad dh\right),$$

$$= \left(P_t^{d2}\cdot h_t^{d2} - b_1^d\cdot h_t^{d2} - b_2^d\cdot (h_t^{d2})^{b_3^d}\right) - \left(P_t^{d1}\cdot h_t^{d1} - b_1^d\cdot h_t^{d1} - b_2^d\cdot (h_t^{d1})^{b_3^d}\right)$$

where PS_t denotes domestic producer surplus in period t, $P_t^{d\,r1}$ and $P_t^{d\,r2}$ denote domestic producers' incentive prices in period t under FACT and FAIR, respectively, and h_t^{d1} and h_t^{d2} denote realized domestic production in period t under each farm program.

The change in domestic taxpayer welfare is calculated as the negative of the change in the sum of deficiency, loan deficiency, and PFC payments made to producers under each farm program.

The change in net domestic social welfare is calculated as the change in the sum of domestic producer and consumer surpluses, assuming all government payments are pure transfers (that is, we assume no transaction costs or costs associated with raising government funds).

3. EMPIRICAL PARAMETERIZATION

When developing their commodity storage model, Williams and Wright (1991) define the point at which domestic demand and supply curves would cross under certainty as the non-stochastic equilibrium. For expositional purposes, they take this point as given and use it to locate baseline demand and supply curves for the model. Unfortunately, such a non-stochastic equilibrium point is not easily inferred from observed data in an empirical application. We proceed by searching for market conditions under which observed market prices and quantities are most likely to reflect approximate non-stochastic equilibrium values.

We assume that both domestic and foreign consumption demand are non-stochastic and essentially stationary over the period of time under consideration (1995-2002). We fit demand curves to market conditions in 1995 (the year in which FAIR was debated) using farm price and consumption quantity data. A summary of model parameter values is presented in Table 10.3.

Unlike demand, both domestic and foreign supply are stochastic, due to stochastic variation in yields. We therefore need to identify a non-stochastic equilibrium price and quantity for each supply relationship. For domestic supply, various factors eliminate many recent years as candidates for the non-stochastic equilibrium. First, average cotton yield appeared to increase considerably over the period 1981-85 (Figure 10.1), so we eliminate this period of significant supply shift from consideration. Next, we take advantage of the existence of the target price program to minimize the influence of price expectations on supply. When farm price is significantly below target price, producers' incentive price is likely approximately equal to the target price. All years from 1986 through 1995 fit this criterion. However, when farm price is below target price and ending cotton stocks are high, the target price is even more likely to reflect producers' incentive price. High stocks put downward pressure on farm price in the following year, making it even more likely that the target price will be the relevant supply price in the coming year. Since 1985, years with relatively high cotton stocks (as characterized by stocks-to-use ratios above 22.5 percent identified by Meyer (1998) as a threshold level for significant effects of stocks on price) include 1986, 1987, 1988 and 1992. Of these years, only 1992 was covered by FACT. Hence, we choose 1992 as a year during which non-price farm program variables reflect the FACT base-case farm program and producers' incentive price, $P_N^{d^r}$, is likely well approximated by the target price. Because 1992 plantings do not produce a harvested crop until 1993, the planned production quantity in 1993, $\bar{h}_N^d$, (equal to (deterministic) 1992 planted acreage multiplied by mean yield) is chosen as the non-stochastic equilibrium quantity. Thus, we choose the 1992 target price and the planned 1993 production quantity as the domestic non-stochastic equilibrium ($P_M^{d^r}$, $\bar{h}_N^d$). U.S. cotton supply parameters (Equations 15-17) are determined using the non-stochastic equilibrium, an estimate of U.S. average variable cost of production in 1992, and an estimate of U.S. supply elasticity taken from the literature, all presented in Table 10.3.

The identification of a non-stochastic equilibrium point for foreign supply is even more problematic than it is for domestic supply, because foreign market distortions due to foreign government interventions vary across countries, making it difficult to determine a situation in which the foreign supply-inducing price is revealed. We chose the same base year, 1992-93, to represent the foreign non-stochastic equilibrium as we did to represent the domestic non-stochastic equilibrium. Because foreign markets are net importers of U.S. cotton, we assume that in non-stochastic equilibrium the foreign supply-inducing price of cotton, $P_N^{f^r}$, must equal the expected U.S. farm price (U.S. futures price) plus average shipping cost (Table 10.3). Foreign planned production in non-stochastic equilibrium, $\bar{h}_N^f$ (Table 10.3), is set equal to the 1993 level of aggregate foreign-planned production. Foreign supply parameters are derived from the non-stochastic equilibrium ($\bar{h}_N^f$, $P_N^{f^r}$), an estimate of foreign shut-down price (b_1^f), and a representative value of foreign supply elasticity (η_f) chosen by the authors from estimates in the literature, are all presented in Table 10.3.

4. MODEL SIMULATION RESULTS

Based on information available to decision makers in 1995, we use the model parameters summarized in Table 10.3 to simulate both domestic and foreign cotton production, consumption, storage, and trade for the proposed duration of FAIR (that is, years 1996-2002). We conduct two simulations, one reflects the FACT policy regime, and the other reflects policy under FAIR. Simulation results are presented in Table 10.4.

This is an ergodic model; that is, initial conditions matter. The initial conditions (aside from differences in policy parameters) are the same for each simulation and consist of the realized values of both domestic and foreign carry-in storage and production (harvest) in 1995. Given these initial conditions, the simulations determine domestic and foreign carry-out storage, consumption, trade, prices, and planting (planned harvest for 1996) in 1995. Although producers make rational planting decisions each year based on known (stationary) yield distributions (domestic and foreign), harvests in years immediately following depend upon both plantings and realized yield disturbances unknown at planting. Model results depend upon which sequences of realized yield disturbances are assumed. A Monte-Carlo analysis of yield sequence distributions is beyond the computational scope of this analysis. Instead, we set each domestic and foreign realized-yield disturbance equal to zero for each year 1996-2002. That is, we assume that both domestic and foreign yields in each year 1996-2002 will equal their respective 1984-95 mean yields. Decision-makers do not know, *ex ante*, what the realized, *ex post*, yield disturbance will be for a given year. Given exogenously specified sequences of yield disturbances (foreign and domestic), the simulations calculate all model variable values for each year 1996 to 2002.

4.1 FACT Solution Projected for 1995-2002

Under FACT (Table 10.4, upper panel), relatively small foreign carry-in stocks and moderate foreign harvests in 1994 lead to relatively small carry-out stocks and large increases in 1994 plantings for harvest in 1995. These large 1994 plantings result in relatively large domestic and foreign harvests in the initial model year of 1995. These large 1995 harvests result in relatively low farm prices, high consumption, large carry-out stocks, low exports, and low plantings in 1995.

In 1996, lower harvests result from the low 1995 plantings. However, the large stocks carried over from 1995 moderate the impacts of the lower harvests on prices and consumption levels. Because foreign harvest falls by much more than domestic harvest in 1996, domestic exports are very large in this year. Planting in 1996 for harvest in 1997 increases somewhat, but is still restrained due to remaining stocks from the large 1995 harvest.

In 1997, slightly higher harvests combined with lower carry-in stocks result in increases in price and decreases in consumption. Carry-out storage in 1997 is reduced in response to relatively high current prices in 1997, with

Table 10.4 Simulation Results Under FACT and FAIR[1]

Year	Domestic carry-in Sd t-1	Domestic harvest hd t	Domestic consumption qd t	Domestic farm price Pd t	Expected domestic farm price Et[Pt + 1(Sd t)]	Domestic exports Z t	Foreign carry-in Sd t-1	Foreign harvest hf t	Foreign consumption qf t	Foreign farm price Pf t
	million bales			$/bale	$/bale	million bales				$/bale
					FACT results					
1995	2.6	17.5	10.3	293.0	325.6	3.0	26.0	75.1	74.9	316.5
1996	6.8	16.8	10.2	302.5	335.4	12.3	29.0	62.0	74.3	326.0
1997	1.1	16.9	9.9	333.0	342.7	8.1	29.0	63.7	72.5	356.5
1998	0.0	17.0	9.8	348.0	349.6	7.2	28.0	64.3	71.6	371.6
1999	0.0	17.0	9.8	348.0	349.6	7.2	28.0	64.3	71.6	371.6
2000	0.0	17.0	9.8	348.0	349.6	7.2	28.0	64.3	71.6	371.6
2001	0.0	17.0	9.8	348.0	349.6	7.2	28.0	64.3	71.6	371.6
					FAIR results					
1995	2.6	17.5	10.3	294.4	327.1	2.9	26.0	75.1	74.8	317.9
1996	6.9	16.3	10.2	305.1	338.0	11.2	29.0	62.8	74.2	328.6
1997	1.8	16.6	10.0	328.4	347.8	8.4	29.0	63.8	72.8	351.9
1998	0.0	16.8	9.7	349.3	350.8	7.1	28.0	64.4	71.5	372.8
1999	0.0	16.8	9.7	349.3	350.8	7.1	28.0	64.4	71.5	372.8
2000	0.0	16.8	9.7	349.3	350.8	7.1	28.0	64.4	71.5	372.8
2001	0.0	16.8	9.7	349.3	350.8	7.1	28.0	64.4	71.5	372.8

[1] *Assumes realized yield each year 1996-2002 equals mean yield 1984-95*

Source: Authors' calculations.

domestic carry-out storage essentially falling to zero. Trade is moderate. With the rebound in farm price in 1997 and low 1997 carry-out, expected 1998 farm price is relatively high, and domestic and foreign 1997 plantings increase in anticipation.

The economy reaches equilibrium in 1998, given the continuing stream of realized yields equal to mean yields. The equilibrium is characterized by moderately high domestic harvest, consumption, and exports, with little domestic storage and a realized domestic farm price of U.S. $348 per bale (72.5 cents/lb), which is slightly below the 1995 domestic target price of U.S. $349.92 per bale (72.9 cents/lb). The equilibrium domestic farm price is significantly above the 1995 loan rate of U.S. $249.22 per bale (51.92 cents/lb). Foreign harvest and consumption are moderate in equilibrium, with foreign storage meeting the (exogenous) foreign stocks-to-use ratio target. The difference between foreign farm price and domestic farm price reflects the per-bale shipping cost for cotton.

It is important to emphasize that the model does not predict that the cotton market generally will reach equilibrium. Rather, only under the particular sequence of domestic and foreign yields assumed in this simulation will the cotton market reach the particular equilibrium described by the results in Table 10.4. However, from the perspective of policymakers in 1995, some sequence of realized yield disturbances must be assumed over the 1996-2002 period to simulate alternative policy outcomes; the reported simulations utilize the straightforward assumption that realized yields in each of these years will equal mean yields from the recent past (1984-95).

4.2 FAIR Solution Projected for 1995-2002

Under the FAIR policy simulation (Table 10.4, lower panel), the same yield sequence is considered under the same 1995 carry-in and harvest initial conditions. However, under FAIR, the target price is removed. Even though FAIR is not implemented until 1996, anticipation of its implementation influences (via rational expectations) storage and planting decisions in 1995. Compared with FACT simulation results, removal of the target price under FAIR reduces domestic plantings in 1995. Reduced plantings in anticipation of FAIR raises 1995 expectations of 1996 prices relative to expectations under FACT. Higher price expectations increase carry-out in 1995, relative to carry-out under FACT. Increased carry-out in 1995 reduces the quantity of cotton available for consumption in 1995, raising 1995 prices and reducing both domestic and foreign consumption and trade in that year. Finally, the increase in expected 1996 prices increases foreign plantings in 1995, relative to foreign plantings in 1995 under FACT.

In 1996 under FAIR, larger carry-in, smaller domestic harvest, and larger foreign harvest result in a net increase in the quantity of available cotton relative to availability under FACT. However, because price expectations for 1997 are higher under FAIR in the absence of a target price, carry-out storage in 1996 is significantly higher under FAIR than it is under FACT. As a result, both domestic and foreign consumption are lower in 1996 under FAIR. The effects of higher 1997 price expectations and 1996 carry-out under FAIR outweigh the effect of higher cotton availability in 1996 under FAIR, resulting in higher 1996 farm prices under FAIR. In the absence of the domestic target-price support under FAIR, domestic plantings are lower in 1996 and foreign plantings are higher, resulting in a moderate overall increase in planting relative to FACT.

In 1997 under FAIR, higher foreign harvest and much higher domestic carry-in outweigh lower domestic harvest to produce a larger quantity of available cotton relative to cotton availability under FACT. Because expectations of 1998 prices are higher in the absence of target prices under FAIR, carry-out storage in 1997 is higher under FAIR than it is under FACT. However, in contrast to 1996, larger carry-out storage in 1997 under FAIR is not sufficient to overcome the effect of larger carry-in storage in that year, and 1997 farm prices are lower under FAIR as a result. With lower prices in 1997 under FAIR, both domestic and foreign consumption are larger relative to consumption levels under FACT, again in contrast to results for 1996. Trade is moderately higher in 1997 under FAIR.

As under FACT, the economy reaches equilibrium in 1998 under FAIR, given the assumed sequence of realized yield disturbances. However, equilibrium under FAIR is characterized by lower domestic harvests, higher foreign harvests, lower domestic and foreign consumption, higher expected and realized prices, and lower trade, relative to FACT. Equilibrium storage is similar under the two policies.

4.3 Welfare Comparisons of FACT and FAIR

The anticipated (*ex ante*) changes in the economic welfare of domestic cotton producers and consumers, as well as the anticipated changes in U.S. agricultural support payments due to a change in cotton policy from FACT to FAIR are presented in Table 10.5. Results are presented for each year from 1995-2002 in terms of the anticipated change in each listed variable caused by a decision made in 1995 to change cotton policy from FACT to FAIR, beginning in 1996.

The anticipated change in domestic consumers' economic welfare is measured by the change in domestic consumers' surplus (Table 10.5, column 3). Given stationary demand, the anticipated change in domestic consumers'

Table 10.5 Domestic Welfare Results: Moving from FACT to FAIR[a]

Year	*Change in domestic farm price*	*Change in domestic farm price*	*Change in domestic consumer surplus*	*Change in domestic producer surplus*	*Change in domestic deficiency payments*	*Change in domestic PFC payments*	*Change in domestic producer welfare*	*Change in domestic government payments*	*Change in domestic social welfare*
	1995 cents/lb	*1995 $/bale*	*1995 U.S. million dollars*						
1995	0.31	1.5	-15.2	25.9	-998.5	0.0	-972.6	-998.5	10.6
1996	0.52	2.5	-26.0	58.9	-797.2	649.6	-88.7	-147.6	32.9
1997	-0.96	-4.6	45.6	-72.3	-285.9	626.4	268.3	340.5	-26.6
1998	0.25	1.2	-11.8	21.5	-31.8	672.8	662.5	641.0	9.7
1999	0.25	1.2	-11.8	21.5	-31.8	649.6	639.3	617.8	9.7
2000	0.25	1.2	-11.8	21.5	-31.8	591.6	581.3	559.8	9.7
2001	0.25	1.2	-11.8	21.5	-31.8	475.6	465.3	443.8	9.7
2002	0.25	1.2	-11.8	21.5	-31.8	464.0	453.7	432.2	9.7
		NPV's[b]:	-46.9	104.5	-2115.7	3592.1	1580.9	1476.4	57.6

[a] *Assumes realized yield each year 1996-2002 equals mean yield 1984-95.*

[b] *Assumes real interest rate r = 0.03.*

surplus depends upon the anticipated change in the domestic market price (Table 10.5, columns 1 and 2). Moving from FACT to FAIR increases the anticipated domestic market price in all years from 1995-2002, except for 1997. As a result, anticipated domestic consumers' surplus is lower under FAIR each year, except in 1997. The net present value (assuming a real interest rate of 3 percent) of the anticipated changes in domestic consumers' surplus over all years—1995-2002—is U.S. –$46.9 million.

The change in anticipated domestic producers' economic welfare (Table 10.5, column 7) is measured by the sum of the anticipated changes in domestic producers' surplus net of government payments (Table 10.5, column 4), deficiency payments (Table 10.5, column 5) and PFC payments (Table 10.5, column 6). Moving from FACT to FAIR increases anticipated domestic producers' surplus in all years from 1995-2002, except for 1997, due to higher anticipated domestic prices under FAIR in all years except 1997. The net present value of the anticipated changes in domestic producers' surplus over all years from 1995-2002 is U.S. $104.5 million. Moving from FACT to FAIR decreases anticipated deficiency payments in all years from 1995-2002. The net present value of the anticipated changes in deficiency payments over all years from 1995-2002 is U.S. –$2.1157 billion. Moving from FACT to FAIR introduces anticipated PFC payments in all years from 1996-2002. The net present value of the anticipated changes in PFC payments over all years from 1995-2002 is U.S. $3.59 billion. Moving from FACT to FAIR decreases anticipated producer welfare in 1995 and 1996, due to large anticipated reductions in deficiency payments under FAIR relative to anticipated PFC payments for these years. However, moving from FACT to FAIR increases anticipated producer welfare in years 1997-2002, due to small, anticipated reductions in deficiency payments under FAIR relative to anticipated PFC payments for these years. The net present value of the anticipated changes in producers' welfare over all years from 1995-2002 is U.S. $1.58 billion.

The change in anticipated total domestic government cotton support payments (Table 10.5, column 8) is measured by the sum of the anticipated changes in deficiency payments (Table 10.5, column 5) and PFC payments (Table 10.5, column 6). Moving from FACT to FAIR decreases anticipated total government payments in 1995 and 1996, but increases anticipated payments in years 1997-2002. The net present value of the anticipated changes in total domestic cotton support payments over all years from 1995-2002 is U.S. $1.48 billion.

Under the assumption that government cotton support payments are pure transfers with no associated transaction costs, the anticipated change in domestic social welfare (Table 10.5, column 9) is measured by the sum of the anticipated changes in domestic consumers' surplus and domestic producers' surplus. Moving from FACT to FAIR increases anticipated domestic social welfare in all years from 1995-2002, except for 1997. The net present value of the anticipated changes in domestic social welfare over all years from 1995-2002 is U.S. $57.6 million.

5. DISCUSSION AND CONCLUSION

Questions have been raised regarding the gains to producers under FAIR. When crop prices are high, taxpayer advocates argue that growers are over-compensated by PFC payments, because deficiency payments would have been low, or non-existent, under prior farm legislation. When commodity prices are low, farmers argue that they would have been better off receiving deficiency payments than they would have been receiving PFC payments under FAIR. Of course, these arguments are based on hindsight, and offer no answers to the question of whether or not the adoption of FAIR was a good *ex ante* decision from the perspectives of various interest groups.

Our analysis shows that the adoption of FAIR was a good *ex ante* decision for cotton growers. At the time adoption was being debated in 1995, farmers would have expected to receive U.S. $1.58 billion more in net present value of producer surplus and government payments under FAIR over the seven-year life of the program. Additional producer surplus due to higher prices under FAIR (caused by decreased output under inelastic demand) would account for U.S. $104.5 million of this amount. An excess of anticipated PFC payments under FAIR, above anticipated deficiency payments under FACT, accounts for the remainder. Because anticipated market price never falls below the loan rate in our simulations, differences in loan deficiency payments are not factors when determining the gains of adoption to producers.

In contrast to the producer gains of adopting FAIR, cotton consumers would have expected, in 1995, to lose roughly U.S. $47 million in net present value of consumer surplus over the seven-year life of the program due to higher anticipated cotton prices under the program. Taxpayers also would have expected to lose under FAIR. The net present value of the anticipated increases in government cotton program payments over all program years is U.S. $1.48 billion.

Despite the anticipated increase in government payments under FAIR, if these payments are treated as pure transfers, then anticipated overall domestic welfare (the sum of anticipated changes in consumer and producer surpluses) increases by U.S. $57.6 million under FAIR. This result is due to the decoupling of program payments from production under FAIR and the anticipated reduction in overproduction that would have occurred under FACT target prices.

It is important to note that, due to the presence of storage, these anticipated welfare values depend on the anticipated paths through time of realized domestic and foreign yields, even though anticipated yield distributions (one domestic distribution and one foreign distribution) are assumed stationary over time. Here, we have made the reasonable assumption that anticipated realized yields each year through 2002 equal mean yields. However, different assumptions regarding the anticipated paths of realized yields produce qualitatively different results. For example, simulation results not reported here show that the assumption of anticipated paths of realized yields, which are one standard deviation higher than mean yields, results in anticipated producer losses and taxpayer savings from adopting FAIR (Dumas and Schmitz, 1999). Future work might employ Monte Carlo-type analyses to calculate expected outcomes over all possible anticipated paths of realized yields.

Results also depend upon the initial, carry-in levels of domestic and foreign storage in 1995. If carry-in storage had been higher (or lower) in 1995, the anticipated changes in economic welfare for each interest group would have been different, even with the same underlying demand and supply curves, and with the same yield distributions and anticipated paths of realized yields. This means that the relative desirability of each policy regime to each interest group, at least in terms of anticipated economic welfare, depends on the year in which the policy choice is made. A policy choice that may be rational for a particular interest group in years with high carry-in may not be rational for that group in years with low carry-in.

Although the present study focuses exclusively on cotton, the methodology developed here could be adapted to investigate the effects of FAIR upon other agricultural commodities. In addition, similar *ex ante* techniques could be used to analyze the anticipated welfare implications of future farm policy reform efforts.

6. ENDNOTES

[i] Welfare measures are known to be sensitive to demand curvature (Williams and Wright, 1991: 106–109), which depends in part on functional form. However, obtaining model estimates for an alternative demand specification requires considerable computational time (approximately three, 24-hour days running 10 Pentium II computers simultaneously) and is beyond the scope of this study. For analysis of the sensitivity of model results to demand specification see Dumas and Schmitz (1999).

[ii] Since the advent of the marketing loan/loan-deficiency payment policy in 1985, the U.S. Commodity Credit Corporation (CCC) has owned a relatively small proportion of U.S. cotton stock.

[iii] Alternatively, we could assume the existence of a futures market with risk-neutral speculators having rational expectations and negligible capital constraints, in which case the risk attitudes of storers would be reflected by their participation in the futures market and would not affect storage behavior.

[iv] Source: USDA/ASCS, Appendix Tables 1 and 16. Various years.

[v] Source: Authors' Calculations based on data taken from USDA/ASCS, Appendix Tables 1 and 16. Various years.

[vi] In this study, we are interested in the domestic U.S. cotton market. The analysis of anticipated changes in foreign welfare resulting from FAIR is beyond the scope of this chapter.

7. REFERENCES

Babula, R.A. 1987. "An Armington Model of U.S. Cotton Exports." *Journal of Agricultural Economics Research* 39(4): 12–22.

Coleman, J., and M.E. Thigpen. 1991. "An Econometric Model of the World Cotton and Non-Cellulosic Fibers Markets." World Bank Staff Commodity Working Paper No. 24, Washington, D.C.

Dow Jones & Company, Inc. 1993. *Wall Street Journal*. Futures Price listings. New York, NY. (April).

Duffy, P.A., and Wohlgenant, M.K. 1991. "Effects of an Export Subsidy on the U.S. Cotton Industry." *Southern Journal of Agricultural Economics* 23 (December): 1–7.

Dumas, C. and T.G. Schmitz. 1999. "*Ex ante/Ex post* Analysis of U.S. Agricultural Policy Reform: A Model of Storage with International Trade under Rational Expectations." Arizona State University Working Paper #99-05, Morrison School of Agribusiness and Resource Management, Mesa, AZ.

Gurkan, A.A. and H. Kasnakoglu. 1991. "The Political Economics of Agricultural Price Support in Turkey: An Empirical Assessment." *Public Choice* 70:277–298.

Hudson, D. 1999. Personal Communication. Department of Agricultural Economics, Mississippi State University, Starkville, MS (21April).

ICAC. (International Cotton Advisory Committee). 1998. "Survey of the Cost of Production of Raw Cotton." Washington, DC (October).

Meyer, L.A. 1998. "Factors Affecting the U.S. Farm Price of Upland Cotton." Special article in: *Cotton and Wool Yearbook 1998*. Washington, DC: ERS/USDA (December).

Microsoft. 1999. Product information available from web site: http://www.microsoft.com.

Monke, E.A., and L.D. Taylor. 1985. "International Trade Constraints and Commodity Market Models: An Application to the Cotton market." *Review of economics and Statistics* 67(1): 98–107.

Mues, C., and P. Simmons. 1988. "The Effects of US Farm Policy on Australian Cotton Revenues." Paper presented at 32nd Annual Conference of Australian Economics Society, La Trobe University, Melbourne, Australia (8–12 February).

Schmitz, T.G., A. Schmitz, and C. Dumas. 1997. "Gains from Trade, Inefficiency of Government Programs and the Net Economic Effects of Trading." *Journal of Political Economy* (April): 637–48.

USBLS (U.S. Bureau of Labor Statistics). 1999. Producer Price Index data, Commodities Index, 1982 = 100. Internet website: http://146.142.4.24/cgi-bin/surveymost?wp (February)

USDA/ASCS (U.S. Department of Agriculture, Agricultural Stabilization and Conservation Service). Various years. *Cotton and Wool Yearbook*. Washington, DC.

USDA/ERS (U.S. Department of Agriculture, Economic Research Service). 1999. "U.S. Cotton Production Costs, 1975–97." Washington DC. Internet website: http://www.econ.ag.gov/briefing/fbe/car/data/table17.wk1

_____. 1996. "1996 FAIR Act Frames Farm Policy For 7 Years." *Agricultural Outlook Supplement*. Washington, DC (April).

_____. 1995. *Cotton: Background for 1995 Farm Legislation*. By E.H. Glade, Jr., L.A. Meyer, and S. MacDonald. Agricultural Economic Report No. 706. Washington, DC.

USDA/ERS/CAD (U.S. Department of Agriculture, Economic Research Service, Commercial Agriculture Division). 1996. *The Cotton Industry in the United States*. E.H. Glade, Jr., L.A. Meyer, and H. Stults, eds., Agricultural Economic Report No. 739. Washington, DC. (July).

USDA/ERS/CED (U.S. Department of Agriculture, Economic Research Service, Commodity Economics Division). 1989. *Cotton: Background for 1990 Farm Legislation*. by H. Stults, E.H. Glade, Jr., S. Sanford, and L.A. Meyer. Commodity Economics Division. Staff Report No. AGES 89-42, Washington, DC.

USDA/FAS (U.S. Department of Agriculture, Foreign Agricultural Service). 1998. *Cotton: World Markets and Trade*. Washington, DC. Internet website: http://www.fas.usda.gov/cotton/circular/1998/98-11/toc.htm.

USDA/FSA (U.S. Department of Agriculture, Farm Service Agency). 1998. *Commodity Fact Sheet: Upland Cotton—Summary of 1997 Commodity Loan and Payment Program*. Washington, DC (September).

_____. 1997. *Fact Sheet: Production Flexibility Contract Data—Fiscal Year 1997*. Washington, DC: USDA/FSA (November).

Williams, J.C., and B.D. Wright. 1991. *Storage and Commodity Markets*. New York, NY: Cambridge University Press.

Chapter 11

Import Rules for Foot-and-Mouth Disease Contaminated Beef

Philip L. Paarlberg and John G. Lee
Purdue University

1. INTRODUCTION

To satisfy the Uruguay Round agreement, the United States proposed a region-based system of sanitary import rules for meat and livestock, whereby, as the risk of importing a disease rises, increasingly strict monitoring, certification, and quarantine rules are triggered. Paarlberg and Lee (1998) present an alternative policy in which the level of a trade barrier is linked directly to the health risk associated with imports. To illustrate how the risk of importing products containing Foot-and-Mouth Disease (FMD) affects the welfare-optimizing tariff, we use a simple model of trade in beef. The level of a welfare-maximizing tariff imposed by the United States is affected both by the risk taken with beef that contains FMD as it moves from one country to another and by the expected loss in U.S. national beef output. Greater risks of importing beef that contains FMD result in larger tariffs against the exporter of contaminated beef. Also, increased expected U.S. output losses cause increased tariffs. When the risk of importing FMD is low and the expected output loss is slight, the discriminatory tariff due to FMD is negligible. When the risk of importing FMD is high and the expected output loss is great, then the discriminatory tariff is prohibitive.

This chapter expands the theoretical and empirical models used by Paarlberg and Lee (1998) by treating output loss as a random variable, and it links imported quantities to the probability of experiencing an FMD outbreak. Whereas our previous model in 1998 considered only two exporting countries, a clean exporter and an infected exporter, this model adds multiple exporting regions each with a different probability of exporting FMD to the United States. The result is a vector of barriers levied against exporting regions that reflect differing probabilities of exporting FMD to the United States.

We begin this chapter with an overview of the U.S. import rules proposed in 1996 for FMD. These rules identify essential components included in the model. Subsequently, a conceptual trade model is developed that incorporates some of the proposed U.S. rules. The conceptual model generates rules for setting import barriers as an alternative policy to the quarantine system, and forms the basis of the empirical model and its results.

2. PROPOSED U.S. RULES ON TRADE FOR NATIONS WITH FOOT-AND-MOUTH DISEASE

In the Uruguay Round under the General Agreement on Tariffs and Trade (GATT), a new Sanitary and Phytosanitary (SPS) Agreement was signed. That agreement requires nations to base the SPS rules on scientific evidence and acceptable risk criteria and to recognize disease-free regions. To comply with the new SPS agreement, in 1996 the United States circulated proposed changes in its import rules for livestock and meat for comment.[i]

Whereas under the old U.S. policy nations were classified as either free or not free of FMD, the proposed rules classify nations and regions according to one of six risk levels. The categorization of an FMD risk in a nation or region can be accomplished using either qualitative criteria or a quantitative assessment of the probability of FMD appearing in the national livestock herd. Five categories of risk have probability thresholds. A region is assumed to have negligible risk (RN) if fewer than one out of one million live animals in that region are expected to become infected. Regions of slight risk (R1) are those in which fewer than one out of 100 thousand animals might be infected. Low-risk regions (R2) have a probability of one out of 10 thousand; moderate-risk regions (R3) have a probability of one out of one thousand; and high-risk regions (R4) have a probability of one out of one hundred. A sixth category is designated for regions with unknown risk (RU) and this category is treated as a high-risk (R4) category.

Higher categories result in increasingly more severe trade restrictions. Imports of animals and meat from RN regions face few restrictions. Live animals must be inspected, but face neither quarantine nor testing restrictions. Fresh, chilled, or frozen meat can be imported into the United States with few restrictions. Animals from R1 regions must be certified that they were born and raised in RN or R1 regions, and must report a negative disease test within 30 days of export to the United States. Fresh, chilled, or frozen meat can be brought into the United States with a similar origin certificate. Animals from regions in higher risk categories face increasingly severe pre-export and post-export quarantines. In addition, they have rules on the length of time that the area in which the animal resided has to have been free of disease, and the distance a disease-free buffer zone is surrounding the area. For meat, rules are added on processing—removal of bone, blood clots, and lymphoid tissue—and on storage temperature, time, and pH levels. Meat from R4 and RU regions retains the cooking requirement of the previous legislation (Section 306 of the Tariff Act of 1930).

3. INCORPORATING THE PROPOSED RULES INTO A TRADE MODEL

This section develops a partial equilibrium trade model for beef to reflect some of the proposed U.S. import rules for FMD. The resulting policy rules are not an additional layer of protection, rather, they are alternatives to the present system. To keep the theoretical and empirical models tractable, the analysis involves only beef while it focuses on alternative policy rules. Hence, it is assumed that trade in live animals is so small that it can be ignored. Therefore, linkages to other sectors such as swine, which are also affected by and can transmit FMD, are ignored. Beef is assumed to be a homogeneous product, except for the FMD risk, which is known by customs officials.[ii] Customs officials are able to identify the origin of the beef, but once it clears customs, its national identity is lost. As this is a partial equilibrium model, the prices of other goods and income are exogenous and these factors are not shown in the model presented below to keep the notation simple. All regions are treated as large regions in the sense that their actions can potentially affect the world price of beef. Construction of the empirical model will show the validity of this assumption.

The world beef market is divided into six regions. Five regions are net exporters whose behaviors are described by excess supply functions:[iii]

$$(1)\quad x_i = x_i\left(p_i\right); \partial x_i / \partial p_i \geq 0 \; \forall i = 1, \cdots 5,$$

where x_i denotes exports of beef by region i, and p_i denotes the price of beef in region i. The sixth region, the United States, is a net importer. Demand for beef in the United States is given by,

$$(2)\quad D_u = D_u\left(P_u\right); \partial D_u / \partial P_u \leq 0,$$

where D_u denotes U.S. beef demand and P_u denotes the U.S. beef price. The supply of beef in the United States, S_u, depends on the price of beef in the United States, and on a variable capturing the severity of an FMD outbreak on beef output, α:

$$(3) \quad S_u = S_u(P_u, \alpha); \partial S_u / \partial P_u \geq 0; \partial S_u / \partial \alpha \leq 0.$$

The severity of an FMD outbreak, α, is treated as having a distribution with a mean, μ, and a variance, σ^2, that is independent of the level of imports. That is, while more imports increase the chance that an outbreak will occur, the severity of a given outbreak is not affected by the total level of imports. Beef imports by the United States, M_u, are the difference between domestic demand and supply:

$$(4) \quad M_u = D_u - S_u$$

Global market clearing is determined by quantity and by price linkages. Imports of beef by the United States must equal exports of beef by the five exporting regions:

$$(5) \quad M_u = \sum_i x_i$$

Although U.S. customs officials know the origin of the beef entering the United States and its FMD risk, U.S. consumers of beef are assumed to be unaware of the origin of the beef they consume. Thus, beef in the United States has a common price regardless of its origin. Because customs officials can identify the national origin given the absence of smuggling or arbitrage to avoid import barriers, the United States has the ability to set discriminatory barriers specific to the different exporting regions.

In the theoretical model, the total trade barrier consists of two parts. The first part is a traditional specific tariff, t_i, levied on imports from country i.[iv] The second component is a premium based on the FMD risk represented by the exporting region, R_i. This premium is similar to an insurance premium in the sense that it compensates the importing country for the FMD risk it incurs by permitting imports from regions with FMD risk. Thus, the price linkages in the model are:

$$(6) \quad p_i = P_u - t_i - R_i; \ \forall i = 1, \cdots 5$$

The five exporting regions correspond to the risk categories (note that the R4 and RU categories are combined since, under the U.S. proposal, RU regions face restrictions equivalent to R4 regions). For each of these categories there is a known probability of a single contaminated metric ton[v] or animal (on a meat basis) crossing the U.S. border. As the quantity of U.S. imports from a particular region with FMD increases, the probability that a contaminated shipment has crossed the border, π_I, also increases:

$$(7) \quad \pi_i = \pi_i(x_i); \partial \pi_i / \partial x_i > 0; \forall i = 1, \cdots 5$$

The probability that an outbreak occurs in the United States from any source, π, is the cumulative effect of the probabilities from the five exporting regions:

$$(8) \quad \pi = (\pi_1, \pi_2, \pi_3, \pi_4, \pi_5)$$

Thus, 1-π is the probability that no outbreak occurs in the United States.

To determine the vector of U.S. barriers levied against the various exporting regions, a welfare function for the United States must be established. This model assumes the goal of U.S. policy is to maximize its consumer and producer welfare plus the tariff and premium revenues less the cost of an FMD control program. Welfare, *G*, in this partial equilibrium model is defined by consumer and producer surplus:[vi]

$$(9) \quad G(P_u, \alpha) = \int_p^k D_u(P_u) dP_u + \int_0^p S_u(P_u, \alpha) dP_u.$$

The tariff and premium revenue, *TR,* is the sum of the specific tariff and premium levied against each exporting region multiplied by the quantity exported to the United States from that region and is assumed to be redistributed by the U.S. Government:

$$(10) \quad TR = \sum_i (t_i + R_i)$$

The cost of FMD control depends on the type of program adopted and the severity of the outbreak. The U.S. Government exogenously determines the type of program used to manage FMD outbreaks—no control, quarantine and slaughter, and vaccination. There is a multiple interaction between the outbreaks and the control program. The different control programs affect the distribution function and alter the expected magnitude of an outbreak. At the same time, more severe outbreaks, given as an increase in α, result in an increased cost for an FMD control program. The cost of type $t(C_t)$ is expressed as:

$$(11)\quad C_t = C_t(\alpha_t); \partial C_t / \partial \alpha_t \geq 0.$$

Given the components of expected U.S. welfare, equations (7) to (11), the expected welfare function under control program $t(W_t)$ is:

$$(12)\quad \begin{aligned} W_t =& \left[1-\pi\left(\pi_1\{x_1\},\cdots\pi_5\{x_5\}\right)\right]\left[G_u(P_u)+\sum_i(t_i+R_i)x_i\right]+ \\ & \pi\left(\pi_1\{x_1\},\cdots\pi_5\{x_5\}\right)\left[G_u(P_u,\alpha)+\sum_i(t_i+R_i)-C_t(\alpha_t)\right]. \end{aligned}$$

Maximizing the U.S. welfare and solving for the tariff and the risk premium facing an individual exporting region, *i*, gives:

$$(13)\quad t_i = x_i / (\partial x_i / \partial p_i)$$

$$(14)\quad R_{it} = -\left[G_u(P_u,\alpha_t)-G_u(P_u)-C_t(\alpha_t)\right]\left(\partial\pi / \partial\pi_i\right)\left(\partial\pi_i / \partial x_i\right).$$

The standard optimal tariff expression in a partial equilibrium model is expression (13) where the slope of the excess supply determines the tariff, t_i. The additional premium for region *i* due to the FMD risk is shown by expression (14) which consists of the expected welfare change due to an FMD outbreak multiplied by the probability of an outbreak being caused by imports from region *i*.[vii] Note that the risk premium facing country *i* depends upon the type of control program, *t*. An FMD outbreak results in a loss of meat supplies causing the price to increase. Consumers experience an unambiguous welfare loss. In the case of producers, there is a welfare loss due to the lost output, but there is a gain due to the higher price. That gain is part of the consumer surplus loss so that the United States, as a whole, suffers a welfare loss in the presence of an outbreak. Thus, $G_u(P_u, \alpha_t) < G_u(P_u)$. If the probability of a region causing an outbreak is zero—as might be the case for an RN region—then no premium on the tariff is added. As the probability of a region causing an outbreak increases, so does the premium. Thus, regions with greater risks face larger total barriers. The larger barriers shift the mix of imports away from the more risky regions.

4. EMPIRICAL MODEL

The empirical model uses the risk classification proposed by the U.S. Department of Agriculture's Animal and Plant Health Inspection Service (APHIS) in 1996 to categorize regions into five risk categories: negligible, slight, low, moderate, and high/unknown. The behavioral equations are assumed to be linear and the elasticities used to calibrate the equations to the 1994-96 base period are from the static world policy simulation (SWOPSIM) model used by the U.S. Department of Agriculture to evaluate the impacts of the Uruguay Round (Sullivan, Wainio, and Roningen, 1989). To determine the slopes and constants for the linear equations, the world beef trade is divided into two types of beef. The clean market consists of the United States and RN regions; the model assumes a price of U.S. $2,000 per ton for those regions. The other type of beef is infected beef traded by the other four regions; the model assumes that the excess supply of infected beef to the United States equals zero with a price of U.S. $1,500 per ton.

As shown by equations (13) and (14), the other key components to determine the welfare-optimizing barriers facing each exporting region are the risks associated with imports of beef from each region, the expected loss in welfare if an FMD outbreak occurs, and the cost of the control program. The first component, the probabilities of FMD appearing in livestock herds, are given for each risk category by APHIS; they range from one out of one million animal units to one out of 100 animal units. In addition, an adjustment must be made for the risk that the commodity remains infected after storage, handling, and processing for export—what APHIS terms the commodity risk. The proposed rules do not give specific commodity risks. Because cooking can destroy FMD and cooked beef may already be imported from regions with FMD, the situation considered here is concerned with fresh beef. While removal of lymphoids and de-boning lowers the commodity risk of fresh beef, without additional information, a commodity risk of one out of 1,000 is used. This means that beef is assumed infected with a probability equal to that of FMD appearing in the herd multiplied by 0.001, and these risks are used for the term, $\partial\pi_i/\partial x_i$, in each equation (14). As a result of this assumption, the premia found in the numerical results may be overstated because they ignore steps that reduce the risk of FMD transmission below one in 1,000. This model also assumes that the probability of an outbreak in the

United States is the sum of the probabilities of obtaining FMD from each risk category.

The second component is the expected loss in U.S. welfare in the event of an FMD outbreak. In this analysis, the scenarios for alternative outbreak patterns over a 15-year period developed by Aulaqi and Sundquist (1979) are used. They consider three alternative FMD control strategies. The first scenario, when FMD is assumed to be endemic with no public control, FMD epidemics affecting 40 percent to 75 percent of susceptible livestock occur every five to 10 years. Thus, for their 15-year period, they experience three epidemics with the first outbreak being the most severe. Between these epidemics, the infection rate drops below 1.0 percent. The average infection rate is 6.3 percent over the 15-year period. If a compulsory vaccination program is introduced following the introduction of FMD, Aulaqi and Sundquist (1979) report an average infection rate of 0.2 percent. The third scenario is eradication in which animals are quarantined and slaughtered following an outbreak until the disease is eliminated. Based on the British experience of an FMD outbreak in 1966 and 1967, Aulaqi and Sundquist (1979) estimate that about 1.0 percent of U.S. livestock would be lost.

While FMD is not always fatal to livestock or is not infectious to humans, this analysis assumes that all livestock stricken with FMD are destroyed and their meat is not marketed. To find the decline in U.S. beef output, the average U.S. cattle and calf inventory for the base period of 103.22 million head is multiplied by the assumed losses of 6.3 percent, 0.2 percent, and 1.0 percent. These animals are converted to beef lost by multiplying by the ratio of beef production to animals slaughtered in the base period. The resulting beef output losses are 2.114 million tons under the endemic scenario; 0.0671 million tons under the compulsory vaccination program; and 0.3355 million tons under the eradication program.

The final term to determine in equation (14) is the cost of the different types of control programs. In all three cases, beef producers are assumed to be compensated for the livestock lost because the producer surplus loss is assumed to be paid by taxpayers. For the policy of eradication, additional administrative and operating costs are incurred. Aulaqi and Sundquist (1979) analyze these costs for the late 1970s. Inflating their cost figures to 1996 prices gives administrative and operating costs of U.S. $500 million, excluding the payments to producers for the animals. Under the vaccination policy, the vaccine costs between U.S. $0.30 and U.S. $0.60 per animal with an additional U.S. $0.20 per vaccination to administer the vaccine. To be effective, at least 90 percent of the herd must be vaccinated twice per year. In this analysis, the higher per-animal vaccination cost figure is used. Given the base herd, the annual vaccination costs are U.S. $165.15 million.

5. MODEL RESULTS

The specific tariffs and premia by risk category are shown in Table 11.1.[viii] The first column gives the welfare optimizing tariffs when there is no expected output loss due to FMD and no risk of obtaining FMD through imports—the no FMD scenario.[ix] For each FMD outbreak and control scenario[x] two columns of results are presented. The first column is the vector of tariffs t_i, when the expected output loss due to FMD is recognized in setting the tariff [equation (13)], but the import risk—equation (14)—is not included. These tariffs are higher than the no FMD tariffs because the output loss raises U.S. beef imports, the term $x_i/(\partial x_i/\partial p_i)$ increases. The second column of results is the premia, which incorporate the FMD risk by region.

Because imports are expected to trigger outbreaks and a fall in U.S. beef supply, even when the risk is not directly incorporated into the tariff setting, the tariffs increase as the expected imports increase. The uniformity of the tariffs levied across the regions is imposed on the model solution for two reasons: (1) In the absence of any FMD risk, the World Trade Organization's (WTO) non-discrimination rule applies; (2) if price discrimination based on each region's excess supply elasticity is allowed, the effects of pure price discrimination and the differences due to risk could be confused.[xi]

Including import risk and the associated expected loss in national welfare further adds to the size of the trade barrier. For the region classified as negligible risk, the premia are small, ranging from U.S. $3.81 per ton under the endemic scenario to U.S. $0.22 per ton under the vaccination program. As FMD risk increases, the premia increase in proportion to the increased probability of importing FMD by region. Thus, in the endemic scenario the pre-

Table 11.1 Beef Tariffs and Risk Premia by Risk Category and Control Program[1]

Region risk	*No FMD tariff*	*Endemic tariff*	*Endemic premia*	*Quarantine-slaughter tariff*	*Quarantine-slaughter premia*	*Vaccination tariff*	*Vaccination premia*
				dollars per ton			
Negligible (RN)	77.34	117.41	3.81	83.70	1.03	78.61	0.22
Slight (R1)	77.34	117.41	38.09	83.70	10.33	78.61	2.21
Low (R2)	77.34	117.41	380.94[2]	83.70	103.28[2]	78.61	22.11
High (R4)	77.34	117.41	38,094.00[2]	83.70	10,327.87[2]	78.61	2,210.87[2]

[1]*Moderate-risk countries are net beef importers, so no excess supply by moderate-risk countries to United States is allowed.*

[2]*Prohibitive tariffs as there are no exports to United States for prices below U.S. $1,500 per ton.*

Source: Authors' calculations using model.

mium for RN regions is U.S. $3.81 per ton, but it is U.S. $38.09 per ton for R1 regions as the risk increases from one in one million to one in 100 thousand.

With calibration of the model so that beef is not supplied to the United States by any region at a price below U.S. $1,500 per ton, several of the barriers are prohibitive. For each FMD control scenario the total barrier levied against R4 and RU regions is prohibitive regardless of U.S. control strategy; therefore, these regions do not supply the U.S. market. For the R2 classification, the total barriers (tariff plus risk premium) under the endemic and quarantine-slaughter scenarios of U.S. $498.35 and U.S. $186.98 per ton are also prohibitive because the U.S. price less the barrier is below U.S. $1,500 per ton. The total barrier on an R2 region under the vaccination scenario of U.S. $100.72 per ton allows that region to supply 96 thousand tons. Thus, only RN and R1 regions are consistent suppliers of beef to the United States.

Policy provides an inducement, or threat, for an exporting region to improve, or control, quality. Under the quarantine-slaughter control program, if an R2 region can adopt procedures to lower its risk classification to the R1 category, the premium paid falls from U.S. $103.28 per ton to U.S. $10.33 per ton. The reduction in risk premium could more than compensate for the cost incurred to lower its risk classification. Alternatively, there is a substantial penalty paid if a region classified as an R1 region fails to maintain that risk classification.

The above results use the 15-year average loss found by Aulaqi and Sundquist (1979) as the expected loss, but as they show, catastrophic losses in individual years can occur. We show the premia under the endemic scenario when the expected loss from FMD used to set the premia is catastrophic (Table 11.2). The catastrophic loss of 30 percent is compared to the 6.3 percent loss used in the earlier analysis. For regions in the RN category, the premium due to an increase in the FMD loss from 6.3 to 30 percent rises from U.S. $3.81 per ton to U.S. $20.96 per ton, while the premium on R1 regions rises from U.S. $38.09 per ton to U.S. $209.59 per ton. Even with

Table 11.2 FMD Premia with Base and Catastrophic Losses under an Endemic (No Control) Scenario[1]

	Percent decline in beef output due to an FMD outbreak:	
Region	*6.3*	*30*
	dollars per U.S. ton	
Negligible (RN)	3.81	20.96
Slight (R1)	38.09	209.59
Low (R2)	380.94	2,095.88

[1]*Because R4 and RU regions face prohibitive barriers in all scenarios, results are not reported.*

Source: Authors' calculations using model.

the large total barriers—U.S. $289.09 per ton for RN regions and U.S. $477.72 per ton for R1 regions—given the catastrophic loss (30 percent), the United States imports large quantities of beef to compensate for the large output lost to the FMD outbreak. Of that beef, only 3.2 percent is imported from R1 countries.

The extent to which the type of control strategy affects the total barrier, given a common expected output loss of 6.3 percent, is shown in Table 11.3 for RN, R1, and R2 regions. For the same expected output loss, the type of control strategy has little affect on the barriers. Under the endemic or no control program, the RN region's barrier is U.S. $121.22 per ton. The barrier rises to U.S. $121.38 per ton under the vaccination program, and increases a bit more to U.S. $121.72 per ton with the quarantine-slaughter policy. A similar pattern occurs for the higher risk categories, but larger steps result from the increased risk. Yet, the increases are not massive across the different control programs. For R1 regions, the largest difference is U.S. $5 per ton. For R2 regions, which are not supplying the United States in this scenario, the largest difference is U.S. $50 per ton.

The point of adopting a control program is to reduce the expected severity of the output drop. That effect is captured in Table 11.1 where the total barriers for the RN, R1, and R2 regions in the quarantine-slaughter and vaccination scenarios are noticeably lower than those under the no-control scenario and closer to the levels of the no-FMD scenario. One issue is whether the additional costs of the quarantine-slaughter and the vaccination programs are compensated by reducing the expected output losses and consequent expected loss in welfare. The expected social loss in the endemic scenario compared to the no-FMD scenario is U.S. $3.8 billion. The quarantine-slaughter policy costs U.S. $500 million dollars, but reduces the expected output loss from 6.3 percent to 1.0 percent. The expected social loss due to FMD in this case, including the U.S. $500 million program cost, is U.S. $1.0 billion. Thus, by spending U.S. $500 million, given these estimates there is a U.S. $2.8 billion reduction in social cost. The vaccination program further reduces the expected loss and social cost—down to U.S. $281 million.

Table 11.3 Total Barriers under Alternative Control Strategies with a Constant Expected Output Loss of 6.3 Percent[1]

Region	*No control*	*Quarantine slaughter*	*Vaccination*
		dollars per U.S. ton	
Negligible (RN)	121.22	121.72	121.38
Slight (R1)	155.50	160.50	157.15
Low (R2)	498.35	548.35	514.86

[1]*Because R4 and RU regions face prohibitive barriers in all scenarios, results are not reported.*

Source: Authors' calculations using model.

However, unlike the quarantine-slaughter policy, which is only used when an outbreak occurs, the vaccination program costs are incurred each year. Given the cost estimates used in this analysis, if outbreaks occur less than every 3.3 years, the vaccination cost is lower than the quarantine-slaughter policy. If outbreaks occur less frequently, the quarantine-slaughter policy is less expensive.

6. CONCLUSIONS

This chapter expands research that links trade barriers to livestock health risks by developing a model for beef trade that captures features of proposed U.S. import rules for beef. Those rules recognize multiple supplying regions, each with a different risk of causing an FMD outbreak in the United States. The model assumes that trade barriers are set to maximize expected welfare given the probability that FMD will be imported. The resulting barriers consist of two terms. One term is the standard optimal tariff facing each exporter. The second term is a premium that recognizes the probability that FMD will be contained in the exporter's beef and the expected loss in U.S. welfare if an FMD outbreak occurs.

The empirical model classifies regions into risk categories based on the system proposed by the United States. There are four possible suppliers of beef to the United States and three possible FMD control strategies in which each control strategy has a different expected output loss and cost. The results show that the premia set are sensitive to the expected output losses and to the risk represented by each category. The premia increase as the expected output loss rises. The total barriers—tariffs and premia—levied against R4 and RU regions are prohibitive, regardless of the type of control program adopted. Under the endemic and quarantine-slaughter policy, even the R2 category faces prohibitive barriers. However, the vaccination policy lowers the expected social loss from an FMD outbreak enough to allow R2 regions to export to the United States.

A given scenario in which the outbreak loss is expected to be catastrophic sharply increases the premia but continues to allow RN and R1 regions access to the U.S. market to replace lost U.S. beef production. Virtually all of the beef imports originate in RN countries.

The estimated barriers are not overly sensitive to the control program when the expected loss is held constant; rather, it is the reduction in the expected losses due to the control program that drives the barrier levels. Of the three control programs evaluated, those that reduce the frequency and the severity of FMD outbreaks are worth the additional costs incurred.

7. ENDNOTE

[i] The United States drafted and circulated for comment its proposed changes to U.S. import rules for meat and livestock (USDA/APHIS, 1996). Following a period of comment, revised rules have been developed. At the time of this research, the revised rules were not yet public so the model follows the initial proposal. The most likely change is in the regionalization, which, in the initial proposal, follows a national system.

[ii] A richer empirical specification would relax several assumptions made here. Animal trade would be considered. Linkages to swine, pork, and feedstuffs would need to be incorporated. Inter-species disease transmission would need to be allowed. Furthermore, the dynamics of the sectors and meat-product differentiation would need to be included. Since the focus of this work is on the regional FMD transmission differences, the complications are left for subsequent research.

[iii] Because the R4 and RU regions are treated equivalently in the regulations, they are aggregated into a single region. Thus, there are five exporting regions in the model instead of six.

[iv] Unlike existing WTO rules, the tariffs are allowed to be discriminatory to fully separate the exercise of market power from the recognition of FMD risk.

[v] Ton is assumed to be one metric ton.

[vi] The example in this chapter is a livestock disease that is not transmitted to humans. Thus, consumer welfare is appropriately measured by conventional metrics, such as consumer surplus. If the disease was a human disease, or if it was a livestock disease that could be transmitted to humans, more careful consideration of consumer welfare would be required. One issue to consider then is the impact of consumer income. Another issue to consider is the interaction of risk and expected consequences on human health. For example, salmonella represents a relatively high risk in the short term, but is not usually fatal. In contrast, bovine spongiform encephalopathy (BSE) carries a low risk, takes several years to develop, and is fatal. Thus, an inter-temporal dynamic model that captures the impact of disease transmission on human health and on income over time is needed.

[vii] In this formulation, it is assumed that the risk, π_i, is independent of, π_j, for $i \neq j$ in the model. Therefore, FMD outbreak risk is independent and static across regions over the period of analysis.

[viii] Both countries assigned to the moderate region are importers. In this homogeneous good model they do not supply the United States and barriers are not reported in the tables.

[ix] The tariffs found are roughly five percent and are below the 31 percent Uruguay Round commitment (Ingco, 1994). Thus, the United States could impose the tariffs generated in this model and satisfy the Uruguay Round commitments.

[x] The prices and quantities in the no FMD scenario differ from those in the 1994-96 base because the base data recognize an FMD risk and prohibit imports from non-negligible risk regions.

[xi] An example of confusion between policy instruments can be found in Krutilla (1991) where an environmental tax substitutes for the failure of the country to use an optimal tariff. To keep the focus on the role of the risk premia, a system of non-discriminatory tariffs is imposed on the empirical model.

8. REFERENCES

Aulaqi, N.A. and W.B. Sundquist. 1979."A Benefit-Cost Analysis of Foot-and-Mouth Disease Spread and Control in the United States," in E.H. McCauley et al. eds., pp. 71–102, *A Study of the Potential Economic Impact of Foot-and-Mouth Disease in the United States.* Staff paper TB-1597. University of Minnesota, St. Paul, MN.

Ingco, M.D. 1994. "How Much Agricultural Liberalization Was Achieved in the Uruguay Round?" Paper presented to the Annual Meeting of the International Agricultural Trade Research Consortium, Washington, DC (15–16 December).

Krutilla, J.D. 1991."Environmental Regulation in an Open Economy." *Journal of Environmental Economics and Management* 20: 127–142.

Paarlberg, P.L. and J.G. Lee. 1998. "Import Restriction in the Presence of a Health Risk: An Illustration Using FMD." *American Journal of Agricultural Economics* 80: 175–183.

Sullivan, J., J. Wainio, and V. Roningen. 1989. *A Database for Trade Liberalization Studies.* Economic Research Service Staff Report No. AGES89-12. Washington, DC: ERS/USDA.

USDA/APHIS (U.S. Department of Agriculture, Animal and Plant Health Inspection Service). 1996. "Importation of Animal and Animal Products: Proposed Rule." Docket No. 94-106-1. *Federal Register* 61(76): 16977–17027 (18 April).

Chapter 12

Trade Distortions in a Free-trade Zone: The Case of Sanitary and Phytosanitary Restrictions

Thomas H. Spreen, John J. VanSickle, and Charlene M. Brewster[1]
University of Florida

1. INTRODUCTION

The intent of the North American Free Trade Agreement (NAFTA) is to create a customs union, or free-trade zone, among the countries of Canada, the United States, and Mexico, in which goods and services can be freely traded. The European Union (E.U.) is even more ambitious in that it not only allows tariff-free trade but has also created a common currency. The Common Market of the South (MERCOSUR) is a free-trade zone that encompasses Brazil, Argentina, Paraguay and Uruguay and is similar in scope to the zone designated in NAFTA.

An individual country within a free-trade zone will often have legitimate concerns regarding both the health of its citizens and the viability of its natural ecosystems. Agricultural trade facilitates the transportation of organisms to new environments, introducing a wide array of pests, microbes, and plant and animal diseases that cause widespread destruction when carried into a country. Differences in climate, vegetation and other factors—including natural predators—influence the ability of a pest, or disease, to become established in a new environment.

In general, government is charged with the responsibility of protecting domestic welfare. Sanitary and phytosanitary (SPS) restrictions are the means by which governments enforce health standards on imports in order to prevent the importation of plant and animal diseases and of pests that could adversely affect domestic producers and domestic consumers. Of the many methods used to enforce SPS restrictions, border inspections and quarantines are the most common.

SPS restrictions are members of the class of trade-related policies known as technical barriers. Technical barriers and quotas form the group of trade restrictions known as non-tariff trade barriers. A recent study by Roberts, Josling, and Orden (1999) summarizes the scope and intent of technical trade barriers, including SPS restrictions. These authors define technical trade barriers:

> . . . regulations and standards governing the sale of products into national markets which have as their *prima facie* objective the correction of market inefficiencies stemming from externalities associated with the production, distribution and consumption of these products. (Roberts, Josling, and Orden, 1999: 3)

What are SPS restrictions? The World Trade Organization Agreement on SPS measures recognizes them as measures aimed at protecting the following:

- human or animal health from food-borne risks;
- human health from diseases carried by plants and animals; and
- animals and plants from pests and diseases.

Food-borne risks may arise from the presence of additives, contaminants, residues, toxins or disease-causing organisms like bacteria. SPS measures can apply to domestically produced food as well as to imported products (WTO, 1999).

The specific aim of sanitary restrictions is to protect human and animal health. These restrictions include the setting of maximum allowable levels of pesticide residues, the allowance of only certain food additives, and the requirement that products come from disease-free areas. Sanitary restrictions are generally enforced through border inspections, but thorough inspections can be quite costly. The importing country must often make a trade-off between exhaustive border inspections and cost; hence, such inspections may be sporadic. As an alternative to border inspections, an importing country may specify a protocol relating to the production, processing, or treatment of a product. This protocol (if enforced by the exporting country) will allow a product to enter the importing country without extensive inspection. The use of protocols is potentially important because it transfers the cost of enforcement onto the exporting country.

Phytosanitary restrictions are regulations that relate to plant health. Countries may choose to impose phytosanitary restrictions on imports to protect against potentially harmful pests from other areas. An example of a potentially damaging pest is an insect that is endemic in the exporting country. In this case, an importing country will choose to protect itself from that pest because of its likely adverse affect on domestic production. Phytosanitary restrictions may also be imposed to protect forests and other wild flora. Similar to the case of sanitary measures, phytosanitary regulations are enforced primarily through border inspections. Again, specific production or processing protocols may be used to allow importation from a country with minimal inspection.

The option to use import protocols to enforce SPS restrictions as opposed to border inspections has two important implications. First, the use of a protocol may serve to facilitate trade. Producers in the exporting country understand that if certain procedures are followed, their products will be allowed to pass into the regulating country. The second implication is that the cost of enforcing the standard is transferred to the exporting country.

One example of the use of protocols in the imposition of a phytosanitary barrier is the export of fresh grapefruit from Florida. Florida is the largest grapefruit-producing region in the world. It accounts for almost 50 percent of world grapefruit production (FAO), and Florida grapefruit producers export grapefruit to both Europe and Japan (FASS). The state of Florida, however, is plagued with periodic outbreaks of the Mediterranean fruit fly, and fruit production in Europe and Japan would be severely affected if the fruit fly were to become established in either country.

Thus, Florida grapefruit growers have agreed to participate in a fruit-fly protocol whereby fly traps are placed in citrus groves. If no flies are found, the grove is certified to be fruit-fly free, and fruit from that grove may enter export channels destined for Japan or Europe. If one fly is found, the grove loses its certification for a fixed period of time, and all fruit from that grove must be sold in the domestic market, or in Canada.[ii] Growers pay a per-acre fee of U.S. $25 to cover the cost of enforcement (Muraro, Hebb, and Stover, 1998). The program is administered by the U.S. Department of Agriculture (USDA).

2. A CONCEPTUAL FRAMEWORK

We present a conceptual model of phytosanitary barriers and their likely impact on the domestic economy (Figure 12.1). In the absence of a phytosanitary barrier, domestic producers in the importing country would be adversely affected if the plant or animal pests were to become established. The adverse effects of pest infestations include increased production costs as producers spend more money to fight pest outbreaks. Producers would also face a decline in the marketability of their product because it would no

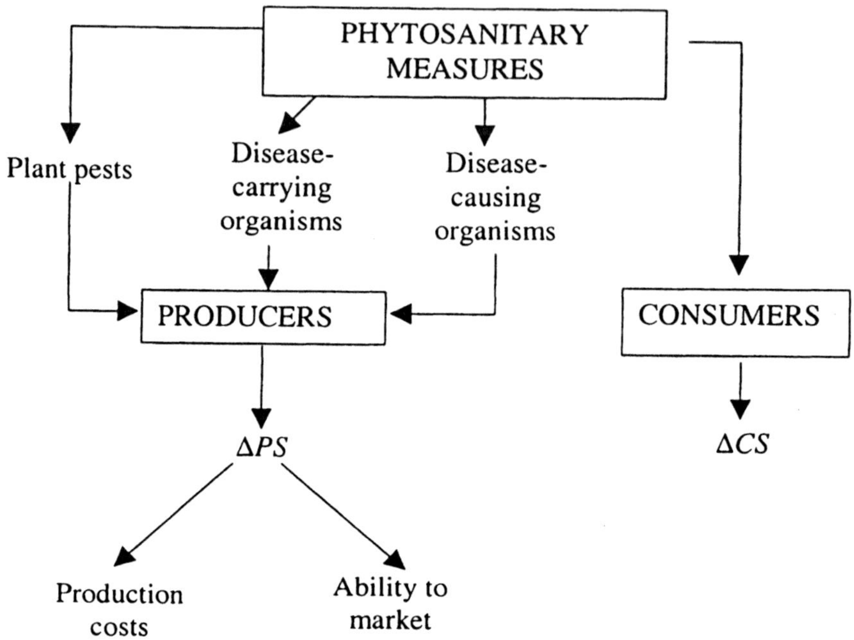

Figure 12.1 Model of Phytosanitary Barriers

longer originate from a certified pest-free zone. Imposition of the phytosanitary barrier prevents harmful pests from being introduced into the local ecosystem; it also reduces or eliminates competition from imports. There will be a resulting change in producers' surplus (ΔPS). Also, consumers are affected indirectly when phytosanitary measures are imposed. There is a loss of consumers' surplus (ΔCS) when the aggregate supply in the domestic market declines.

We present a conceptual model of sanitary barriers and their likely impact on the domestic economy (Figure 12.2). The imposition of sanitary measures will have a direct impact on both producers and consumers. Beneficial producer impacts include protection from harmful diseases and the reduction of import competition. An increase in the quality of the food supply may translate directly into noticeable consumer health benefits (ΔH). Consumers are no longer exposed to the risk of chronic diseases, such as cancer, or the acute effects of poisoning, both of which may be caused by

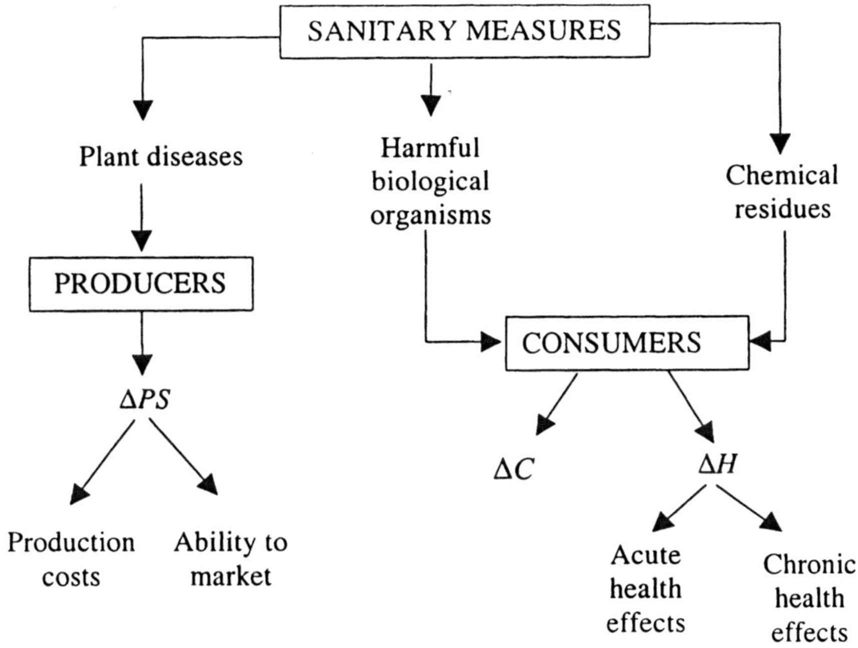

Figure 12.2 Model of Sanitary Barriers

by pesticides and food additives. Consumers may also realize an indirect effect through reduced aggregate supply (ΔCS).

3. ANALYSIS OF SPS BARRIERS

Taylor and Howitt (1993) provide a useful theoretical framework for analyzing changes in social welfare that are associated with resource policies. In that framework, they assume that infinitesimal changes in the policy variable are possible. But in this case, the imposition of an SPS barrier is treated as a distinct and discrete activity. This assumption, however, does not essentially change the analysis if a variable, B, is defined. Assuming that $B = 1$ indicates the presence of an SPS barrier and that $B = 0$ indicates the absence of such a barrier. Changes in net social welfare, W, when the barrier is enacted ($B = 1$), will be:

$$(1) \quad \Delta W = \Delta CS \frac{\Delta W}{\Delta CS} + \Delta PS_Y \frac{\Delta W}{\Delta PS_Y} + \Delta PS_X \frac{\Delta W}{\Delta PS_X} + \Delta EC \frac{\Delta W}{\Delta EC} + \Delta H \frac{\Delta W}{\Delta H},$$

where W is aggregate social welfare, PS_y and PS_x denote producers' surplus products and input markets, respectively, CS denotes consumers' surplus,[iii] EC denotes the enforcement costs associated with the imposition of the proposed SPS barrier, and H is some index of human health.[iv] Assuming that the impacts on input markets are minimal, equation (1) simplifies to:

$$(2) \quad \Delta W = \Delta CS \frac{\Delta W}{\Delta CS} + \Delta PS \frac{\Delta W}{\Delta PS} + \Delta EC \frac{\Delta W}{\Delta EC} + \Delta H \frac{\Delta W}{\Delta H},$$

where PS represents producers' surplus in the single-product market under consideration. The imposition of a phytosanitary barrier would likely cause consumers' surplus to decline, producers' surplus to increase, enforcement costs to increase, and a zero effect on human health. Social welfare is negatively related to increased enforcement costs, because these costs are supported through public funds. Therefore, on the right-hand side of equation (2), the first term is negative, the second term is positive, the third term is negative and the last term is zero. It follows that the phytosanitary regulation is welfare-improving if:

$$(3) \quad \Delta PS \frac{\Delta W}{\Delta PS} > \Delta CS \frac{\Delta W}{\Delta CS} + \Delta EC \frac{\Delta W}{\Delta EC}.$$

There is an improvement in social welfare if the increase in welfare derived from the benefits accruing to producers exceeds the loss of welfare to consumers through reductions in consumers' surplus and increased enforcement costs.

The acknowledgement of the existence of linkages between consumers' and producers' surplus changes, and changes in the aggregate welfare function is consistent with the argument postulated by Paarlberg and Abbott (1986). These authors argue that policymakers may place different weights on the contributions of producers' surplus and consumers' surplus to welfare. In expression (3), the terms $\frac{\Delta W}{\Delta CS}$ and $\frac{\Delta W}{\Delta PS}$ capture the weighted change in the government welfare function that results from changes in consumers' surplus and producers' surplus, respectively. The fact that these terms may differ in magnitude explains, in part, why policy decisions that favor producers over consumers (or vice-versa) are made.

As in the case of phytosanitary barriers, the imposition of a sanitary barrier would likely reduce consumers' surplus, increase producers' surplus and increase enforcement costs. In addition, the sanitary barrier may also positively affect human health if pesticide residues, or food additives, are reduced, or if it prevents the introduction of a harmful disease that directly, or indirectly, affects humans. The sanitary barrier is welfare-improving if:

$$(4) \qquad \Delta PS \frac{\Delta W}{\Delta PS} + \Delta H \frac{\Delta W}{\Delta H} > \Delta CS \frac{\Delta W}{\Delta CS} + \Delta EC \frac{\Delta W}{\Delta EC} .$$

The measure is welfare-improving if the benefits accruing to producers plus the increase in welfare attributable to enhanced human health exceed the loss in welfare resulting from decreases in consumers' surplus and increased enforcement costs.

3.1 Disaggregating Changes in Producers' Surplus

Consider the change in producers' surplus more closely. In the case of both sanitary and phytosanitary barriers, it is possible to identify at least three beneficial sources of change in producers' surplus that result from the imposition of a barrier. The most obvious source is the reduction in availability of imports when the SPS barrier is imposed. Imports will decrease because exporters will be unable to achieve the standards required by the SPS barrier, or their costs will increase as a result of the additional technology required by the SPS barrier (Figure 12.3). Excess supply, *ES*, in the exporting country (or countries) shifts, causing the aggregate supply, *AS*, in the importing country to shift as well. As a result, domestic prices increase, and consequently, domestic production also increases.

Domestic producers may also benefit if the SPS barrier results in the elimination of a harmful pest (or disease), or if the SPS barrier prevents the introduction of such a pest or disease. This effect is the so-called supply-shifting phenomenon described by Roberts, Josling and Orden (1999). If a new pest (or disease) is introduced through imports, domestic producers would be adversely affected either through reduced yields or through increased production costs as additional inputs are required to counteract the

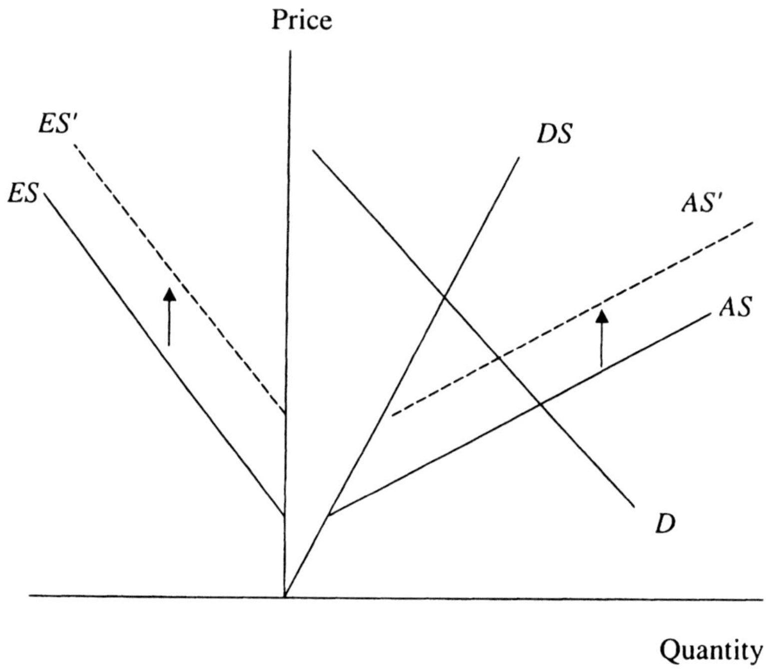

Figure 12.3 Inward Shift in Import Supply Results from the Imposition of an SPS Barrier

adverse effects of the pest, the disease, or both[v] (Figure 12.3). In the absence of an SPS restriction, the domestic supply curve, *DS*, shifts as a consequence of increased production costs and/or lower yields. As domestic producers are less competitive compared to their import competitors, they lose market share, and hence, their surplus is diminished.

The third possible benefit to domestic producers that arises from the imposition of an SPS barrier occurs when the domestic industry is an exporter. It is common for countries to import a horticultural crop during the domestic producers' off-season, and to export that crop during the domestic harvest season. For example, the United States imports fresh apples and pears from Chile in the spring and early summer when the only available domestic supply comes from cold storage of the previous fall harvest. At the peak of the U.S. harvest season in late summer and fall, however, apples grown in the United States are exported to Canada and Mexico. The introduction of a harmful pest or disease from off-season import suppliers may, in turn, affect the ability of domestic producers to export to third countries.

These three points may have helped domestic producers to successfully lobby for the enactment of SPS barriers. It appears to be fairly easy for domestic producers to convince policymakers that allowing unrestricted imports would be detrimental to the domestic industry for any, or all, of the arguments presented above. Orden and Romano (1996) refer to this process as the capture of the policy process by the domestic industry. By exaggerating the potential negative impacts of imports, interested parties may be able to convince policymakers that the imposition of an SPS restriction ($B=1$) is welfare improving. This will be the case when policymakers believe that the left-hand side of equation (3) or of equation (4) is greater than the right-hand side.

Consumer groups may also seek to influence the process by trying to convince policymakers of the potential health impacts associated with imports. In this case, improvements in welfare associated with the sanitary barrier are related to improvements in health, rather than to improvements in producers' surplus. In this manner, although producers and consumers may have quite different agendas, they form an alliance to promote the imposition of a sanitary barrier.

3.1.1 Stochastic Impacts and Their Measurement

Although the previous discussion assumed that changes in welfare could be determined with certainty, it is likely that some components of welfare may be measured with error, or may be inherently stochastic. Consider the right-hand side of equation (3). It is likely that the enforcement costs associated with a particular phytosanitary regulation can be estimated in a straightforward manner. The government will simply determine the resources that will be allocated to enforcement. Therefore, it can be reasonably assumed that enforcement costs are non-stochastic and are measured without error. It should be noted, however, that the effectiveness of the enforcement is unknown and that there is a trade-off between the quantity of resources allocated to inspection and the efficacy of the enforcement process.

The measurement of the likely impact of imposing SPS restrictions on consumers is tied directly to the quality of economic models. Although the theoretical tools used to analyze the impact of restrictions on aggregate consumption and equilibrium price are well developed, the empirical measurement of these effects is an inexact science. A long list of the problems that will likely be encountered in such an empirical study is not compiled in this chapter, but complicating factors include product heterogeneity, regional differences in tastes and preferences, and inexact specification of demand models. Economists, however, are likely to gloss over these issues and will usually assume that consumer impacts are both non-stochastic and measured without error.

The measurement of producer effects is often the most problematic. As discussed in the previous section, producer effects can be disaggregated into three groups: (1) domestic-market effects; (2) export-market effects; and (3) ecological effects. Measurement problems associated with domestic-market effects are similar to those associated with the impact on consumers; the type of empirical economic model used will vary widely depending on the nature of the proposed restriction and on the availability of data. The measurement of export-market effects will also be dependent on a valid empirical-economic model. It is clear, however, that ecological effects will be both stochastic and subject to large measurement error.

Therefore, a better representation of equation (3) is that the proposed phytosanitary restriction is welfare-improving in the sense of first-degree stochastic dominance if:

$$(5)\quad \int_0^A \left[\Delta PS \frac{\Delta W}{\Delta PS} \right] dx \geq \Delta CS \frac{\Delta W}{\Delta CS} + \Delta EC \frac{\Delta W}{\Delta EC},$$

for all A, with the left-hand side of equation (5) strictly greater than the right-hand side for some A. Similarly, a sanitary restriction is welfare-improving in the sense of first-degree stochastic dominance if:

$$(6)\quad \int_0^A \left[\Delta PS \frac{\Delta W}{\Delta PS} + \Delta H \frac{\Delta W}{\Delta H} \right] dx \geq \Delta CS \frac{\Delta W}{\Delta CS} + \Delta EC \frac{\Delta W}{\Delta EC},$$

with the left-hand side of equation (6) strictly greater than the right-hand side for some *A*.[vi]

The measurement of risks to human health associated with the consumption of contaminated foodstuffs or of undesirable production processes is relatively new to the economic literature. Lichtenberg and Zilberman (1990) note that "these risks are often quite subtle and, therefore, difficult to detect or verify reliably, and quantitative estimates of risk are typically subject to considerable uncertainty." Harper and Zilberman (1992) examine the problems associated with formulating environmental policy in the presence of uncertain human-health risks. They examine two different decision rules to evaluate the trade-off between health and economic welfare, and they note that these rules lead to opposite policy conclusions. As a result, the authors recommend the adoption of a safe minimum standard approach (SMS) in which economic costs and benefits are weighed only after some minimum acceptable level of health safety has been met.

Nyamusika et al. (1994) use a different approach in weighing the costs and benefits of animal-disease control. In their study, the authors use an epidemiological-model, bovine-respiratory disease (BRD) to capture the uncertainty associated with the spread of the disease. They estimate the economic returns associated with a vaccination program.

4. SPS BARRIERS IN ACTION

The United States and Mexico are major producers of citrus crops, ranking second and fourth respectively in world citrus production (FAO). It is not surprising, therefore, that trade in citrus products was a contentious issue during the negotiation of NAFTA, and has continued to be so after the implementation of NAFTA. The two countries are similar in that each has a large domestic market for citrus products, and that both also compete in export markets. They also share the common practice of using phytosanitary barriers as means to restrict fresh citrus imports.

Florida is the largest citrus-producing state in the United States. It is the world's second largest producing area for orange juice (after Sao Paulo, Brazil) and is also the largest exporter of fresh grapefruit. Citrus producers in Florida are periodically affected by outbreaks of the Mediterranean fruit fly. The fruit fly is a particularly insidious pest in that it has a wide array of hosts, including citrus, tomatoes, peppers, and most stone fruits.

As discussed earlier in this chapter, Florida grapefruit growers subscribe to a fruit-fly program administered by the USDA. For a per-acre flat fee of U.S. $25, the USDA places and monitors fruit-fly traps in commercial citrus groves (Muraro, Hebb, and Stover, 1998). If no fruit flies are found, then the grove is certified fly-free. Both the European Union and Japan accept this protocol and allow the importation of Florida grapefruit from those growers whose groves have been certified. Alternatives to this approach include fumigation (with methyl bromide) and cold storage. However, both of these alternatives are expensive and result in loss of fruit.

Mexico is the third largest producer of oranges in the world. It also has significant plantings of tangerines and grapefruit. Until the mid-1980s, Mexican tangerine producers in Nuevo Leon were successful in exporting tangerines to the United States. Continuing outbreaks of the Mexican fruit fly, however, resulted in the United States imposing a protocol that all shipments of fresh citrus from Mexico be fumigated with methyl bromide.[vii] From the Mexican perspective, a particularly galling aspect of U.S. policy is that USDA inspectors must be present to oversee the fumigation process. Yet the cost of these inspectors is borne by Mexican citrus growers.

After the imposition of the required methyl-bromide treatment, the number of shipments of fresh citrus from Mexico to the United States dropped dramatically, from 25,000 metric tons in 1983 to 4,000 metric tons in 1985. Even after the implementation of NAFTA, which effectively eliminated U.S. tariffs on fresh-citrus imports, Mexico's exports to the United States remained low.

Before 1994, Mexico imposed a 20 percent *ad valorem* tariff on all fresh-citrus imports, which effectively kept Florida grapefruit out of the Mexican market. With the coming of NAFTA in 1994, Florida grapefruit growers expected the market for grapefruit in Mexico to open up. When Florida growers attempted to export fresh grapefruit to Mexico, however, they found that the Mexicans had imposed a reciprocal phytosanitary barrier on Florida grapefruit. Because of the presence of the Mediterranean fruit fly in Florida, all fresh-citrus exports, including grapefruit exported from Florida to Mexico, were to be fumigated with methyl bromide. Furthermore, this fumigation was to be supervised by Mexican inspectors, who, in turn, were to be paid by Florida growers.

Historically, Mexico has been a small market for grapefruit, and because of this, Florida growers have not elected to follow the protocol imposed by the Mexican government. Hence, the United States has not exported any grapefruit to Mexico.

It is worth noting that the USDA certified several states in Western Mexico to have met the criterion of fruit-fly-free status. These states are now free to export fresh oranges and tangerines into the United States. However, the large citrus-producing states in Eastern Mexico—including Nuevo Leon, Tamaulipas, San Luis Potosi and Veracruz—remain under the methyl-bromide fumigation requirement.

5. CONCLUDING REMARKS

In this chapter, we presented and discussed issues related to SPS barriers. A model from the perspective of government policymakers was developed and analyzed. The model suggests that government policymakers who seek to improve aggregate domestic welfare (encompassing human and animal health concerns), and who seek to preserve their domestic environment, will choose to impose SPS barriers if the benefits derived from these barriers outweigh their costs. The benefits associated with a phytosanitary barrier are those accruing to domestic producers. These benefits are gained through reduced import competition and through the probability that adverse pests and diseases will not be introduced into domestic-production areas. The costs associated with a phytosanitary barrier are those borne by consumers, including a reduced aggregate supply and increased enforcement costs. The importing country may be able to shift the burden of enforcement cost onto its import-supply countries. In the case of sanitary barriers, the analysis is similar except that sanitary barriers encompass micro-organisms that may adversely affect human and/or animal health.

A major problem in the analysis of the economic impact of SPS barriers is that both the costs and benefits associated with a particular barrier may be subject to measurement error, or may be inherently stochastic. Both of these issues cloud the regulatory process. For example, the likelihood that a particular pest could be introduced into the importing country's ecosystem through the imports of a particular commodity may be subject to large measurement error. In this case, domestic producers may exaggerate either the probability of infestation or the cost impact of infestation. In this way, domestic producers can capture the regulatory process and, through the imposition of SPS barriers, limit competition from imports. In the case of human-health issues, consumer groups may join domestic producers to form a coalition. This coalition will pursue the implementation of a sanitary barrier that serves to promote the varying interests of both groups.

For the fresh-citrus trade between the United States and Mexico, it is the issue of enforcement costs that serves to enhance the impact of a legitimate phytosanitary barrier on consumers in both countries. While citrus producers in Florida have a protocol related to fruit flies that is satisfactory to both Japan and the European Union, Mexico is insisting that its inspectors be present in Florida to oversee the satisfactory completion of the protocol for the Mediterranean fruit fly before fresh Florida grapefruit will be allowed into Mexico.

As the push for trade regimes that are less dependent upon tariffs, quotas and domestic subsidies continues throughout the world, it is likely that the battleground for future trade disputes will be related to SPS issues. This chapter will hopefully help in future dispute resolutions that are based upon good science, not upon subjective criteria.

6. ENDNOTES

[1] Thomas H. Spreen, John J. VanSickle and Charlene Brewster are professor, professor and postdoctoral research associate, respectively, Food and Resource Economics Department, University of Florida, Gainesville, FL. The authors gratefully acknowledge the technical assistance of Suzanne Thornsbury, University of Florida, Indian River Research and Education Centre, Fort Pierce, FL; Richard Kinney of the Florida Fresh Fruit Shippers

Association; Charles Moss of the University of Florida; and the editorial assistance of H. Carole Schmitz, Jason Snart, and Kim Box.

[ii] California, Texas, Arizona and Hawaii will not accept grapefruit from groves that lose their fruit-fly certification. In these cases, fruit is sent to a post-harvest fumigation chamber, where it is exposed to methyl bromide.

[iii] Just and Hueth (1979) define producers' surplus and consumers' surplus relative to general equilibrium supply and demand curves. They show that producers' surplus relative to the general equilibrium supply curve incorporates profits in both factor and product markets. They also show that consumers' surplus relative to the general equilibrium demand curve is consumers' surplus in both product market plus profits in all intermediate markets (Taylor and Howitt, 1993). In this case, changes in one product market are examined, assuming a negligible impact in other product and input markets; hence, we adopt a partial equilibrium rather than a general equilibrium framework.

[iv] If the proposed SPS barrier affects more than one commodity, then the producers' and consumers' surplus associated with all affected commodities would be included on the right-hand side of the equation.

[v] Or, in the case of livestock production, decreased production would result from lower weight gain or from higher death losses.

[vi] The left-hand side of this expression is a gross simplification because the domains of the two integrals are different. Stochastic dominance in a multivariate framework is much more complicated and is not investigated here.

[vii] A notable exception is the export of Persian limes. Persian limes are not susceptible to the fruit fly and, hence, are not subjected to the fumigation requirement.

7. REFERENCES

FASS (Florida Agricultural Statistics Service). Various issues. *Citrus Summary.* Orlando, FL: FASS.

FAO (Food and Agriculture Organization of the United Nations). Various issues. *Citrus Statistics.* Rome, Italy.

Harper, C. and D. Zilberman. 1992. "Pesticides and Worker Safety." *American Journal of Agricultural Economics* 74: 68–78.

Just, R. and D. Hueth. 1979. "Welfare Measures in a Multimarket Framework." *American Economic Review* 69: 947–954.

Lichtenberg, E. and D. Zilberman. 1990. "Efficient Regulation of Human Health and Safety Under Uncertainty: California Water Quality Case Studies," in D. Zilberman and J. Siebert, eds., *Economic Perspectives on Pesticide Use in California.* Department of Agricultural and Resource Economics, University of California, Berkeley, CA.

Muraro, R.P., J.W. Hebb, and E.W. Stover. 1998. "Budgeting Costs and Returns for Indian River Citrus Production, 1997-98." Economic Information Report 98-5, Food and Resource Economics Department, University of Florida, Gainesville, Florida (August).

Nyamusika, N., T. Spreen, O. Rae, and C.B. Moss. 1994. "A Bioeconomic Analysis of Bovine Respiratory Disease Complex." *Review of Agricultural Economics* 16: 39–53.

Orden, D. and E. Romano. 1996. "The Avocado Dispute and Other Technical Barriers to Agricultural Trade Under NAFTA." Invited paper presented at the NAFTA and Agriculture: Is the Experiment Working? symposium, San Antonio, Texas (November).

Paarlberg, P.L. and P.C. Abbott. 1986. "Oligopolistic Behavior by Public Agencies in International Trade: The World Wheat Market." *American Journal of Agricultural Economics* 68(5): 28–42.

Roberts, D., T.E. Josling, and D. Orden. 1999. "Technical Barriers to Trade: An Analytical Framework." Technical Bulletin No. 1876, Economic Research Service, U.S. Department of Agriculture, Washington, DC.

Taylor, C. and R. Howitt. 1993. "Aggregate Evaluation Concepts and Models," in G. Carlson, D. Zilberman, and J. Miranowski, eds., *Agricultural and Environmental Resource Economics*. Oxford, UK: Oxford University Press.

WTO (World Trade Organization). 1999. "Understanding the WTO Agreement on Sanitary and Phytosanitary (SPS) Measures." Internet website: http://www.wto.org/wto/goods/spsund.htm (April).

Chapter 13

Sanitary and Phytosanitary Issues: Where Does the WTO Go From Here?

Suzanne D. Thornsbury
University of Florida

1. INTRODUCTION[i]

The Agreement on the Application of Sanitary and Phytosanitary Measures (SPS Agreement) governs regulations designed to mitigate potential negative externalities associated with the movement of products across national borders that might adversely impact human, animal, or plant health or life. When legitimate externalities or other market failures are addressed, such technical barriers have the potential to increase national welfare, even without consideration of terms-of-trade effects. All nations employ a wide range of standards and regulations to govern sales of agricultural products in national markets, the majority of which are considered justified commercial limitations.[ii] Governments may also impose technical barriers to isolate domestic producers from international competition. In these cases, under small-country assumptions, technical barriers are welfare-decreasing policies.

In contrast to the Agreement on Agriculture, which sets numerical limits and commitments in the areas of market access, domestic support, and export subsidies, provisions of the SPS Agreement provide guidelines for government behavior when implementing technical measures. The goal of the SPS Agreement is to protect the legitimate rights of importing countries with respect to national health and safety without providing a loophole for countries to avoid other trade-liberalizing disciplines of the Uruguay Round Agreements. Guidelines are designed to help identify when such barriers are disguised protection, but the legitimacy of specific SPS measures continues to be evaluated on a case-by-case basis.

After approximately five years of implementation, it is possible to compare the *a priori* expectations for the SPS Agreement in regard to freer agricultural trade with results to this point. Implementation and interpretation of the SPS Agreement is a dynamic process that continues through actions and findings of both World Trade Organization (WTO) members and Committees. Earlier evaluations of specific provisions and WTO panel actions are documented in the literature and will be drawn on below.[iii] The positions of WTO members, as they prepare for the next round of negotiations, can then be evaluated within this context.

2. *A PRIORI* EXPECTATIONS

At the time of ratification, the Uruguay Round Agreements were generally perceived to include significant steps toward freer trade in agricultural products, although there were some concerns over the ability of the SPS Agreement to achieve the desired results. In addition to the institutional strengthening components of the General Agreement on Tariffs and Trade (GATT) itself, the use of science to provide criteria for evaluation, plus transparency and regionalization clauses, were viewed as critical components of the SPS Agreement. Difficulties in interpretation and application of science, loss of national sovereignty, and potential adverse reactions to institutional strengthening were areas of concern.

Uniform harmonization within international standards is not an explicit requirement of the SPS Agreement, yet it was obviously a key to implementing success (Josling, 1998).[iv] Recognizing that the international guidelines may not reflect the preferences and/or needs for externality mitigation of every nation, the SPS Agreement also allows members to set a higher level of protection if it is based on a risk assessment that incorporates available scientific evidence. The use of science to provide criteria for policy evaluation is a unique feature of the SPS Agreement that arguably holds governments to a higher standard of accountability when compared to criteria used as the basis for implementing other, non-SPS, technical measures such as those covered by the Technical Barriers to Trade (TBT) Agreement (Roberts, 1998). In addition, scientific criteria were viewed as means to provide greater stability of expectations among trading partners concerning trade and in designing domestic regulations (Bhagwati, 1996).

A more transparent process, implemented, in part, through the equivalency and notification guidelines, was also viewed as a significant step toward freer trade. The opportunity for public scrutiny of regulations could raise the political costs of using technical regulations as disguised protection for domestic agriculture (Sykes, 1995). Under the SPS guidelines, members must accept the equivalence of other members' SPS measures if they achieve the same level of risk protection. In order to determine if the level of protection is the same, members are encouraged to provide access for testing, inspection of facilities, or other relevant procedures (WTO, 1999: Art. 4, Para. 1), again increasing the potential for exposure of standard-setting processes and enforcement.

Notification requirements provide additional opportunities for public comment before regulatory changes are enacted and are seen to be a catalyst for effective implementation of the SPS Agreement. Posting of measures under consideration is required, if the proposed change puts in place restrictions that are not substantially the same as an international standard, and if the measures may have a significant impact on trade (WTO, 1999: Art.7). Hence, expectations were that the SPS Agreement would enhance *ex ante* opportunities for members to comment on proposed regulatory changes. The transparency requirements are obligatory for all members, including least-developed countries, except when urgent problems arise.[v]

Finally Article 6 of the SPS Agreement obliges members to adopt their regulations to conditions in regional areas, not necessarily constrained by geographic areas defined by national borders. In particular, members are obligated to recognize the existence of pest- or disease-free areas within an exporting nation or within regional conditions that might exist across several countries. *A priori*, the regionalization criterion was viewed as a positive step when opening up previously restricted areas to trade.

Creation of the WTO and of revised dispute settlement procedures were seen as critical components of institutional strengthening that would impact the governance of sanitary and phytosanitary restrictions. The concept of the GATT as a single undertaking is new with the Uruguay Round Agreements. Prior to 1995, members have been able to sign the basic GATT Agreement without signing numerous side-agreements (such as the 1979 TBT Agreement) that were also negotiated among subsets of members. Since 1995, all members have been bound by the full set of GATT/WTO obligations, thus helping to clarify the legal relationship between the side agreements and obligations of GATT.

One of the primary goals of the Uruguay Round negotiations was to establish a more efficient and credible dispute settlement mechanism based on an agreed-upon set of procedures.[vi] The changes in dispute resolution have moved GATT somewhat from a system of soft law, in which legal norms do not effectively compel compliance, to one of hard law, in which there is a relatively high expectation of compliance with legal norms (Abbott, 1997). This may increase the likelihood of compliance and the credibility of the enforcement mechanisms. The strengthened dispute-settlement procedures are likely to be of particular relevance to technical barriers, since the SPS Agreement provides guidelines for government behavior, as opposed to numerical limits on policy adoption.

In addition to the positive *a priori* expectations for implementation of the SPS Agreement, there were some concerns over the ability of the SPS Agreement to achieve desired results. While the use of science to provide criteria for policy evaluation in the Uruguay Round SPS Agreement is more stringent than criteria negotiated in previous GATT Rounds, there is substantial room for interpretation by members. Other than requiring a risk assessment, the SPS Agreement does not provide guidelines for identifying acceptable scientific procedures and concurrence may be difficult to achieve. The results of a risk assessment depend critically on probabilities, both objective and subjective, associated with specific events and outcomes. Decision makers must choose among policy options when they are uncertain about the probabilities to associate with specific outcomes, and, thus, often exhibit a cautious approach and can be shown to prefer situations where they have more knowledge. In particular, probabilities for low-probability, high-consequence events are often over estimated. Over time, criteria in use will change as scientific and risk-assessment methodology evolves, increasing the potential for disagreement between opposing scientific views. The SPS Agreement provides little direction in cases of scientific ambiguity, except through allowing temporary measures to be enacted while more evidence is accumulated (James, 2000).

Differences between scientific risk assessment and public perception may prove difficult to resolve as WTO members debate the role of consumer sovereignty in setting standards. Additional *a priori* concerns over potential losses in national sovereignty include the ability to set acceptable regulatory standards by consensus in an international forum. Critics fear that, through negotiation, internationally adopted standards could reflect the least-common denominator, or lowest level of protection. The emphasis on reaching consensus might also act to ensure delay or avoidance of discussion on controversial matters. Governments may assert that their own interests are not represented adequately in international decision making, particularly in developing countries, where the resources to participate in multiple meetings and organizations are limited. Additionally, governments may argue that national protection levels should be set within their boundaries and without the negotiated consideration of other interests, which is a part of any international decision.

Finally, there was some concern that institutional strengthening mandated during the Uruguay Round might not enhance the long-run movement toward freer trade. The prospect of an increase in the number of disputes could cause governments to resist making new commitments or could lead to a breakdown in negotiations (Abbott, 1997). Further, the new hard laws were not always designed to decrease trade barriers; for example, the SPS Agreement specifically authorizes members to set their own acceptable level of risk, which could potentially legitimize reduced trade.

Since technical barriers implement trade restrictions that may be deemed legitimate by the WTO, the dispute settlement process is likely to be increasingly important in resolving controversy over specific issues. Panel rulings and statements will help clarify the acceptable interactions of science, economics, and politics when determining the role of technical barriers in agricultural trade. However, as the WTO still depends on the consensus and support of member nations, the dispute settlement procedures may not ultimately be the constraining factor for government behavior. Individual government commitments to freer trade and the underlying principles of GATT will be critical factors for success.

3. IMPLEMENTATION

Implementation of the SPS Agreement, after approximately five years, has produced mixed success. Progress has more often been subtle, while setbacks have been more widely publicized. As a mechanism to move toward freer trade in agricultural products, the SPS Agreement has been arguably successful. Movement toward international standard setting, regionalization, and transparency criteria has generally proven to be effective, albeit with continuing progress underway. Concerns remain over the role of politics when setting standards, and over the ability of developing countries to fully participate in the process. Debates and controversies over dispute settlement and consumer sovereignty remain highly publicized.

There are no readily available quantitative measures indicating the number of members who have revised domestic regulations to comply with international standards or the resulting level of market access achieved. The relevant standard-setting bodies, however, have been increasingly active in cooperative roles with WTO. In an internal review, the SPS Committee specifically recognized (1) the intensification and simplification of the Alimentarius Commission (Codex) standard-setting procedures; (2) increased activity in the Office International des Epizooties (OIE); and (3) revision of the International Plant Protection Convention (IPPC).

As anticipated, the regionalization criterion has opened up some previously restricted areas to agricultural trade. For example, prior to the Uruguay Round Agreements, cattle, swine, sheep, and some meat from countries where foot-and-mouth disease was present were banned from entry into the

United States. Revised, post Uruguay Round, U.S. regulations will allow imports to be made from low-risk regions within a country. As a result, Argentina was able to export these products to the United States on a limited basis.[vii] In its review of the operation and implementation of the SPS Agreement, the SPS Committee noted the critical importance of continuing to recognize pest- or disease-free areas, or to recognize areas of low-pest or low-disease prevalence for international trade in agricultural products.

Transparency of WTO Member policies has been enhanced. There were over 1100 SPS notifications filed during the first five years of implementation, providing *ex ante* opportunity for members to comment on proposed regulatory changes. While initially skewed toward developed countries, an increasing number of members have begun to notify their SPS measures. By March 1999, 59 members had submitted notifications, 91 members had established National Notification Authorities, and 100 members had established National Enquiry Points (WTO/SPS, 1999a). For calendar year 1999, the SPS Committee reported the circulation of 95 SPS notifications, including 32 from high-income, 59 from middle-income, and four from low-income countries (Table 13.1). Notifications from middle- and lower-income countries have been steadily increasing over the implementation period.[viii] The ability to receive notifications in a timely manner has also been enhanced, as members are encouraged to publish proposed regulatory changes at self-designated sites on the World Wide Web.

Of the total number of SPS notifications prior to 1999, 83 were challenged in the informal forum provided by the SPS Committee that noted, in its internal review, substantial time was devoted to the discussion of specific implementation problems. A number of SPS-related trade issues were resolved through informal discussion within the Committee. Additional disputes were resolved through bilateral negotiations. For example, between October 1996 and September 1998, USDA's Animal and Plant Health Inspection Service reported that 112 restrictions related to plant and animal health were resolved through bilateral technical exchange. Existence of informal resolution mechanisms are particularly critical for developing coun-

Table 13.1 SPS Notifications Circulated, 1999

WTO members' income	*Measures notified*	*Members notifying*
	number	*percentage*
High	32	42
Upper-middle	25	54
Lower-middle	34	38
Low	4	10
Total	95	n.a.

Source: WTO/SPS (2000) and author's calculations.

tries, which would be less likely to undertake a more expensive and time-consuming formal process.

Enactment of the SPS Agreement has put pressure on member countries to examine their own technical standards. One strength of the agreement is the focus on how countries set standards, and a more precise definition of what can be considered justifiable (Josling, 2000). Broad-based regulatory review has occurred among several countries, and discussions continue over acceptable processes for setting standards (Roberts, 1998).[ix] Wide-ranging reviews of compliance are underway in numerous developing and least-developed countries, where the scientific research tradition and facilities may historically have not been as strong. The marginal cost of setting standards, particularly those that are complex in nature, can be very high in such countries. The SPS Agreement has been of benefit by inducing members to examine and/or develop their own scientific infrastructure and human capital.

Another important contribution of the SPS Committee is its role in the clarification of the Agreement through rulings and interpretations. Extensive discussion on particular issues has helped draw attention to specific trade concerns and to avoid potential conflicts (WTO/SPS, 1999a). For example, the ability of members to set a prohibitively low level of acceptable risk will be limited by the non-discrimination policy of GATT, which requires that domestic and imported products be subject to the same requirements. Non-discrimination has provided additional impetus for members to enact consistent standards where none existed, limiting opportunities for sale of substandard goods, both imported and domestic.

Despite progress made through implementation of the SPS Agreement, there still remain areas of concern. The legitimacy of specific technical barriers remains subject to challenges on scientific or other procedural grounds (for example, not least-trade restrictive). Empirical evidence concerning the extent of questionable technical measures in international agricultural trade is very difficult to assess, because there is no comprehensive data source available. A 1996 USDA survey of technical barriers to U.S. agricultural exports identified 302 questionable barriers imposed by 61 countries and two regional trading blocks, which had an estimated trade impact of U.S. $4.9 billion.[x] Since the cross-sectional survey was not repeated, there is no direct evidence concerning either escalation or reductions in these numbers.

Among all countries, formal WTO complaints related to nine distinct SPS issues have been filed under the dispute settlement procedures since 1995, whereas during the 47 prior years of GATT there were virtually no trade disputes over SPS measures that reached a formal-dispute settlement. Three of the recent disputes have advanced to WTO panel and Appellate Body rulings: the E.U.-U.S./Canada Hormones case, the Australia-Canada

Salmon case, and the Japan-U.S. Varietal Testing case (Roberts, 1998). Outcomes of the three cases have been somewhat disappointing in terms of increasing agricultural trade. In the Hormones dispute, the complainants were authorized to introduce retaliation in the form of increased tariffs on E.U. exports after the European Union failed to either modify its measure to comply with the final Appellate Body ruling or to offer acceptable compensating trade concessions. In the Salmon case the revised measures notified by Australia have been challenged by Canada in regards to their compliance. The degree of market access created by the Varietal Testing ruling will not be known until Japan notifies its intended changes.

The loss of national sovereignty remains a publicly visible source of opposition to the WTO (and, by inclusion, to the SPS Agreement); however, it is difficult to quantify this complaint. Roberts, Orden, and Josling (1999) argue that Appellate Body rulings in the Hormones case provide evidence against such a claim. The three rulings have upheld member obligations to base technical regulations on scientific-based risk assessment while allowing national standards that are more stringent than those set internationally. In addition, the right to invoke the precautionary principal as a means of mitigating unfamiliar risks on a temporary basis was reinforced. Although the rulings provided clarification of SPS guidelines, continued non-compliance by the European Union, and enforcement of authorized sanctions has highlighted the continued debate over consumer sovereignty and science.

An additional concern remains about the ability of developing and least-developed countries to participate fully in the benefits of the SPS Agreement. Lack of opportunities to participate in several international standard-setting bodies, as well as in WTO and in numerous committees, due to resource constraints, were concerns *a priori*. Membership lists for WTO, Codex, OIE, and the IPPC indicate that both the number and the percentage of members in each of the organizations were higher for non high-income countries in almost every case (Table 13.2). However, membership in an organization is not always indicative of active participation and decisions undertaken will also have trade impacts on countries that do not participate.

Table 13.2 Membership in WTO and International Standard-setting Bodies, 1999

	Organization				
Members' income	*WTO*	*Codex*	*OIE*	*IPPC*	*Total*
High	36	33	26	33	50
Upper-middle	24	25	23	31	36
Lower-middle	32	40	32	46	57
Low	42	54	26	53	63

Source: WTO/SPS (1999b) and author's calculations.

4. FOCUS ON CITRUS

Citrus is a relatively small (less than two percent) component of agricultural cash receipts in the United States, yet it is critically important to the agricultural economies of selected states, notably to Florida (Table 13.3). On average, approximately 10 percent of U.S. citrus production entered export markets during the 1990s. The percentage of international sales was higher for selected commodities, with shares approaching 27 percent in 1990 and 1991 for grapefruit.

The 1996 USDA survey results included 18 questionable technical barriers applied to all citrus, or to specific citrus products. The estimated trade impact was U.S. $103.94 million. Five additional barriers were identified as applied to fresh fruit, and three barriers were identified as applied to all fresh fruit and vegetables, for a total cost of U.S. $42.4 million. Again, since the survey was not repeated, there is no way to determine if these numbers have increased or decreased since 1996. However, there have been at least 12 WTO notifications of changes in technical standards that impact the citrus trade, including eight from the United States, indicating that members have been making modifications to technical regulations as they have been applied to citrus.

Expanding trade in citrus has not been free of SPS controversies. In 1998, the Government of Argentina officially requested consideration to allow the export of fresh grapefruits, lemons, and oranges from four states in Argentina into the United States. The national plant-protection organization of Argentina (SENASA) proposed that the citrus-growing area in the four Argentinean states of Catamarca, Jujuy, Salta, and Tucuman is free of canker [(*Xanthomonas axonopodis* pv. *citri* (Hasse)].[xi] A 1998 economic assessment found that Argentinean lemons are the product most likely to be exported to the United States and that, due to different growing seasons, shipments were expected to complement, rather than to compete with, U.S. production (USDA/APHIS, 1998).

Table 13.3 Citrus Cash Receipts, Selected U.S. States, 1996

State or country	*Cash receipts*	
	U.S. $1,000	*percent of total*
Florida	1,607,570	23.77
California	897,241	3.64
Arizona	55,340	2.34
Texas	21,220	0.16
United States	2,581,371	1.31

Source: USDA/ERS (1999) and author's calculations.

The possibility of easing the ban on the import of Argentinean citrus to the United States has raised the anticipation that U.S. citrus might gain access to Argentinean markets, which were previously closed to any region not completely free of fruit flies. U.S. officials have long contended that there is no need for such a ban, and that a less restrictive import protocol could be applied, since fruit flies are also present in Argentina. A comparison of relative prices suggests the possibility of access to the Argentinean market for Florida grapefruit.

In August 1998, the U.S. formally notified the WTO of regulatory modifications that would recognize an area within Argentina as free from citrus canker (WTO/SPS, 1998). Imports were permitted under conditions designed to prevent canker infestations from the other citrus-growing regions of Argentina and infections from additional citrus diseases, specifically sweet orange scab (*Elsinoe australis* (Bitancourt and Jenkins)), black spot [*Guignardia citricarpa* (Kiely)], and other plant pests. However, continued negotiations over the definition of appropriate conditions of entry and a subsequent outbreak of citrus canker in regions of U.S. production delayed resolution and no product was transferred in either direction until 2000.

Additional negotiations concerning risks associated with citrus canker are continuous between the United States and the Republic of Korea. The United States allows imports of citrus from the Cheju region of Korea, except into the five major citrus-producing states (California, Florida, Arizona, Louisiana, and Texas), under a set of specified phytosanitary safety measures. [xii] The Korean government has questioned the legitimacy of geographic limitations and has requested permission to expand imports into the additional states. USDA/APHIS is undertaking a risk assessment but has clearly stated that destination is a risk factor with the probability of disease establishment in production areas to be quite high (WTO/TPR, 2000). Simultaneously, the U.S. and Korea are engaged in consultations regarding mandatory fumigation and quarantine requirements on citrus entering Korea. USDA officials have argued that the treatments shorten the shelf life of fresh fruit, and they have requested exemption for fruit originating in regions of California that have been declared fruit-fly free.

5. POSITIONS FOR THE NEXT ROUND

Article 20 of the Agreement on Agriculture committed members to initiate continuing negotiations one year before the end of the implementation period. The mandate for negotiations includes taking into account: (1) experience to that date from implementing the reduction commitments; (2) effects of the reduction commitments on world trade in agriculture; (3) non-trade concerns such as special and differential treatment to developing-country members; (4) objective to establish a fair and market-oriented agricultural trading system; and (5) further commitments necessary to achieve the long-term objectives listed in the preamble to the Agreement.

In preparation for the next round of negotiations, a series of three Ministerial Conferences were held; the most publicized occurred November 30 to December 3, 1999, in Seattle. Although the 1999 WTO Ministerial meetings in Seattle ended without agreement on an agenda for the next round, a preliminary agenda for agriculture was drafted. In this early draft there was no consensus to open, or to renegotiate, the SPS Agreement, and the preliminary positions of specific countries do not indicate a strong movement to do so. The European Union has raised the possibility of renegotiation and their position on agriculture specifically mentions the SPS Agreement, particularly as it relates to food-safety issues, and asks for confirmation that non-discriminatory science-based measures, designed to achieve the determined level of safety, are in conformity with the SPS Agreement (WTO/GC, 1999).

However, as in the initial Uruguay Round negotiations, discussion with regard to other agricultural trade and institutional issues may have significant impacts on interpretation and further implementation of the SPS Agreement. Debate over participation from developing and least-developed members is continuous. In review of implementation issues, the SPS Committee recognized a continuing need for enhanced technical assistance through the international organizations as well as for continued bilateral exchange. An increasing concern of least-developed country members is their declining share in world trade, investment, and output (WTO, 2000). The ability to implement and/or to document compliance with SPS regulatory restrictions has been suggested as a contributing factor to this marginalization. Roberts, Orden, and Josling (1999) caution against the extensive use of claims for special and differential treatment by developing countries. An overuse of such claims could reinforce perceptions that the products are unable to meet standards, and may actually hamper market access.

The use of biotechnology would likely, at least in part, be governed by the disciplines of the SPS Agreement. Prior to the Ministerial meeting in Seattle, the Canadian government proposed the establishment of a working group on biotechnology. Such a group would be a cross-committee and, thus, could incorporate issues governed by multiple agreements. While the working group was not launched, biotechnology will be a contentious topic of debate in the future for any agenda on agriculture. The debate over consumer sovereignty and risk will likely be at the center of discussions on biotechnology with significant impacts on interpretation and enforcement of the SPS Agreement.

6. CONCLUSION

The Uruguay Round negotiations ended with substantial institutional reforms, both for GATT, in general, and with regard to agricultural trade, in particular. Creation of the WTO provided a formal organization to oversee GATT Agreements and strengthened compliance and enforcement mechanisms. The Agreement on Agriculture brought agricultural trade discipline more in accordance with the underlying principles of GATT by requiring specific numerical commitments on market access, domestic support, and export subsidies. Technical barriers to trade were explicitly addressed in the SPS and TBT Agreements, designed to recognize the right of governments to limit trade through technical barriers as a mechanism to correct negative externalities, while seeking to limit the ability to intentionally misuse such measures. The SPS Agreement was seen, *a priori*, by most as a positive step toward freer trade in agricultural products.

As WTO members drafted their positions for the Ministerial meetings in Seattle, there was not a strong agenda to reopen the SPS Agreement for further negotiations. In an internal review, the SPS Committee concluded that the agreement had contributed to improving international trading relationships. Increasing use of the transparency requirements; adaptation of regulations to regional conditions; progress in the recognized standard-setting bodies; and continued conflict resolution through informal negotiation and discussion are areas of subtle and steady progress toward freer trade. Areas of concern include enhanced participation by developing countries, clarification of the relationship between consumer sovereignty and science in the process, and enforcement of formal Dispute Settlement Body rulings. A consensus seems to be emerging that the SPS Agreement has been successful, and that future progress can be best achieved through continued refinement of the existing structure, the work of the SPS Committee, and the WTO dispute settlement procedures.

7. ENDNOTES

[i] The author gratefully acknowledges the research assistance of Tara M. Minton.

[ii] The 1999 Secretariat Report on U.S. Trade Policy Review noted 50,000 U.S. standards in existence to protect consumers, public health, safety, and the environment (WTO/TPR, 2000).

[iii] For example, Roberts (1998) and James (2000) provide assessments of the U.S.-E.U. dispute over hormone-treated beef and over implications of WTO dispute panel rulings for interpretation of the SPS Agreement.

[iv] Recognized organizations for international standard setting are the Codex Alimentarius Commission (Codex) for food additives, veterinary drug and pesticide residues, contaminants, methods of analysis and sampling, and codes and guidelines of hygienic practices; the International Office of Epizootics (OIE) for animal health; and the International Plant Protection Convention (IPPC) for plant health.

[v] Allowances are made for *ex poste* emergency notifications (Annex B) and provisional measures in cases where relevant scientific evidence is insufficient (WTO/SPS, 1999b: Art. 5.7).

[vi] In order to achieve this goal, the previous consensus-based dispute settlement system was converted to a quasi-judicial system governed by the WTO. Under the new system, members have the right to initiate a panel hearing without full consensus; there are strict time limits set for each stage in the dispute process; panel reports can only be rejected by unanimous vote; and panel rulings are to be binding on all members.

[vii] In 2000, foot-and-mouth disease was discovered in parts of Argentina, Brazil, and Uruguay. The protocol established identified the problem before the product was transferred and a temporary ban was imposed.

[viii] Roberts, Orden and Josling (1999) reported notifications from 81 percent, 54 percent, 44 percent, and 6 percent of high, upper-middle, lower-middle, and low-income countries from 1995 to June 1999, respectively.

[ix] The debate over the role of economics, particularly cost-benefit methodology, in setting technical standards continues. Kopp, Krupnick, and Toman (1997) and James (2000), among others, provide discussion of these issues.

[x] Barriers identified in the USDA survey include an expert consensus-based view of 1996 measures that decreased, or potentially decreased, U.S. agricultural exports to specified export markets and appeared to violate at least one of the disciplines of the WTO Agreements. Of the 302 measures identified, over 86 percent were considered questionable because they potentially violate a principle of the SPS Agreement, 9 percent were potentially disciplined by the TBT Agreement and 5 percent served more than one purpose and so might be covered by multiple Agreements, or appeared to violate another provision of GATT 1994 (Thornsbury et al., 1999).

[xi] Citrus canker is a bacterial disease that effects most commercial citrus varieties, causing lesions on leaves, stems, and fruit and eventual tree decline. There is no known human safety risk associated with citrus canker, yet the disease will eventually make citrus plantings non-productive and has an immediate negative impact on the quality of fruit entering the fresh market.

[xii] In 1998/99 31 metric tons (MT) of tangerines were imported to the U.S. from Korea. Volume rose to 377 MT in 1999/2000. Over the same time periods 19,000 MT and 27,516 MT of oranges, 2,585 MT and 1,215 MT of lemons, and 2,197 MT and 690 MT of grapefruit were exported from the U.S. to Korea (USDA/FAS, 2000).

8. REFERENCES

Abbott, F.L. 1997. "The Intersection of Law and Trade in the WTO System: Economics and the Transition to a Hard Law System," in *Understanding Technical Barriers to Agricultural Trade*, D. Orden and D. Roberts, eds., Chapter 3, St. Paul, MN: University of Minnesota, Department of Applied Economics.

Bhagwati, J. 1996. "Trade and the Environment: Exploring the Critical Linkages," in *Agriculture, Trade and the Environment.* M. Bredahl, N. Ballenger, J. Dunmore, and T. Roe, eds. Boulder, CO: Westview Press.

James, S. 2000. "An Economic Analysis of Food Safety Issues Following the SPS Agreement: Lessons from the Hormones Dispute." Policy Discussion Paper No. 0005, Centre for International Economic Studies, University of Adelaide, Adelaide, South Australia (February).

Josling, T. 1998. *Agricultural Trade Policy: Completing the Reform.* Washington, DC: Institute for International Economics.

Josling, T. 2000. "Issues in the WTO Agricultural Negotiations: An Overview." Paper presented at the Global Agricultural Trade in the New Millennium Conference, New Orleans, Louisiana (25–26 May).

Kopp, R., A. Krupnick, and M. Toman. 1997. "Cost-Benefit Analysis and Regulatory Reform: An Assessment of the Science and the Art." Discussion Paper 97-19, Resources for the Future, Washington, DC (January).

Roberts, D. 1998. "Preliminary Assessment of the Effects of the WTO Agreement on Sanitary and Phytosanitary Trade Regulations." *Journal of International Economic Law* 2: 377–405.

Roberts, D., D. Orden, and T. Josling. 1999. "WTO Disciplines on Sanitary and Phytosanitary Barriers to Agricultural Trade: Progress, Prospects, and Implications for Developing Countries." Invited Paper presented at the 1999 World Bank Global Conference on Agriculture and the New Trade Agenda from a Development Perspective: Interests and Options in the WTO Negotiations, Geneva, Switzerland (1–2 October).

Sykes, A.O. 1995. *Product Standards for Internationally Integrated Goods Markets.* Washington, DC: The Brookings Institute.

Thornsbury, S., D. Roberts, K. DeRemer, and D. Orden. 1999. "A First Step in Understanding Technical Barriers to Agricultural Trade," in G.H. Peters and J. Von Braun, eds., pp. 453–463. *Food Security, Diversification, and Resource Management: Refocusing the Role of Agriculture?* Brookfield, VT: Ashgate.

USDA/APHIS (U.S. Department of Agriculture/Animal and Plant Health Inspection Service). 1998. *Proposed Rule for Importation of Grapefruit, Lemons, and Oranges from Argentina, Environmental Assessment.* Washington, DC: USDA/APHIS (November).

USDA/ERS (U.S. Department of Agriculture/Economic Research Service). 1999. "U.S. and State Farm Income Data." Internet website:
http://www.econ.ag.giv/briefing/farmincome/finfidmy.htm

USDA/FAS (U.S. Department of Agriculture, Foreign Agricultural Service). 2000. *Republic of Korea, Citrus, Semi-Annual Report.* GAIN Report KS0054, Washington, DC.

WTO (World Trade Organization). 2000. *Declaration of the Ministers of Trade of the Least-Developed Countries, Seattle, 20 November 1999.* WT/L/343, Geneva, Switzerland (9 February).

_____. 1999. The Legal Texts: The Results of the Uruguay Round of Multilateral Trade Negotiations. Cambridge, UK: Cambridge University Press.

WTO/GC (World Trade Organization, General Council). 1999. *E.C. Approach on Agriculture.* WT/GC/W/273, Geneva Switzerland (27 July).

WTO/SPS (World Trade Organization, Committee on Sanitary and Phytosanitary Measures). 2000. *Documents and Notifications Circulated During 1999.* G/SPS/GEN/171, Geneva, Switzerland (15 March).

_____. 1999a. *Review of the Operation and Implementation of the Agreement on the Application of Sanitary and Phytosanitary Measures.* G/SPS/12, Geneva, Switzerland (11 March).

_____. 1999b. *Membership in WTO and International Standard-Setting Bodies.* G/SPS/GEN/49/Rev.1, Geneva, Switzerland (7 May).

_____. 1998. *Notification.* G/SPS/N/USA/132, Geneva, Switzerland (28 August).

WTO/TPR. (World Trade Organization, Trade Policy Review Body). 2000. *Trade Policy Review: United States.* WT/TPR/M/56/Add.1, Geneva, Switzerland (9 March).

PART III

FOREIGN-DIRECT INVESTMENT TRADE AND VERTICAL CONTRACTING

Chapter 14

Foreign Direct Investment and Vertical Contracting in the Agri-Food Sector of Transition Economies

Hamish R. Gow[1,2] and Johan F. M. Swinnen[1,3]
[1]*Universiteit Leuven,* [2]*University of Illinois, and* [3]*European Commision*

1. INTRODUCTION

Total flows of foreign direct investment (FDI) into transition countries (TCs) increased significantly during the early 1990s.[i] FDI has a positive impact on industry in TCs because it provides capital essential for restructuring and modernization, as well as bringing managerial and technological skills, which are in short supply within the region (OECD, 1998). Agri-food sector FDI has accounted for a substantial share of total FDI. The vast majority of agri-food FDI has been directed to the agro-industry, not to primary agriculture, but FDI is still having a major impact on primary agriculture.

Agricultural transition in Eastern Europe and the former Soviet Union has been characterized by declining output and decapitalization of the production system.[ii] An essential factor in this transition has been the break-up of the system of contracting (and contract enforcement) in the agri-food chain, which had been vertically integrated and centrally planned.[iii] With imperfectly developed market and legal institutions, enforcement problems within the agri-food sector have caused contractual disruptions. In the absence of credible and enforceable contractual arrangements, the opportunity exists for one of the parties involved in the contract to attempt to extract a greater share of the appropriable quasi-rents accruing to the relationship-specific investment by renegotiating the contractual terms *ex post* (that is, holding up the transaction). These hold-up problems cause under-investment in relationship-specific investments (Klein, Crawford, and Alchian, 1978; Williamson, 1985).[iv] An important form of hold-up problems in TCs are delayed payments resulting from a combination of agriculture-specific characteristics and transition-specific problems (Gow and Swinnen, 1998).

Empirical observations suggest that FDI in upstream and/or downstream industries has provided a solution to the contracting problems, and has induced contract enforcement and access to inputs for the farms through private contract enforcement mechanisms under the form of innovative vertical contracting. While many studies have analyzed the impact of FDI (Bende-Nabende, 1998; Borensztein, de Gregorio, and Lee, 1998; Brenton and di Mauro, 1999), including the spillover effects on other sectors[v] (Lall, 1980; Watanabe, 1983; Blomström and Kokko, 1997), the impact of FDI in TCs discussed here goes beyond these traditional arguments, since FDI induces private contract enforcement and contractual convergence across sectors.

To explain the impact of FDI in transition countries, we use a neo-institutional economics-contracting model based on Klein's (1996) model of contract enforcement and hold-up problems. We show why the reforms have induced massive contract disruptions and how FDI-induced vertical contracting has been successful in providing private enforcement of contracts during the transition period. We present empirical evidence on FDI-induced vertical contracting, including a case study. The case study illustrates the dramatic potential aggregate impact of FDI (including horizontal and vertical spillovers) on investments, productivity, output, and trade, both for the processed product and for the raw material.

2. FOREIGN DIRECT INVESTMENT IN TRANSITION COUNTRIES

Between 1991 and 1996, the amount of FDI into TCs more than quintupled. The main recipients of FDIs have been the Czech Republic, Hungary, Poland, and Russia. These four countries together accounted for about 80 percent of total FDI flows into the region in 1996. The vast majority of investors are from Organization for Economic Cooperation and Development (OECD) countries; geographic proximity is a major determinant of investment. For example, German companies have been important investors in the Czech Republic and Poland; Scandinavian firms in the Baltic states; and Greek investors in Bulgaria (OECD, 1998).

The share of total FDI going to the agri-food sector ranged from 7 percent in Croatia to about 25 percent in Bulgaria (Table 14.1). On a per-

Table 14.1 Cumulative Inflows of Agri-food Foreign Direct Investment (FDI) in Transition Countries by Subsector, 1990–97

	Milling and baking	*Sugar and confec-tionery*	*Veg. oils and fats*	*Fruit and vege-tables*	*Milk and dairy*	*Meat and ready meals*	*Alcoholic beverages*	*Soft drinks*	*To-bacco*	*Other pro-cessed food*	*Total food pro-cessing*	*Per capita U.S.$*	*Share of agri-food FDI in total FDI*
Albania[a]	–	–	–	–	4.0	–	–	10.0	-	–	14.0	4.0	–
Bulgaria[a]	46.8	-18.6	6.8	27.7	12.3	–	37.0	40.9	0.9	–	191.0	22.7	25
Croatia	–	–	–	3.0	–	–	33.0	31.2	–	0.4	68.0	15.1	7
Czech Republic[a]	49.0	23.0	32.0	–	3.0	4.0	157.0	237.0	420.0	72.0	997.0	96.8	8
Estonia	9.0	–	–	5.0	11.0	1.0	16.0	15.0	17.0	11.0	85.0	57.4	11
Hungary[a]	41.8	173.0	–	230.5	167.3	9.6	123.9	21.8	32.0	32.4	832.2	80.8	–
Lithuania	12.0	23.0	–	–	15.0	2.0	25.0	–	64.7	8.2	150.0	40.4	16
Poland	107.6	765.9	110.6	28.2	123.1	70.8	226.6	454.2	730.0	298.2	2915.0	75.7	14
Romania[a]	–	42.4	–	–	–	–	166.0	321.0	–	0.5	239.9	10.6	18
Russia	65.0	692.7	–	–	138.6	57.0	121.9	294.2	90.0	–	1459.4	9.9	14
Ukraine[a]	–	42.5	91.5	4.4	50.0	50.0	6.8	116.0	–	–	361.2	7.1	22
Total	331.2	1781.1	240.9	298.8	524.2	194.4	913.2	1251.3	1354.6	422.7	7312.8	24.2	–

[a]*Estimates*

Source: OECD (1998)

capita basis, the Czech Republic, Hungary and Poland stand out as the recipient countries with the highest agri-food FDI.

The vast majority of agri-food FDI has been directed to the agro-industry, rather than to primary agriculture. At most two percent to three percent of total FDI in Romania and the Ukraine has been directed to primary agriculture. Within agro-industry, most FDI has been directed into the sugar and confectionery, tobacco and soft drink subsectors. Alcoholic beverages, and milk and dairy production also attracted substantial FDI. However, meat processing, for example, has received relatively little investment from foreign firms.

3. THE IMPACT OF FDI: DIRECT EFFECTS AND SPILLOVERS

3.1 General

FDI provides much of the capital and managerial and technological skills needed for the restructuring and modernization of agro-industries in transition countries. Also foreign managers can often more easily transfer tacit know-how than can local managers who are not yet familiar with the workings of private enterprise (Teece, 1986).

It is well known that FDI not only benefits the recipient company, but it can also affect the behavior and performance of companies with whom it competes (so-called horizontal spillovers) and/or exchanges products, either as a supplier or as a purchaser (so-called vertical spillovers). For example, the foreign-owned firms force local competitors to improve their operations, increase their managerial efforts, and adopt new marketing and contracting techniques. Furthermore, the demonstration of new technologies increases the know-how of local firms, and, in combination with increased competition, forces them to operate more efficiently (Blomström and Kokko, 1997).

Some of the spillovers from FDI result from cooperative linkages between the multinational corporations' (MNC) foreign affiliates, and local suppliers, because suppliers are forced to meet the higher standards of quality, reliability, and speed of delivery (Watanabe, 1983). Spillovers also occur when local firms benefit from the MNC affiliate's superior knowledge of product, or process, technologies or markets, without incurring the initial setup and development costs that may exhaust the gains from the improvement. MNCs can affect the economic welfare of input suppliers in three ways: (1) the quality of goods and services that they buy, (2) the influence they may exert upon the terms of procurement, and (3) the impact they may have upon the technological capability, managerial initiative, and organizational competence of their suppliers (Dunning, 1993). Direct spillover effects come from MNCs who contribute to (1) raising the productivity and efficiency of local suppliers as they help prospective suppliers set up their production; (2) providing technical assistance or information for raising the quality of the suppliers' products; or (3) facilitating innovations that provide, or assist, in the purchase of raw materials and intermediaries; (4) providing training and helping with management and organization; and (5) assisting suppliers to diversify by finding additional customers (Lall, 1980).

3.2 Impact in Transition Countries

Empirical observations suggest that spillover effects are very important in transition countries. For example, while the vast majority of agri-food FDI has been directed to the agro-industry segment of the supply chain, rather than to primary agricultural production, FDI has still had a major impact on growth and efficiency at the farm level. The beneficial impact of FDI has arisen from vertical contracting between farmers and the foreign affiliates in the upstream and downstream industries. The inflow of FDI into the agri-food chain, when accompanied by innovative vertical contracting and financing, has successfully reduced financial constraints and stimulated investment and technology adoption at the farm level, which has resulted in substantial quality and yield improvements for contract producers (Gow and Swinnen, 1998).

Additionally, FDI spillovers have occurred through changing management and contracting practices both within the initially affected sector, as well as across adjacent sectors. Firms have been observed imitating these contractual relationships as they compete for the same primary producers and as they compete for their fixed factor resource, which is land.

4. TRANSITION AND CONTRACT DISRUPTIONS

4.1 A Model of Contract Hold-up Problems in Transition[vi]

Contracts are naturally incomplete, because economic agents find it difficult and expensive to foresee and plan for all possible contingencies, as well as to enforce these contracts, especially when outcomes are unobservable or non-verifiable by a third party (Hart, 1995). Contractual incompleteness thus often results in parties exposing themselves to *ex post* costs and hazards related to their sunk investments in relationship-specific assets; that is, the occurrence of hold-up problems.

There are two mechanisms to reduce the likelihood of a hold-up problem: private sanctions and legal (court) enforcement. Private sanctions include both the losses that result from termination or from nonrenewal of the contract or relationship (that is, the future quasi-rents of the relationship-specific investments), and the damage to the reputation of the party that holds up the transaction. (The latter includes future transacting parties that impose an increased cost of doing business on the reneging party by demanding more explicit and/or favorable contractual terms and by preferring written contracts to verbal promises.)

Traditional contract theory usually considers that court enforcement and private enforcement are alternatives. However, Klein emphasizes a fundamental complementarity between court enforcement and private enforcement. Contractual terms are used "to economize on the amount of private enforcement capital necessary to make a contractual relationship self-enforcing by merely 'getting close' to the desired performance in a wide variety of circumstances (without creating undue rigidity) and to let the threat of private enforcement move performance the remainder of the way to the desired level" (1996: 455–456).

It is sometimes not viable to use legal-dispute mechanisms due to a combination of litigation costs, ineffective contract law, poor third-party verifiability, and the potential loss of the only suitable trading partner for that commodity. This is especially true in TCs. For example, the agricultural-processing sector is often characterized by geographical monopsonies. Therefore, the potential loss of their sole trading partner can impose high costs on a production enterprise, especially when the relationship-specific investment has already been sunk. Further, the legal and judicial systems are still in their embryonic stages of development, hence, outcomes of any court decision are highly uncertain and are nontransparent.

Firms may also prefer incomplete contracts. Strict specification of the contractual terms may produce unwanted rigidity. For example, once contractual terms are written down, one of the parties may decide to hold up the transaction by enforcing the literal terms, even if these run against the initial intentions of the contract.[vii] With incomplete contracts, parties gain greater flexibility to opt out of the contractual arrangements if future market conditions deviate substantially from expectations.[viii] Thus, transacting parties may intentionally elect to leave specifications out of the contract, opting, instead, to use private sanctions, as opposed to courts, to enforce the contractual arrangements.

4.2 When do Hold-up Problems Occur?

At each point in time, both partners in a contract consider the costs and benefits of holding up the contract. A hold-up problem will occur when benefits are greater than are the costs to one party. Klein (1996) argues that, *ex ante*, transacting parties will only engage in a transaction if they expect that both parties will honor the (implicit) contract (that is, if (for both parties) the expected cost of contractual breach is higher than the expected benefits). The main reason hold-up problems occur is that unanticipated changes in the external environment affect the cost/benefit ratio sufficiently to make it optimal for one party to hold up the contract.

Consider the following example. A producer of product x needs to invest in production facilities specific for a certain delivery to a processor. To prevent a hold-up problem by the processor after the producer makes the investment, both parties agree to a contract that specifies product characteristics (that is, quality, quantity, and a fixed price). Assume that the price is set at the expected market price P_0 (Figure 14.1). Once the contract has been agreed upon, the actual market price, P, may deviate from the contracted price P_0. If $P > P_0$, the contract provides unanticipated rents to the processor and the benefits of breaching the contract increase for the producer, since he could get a higher price by selling his product in the market (and vice versa). The producer's benefits of breaching the contract, $H_A(p)$, increase with an increase in the wedge between the actual price and the contracted price. At some price, P_A, the benefits $H_A(p)$, will become larger than the costs for the producer of breaching the contract, K_A, where K_A is the sum of the reputation losses and capital costs. Analogously, P_B represents the market price below which is optimal for the processor to breach the contract and to buy his supplies in the market.

If the market price stays within the $(P_B, P)_A$ price range, the contract will be honored, otherwise it will not be honored. The range (P_B, P_A) is therefore called the self-enforcing range of the contract (Klein, 1996). The self-enforcing range measures the extent to which market conditions can change without precipitating a hold-up problem by either party.[ix] Changes in market conditions may alter the value of specific investments and, therefore, the benefits of a hold-up problem. Yet, as long as the relationship remains within the self-enforcing range whereby each transactors' benefits of a hold-up problem is less than his or her costs, a hold-up problem will not take place.

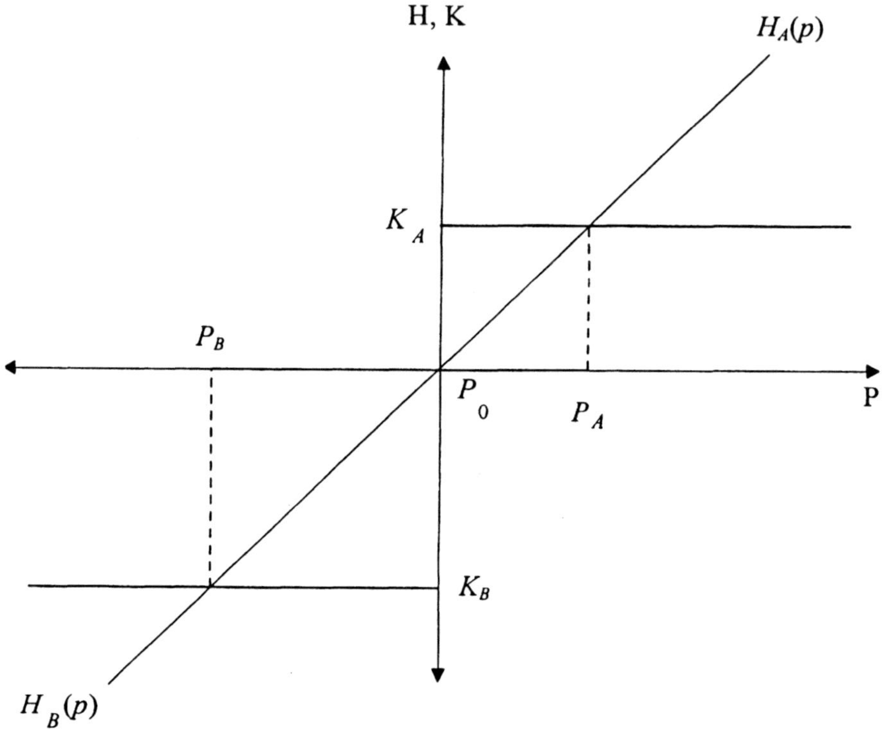

Figure 14.1 The Self-enforcing Range in Contracts

In this framework, hold-up problems only occur when a sufficiently large unanticipated event shifts a contract outside the self-enforcing range. A hold-up problem, thus, would never occur in a fully anticipated world. If the transactor had anticipated the possible occurrence of the present market conditions and had recognized the potential for hold-up problems to occur, it is unlikely that he would have undertaken the initial investment to begin with. Otherwise, he would have insisted upon different specifications in the contract. Alternatively, since the magnitude of the private sanctions (reflected in K_A and K_B) affects the size of the self-enforcing range, a different distribution of private sanctions could also prevent a breach of contract.

4.3 Transition and Contract Hold-up Problems

Prior to the initial reforms in Central and Eastern Europe, most of the agri-food supply chain was centrally planned and vertically integrated.[x] The central authority provided contractual enforcement, and transacting parties faced a low (or zero) probability of having a hold-up problem. The reforms caused several institutional changes, which lead to widespread hold-up problems.

First, economic reforms split the agri-food chain into autonomous enterprizes.[xi] Second, contractual terms were no longer enforced by the legal system or by the central planning authority, and a new legal enforcement mechanism was absent or ineffective.[xii] Third, since the transacting parties had no previous experience with hold-up problems, private-enforcement levels were left unchanged and producers continued making relationship-specific investments. Fourth, the reforms brought unanticipated shocks that dramatically changed market conditions, which shifted the processors outside the self-enforcing range.

As market conditions moved outside the self-enforcing range, hold-up problems were not prevented by court actions. The lack of transparency, or the absence of a suitable legal system, has complicated court enforcement in TCs.[xiii] Once hold-up problems occurred, they reinforced themselves because the private enforcement capital of processors had declined as their reputation and the relationship-specific investments by both partners decreased, which further reduced the self-enforcing range.

4.4 Transition Hold-up Problems in the Agri-food Chain

Within the agri-food chain these hold-up problems have typically been characterized by long payment delays for delivered product (that is, accounts payable) by downstream processors. They have effectively provided processors with an interest-free loan for the length on the delay, and have caused additional financial strain upon the already distressed producers. For example, in the Slovak Republic, the length of the delay is (strongly) negatively correlated with the profitability of the farm. In 1994, the average number of days of delays for collection of accounts receivable were 80.97 and 110.76 for profitable and unprofitable farms, respectively. The delays were similar in 1995, with averages of 82.15 and 112.70 for profitable and unprofitable farms, respectively (OECD, 1997).

Delayed payments are not only an important constraint for farms, but also for other companies in the agri-food chain. A recent survey among food processors in the Czech Republic, Hungary, and Slovenia concluded that late payments were considered the single most important obstacle to company growth in the Czech Republic and Slovenia, while it ranked number 3 out of 12 causes in Hungary (Table 14.2). Considering that these are some of the most advanced TCs, one can imagine that this problem is at least as important in other TCs.[xiv]

Table 14.2 Importance of Various Barriers to Growth for Food Processors

	Czech Republic	*Hungary*	*Slovenia*
Type of barrier	*Average survey scores (1 – 4[a])*		
Exchange rate instability	2.16	2.52	2.04
Inflation	2.41	3.17	2.69
The level of interest rates	3.19	3.15	2.77
Access to credit	2.45	2.43	2.38
Activities of organized crime and gangsters	1.58	2.03	1.73
Government price controls	1.74	1.57	2.67
Other government intervention	1.25	2.62	2.67
Your company having high levels of debt	2.48	2.02	2.23
Late payment by customers	3.44	2.70	3.54
Enforcement of bankruptcy laws	2.10	1.29	2.21
Activities of state monopolies	1.90	1.91	1.83
Problems with privatization	1.61	1.49	0.94
Exchange rate instability	2.16	2.52	2.04

[a] 1 describes no problem, 4 describes a major barrier.
Source: Gorton, Buckwell, and Davidova (1999).

The response of farms was to internalize their exchange transactions through vertical integration, to shift exchange to spot markets, or to terminate their activities and wait for better market conditions. Combined, these factors have contributed to the decrease in agricultural output, to the decline in fertilizer use, and to investments in livestock and tractors, but with minimal changes in the use of arable land (Gow and Swinnen, 1998).

5. FOREIGN DIRECT INVESTMENT, VERTICAL CONTRACTING, AND PRIVATE-CONTRACT ENFORCEMENT

Empirical observation suggests that FDI in upstream and/or downstream industries has provided at least one solution to contracting problems, and has induced contract enforcement and access to inputs for the farms through private contract-enforcement mechanisms under the form of innovative vertical contracting. More specifically, FDI has provided food processors with the ability and incentives to shift the self-enforcing range to better match the transacting parties' expectations of future market conditions, thereby sustaining the contracts. Through the provision of innovative and credible contracts with producers, enforced by private capital, such processors have been able to reduce the probability of a hold-up problem and have stimulated increased relationship-specific investment by producers.

FDI-induced vertical contracting between agribusiness and farms has taken many forms, but it has included conditions for product delivery and payments as well as for input-support programs for farms. Tables 14.3 and

14.4 summarize such programs that various agri-businesses have implemented after FDI inflows in Central and Eastern Europe (Table 14.3) and in the former Soviet Union (Table 14.4). These programs include pre-financing of inputs (chemical, fertilizer and seeds), interest subsidies on machinery investments, and technical support for technology adoption and management.

6. A CASE STUDY: JUHOCUKOR A.S.

Juhocukor a.s. is the largest sugar processing company in Slovakia and was taken over by a foreign company (Eastern Sugar BV[xv]) in 1993. Following the takeover, Juhosukor's sugar output increased by 212 percent and sugar-beet deliveries increased by 157 percent between 1993 and 1997 (Figure 14.2). This coincided with dramatic increases in investments, productivity, and output at the farm level, as well as strong spillover effects to other sectors.

Before the takeover, Juhocukor faced many problems. Besides the standard ones—such as financial constraints, under-utilized processing

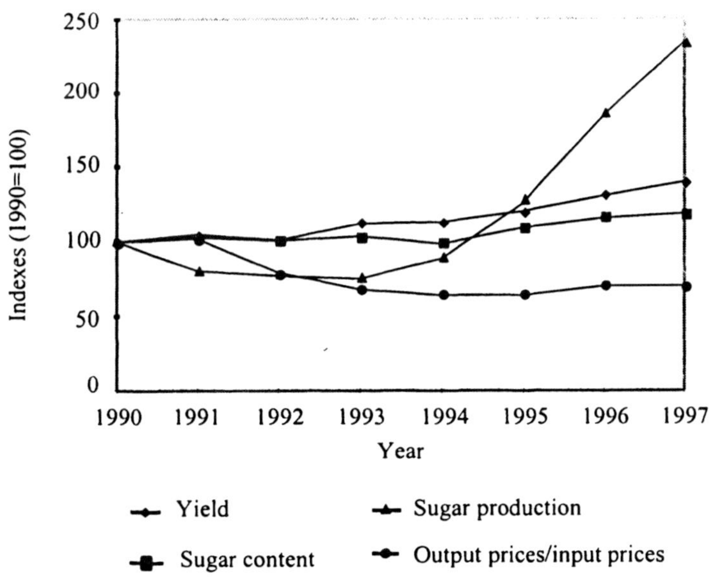

Figure 14.2 Production Indicators for Juhocukor a.s. and their Producers

Table 14.3 FDI Induced Vertical Contracting and Support to Producers by Agribusiness in Central and Eastern Europe

Foreign investor	*British/French sugar processor*	*British sugar processor*	*Belgium beer producer*	*Belgium maltster*	*European seed merchant*
Local firm or partner	*Local sugar processor*	*Local sugar processor*	*Local brewery*	*Local malting plant*	*Foreign food manufacturers*
Activity	*Sugar production*	*Sugar production*	*Malting*	*Malting*	*Seed manufacturing*
Countries of investment	*Hungary* *Czech Republic* *Slovakia*	*Poland*	*Romania* *Hungary*	*Croatia*	*Central and Eastern Europe*
Producer type	*Sugar beets*	*Sugar beets*	*Barley*	*Barley*	*Crops*
Types of producer support					
Production and equipment training	Yes	Yes	Yes	Yes	Yes
Credit access programs	Yes	Yes	Yes	Yes	Yes
Input provision and facilitation	Yes	Yes	Yes	Yes	Yes
Machinery procurement	Yes	Yes	—	—	Sometimes
Agronomical support	Yes	Yes	Yes	Yes	Yes
Veterinary support	—	—	—	—	—
Harvest and handling support	Yes	Yes	Yes	Yes	Yes
Quality control	Yes	Yes	Yes	Yes	Yes
Transportation	Yes	Yes	Yes	Yes	Yes
Specialized storage	Yes	Yes	Yes	Yes	Yes
Business and financial management support	Yes	Yes	Yes	Yes	Yes
Market access	Yes	Yes	Yes	Yes	Yes
Timely payments	Yes	Yes	Yes	Yes	Yes

Source: Authors' compilation.

Table 14.4 FDI Induced Vertical Contracting and Support to Producers by Agribusinesses in the Former Soviet Union

Western investor	*U.S. combine producer*	*Swiss confectionary producer*	*U.S. dairy processor*	*U.S. food catering firm*	*U.S. nut processor*	*U.S. agri-chemical manufacturer*
Foreign partner	*Ag. equipment dealer*	*Dairy plant*	*Dairy plants*	*Food catering enterprise*	*Facilities owners*	*Ag. input distributors*
Activity	*Ag. equipment leasing and sales*	*Confectionary*	*Dairy products*	*Fast food service*	*Walnut processing*	*Fertilizer and pesticide sales*
Countries of investment	*Ukraine*	*Russia*	*Ukraine Moldova Kazakhstan*	*Russia Ukraine*	*Moldova*	*Ukraine Russia*
Producer type	*Grain and oilseed*	*Dairy producers*	*Dairy producers*	*Vegetable producers*	*Walnut growers*	*Crop producers*
Types of producer support						
Production and equipment training	Yes	Yes	Yes	Yes	Yes	Yes
Credit access programs	Yes	—	—	—	Yes	Yes
Input provision and facilitation	Yes	Yes	Yes	Yes	Yes	Yes
Machinery procurement	—	—	Yes	Yes	Yes	Yes
Agronomical support	Yes	Yes	Yes	Yes	Yes	Yes
Veterinary support	—	Yes	Yes	—	—	—
Harvest and handling support	Yes	Yes	Yes	Yes	Yes	Yes
Quality control	Yes	Yes	Yes	Yes	Yes	Yes
Transportation	Yes	Yes	Yes	Yes	Yes	Yes
Specialized storage	Yes	Yes	Yes	Yes	Yes	Yes
Business and financial management support	Yes	Yes	Yes	Yes	Yes	Yes
Market access	Yes	Yes	Yes	Yes	Yes	Yes
Timely payments	Yes	Yes	Yes	Yes	Yes	Yes

Source: Authors' compilation.

facilities, poor quality control, outdated processing equipment, and low-quality inputs—it had gained a bad reputation for not paying farms within a reasonable time after delivery of sugarbeets. These delayed payments had worsened problems farms already had with severe cash flow and profitability. As a consequence, farmers no longer wanted to invest in sugar-beet production for Juhocukor: sugar-beet deliveries to Juhocukor declined from 315,000 tonnes in 1990 to 214,000 tonnes by 1993, its contracted hectares decreased from 7,800 hectares to 6,000 hectares, and its sugar production decreased from 32,000 tonnes to 24,000 tonnes over the same period.

In the framework of our contract model, it was the case that the reform-induced changes in the institutional and market conditions and the erosion of Juhocukor's reputation by previous hold-up problems had shifted the expected market conditions outside the self-enforcing range, and had made private contract enforcement impossible under the existing circumstances. In order to stimulate farms to invest in high-quality beet production and to deliver it to the company after the FDI takeover, Juhocukor introduced several programs that were intended to make the contracts self-enforcing and were intended to reduce the likelihood of a hold-up problem, as perceived by the beet producers, by restructuring the private enforcement capital (that is, the distribution of costs and benefits of a contract hold-up problem). In addition, the programs were intended to ease producers' access to necessary inputs.

First, to improve its reputation, the company started paying contracts at the time of product delivery to the factory (it was the only company providing timely payments). Second, an input provision and investment facilitation program assisted farmers with the purchase of inputs (such as seeds, fertilizer, and chemicals), and assisted farmers when accessing credit. The investment and financial facilitation packages were initially provided on an *ad hoc* basis, but were formalized in 1995 into a program developed with Polnobanka (the main bank lending to agriculture in Slovakia) for the financing of machinery investment and working capital. The program provided Polnobanka with a guarantee for the repayment of both the principal and interest on each loan, and provided the contract growers with an interest-rate subsidy between three percent and seven percent.

Second, through accepting the residual claim on these investments via the guarantees, Juhocukor effectively increased its own costs of not honoring the contracts, thereby making its honoring of the contract more likely and its promises to do so more credible.[xvi] At the same time, the program reduced the costs of a Juhocukor hold-up problem for producers. In the event of non-payment, beet producers would not have to pay for their purchased inputs and would only lose returns to their labor and personal capital contributions. By simultaneously reducing the likelihood of a hold-up problem by

Juhocukor and by limiting the costs of it for producers, the programs induced farms to invest in sugar beets.

Third, a technical support and extension program (which included agronomical support, soil testing, integrated pest management, production and managerial advice), expanded the amount of non-salvageable relationship-specific investment that Juhocukor had tied up in the contract, which increased the amount of capital Juhocukor had committed and thereby enlarged the self-enforcing range of the contract.

Finally, Juhocukor attempted to restore its damaged reputation through an extensive media and public-relations campaign. The campaign informed potential producers about the contracts and programs, and signaled to producers that Juhocukor was willing to publicly risk its reputation to back these contracts. As their reputation increased over time, so did the amount of private enforcement capital that Juhocukor had committed to the contract.

7. IMPACTS OF FOREIGN DIRECT INVESTMENT AND OF VERTICAL CONTRACTING

7.1 Direct Effects and Spillovers

The aggregate impact of FDI, including the horizontal and vertical spillovers, can be quite dramatic. It is difficult to obtain reliable and disaggregated data on these effects. In the case of Juhocukor, for which detailed information is available, the combined impact was dramatic. At the farm level, the programs induced large increases in the production and technical efficiencies of sugar-beet production: average yields increased from 33 tonnes per hectare with 13 percent sugar content in 1993 to about 45 tonnes per hectare with 16 percent sugar content for the 1997 season. Further, after the introduction of the finance and investment programs in 1995, contracted hectares increased from around 6,000 tonnes to 9,500 tonnes. [(Note that the terms of trade remained stable from 1993 to 1997, which indicates that something, apart from price change, is driving these results (Figure 14.2).] In combination, the increase in farm yields and contracted hectares resulted in an increase in Juhocukor's sugar production from 24,000 tonnes in 1993 to an estimated 75,000 tonnes in 1997.

The impact of FDI on total Slovakian sugar production was even heavier, since competition induced other sugar companies to imitate Juhocukor's contractual arrangements, none of which were taken over by a foreign company until 1998. With a one- or two-year delay, the result was increases in output and productivity in the other sugar companies and their suppliers as well (Figure 14.3).

The aggregate impact of FDI on total Slovakian sugar production and trade is summarized (Figure 14.4). Total sugar-beet production fell from 1.87 million tonnes in 1989 to 1.1 million tonnes by 1993. With spillover effects reflected in output after 1995, aggregate sugar output increased from 140,000 tonnes to around 250,000 tonnes between 1995 and 1998. The trade impacts were significant because sugar imports declined by 50 percent over this period (from 75,000 tonnes to around 35,000 tonnes), despite a strong growth in (officially recorded) consumption from 210,000 tonnes to 275,000 tonnes.

8. CONTRACTUAL CONVERGENCE AND MORE SPILLOVER EFFECTS

Besides the spillover effects to supplying farms and to competing companies, spillover effects occurred across commodity sectors. Contractual convergence emerged from competition among contractual arrangements (Eggertsson, 1990). Different contracts converged to a new equilibrium as firms began imitating other firms' (initially experimental) contractual arrangements once they are judged to be successful.

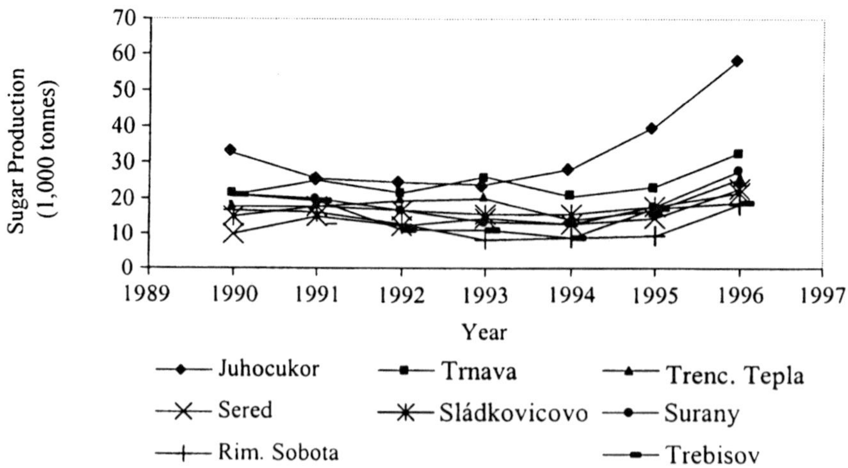

Figure 14.3 Sugar Production by Processing Company in Slovakia

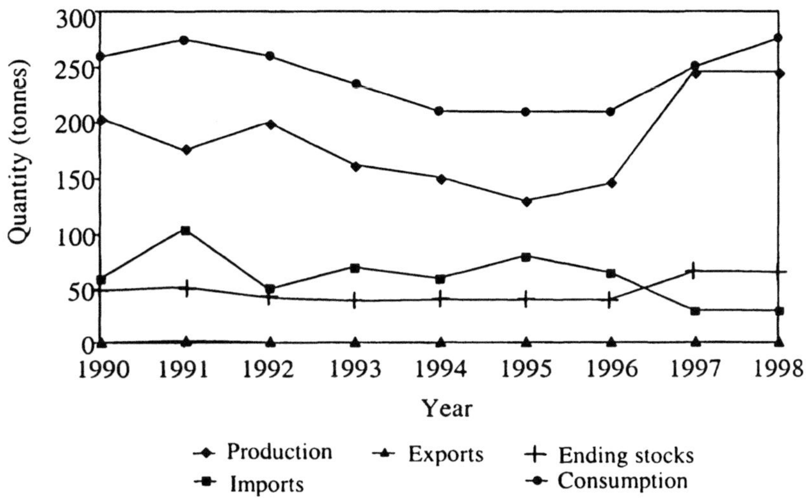

Figure 14.4 Sugar Production, Consumption, and Trade in Slovakia

In the Juhocukor case, it was observed that firms competing for the same farm resources (that is, land) were forced to offer similar contractual arrangements, thereby causing contractual convergence. The contractual convergence and subsequent investment and technology adoption at the upstream production level has not been confined solely to the directly affected commodity sector. The adoption of similar contractual arrangements has also been observed in the adjacent commodity sectors, such as oilseeds, as these firms have also had to compete for the same factor inputs (such as land), and have evidenced similar results.

Similar convergences and impacts have followed FDI in other sectors, especially when there has been a requirement for high-quality inputs by the downstream processing firms. For example, Palma Tumys (Henkel) in the Slovakian oilseed sector provided similar contracts and associated programs. These programs have been equally successful. Other Slovakian sectors that were similarly affected include the brewing barley sector with Zlaty Bazant (Heineken); corn starch and isoglycose with the Amylum group; and the dairy sector with Rajo (Schärdinger) and Majcichov (Farmco).

Finally, another contractual spillover effect observed is that, following the success of the contract-based lending that Polnobanka developed with Juhocukor and Palma-Tumys, Polnobanka has now developed a standard series of credit lines which enable downstream enterprises to provide farmers with advance payments for contracted supplies. The schemes use draft loans with the future harvest acting as the collateral and operate in a similar way to the scheme previously discussed.

9. CONCLUSIONS

FDI in the agri-food sector has increased significantly since the beginning of the reforms in TCs. The impact of FDI on the agri-food sector of TCs appears to be significantly larger than was once thought. FDI has had important effects on productivity, output, and trade in the industry in which it occurs as well as in the upstream and downstream market segments, as it has induced increased investment and technology adoption across the sector.

The spillover effects at the farm level results from FDI-induced vertical contracting. As part of this vertical integration, agribusiness enterprizes taken over by foreign companies have, in many cases, implemented producer-support programs, including financial and investment assistance programs and extension support that have had a substantial positive impact.

This chapter has attempted to explain why FDI programs have contributed to an increase in farm output and productivity. It is argued that these programs provide private enforcement capital to enforce contracts between the company and the farm. Contract disruptions and hold-up problems—mostly in the form of payment delays—have become pervasive in TCs' agri-food sectors and have caused sub-optimal resource allocations, reduced investment, and declines of output and exports. In the absence of contract enforcement mechanisms due to transition problems, private enforcement mechanisms are crucial to enforce contracts, and to provide credible incentives for investments.

Vertical contracting and the associated support programs increase the costs of contract breaches for the FDI company, making contract breach by the company less likely, and reducing the cost for the farm. The support programs, therefore, increase the private enforcement capital to the contract and improve incentives for firms to make contract-specific investments. At the same time, by providing guarantees to banks and to the companies that supply inputs to the farmers, the flow of farm inputs increases. In combination, both appear to have stimulated the growth and productivity increases at the farm level.

The empirical results from a case study of FDI in a Slovakian company show dramatic positive effects of FDI on output, yields, and quality. Empirical evidence from the former Soviet Union and Central and Eastern Europe suggest that the effect of this FDI-induced vertical contracting may be dramatic. Also, empirical evidence further suggests that the introduction of the contractual arrangements has caused spillover effects to other firms and even to other sectors. Competing firms have imitated the contractual arrangements inducing positive impacts beyond the direct FDI-induced vertical contract.

10. ENDNOTES

[i]Brenton and di Mauro (1999) argue that E.U. FDI in advanced transition countries is consistent with expectations based on a gravity model, but that FDI in less advanced countries in the region is considerably below its potential level.

[ii]The focus in this paper is on the transition countries of Eastern Europe and the former Soviet Union, although some of the arguments also apply to transition countries elsewhere in the world. For a comparison of transition in Eastern Europe, the former Soviet Union, and East Asia, see Macours and Swinnen (2000b, 2001).

[iii]Other causes include declining terms of trade for agricultural producers, financing problems due to a reduced supply of agricultural credit, market uncertainty with high inflation, disruptions caused by farm restructuring, privatisation and land reforms, and the weather (Macours and Swinnen, 2000a).

[iv]Hold-up problems are defined as "the general business problem in which each party to a contract worries about being forced to accept disadvantageous terms later, after it has sunk an investment, or worries that its investment may be devalued by others." (Milgrom and Roberts, 1992, p.136).

[v]Blomström and Kokko (1997) suggest that the most significant channels for the dissemination of modern, advanced technology are external effects or spillovers from FDI. They occur when the multinational corporation (MNC) affiliate cannot reap all of the productivity or efficiency benefits that follow in the host country's local firms as a result of their entry or presence. These spillovers may occur when local firms copy some technology used by the MNC affiliates, when they are forced to a more efficient use of existing technology and resources, or when they search for new and more efficient technologies as a result of the increased competition following market entry of the MNC affiliate.

[vi]See Gow, Streeter, and Swinnen (2000), and Gow and Swinnen (forthcoming) for a more extensive analysis and discussion of the theoretical model.

[vii]See Klein, Crawford, and Alchian (1978), and Klein (1996) for an extensive discussion of these issues related to the *Fisher Body v General Motors* case.

[viii]This is one of the likely reasons for the popularity of spot markets for exchange markets in transition countries. Given the great uncertainty about future market conditions, transactors prefer the flexibility provided by spot markets, because they are not locked into an adverse situation.

[ix]For an analysis of the conditions under which contracts are self-enforcing see Telser (1980) or Bull (1987).

[x]For example, prior to the initial reforms in the Slovak Republic both the upstream and downstream industries were composed of large state-owned companies, one per sector and, in certain cases, the same firm was on both sides of the market (for example, the Agricultural Supplies and Procurement Organization (PZN)). This enabled the state to gain total control of the sector, and it meant that producers effectively faced monopolies on both sides of the market. Additionally, all production and resource allocation decisions, as well as production and price targets, were set centrally by the state. The artificially low

and administratively set commodity prices required that agriculture had to be heavily compensated through an extensive range of subsidies. In addition, massive consumer subsidies were provided for basic foodstuffs. These factors stimulated excessive production and consumption of agricultural products, as well as the establishment of overcapacity in the processing sector (OECD, 1997).

[xi]For example, beginning in 1991 the Czechoslovakian (later Slovak) government initially switched the previously vertically integrated agri-food processing sector comprised of 30 firms and 188 processing plants into 197 separate and autonomous state-owned enterprises. Later, these firms were either privatized during the first voucher privatization, or were sold to selected strategic purchasers in the second stage of privatization (OECD, 1997).

[xii]Many of the Central and Eastern European governments have drafted suitable legislation on contracting, for example, prompt payment laws in the Slovak Republic and Slovenia. However, it takes time for these laws to pass through parliament, thus, until these laws are enacted, transacting parties will have to rely on other enforcement mechanisms.

[xiii]For example, in Slovakia during 1994-95, delays in the payment for delivered milk by dairy companies of up to 6 months or more were not uncommon, yet farmers continued producing milk and delivering it to the processing plants without taking legal action. The Slovak Ministry of Agriculture has recognized this problem and has begun the process of getting a prompt payment law passed through parliament, which requires dairy companies to pay for delivered milk within 14 days. However, the jury is still out regarding if, and how, they will enforce this law.

[xiv]In fact, the shift to barter trading in Russia and other NIS can be interpreted as an institutional response to extreme payment arrears that exist in these countries (Melyukhina and Khramova, 1999; Wandel, 1999).

[xv]Eastern Sugar BV is a 50:50 joint venture between Tate and Lyle and Saint Louis Sucre, created for their sugar activities in Central and Eastern European countries with investments in the Czech Republic, Hungary, and Slovakia.

[xvi]Notice that this program does not increase the self-enforcing range of the contract. Rather, by reallocating the private enforcement capital from the producers to itself, Juhocukor shifted the self-enforcing range and made it less likely that market conditions would fall outside the range (Gow, Streeter, and Swinnen, 2000).

11. REFERENCES

Bende-Nabende A. and J.L. Ford. 1998. "FDI, Policy Adjustment, and Endogenous Growth: Multiplier Effects from a Small Dynamic Model for Taiwan, 1959–1995." *World Development* 226(7): 1315–1330.

Blomström, M. and A. Kokko. 1997. "How Foreign Investment Affects Host Countries." *Policy Research Working Paper 1745*. Washington DC: International Economics Department, The World Bank.

Borensztein, E., J., De Gregorio, and J.W. Lee. 1998. "How does foreign direct investment affect economic growth?" *Journal of International Economics* 45: 115–135

Brenton, P. and F. di Mauro. 1999. "The Potential Magnitude and Impact of FDI Flows to CEECs." *Journal of Economic Integration* 14(1): 59–74.

Bull, C. 1987. "The Existence of Self-enforcing Implicit Contracts." *Quarterly Journal of Economics* 102(1): 147–159.

Dunning, J. 1993. *Multinational Enterprizes and the Global Economy*. Reading, MA: Addison-Wesley Publishing Co.

Eggertsson, T. 1990. *Economic Behaviour and Institutions*. Cambridge, UK: Cambridge University Press.

Gorton, M., A. Buckwell, and S. Davidova. 1999. "Transfers and distortions along the CEEC food supply chains." Paper presented at the FAIR conference on Agricultural Implications of EU-CEEC Accession, Leuven, Belgium (March).

Gow, H.R. and J.F.M. Swinnen. "Private Enforcement Capital and Contract Enforcement in Transition Economies." *American Journal of Agricultural Economics* (forthcoming).

_____. 1998. "Agribusiness Restructuring, Foreign Direct Investment, and Hold-up Problems in Agricultural Transition." *European Review of Agricultural Economics* 25(4): 331–350.

Gow, H.R., D.H. Streeter, and J.F.M. Swinnen. 2000. "How Private Contract Enforcement Mechanisms can Succeed where Public Institutions Fail: The Case of Juhocukor a.s." *Agricultural Economics* 23: 253–265.

Hart, O. 1995. *Firms, Contracts and Financial Structure*. Oxford, UK: Clarendon Press.

Klein, B. 1996. "Why Hold-Ups Occur: The Self-Enforcing Range of Contractual Relationships." *Economic Inquiry* 36 (July): 444–463.

Klein, B., R.G. Crawford, and A.A. Alchian. 1978. "Vertical Integration, Appropriable Rents, and the Competitive Contracting Process." *Journal of Law and Economics* 24: 297–326.

Lall, S. 1980. "Vertical Interfirm Linkages in LDCs: An Empirical Study." *Oxford Bulletin of Economics and Statistics* 42: 203–226.

Macours, K. and J.F.M. Swinnen. "Patterns of Agrarian Transition." *Economic Development and Cultural Change* (forthcoming).

_____. 2000a. "Causes of Output Decline in Economic Transition: The Case of Central and Eastern European Agriculture." *Journal of Comparative Economics* 28 (1): 172–206.

_____. 2000b. "Impact of Reforms and Initial Conditions on Agricultural Output and Productivity Changes in Central and Eastern Europe, the Former Soviet Union, and East Asia." *American Journal of Agricultural Economics* 85(5): 1149–1155.

Masten, S.E. 1996. ed. "Introduction." *Case Studies in Contracting and Organization*. Oxford, UK: Oxford University Press.

Melyukhina and Khramova. 1999. "New Players in the Food Marketing Chain in Russia," in P. Wehrheim and J. Von Braun eds., *Russia's Food Economy: Toward Truly Functioning Markets* (forthcoming).

Milgrom, P. and J. Roberts. 1992. *Economics, Organizations, and Management*. Englewood Cliffs, NJ: Prentice Hall.

OECD (Organization for Economic Cooperation and Development). 1998. *Agricultural Policies Markets and Trade in Transitional Economies: Monitoring and Evaluation, 1998*. Paris, France: Centre for Cooperation with the Economies in Transition, OECD.

_____. 1997. *Review of Agricultural Polices: Slovak Republic*. Paris, France: Centre for Co-operation with Economies in Transition, OECD.

Pirrong, S.C. 1993. Contracting Practices in Bulk Shipping Markets: A Transaction Cost Explanation. *Journal of Law and Economics* 36(October): 937–976.

Swinnen J. and H. Gow. 1999. "Agricultural Credit Problems and Policies during the Transition to a Market Economy in Central and Eastern Europe." *Food Policy* 24(1): 21–47.

Teece, D. 1986. "Transactions Costs Economics and the Multinational Enterprise: An Assessment." *Journal of Economic Behavior and Organization* 7(March): 21–45.

Telser, L.G. 1980. "A Theory of Self-enforcing Agreements." *Journal of Business* 53(1): 27–44.

Wandel. 1999. "Vertical integration in the Russian Agro-Food Sector," in P. Wehrheim and J. Von Braun, eds., *Russia's Food Economy: Toward Truly Functioning Markets* (forthcoming).

Watanabe, S., ed. 1983. "Technological Linkages through Subcontracting in Mexican Industries." *Technology Marketing and Industrialization: Linkages between Small and Large Enterprises,* New Delhi, India: Macmillan.

Williamson, O.E. 1985. *The Economic Institutions of Capitalism.* New York, NY: Free Press.

Chapter 15

The Impact of Food Industry Globalization on Agricultural Trade Policy

Tim Josling
Stanford University

1. INTRODUCTION

It has long been an oddity that the food industry has been so quiet when it comes to influencing agricultural policy. Of course, some industry groups have made their position known on certain aspects of agricultural legislation, but, in general, the agri-food processing sector and the distribution sector have not exerted a major influence on policy. Academic interests have reflected this reality; the vast majority of analytical treatments of the politics of agricultural policy and agricultural trade have focussed on the influence of the farm lobby and their supporters in legislatures. Conventional wisdom considers consumer interests weak and diffused, and, hence, they have rarely received much political attention. Farm policy anomalies, such as the fact that massive transfers are made to relatively few small businesses from consumers or taxpayers, are explained in terms of the over-representation of the rural vote, the identification of political interest with production rather than consumption activities, or public sympathy for what is perceived as a poor and backward sector.

The food industry fits awkwardly into these explanations. It is both the distributor of the output from farms and also the consumer/processor of most raw materials. Are the food companies extensions of the farm sector, and, hence, on the side of the producer? Or are they surrogate consumers, acting on behalf of families who do not buy directly from the farm? Or are they merely neutral intermediaries with no particular views on policy other than to see the food chain function? Do they gain or lose from farm protection?

Are they in favor of freer trade, or do they benefit from the profits to be had under the cover of trade barriers?

One can make a case that the food processing industry, along with the suppliers of raw materials, has generally been supportive of producer groups in the past, though not terribly vocal about that support. Some of the parts of the food industry, in particular those nearer the consumer end of the chain, have staked out somewhat different positions. But the identification of food processing with producer interests has been the dominant trend. The driving notion behind this chapter is that the set of changes that are often grouped together as globalization is shifting the food sector from a relatively passive friend of the producer to a potentially vital ally for the consumer.

Before attempting to explain why such a change might be taking place, consider its possible ramifications for farm policies and for agricultural-trade policies. The food sector itself is considerably larger than the farming industry. The producer-bias in farm policies that has been universally noted for the industrial countries seems, itself, to rest upon the quiescence of the food sector. A more aggressive food-processing industry could conceivably reshape the politics of agriculture. Of course the farm lobby would remain, but a food-processing industry that used its political clout to counter that of the farm groups could potentially prevail. This could have a dramatic impact on the evolution of farm policy, particularly in Japan and the European Union. And it could change markedly the arguments, and indeed the agenda, in the next round of World Trade Organization (WTO) agricultural talks.

The case for the growing significance of the food industry is expressed in this chapter in the form of a series of qualitative statements. The food industry is a massive part of the economy and qualifies for the label of big business. It has also become a global sector, moving away from its local roots. Its ties to the farm sector are weakening, in part because of structural change. In particular it is becoming less involved in the implementation of farm policy. Instead it is increasingly becoming oriented toward the consumer, and toward setting standards for food quality. This new food sector tends to favor a liberal trade and investment regime, though more open global markets may, in turn, require global regulations. As a result, the food industry is expanding the agenda for the current WTO agricultural talks, and will have a considerable voice in the outcome. Though each of these statements is open to empirical analysis, no organized quantitative evidence is given here. Rather, evidence assembled here is anecdotal. If the underlying thesis is plausible, however, the topic deserves much more careful research.

2. THE FOOD INDUSTRY HAS BECOME BIG BUSINESS

The food chain, from the supplier of inputs to the farmer to the retailing of food through restaurants or stores, accounts for roughly 25 percent of the private economy of modern industrial countries, and somewhat more for developing economies. Food processing, itself, is one of the largest of the manufacturing and distribution sectors in most countries, accounting, for example, for one sixth of all market and distribution activity in the United States. Sales from the processed food sector alone are well over one half trillion dollars in the United States—more than double the size of farm cash receipts. Value added in the processing, wholesaling and retailing of food was U.S. $372 billion in 1994, compared to U.S. $85 billion for the farm sector (Henderson, Handy, and Neff, 1996).

The size of some of the individual food firms is quite impressive. Evidence from the *Financial Times* survey of the top 500 firms confirms the economic, and presumably the political, clout of these enterprises (*Financial Times*, 1999). The top ten food firms in Europe (ranked by capitalized value) employed 1.13 million workers and had a turnover of U.S. $248 billion. In the United States, the top ten food firms had a turnover of U.S. $140 billion and employed 700 thousand people. Even in Japan, a much smaller economy than in Europe or the United States, the top ten food firms had a turnover of U.S. $85 billion and employed 162 thousand workers. There is little doubt that the major food firms are large enough to get the attention of politicians.[i]

As important as the size of the food sector is its composition. The structure of the food chain has been changing rapidly, and the relationship between the parts has been shifting.[ii] These changes have primarily involved concentration, leading to the emergence of some relatively large firms. This has been particularly true in some of the input industries, such as the seed sector, in which the U.S. market is dominated by a few firms such as Pioneer, Monsanto and Novartis.[iii] The trend toward concentration is reinforced by the trends of new technology. A prominent example is the adoption of seed from transgenic plants, notably in the United States and Argentina. Life-science companies such as Dow, DuPont, Novartis, and Monsanto have, in effect, been buying their way into the seed business as a way of augmenting the return on scientific research conducted in their laboratories. Such concentration may not be bad in itself, as the companies, at the moment, still have to compete with retained seed, landrace crosses, and seed produced locally. But the potential exists for the exercise of market power by life-science companies, extracting considerable rents for genetic material in plants.

At the other end of the chain, supermarkets have come to dominate the retail distribution of food, at least in several European countries.[iv] This domination has been of serious concern to farmers and their organizations, though there is no evidence that returns to farmers have suffered.[v] Some of this concentration has arisen from the realization of economies of scale in the distribution system, but much of it may be due to changing habits of consumers and the modification of laws that used to protect the small retail firm from competition. The main effect seems to have been a renewed emphasis on quality, which may prove rather beneficial when preparing producers for the rigors of global competition. But the potential presumably exists for oligopsonistic exploitation of the farming sector by powerful retailers.

The trend toward concentration in the United States is apparent in the major role that fast-food outlets play when setting standards for poultry, beef and potatoes. But the cereal and oilseed markets are also characterized by a relatively small number of players in the markets into which farmers sell. The recent purchase of Continental Grain's handling facilities by Cargill has further reduced choice to farmers. Once more, the argument in this chapter is not so much that these developments are undesirable or have hurt farmers, but that they will begin to have an impact on the politics of agricultural and trade policy.

3. THE FOOD INDUSTRY HAS BECOME A GLOBAL SECTOR

The agriculture and food industries have now joined the list of global sectors, supplying a worldwide marketplace from a mutually supportive network of institutions that transcend national boundaries. This process of globalization itself has strong implications for trade policy. It gives these industries a cosmopolitan outlook on policy issues, which is in contrast to the inward focus of national agricultural sectors.

One measure of the international scope of the food industry is the extent of its foreign direct investment (FDI). There has been a significant growth of such investment over the past two decades, complementing the growth in trade. In 1993, sales of foreign affiliates of U.S. food firms amounted to U.S. $100 billion, or about five times the size of direct exports of processed foods from the United States of U.S. $24.4 billion (Henderson, Handy and Neff, 1996).[vi] Sales of U.S. affiliates of foreign firms totaled U.S. $45 billion in that year—more than three times the level of imports into the United States of processed food. In addition, U.S. food-service companies contributed to globalization by selling U.S. $19 billion of food products abroad.[vii] Once again there is a marked difference between the FDI pattern of the United

States (in which food processing, fast-food outlets and soft-drink firms dominate), and the European Union (in which retail firms and alcoholic beverages tend to be more significant).

In many respects, agriculture has always been a global sector. Trading in foodstuffs is as old as civilization. The great expansion of commerce in the 16th and 17th centuries was based upon the discovery of geographically specific commodities, minerals, and tropical plant products. By the 18th and 19th centuries, manufactured goods began to accompany raw materials in global trade, but agricultural and mineral exports were still dominant. Even into the first half of the 20th century, as the industrial countries began to emerge as major trading powers, many countries in the world still made a living from selling agricultural and food commodities to overseas markets. To them, globalization of agriculture was always a fact of economic life.

Not all countries participated fully in the global food system. Some preferred to develop their own food processing industries based on locally produced raw materials. Nor was the trade system truly global. Much agricultural trade followed colonial paths, from the remote producer to the processor in the metropolis. As in many other areas of international economic life, the last half of the century has seen a slow march toward the restoration of the international agricultural and food systems that existed at the turn of the century.

4. THE FOOD INDUSTRY IS WEAKENING TIES TO THE FARM SECTOR

One of the major consequences of the globalization of the food industry is that the link between the farm sector and the marketing and processing sectors is changing. One aspect of this is the trend toward contract farming and other types of pre-selling of output. This would seem to strengthen the relationship between producer and processor, but in fact it makes the relationship more asymmetric. Farmers are, in some sectors, managers of a farming process that is, in large part, planned by input suppliers and/or by output purchasers. This structure has gone furthest, perhaps, in the poultry industry, in which the process is closely controlled from the hatching of eggs to the slaughtered bird. Feed and antibiotics are either provided for the farmer or their choice is controlled. Pig production is also moving in this direction, in part for quality control. Even the dairy sector is becoming structured in the same way. Milk, in both Europe and North America, has long been produced under a quota system that virtually eliminates any flexibility in marketing. Now inputs and feedstuffs are becoming more regulated, not by government but by the dairy that purchases the milk. Crop production was always the archetype of independent farm operation, planting seed purchased from a merchant and selling the harvest to another firm. Increasingly, with designer seeds, the input supplier and the crop purchaser are part of the same entity. Traceability of origin and separability of output are essential to the development and spread of this new technology (Kalaitzandonakes and Maltsbarger, 1998; Marsh, 1998; Schnittker, 1998).

The contract system is an integral part of the management of the food supply chain that in turn is stimulated by the pressures of competing in a global market.[viii] Though some markets are large enough to warrant these marketing and management changes strictly for internal purposes, it is the stimulus of international competition that seems to be driving the move at present. Within self-contained national markets, such contracting might strengthen the links between farmer and processor.[ix] But trade changes this situation rather dramatically. Rather than having contracting reinforce the link between producer and processor, under a liberal trade regime the processor tends to treat local producers and distant producers as alternative and equally viable sources of supply. The link between the economic interest of the processor and that of the local producer is weakened. The processor begins to look around for the most profitable source of supply, even if that means contracting with overseas producers. Pressures upon agricultural trade policy from the processor would shift toward the further reduction of import barriers.

5. THE FOOD INDUSTRY IS BECOMING LESS INVOLVED IN THE MANAGEMENT OF AGRICULTURAL POLICY

A major reason why the food industry in many developed countries has been so close to the producer stems from the symbiotic relationship between the industry and public policy toward the farm sector. Since the 1930s, farm policies have operated largely through the regulation of commodity markets. This has necessitated the cooperation of processors and distributors. In some cases, the function of collection and distribution was taken over by parastatals, providing a direct link between policy and marketing. In other cases, the link was less direct but equally effective. The *quid pro quo* was that governments tended to build fixed margins into policy prices and to regulate competition. The agricultural production process was often unfettered by direct controls, but the marketing system was a handmaiden of policy.

The situation in the European Union demonstrates this symbiosis. E.U. sugar policy is essentially a system of production quotas, administered through the beet crushers, and protected from imports by prohibitive tariffs.

Selected foreign suppliers also receive quota access to the E.U. market. The prices are set for sugar beets, raw sugar, and white sugar. The crushing margin and the refining margin are administratively defined, and representatives from the industry spend considerable effort in ensuring that the E.U. Commission takes into account cost increases and other influences on margins. Moreover, sugar quotas are not usually traded among E.U. countries; they are, in fact, allocated initially as national quotas. The sugar-processing industry has therefore very little incentive to push for a change of regime, whether it is the removal of quotas or the liberalization of trade. And it is not a coincidence that the sugar sector tends to be cohesive and effective in resisting change.[x]

The changes in farm policy over the last decade have altered the landscape dramatically. First, the gradual decoupling of policies relieves the processing industry of the need to administer policy. The direct payment check does not need to go through a processor or wholesaler. Second, the restrictions on export subsidies mean that volumes are constrained whenever the sector is in surplus. For those parts of the food chain that rely on volume for profits, this makes a more liberal market quite attractive. Wholesalers and processors can benefit from imports as well as from domestic production. The third factor may be the most important of all: food tariffs have been reduced as a result of the trade negotiations. Where these tariffs are not adjusted for input costs there will be a squeeze on profits.[xi]

6. THE FOOD INDUSTRY IS INCREASINGLY ORIENTING ITSELF TOWARD THE CONSUMER

The phenomena of contract farming and of integrated supply chains are profoundly altering the structure of the food industry. In part, these changes reflect the development of new technology, and the trends are likely to continue as the biotech revolution develops. The successful introduction of genetically improved varieties requires transparency and traceability in the supply chain, which both rely upon advances in information technology to allow the flow of information up and down the chain. Satellite communications allow for even the most remote part of the chain to be in instant touch. The combination of easily accessed information and strict quality control, made possible by technical advances and structural change, provides the basis for a global industry. The key change, however, may not have been in the technology *per se* but in the reorientation of marketing strategies that accompanied globalization. In short, the consumer has been placed in the driver's seat of the global food system.

The emergence of the consumer as the driving force behind the changes in the food sector is a reflection of both demographic and income trends. Slow population growth in affluent countries, particularly in Western Europe, implies very little market growth for staple foods. The market is, however, expanding rapidly in those countries that are experiencing the development of an urban middle class. Processed foods tend to be purchased in place of local staples. Mass communication tends to lead to the broadening of tastes to include foods from other cultures. In developed countries, the consumer is willing to pay for quality, convenience, and other attributes, including geographical characteristics and particular methods of production (Urban, 1998). These value-added elements, in turn, allow product differentiation and advertising, promoting a market structure of monopolistic competition. Meeting and cultivating consumer tastes for specific foods are far from selling minimally differentiated commodities on global markets. It is this shift that underpins the change in attitude of the food sector to farm policy.

One rough indication of this systemic change is the growth of trade in high value-added products, a growth that has been much greater than that in homogeneous bulk products. In 1985, trade in high value-added products was barely 50 percent of total agricultural trade. It is estimated that, by the year 2000, this share will be around 75 percent of agricultural trade. Part of this growth is due to the effects of rising incomes, allowing consumers to shift away from unprocessed foods. But much of the growth in high value-added goods is due to increasing product differentiation as producers and food retailers attempt to convince consumers of the merits of particular geographical locations, recipes, and brand names. Goods that were once considered non-tradable have found a place in foreign markets for ethnic and exotic foods. Product differentiation, segmentation of the market and quality attribution, and the growth of non-traditional trade, are the keys to the growth of agricultural exports for many countries. Europe itself is enjoying a minor export boom in differentiated commodities by breaking away from its traditional focus on a few undifferentiated products (such as, wheat, sugar, skimmed milk powder, and butter) that could only be sold with heavy subsidies.

7. INCREASINGLY, THE FOOD INDUSTRY IS SETTING THE STANDARDS FOR FOOD QUALITY

The agricultural-trade agenda is largely driven by those who seek access to wider markets, along with those who seek less expensive sources of raw material. However, there is another set of trade-policy issues that has emerged in the process of globalization. These issues center upon the question of standards, which often act as impediments to trade. As traditional trade barriers fall, these other trade impediments become visible, like rocks in an ebbing tide. Many impediments are the result of different regulatory regimes that developed at the time of autarchic national markets. The global economy can survive with significantly different regulations in different countries, but harmonization cuts costs and encourages exploitation of scale economies. On the other hand, legitimate differences may exist among the objectives of regulation, and among the sensitivities of national markets. The resolution to this dilemma is still being worked out in the world of trade policy. The agri-food system happens to be at the center of the controversy and has a lot at stake in the outcome.

Public policy on food standards has been developing rapidly, both at the national and the international level. One set of conflicts arises from differences in food-trade regulations, which are imposed for the ostensible reason of protecting plant, animal, or human health from disease or from other afflictions. The fear is often expressed that such regulations do little more than protect the livelihood of local producers who would otherwise be unable to compete. The spotlight has been turned on the elimination of such back-door protective measures. The Agreement on the Application of Sanitary and Phytosanitary Measures (the SPS Agreement), adopted as part of the Uruguay Round package, attempted to make it easier to distinguish between legitimate (science-based) regulations and those that appear to be simply protecting producer interests. It ruled out those regulations that reflect irrational consumer fears not based on scientific evidence (Roberts, Josling, and Orden, 1999; WTO, 1995). With the unfortunate consequences of the Bovine spongiform encephalopathy (BSE) outbreak in the United Kingdom fresh in people's minds, and the poor performance of public officials when handling the beef crisis, the issue of consumer concerns has come to the fore.

The reaction of the authorities to such consumer concerns has been to review food-safety policy "from plough to plate" (E.U. Commission, 1997; FDA, 1997; MAFF, 1998; USDA, 1997). One aspect of this comprehensive approach is the widespread adoption of the concept of Hazard Analysis of Critical Control Points (HACCP). This approach is based upon the identification of those stages in the food chain at which contamination can take place, and upon the focus of remedial controls regarding those points. Supply chains of a contractual nature have to develop their own type of HACCP procedure as an overall part of quality control.

The case of the potential conflicts over food from transgenic plants has brought these matters to public attention.[xii] Recent concerns about both the environmental impact and the possible health effects of genetically modified organisms (GMOs) have sparked widespread debate upon the appropriate public regulatory framework for the introduction of GMOs into the food supply (Bureau, Marette, and Schiavina, 1997). The remarkably rapid adoption of modified seeds, particularly for corn and soybeans in the United States, Canada, Argentina, and Australia, has put pressure on the importing countries to speed up their approval process so as to prevent the emergence of active trade conflicts. Importing country governments are being pressured by consumer and environmental groups who are asking that the introduction of genetically engineered foods be slowed or halted while more extensive tests are completed. This public-policy dilemma has a direct impact on the private firms at the heart of the new global food system. On the one hand, the advances in biotechnology appear to offer a plausible (and profitable) solution to meeting the expanded demand for food. On the other hand, the consumer, who is assumed to be king in the global marketplace, is showing some reluctance to accept the new technology. This places the resolution to the problem squarely in the hands of the private sector.

Though public agencies must always take the lead when protecting health and safety, private standards also have an important place in the global system of commerce. Firms selling differentiated goods in overseas markets rely on repeat business. No longer can producers hide behind the anonymity of international trade. Once again, supply chains have pioneered the way by developing systems of producer accountability and supply traceability. In the modern food chain, many consumers like to know where foods come from. This means that they are willing to pay for some amount of information, and they will remember if they are dissatisfied. Supply chains can cater to such consumers as well as to those who are less concerned about the origins of their food ingredients.

In the case of GMOs, the private food firms have begun to develop strategies to deal with the problem of low consumer acceptance in major (mostly European) markets. By announcing their own policies, ranging from avoiding all GMO foods in its own-label products (for example, the Iceland and Asda chains in the United Kingdom) to avoiding selling meat from animals that had been fed GMOs (two other U.K. chains), the private sector is trying to maintain both consumer trust and its own access to low-cost supply sources (Grant, 1999). If the consumer trusts the supermarket more than he or she trusts the government, then this strategy could work. It might turn out to be a good example of self-regulation as a viable way of dealing with the problem of standard setting in the global marketplace.

8. THE FOOD INDUSTRY WANTS A LIBERAL TRADE AND INVESTMENT REGIME

The key question for agricultural trade is whether or not the food sector, as it goes global, will add its weight to the pressures for trade liberalization. The structural changes underway in agriculture and the food sector may impel the movement toward freer markets. In the past and in case markets were over-supplied, it was possible to maintain traditional farming systems by utilizing a combination of basic research into yield-increasing technology; of guaranteed markets for undifferentiated raw materials; of fixed and profitable margins for processing activities (which tended to locate near the production base); and of various parastatal marketing agencies. Trade policy acted as handmaiden to these domestic policies. Tariffs and non-tariff barriers were used by governments as instruments to protect domestic markets from competition from abroad. Globalization brings new challenges and requires new policy approaches. Moreover the old policies often get in the way of those that are needed for the new food system. Nowhere is this more evident than in the trade policies.

Trade policy is sensitive to the interests of companies that are engaged in trade and in investment abroad. Globalization of the food sector involves more companies in the operation of agricultural-trade policy. When an international fast-food outlet in a foreign country sells a piece of chicken or a hamburger, the transaction is governed by the rules and policies set down for agricultural trade. More often than not, the outlet will have purchased some of the ingredients abroad. In the case of food and agricultural products, investment is complementary to, rather than a substitute for, trade. Though some investment is obviously in place to jump a trade barrier, most investment seems to be part of the search for new markets, taking advantage of economies of scale by spreading managerial and financial assets over a wider area. Even when this does not happen, trade can increase as a result of the changes in taste brought about by the spread of international cuisine.

The concept of a food system that goes beyond freer trade for raw materials is interesting in a broader context. It offers the possibility, however improbable, of seizing the trade agenda from the agricultural lobbies. Governments in developing countries could, in effect, set up a modern, low-cost food sector through FDI and through liberal market access. For many governments to achieve this without having to go through the painful stages of high protection of basic foods, followed by the difficult transition to competitiveness, would be political bliss. Those developing countries could bypass the farm problem that has plagued Europe and the West. The new global food sector may succeed in shifting the trade agenda from U.S.- and

E.U.-competitive rivalries and squabbles over farm support levels to the role of agriculture and food trade in development policy.

9. THE FOOD INDUSTRY ITSELF MAY NEED REGULATING AT THE GLOBAL LEVEL

So far, the trend toward a global food market has raised few fundamental regulatory questions beyond those of disparate food-quality standards. In part, as a result of the focus on farm level issues, little attention has been given to whether or not new rules are needed to ensure that the performance of the global system is in line with international ideals. But this lack of attention is likely to change as the process of trade liberalization proceeds.

One aspect of global-market regulation that may emerge sooner, rather than later, is that of competition policy. At one level, it is clear that a global trade system will eventually need global competition laws. This apparently uncontroversial conclusion has had little effect, so far, on trade-policy discussions. While some are calling for full-scale negotiations on international competition policy, many maintain that the most that can be done is to make sure that each trading country has its own anti-trust policy in place (MacLaren and Josling, 1999). But this minimalist approach is unlikely to be satisfactory by itself. The best policy for curbing misuse of market power in any one country is an open trade system. The very openness of the trade system, however, allows large firms to develop market power in the world market. Global competition policy should be more about market power in world markets than about enforcing competition policy in each national market.

One emerging global competition issue concerns the concentration of market power in the agri-food distribution chain.[xiii] This issue has two separate, but related, aspects. One is the use of market power by public agencies or by parastatals given the ability to act in a restrictive way. The issue of state trading is coming to the fore in trade talks. At one extreme is the concern among non-state-trading countries that state-trading countries can gain an unfair advantage through hidden export subsidies and import barriers. At the other extreme are fundamental systemic issues such as the behavior of the state-trading entities in China, and, indeed, the extent to which the government controls, albeit indirectly, all trade decisions in that country. As a major player in agricultural trade markets, the terms under which China will enter into the WTO will have a significant impact on the rules that can be set for other countries with parastatal agencies active in the market.

The issue of competition is at the heart of another potential problem facing the agri-food system. Concentration of economic power is not confined to public agencies that are given monopoly rights in importing or exporting.

Private firms can have significant market power to influence prices. Should there be any rules relating to the use of market power in international markets? What are the dangers that the rules are trying to prevent? Is the withholding of supplies to raise the price of commodities a problem? This seems relatively unlikely in the case of basic foods, but could happen with vital supply components, such as seeds. Or is the problem one of dumping and market disruption? The incorporation of anti-dumping rules in a set of more comprehensive competitive regulations is supported by many trade economists. Whatever is agreed to will have significant implications for global agriculture.

10. THE FOOD INDUSTRY IS EXPANDING THE AGENDA FOR TRADE TALKS

The main focus of international trade policy has traditionally been the conditions of entry into markets. As globalization has progressed, so has the scope of trade rules expanded. The new trade-policy environment contains a number of different elements. These include the health, safety, and environmental rules that ensure quality and acceptability in discriminating markets; codes for the treatment of FDI; the regulation of conditions of competition; and the codification of the rights granted to the owners of intellectual property.[xiv] These various facets of trade policy are shown in Table 15.1 together with the main elements of the global-food system.

The traditional agriculture agenda is dominated by disputes over trade liberalization and domestic farm-policy reform (Josling, 1998). This will again be the main focus for the next WTO round of agricultural negotiations. These issues revolve around trade rules for market access for farm goods. Other parts of the food chain also have market-access interests, as discussed above, with pressure for freer access to raw materials coming from processing concerns. Similarly, questions of supply security for developing countries will be on the agenda. With respect to negotiations over standards, each part of the food chain is likely to be engaged. It is not always easy to see where particular issues will be addressed in the negotiations, but the issues of potential environmental damage from GMOs will surface somewhere and will have to be faced. Similarly, at the other end of the chain, consumer-confidence issues will haunt the negotiators until such time as they can find an acceptable way to handle the controversy.[xv]

The possible interface between competition regulation and the agricultural and food system was discussed above. Here it is only necessary to point out that the issues spread across all the stages of the food chain. The risk of market failure due to excessive concentration of economic power seems to

Table 15.1 Interface Between Trade Policy and Elements of the Food-supply Chain

	Inputs	*Farming*	*Processing*	*Distribution*	*Consumer*
Trade rules for market access	Conditions of access to input supplies	Trade liberalization, domestic policy reform	Search for low-cost supplies. Issues of effective protection of processing activities	Intra-firm trade in supply chains	Food-security issues and supply assurance
Health, safety, and environmental standards	Potential damage from transgenic processes	Environmental hazards of chemical farming	Runoff from processing facilities	Hazard control and accountability in the supply chain	Food safety, confidence in standards agencies, role of science
Investment codes		Foreign ownership of land	Globalization of processing activities	Investment in supply chain	Investment in retail system
Competition regulations	Monopoly control of genetic material	Role of single-desk sellers of farm goods	Processor contracts	Control over distribution by public and private monopolies	Retail sector concentration
Intellectual property rights	Patents on biotechnology practices and materials			Geographical labels	

Source: Author.

be most obvious in the seed sector, but the more domestic issues of parastatal marketing agencies, limited choice of processors, and retail concentration are likely to spring up in trade discussions. Whether they will be handled in the context of agricultural talks seems less certain.

The global-trade system, whether it is in agri-food products or in automobiles or computers, depends on investment. Capital accumulated in one country is invested in others, to the mutual advantage of both economies. But global investment also requires rules, and these are not yet fully developed. Several issues are at stake in the area of investment. One is the assurance by the investor that the assets owned by foreigners will not be expropriated, that earnings from investments can be taken out of the country, and that there will not be undue restrictions (such as requirements to use domestic inputs or to export a share of outputs) upon the foreign operation. Firms have alternatives, and countries maintaining policies that are not investment friendly may lose the opportunity to participate in the global division of labor. The global reach of food retailing and processing similarly requires the assurance that facilities abroad will not be expropriated and that undue restrictions will not be placed on the repatriation of earnings. Supply chains also need the environment of predictability that comes from an open-investment policy.

Among the newer aspects of international trade policy is the establishment of rules regarding intellectual property. The emergence of international rules predates the GATT Uruguay Round, with the establishment of the World Intellectual Property Organization (WIPO). But there was insufficient incentive for countries without intellectual property protection to join. The breakthrough came in the Uruguay Round when the negotiating countries signed the Trade-related Intellectual Property agreement (TRIP). TRIP brought a degree of harmonization to the disparate treatment of patents, copyrights and trademarks in various trading countries.

One important area in which the rules on intellectual property are significant is the input industries. The seed sector, in particular, has already made use of such international facilities to try to reclaim some revenue from farmers. The ability to patent plant varieties has been a controversial topic for some years. Now it is possible to patent particular manipulations of genetic material. This would give a much greater chance for companies to license new varieties to farmers.[xvi] This is of concern among some who fear that the highly concentrated seed industry could extract considerable profits from farmers world wide, as they would have to pay from season to season for planting even their own retained seed. Of course, one could foresee some difficulty when enforcing such laws, but firms could, in principle, use so-called terminator genes that would prevent a plant from reproducing.[xvii]

The other key aspect of intellectual property rights that impinges upon the global agri-food system is that of the appellations or guarantees of geographical origins. It is widely held that such geographical labels help the consumer to pick a brand on which they can rely. It is also possible that the same useful information can have the effect of inhibiting competition and earning scarcity rents for the holder of the patent. But, regardless of the merits of particular types of labeling, some form of brand identification is an important part of the new consumer-driven food system. There will be attempts to widen the scope for geographical protection in the upcoming talks.

11. THE FOOD INDUSTRY WILL HAVE A SIGNIFICANT SAY IN THE NEXT WTO ROUND

Agricultural trade policy used to be dominated by farm groups and by those arguing for more protection. One major shift in the 1980s was the growing involvement of some of the multinational-food firms in the agenda-setting aspect of trade negotiations. This trend is likely to continue. First, the processing sector has a strong incentive to look for low-cost supplies. There is, therefore, the incentive to lobby governments for the ability to import those supplies from world markets so as to remain competitive with firms located in countries in which prices are lower. In many cases, the low-cost food suppliers are overseas, as are the main competitors in the global marketplace. Hence, one would expect continued pressure from the food industry to allow raw-material prices to fall to world-market levels over a period of years. Given the disinclination of governments to support these prices indefinitely, the inclination of the food industry will come to dominate in the end. The unresolved question is the pace of such a reduction in protection.

The tendency for the international-food companies to search for low-cost supplies will be reinforced by the pressure from those firms that are already operating in several countries. For these firms, including those in the distribution and retailing business, international trade is often intra-firm trade. Any restriction on the movement of food items within the firm will tend to cause problems for the firm, and hence will be resisted.[xviii] But just as the intra-firm movement of goods can be thwarted by government regulations, so too can the contractual obligations of firms that have come together in other forms of alliance. One would expect that those firms that have been pioneering supply chains, linking producers in one country to wholesale and retail outlets in another, would also find government restrictions on trade irksome. Thus, one might expect these supply chains to add their voice to pressures for trade liberalization in the next round of WTO negotiations.

All these changes are in the direction of a more sophisticated agricultural industry, aware that the future depends on satisfying a variety of consumer tastes and competing for the consumer dollar with other goods and services. As more actors become involved in the political process, the center of gravity will shift perceptibly away from the primary producer. Policy will become less commodity focussed as the emphasis switches to adding value to the raw material and marketing the final product. These changes are crucial to the future of agricultural trade policy. In a situation in which the market is an administered price supported by public purchasing agencies, free trade poses a real threat. In a world in which farmers produce for the market, improvement in access to overseas markets compensates, in part, for more competition on the domestic market. A freer agricultural market no longer means a collapse of prices and mass rural depression. Today it is more likely to spark rural entrepreneurship and healthy market development based on response to the changing food habits of middle-class consumers.

12. ENDNOTES

[i] There are prominent cases of processors apparently influencing farm policy. One such example is that of Archer-Daniels Midland (ADM), supermarket to the world, and the largest corn wet-miller in the United States. ADM is purported to have used its influence to preserve the program of sugar price supports. The profitability of the high-fructose corn syrup (HFCS), a by-product of the wet-milling process, depends upon keeping domestic sugar prices high. A more recent example is that of the banana marketing company, *Chiquita Brands International*, whose Chief Executive Officer reportedly persuaded the U.S. government to support the Central American banana producers in their WTO complaint against the E.U. regime. Though this dispute involves issues well beyond traditional agricultural policy, the E.U. banana regime is, in part, a policy for domestic producers, and the tariff-rate quota system at the heart of the dispute is covered by the rules of the Uruguay Round Agreement on Agriculture.

[ii] Comments on recent developments relate broadly to the past fifteen years and to Western Europe, the United States, and Canada, and increasingly to Latin America. Asian developments are somewhat out of step, in part because of regulatory restrictions on the food sector and because of closer government involvement in the processing and marketing functions.

[iii] On March 15, 1999, DuPont announced its purchase of Pioneer Hi-Bred, the world's largest seed company. This gave the combined firm more than 40 percent of the U.S. corn seed sales and 20 percent of the soybean seed market. Monsanto holds a commanding 80 percent share of the cottonseed market (*Economist*, 1999). There are, of course, still many small firms in the seed industry, so there is no evidence, at present, that prices have been forced up by lack of competition.

[iv] This is most noticeable in the United Kingdom, where three chains sell most of the food products, as well as in the Netherlands, Belgium, and France (Hughes, 1993). In the United States, which pioneered the supermarket, the share of the top five chains has increased from 19 percent in 1992 to 33 percent in 1999 (Cotterill, 1999). Market analysts expect this to increase to nearly 60 percent within the next five years with mergers, though some see a different trend, as the proliferation of food sales from other types of retail outlets, such as gas stations, competes with grocery stores.

[v] There is, of course, a body of political literature that focuses upon the presumed negative impacts of corporate control of agribusiness upon small farmers and upon rural areas. These concerns have been accentuated by the globalization trend discussed in this chapter. Nevertheless, it is difficult to find convincing empirical studies to indicate such negative links. Those countries that have discouraged corporate activity in the food chain do not have a particularly good record in preventing rural distress and in maintaining a healthy environment for the small farmer to prosper.

[vi] The measure of sales of foreign affiliates rather than FDI flows or the asset value of such investments, is convenient because it is easier to compare with trade flows. It overstates the significance of FDI somewhat, as it implicitly assumes 100 percent ownership of for-

eign affiliates by the investor. However, it also omits other types of international linkages, such as joint ventures, which do not involve the transfer of ownership, and of portfolio investment, which usually does not imply control over commercial activities.

[vii] One firm, McDonald's, accounted for U.S. $11 billion of these sales in 1994, representing 43 percent of the total sales of the company (Henderson, Handy, and Neff, 1996).

[viii] For a discussion of the emerging branch of management science that focuses on the supply chain see Beers, Beulens, and van Dalen (1998). The development of supply chains is determined by such factors as technological spillovers, increased demand for variety, the enforcement of property rights in more countries, common commercial codes, and the ease of communication of ideas and management expertise.

[ix] An example of this might be the apparent common interests of dairy farmers and creameries in Canada, expressed in support for the marketing board structure, and of supply management.

[x] Other examples would include the dairy sector in Europe, in which both quotas and margins are administratively fixed. Even the cereal policy is run, in part, through the grain wholesalers, and processed grain products are protected by derived tariffs and based on the grain-import levies.

[xi] In the European Union there is, in the case of cereal and sugar derivatives, a direct link between the protection of the raw material and the derived product.

[xii] Unlike the issue of beef hormones, no actual trade dispute has erupted. The volcano is, however, smoking, and the rumbles of a potential dispute can be clearly heard.

[xiii] There has, of course, been a radical critique of the role of multinational corporations in world trade for many years (see Lehman and Krebs in Mander and Goldsmith (eds.), (1998), for a recent restatement of this case). Economists have tended to reject such critique as having a mainly political focus. But in doing so there is a danger that real issues of competition are overlooked. The full benefits of liberalization, as with privatization in the domestic economy, rest upon the existence of contestable markets.

[xiv] More contentious are the issues of environmental regulations and labor laws, and their interface with trade rules. These issues are not pursued here, though they have some tangential bearing on the globalization of the food industry. They are more central to the areas of fishing, forestry, and textiles than to agriculture and food processing.

[xv] The basic problem is that GMOs raise so many different issues, few of which fall neatly under current rules. The SPS Agreement deals with regulations concerned with animal, plant and human health, but the objection to GMOs is much wider than a health concern. The Technical Barriers to Trade Agreement covers labeling, but does not have the SPS science test for regulations. The Trade-related Intellectual Property (TRIPS) Agreement deals with the patenting issue of genetic material. A separate set of discussions has been going on in the context of the Biodiversity Treaty (Barton, 1996) to cast a Biosafety Protocol that allows countries to block imports of GMO products. But the relation between this initiative and the WTO still needs to be clarified. A WTO working group on GMOs has been suggested to prevent the issue from spilling over into too many other parts of the negotiations.

[xvi] Though plant breeders' rights have been recognized since the 1930s in the United States, it has proved impossible to patent improvements that come through selection in the field (landrace crosses) and not easy to see the justification for doing so. But when the improvement comes in the laboratory, as a result of using particular genetic material in a biotechnological process, the case for restricting unlicensed use increases. There are, however, those who argue strongly against any patenting of life-forms (see Kimbrell, in Mander and Goldsmith (eds), 1998).

[xvii] The other side of this coin is the possibility of the introduction of apomixis: the ability of plants to reproduce without seed. If this were to be feasible, there would be some restraints on the ability of the seed companies to extract rents for the use of patented technology.

[xviii] When European-food firms owned production facilities in overseas countries as a part of the colonial-food system, duties and other restrictions were rare.

13. REFERENCES

Barton, J. 1996. "Biotechnology, the Environment, and International Agricultural Trade." *The Georgetown International Environmental Law Review* 9(95): 126–141.

Beers, G., A. Beulens, and J. van Dalen. 1998. "Chain Science as an Emerging Discipline." Paper presented to the Third International Conference on Chain Management, Wageningen, Netherlands (28–29 May).

Bureau, J.C., S. Marette, and A. Schiavina. 1997. *Non-Tariff Trade Barriers and Consumers' Information: The Case of E.U.-U.S. Trade Disputes on Food Products*. Institut National de la Recherche Agronomique, Department d'Economie, Unité de Grignon (July).

Cotterill, R.W. 1999. "Food Industry Consolidation: The Uncharted Global Venture." *Food Marketing Policy Center Newsletter*, University of Connecticut, Storrs, CT (April).

The Economist. 1999. "In the Mill." Staff paper. (20 March: 64–65).

European Commission. 1997. *The General Principles of Food Law in the European Union, (Green Paper)*. COM(97): 176 (3 December)

FDA (Food and Drug Administration). 1997. *Food Safety from Farm to Table: A National Food Safety Initiative*. Report to U.S. President Clinton, Washington, DC: FDA (May).

Grant, W. 1999. "Biotechnology: A Source of Tensions in the U.S.-EU Trade Relations" (unpublished).

Financial Times. 1999. "FT 500." Staff report (28 February).

Henderson, D.R., C.R. Handy, and S.A. Neff, ed. 1996. *Globalization of the Processed Foods Market*, Agricultural Economic Report No. 742. Washington, DC: USDA/ERS.

Hughes, D., ed. 1993. *Breaking with Tradition: Building Partnerships and Alliances in the European Food Industry*. Wye, UK: Wye College Press.

Josling, T. 1998. *Agricultural Trade Policy: Completing the Reform*. Washington, DC: Institute for International Economics (April).

Kalaitzandonakes, N. and R. Maltsbarger. 1998. "Biotechnology and Identity-preserved Supply Chains." *Choices* (Fourth Quarter): 15–18.

MacLaren, D. and T. Josling. 1999. "Competition Policy and International Agricultural Trade." (unpublished)

Mander, J. and E. Goldsmith, eds. 1998. *The Case Against the Global Economy*. San Francisco, CA: Sierra Club Books.

Marsh, J. 1998. "The Economics of New Technology," in *Global Agribusiness*, Premier Issue (December).

MAFF (Ministry of Agriculture, Fisheries, and Food). 1998. *The Food Standards Agency: A Force for Change*. Washington, DC: The Stationery Office.

Roberts, D., T.E. Josling, and D. Orden. 1999. *A Framework for Analyzing Technical Trade Barriers in Agricultural Markets*, Technical Bulletin No. 1876, USDA/ERS, Washington DC (March).

Schnittker, J.A. 1998. "An Agricultural Revolution with Implications for Sustainability." *Choices* (Fourth Quarter).

USDA (U.S. Department of Agriculture). 1997. *The Food Safety Enforcement Enhancement Act of 1997*. Washington, DC: USDA.

Urban, T.N. 1998. "Beyond Industrialization: The Prescription Food System." *Choices* (Fourth Quarter): 43–44.

WTO (World Trade Organization). 1995. *The Results of the Uruguay Round of Multilateral Trade Negotiations: The Legal Texts*. Geneva, Switzerland: WTO.

Chapter 16

International Trade and the Firm

Timothy G. Taylor and James L. Seale, Jr.
University of Florida

1. INTRODUCTION

Rapid post-war increases in international trade, major policy events such as the Uruguay Round of the General Agreement on Tariffs and Trade (GATT) negotiations, and the proliferation of regional trade agreements [for example, the Common Market of the South (MERCOSUR), the North American Free Trade Agreement (NAFTA) and the Single European Market (SEM)] have produced considerable discussion and debate on international trade, on economic integration, and on an avalanche of theoretical and empirical research. Empirical research has focused upon assessing the likely impacts of these recent international changes on signatory countries, in general, and on the most likely affected industries, in particular.

Much of the theoretical research done in international trade that underpins this empirical work was stimulated by the need to develop theoretical frameworks amenable to empirical implementation. However, theoretical research has also been stimulated by the difficulties of neo-classical trade theories, such as the Heckscher-Ohlin-Samuelson (HOS) model, that has difficulties adequately explaining observed trade patterns. In particular, neoclassical theory seems unable to explain intra-industry trade (IIT) and the observed volume of trade between countries with seemingly similar resource endowments.

In response to these problems and difficulties, new trade theory evolved in the 1980s as an alternative theoretical framework. New trade theory dispensed with the question of why certain countries produce and export what they do, and sought to explain why they would (or would not) continue to produce and export what they do. In order to accomplish this, new trade theory appealed to the assumptions of imperfectly competitive markets, of economies of scale, and of product differentiation (Helpman, 1999). While new trade theory has had some success closing the gap between theory and empirical observation, a considerable gap remains.

In this regard, many possible lines of inquiry have been pursued. The main line of attack used in the new trade theory has been to relax some aspects of the perfect competition assumptions embedded in neoclassical trade theory. Another line of inquiry, recently promoted by Krugman (1991) and others, draws on economic geography and focuses upon transportation costs, agglomeration, and market size. A third line of inquiry, which has received less attention, focuses on the theoretical treatment of the firm within trade theory. In both neoclassical and new trade theory, the firm is generally considered to be a single plant, single-product entity[i] competing against similarly defined foreign entities. The inner scope and boundaries of the neoclassical firm is left in the proverbial economic black box.

Examination of the post-war evolution of international trade suggests that ignoring the inner structure and organization of firms is especially problematic. Rugman (1998) notes that over the past decade "international trade and investment [has been] dominated by a small set of multinational enterprises," most of which are headquartered in Europe, Japan, and the United States. Hummels, Rappaport and Yi (1998) indicate that in 1995 there were more than 39,000 parent firms with about 279,000 foreign affiliates comprising an estimated total foreign direct investment (FDI) stock of more than U.S. $2.7 trillion.

Reported in Abbott and Solana-Rosillo (1998) data provide additional evidence about the United States. In 1993, U.S. manufactured exports[ii] were valued at U.S. $464.8 billion of which U.S. $366 billion were classified as direct exports. Of this, 21 percent (U.S. $98.7 billion) were associated with intra-firm exports. Sales by mostly foreign owned affiliates (MOFA) significantly exceeded direct exports and were valued at U.S. $622.7 billion.

A similar story exists for the food and kindred products sector. Total exports were valued at U.S. $19.6 billion with direct exports amounting to U.S. $16.2 billion in 1993. Intra-firm exports were proportionately smaller than for the manufacturing sector as a whole, accounting for 13.2 percent (U.S. $3.4 billion) of total exports. However, sales by MOFA were U.S. $79.2 billion, considerably greater in proportionate terms to that observed for the manufacturing sector in general.

Though these data point to the importance of multinational enterprise and FDI in international commerce, they do not tell the entire story. Feenstra (1998) recently noted that trade as a percentage of gross domestic product (GDP) in 1990 was not very different from that observed just before World

War I. However, the structure and composition of trade has changed significantly. Merchandize trade, as a percentage of merchandize value-added, has increased dramatically reflecting the increasing degree to which trade is composed of intermediate goods, or is conducted under various forms of vertical and horizontal coordination of which multinational firms represent only one. Much of this coordinated activity is not captured in traditional measures of intra-firm trade or in FDI.

As these trends and data show, an increasing amount of post-war international trade is occurring within the neoclassical black box or under various forms of coordination as opposed to arms-length market transactions. Thus, the trade models that view world trade as occurring through single-product entities (neoclassical firms) trading at arms length seem to be at odds with the actual economic structure and composition of world trade. Thus, the usefulness of such models comes into question. Indeed, one must wonder whether the underlying assumptions of the new trade theory are not based more on attempts to overcome the shortcomings of the treatment of the firm in these neoclassical theoretical models than they are at attempts to more accurately reflect reality (Abbott and Solana-Rosillo, 1998).

This does not, of course, change the importance of traditional policy instruments, but it does suggest that the manner in which they function may considerably differ from those predicted by both neoclassical and new trade theory. Indeed, as noted by Markusen (1995: 169), "the policy and normative analysis that comes out of the new trade theory may be significantly off base."

In this chapter, we argue that viewing trade from the perspective of firm growth and managed coordination of the value chain may provide a useful framework for analyzing trade and globalization issues. From this perspective, we examine the role of the firm in traditional international trade models. We also discuss the alternative theories of the firm that potentially explain the evolving patterns of trade and globalization. While this chapter does not seek to develop a comprehensive theoretical framework, it highlights and pulls together various strands of research in the hopes of stimulating a more focused attempt to create such a theoretical model.

The next section examines the role of the firm in neoclassical economics and contrasts this with alternative constructs of the firm that have evolved from the business strategy-analysis literature. Using these constructs, we attempt to explain the evolution of firm boundaries both horizontally and vertically along the value chain, and suggest that the key to explaining the current patterns of trade and globalization may rest by explaining the location and forms of coordination across production activities both horizontally and vertically. In Section three, we examine the treatment of the firm in both neoclassical and new trade theory. In addition, the strengths and weaknesses of these trade theories are contrasted with models of the multinational business firm. In the final section, we offer some conclusions and suggestions for future directions for research.

2. CONCEPTS OF THE FIRM

Within the discipline of economics, the concept of the firm is a rather curious construct despite its being one of the fundamental components of economic theory. As noted by Penrose (1995: 10), "the firm is by no means an unambiguous clear-cut entity separable from other objects, and it is difficult to define except with reference to what it does and what is done within it." She goes on to note that of all the theoretical concepts used in economic theory, the theory of the firm is "the most often misunderstood and misapplied by both its defenders and its attackers."

There are many theories of the firm, but the present analysis lumps these theories into two categories: (1) neoclassical theories of the firm, and (2) the large body of literature concerned with the growth of the business firm and its boundaries. In neoclassical theory, the firm is best understood according to its fundamental role. The primary functions of this theory are: (1) to acknowledge the existence of economic entities that produce something for others' consumption; (2) to ensure these entities are consistent with some notion of technology; and (3) to provide a vehicle for resource allocation and the determination of equilibrium wage rates in general-equilibrium models. In accomplishing this task, the firm is largely a mathematical (mechanical) construct defined in terms of technology. Indeed, the neoclassical firm can essentially be viewed as an embodiment of a production technology that Shephard (1970: 13) defines as "certain means, arrangements of these means and the uses of materials and services by which goods and services may be produced." Because of its relationship to technology, the firm, in theory, is generally synonymous with a production function.[iii] The theoretical properties that define it are derived from underlying assumptions of a production technology[iv], and from some type of assumption concerning optimizing behavior.

The neoclassical firm has proven to be an excellent workhorse in many contexts. As a mathematical form, it can be represented in various equivalent ways and facilitates model building for static and dynamic analysis. It also captures the essential notion of decision making at the margin and permits a wide range of optimizing behaviors to be explored. Within general-equilibrium models, and as a theoretical basis for developing demand equations for inputs, the neoclassical firm has been most useful.

Difficulties arise, however, when the neoclassical firm is used to describe the "flesh and blood organizations that businessmen call firms" (Penrose 1995: 13). One could, of course, argue that the goal of economic theory and models is not to model such flesh and blood aspects of firms but to capture fundamental economic forces and their interrelationships. Such a view is potentially problematic, however. As noted by Goldin (1995: 201) in a recent *Journal of Economic Perspectives* article on Cliometrics, "economists have recently become aware that the rules of economic and social interaction determine economic outcomes more than the stock of resources and the level of technology." This notion was further supported in a recent *Wall Street Journal* article calling into question mechanical models of business and economy (Petzinger, 1999).

This distinction between technology (as embodied in the neoclassical firm) and social interaction (for example, management) is nicely captured in the assumption of constant returns to scale (CRTS). The assumption of CRTS has generally been elevated to the status of a stylized fact in many economic models. Given the neoclassical firm, such an assumption is quite reasonable. In the long run, when all factors are freely variable, CRTS merely implies that the firm is perfectly replicable. As is well known, however, this creates problems in models based upon perfectly competitive markets in which some source of diminishing returns is needed.[v] The most convenient source of such diminishing returns has been management. While this argument has some appeal, many studies (see Goldin, 1995, and the studies cited therein) have called the reasonableness of this notion into question. Indeed, one could argue that management, broadly defined, is a major source of increasing CRTS rather than a source of diminishing returns.

This example calls attention to one of the major shortcomings of the neoclassical theory of the firm, namely its failure to consider the scope and organization of the firm, in general, and the function of management, in particular. The neoclassical firm is indeed a black box with regard to these issues. If Goldin (1995) and Petzinger (1999) are correct in their assessment of the drivers of economic change, alternative concepts of the firm are required. Such concepts may be found in the business-strategy literature and may be derived from Coase (1937) and others.

The focuses of Coase's seminal research were the questions: why do firms exist? and what defines their boundaries? As noted by Demsetz[vi] (1995), the existence of firms can be attributed to a number factors other than to Coase's transaction costs. Regardless of the motivation for the existence of firms, a useful conceptual starting point is to consider the firm as a collection of resources tied together by an organizational structure. In this context, what distinguishes a firm from a non-firm is that resources within the firm are subject to managed coordination rather than to arms-length market transactions. The focus of the theory of the firm thus turns to the analysis of firm boundaries and to organization.

Numerous theories have offered to explain both the evolution of firm boundaries and the forms of coordination. However, to a large extent, all are variations and refinements of the fundamental notion of transaction-cost economics that analyzes trade-offs between the costs associated with using the market (transaction costs) and the costs associated with managed coordination (agency costs).[vii]

Much of modern transaction cost economics has been attributed to Williamson (1979, 1985). The main focus of his work is on transaction costs associated with asymmetric information in bilateral bargaining relationships, and the existence of relationship-specific assets. Holmstrom and Roberts (1998) note the three critical elements to transaction-cost explanations of Williamson's firm boundaries are frequency of transactions, uncertainty, and asset specificity. The basic argument is that higher degrees of uncertainty and asset specificity favor a hierarchical relationship as the most efficient means of organizing activity. Higher frequencies of transactions enable fixed-cost spreading and favor internalization (integration).

A related, but somewhat different slant to the determination of firm boundaries, pioneered by Grossman and Hart (1986), has been coined as the modern property rights approach. A key part of this model has been the so-called hold-up problem. The classic hold-up problem is characterized by a situation in which one party must make an investment in order to enter into a transaction with some second party. The investment is relationship specific so that its value inside the transaction is greater (perhaps much greater) than it is to some other use. Because of uncertainties surrounding the transaction, it is impossible to develop a complete contract. The existence of an incomplete contract *ex post* creates incentives for negotiation over the distribution of the quasi-rents. In essence one party to the transaction attempts to hold up the other. The inefficiencies associated with this situation can be overcome by several governance structures involving integration or asset ownership.

A third approach to understanding firm boundaries that has been widely used within the context of the multinational enterprise is Dunning's (1977, 1981) Ownership, Location, and Internalization (OLI) framework. The basic premise of this framework is that there are cost disadvantages associated with the conduct of international business. For multinational firms to exist there must be some advantages that more than offset disadvantages. The OLI framework places these advantages into one of three categories: (1) ownership advantages; (2) location advantages; and (3) internalization.

Ownership advantages can be either firm or product specific. Firm-specific advantages include such factors as patents, technological-process knowledge, brand name, and reputation. Especially significant are intangible assets, such as knowledge, other information-based assets, or idiosyncratic characteristics of management. Many of these types of assets have a public good character. Whether one uses transaction costs or Grossman-Hart asset ownership analysis, the presence of such assets clearly favors creation of a multinational hierarchy.

Location advantages refer to factors that make production in the foreign market preferable to domestic production and export. Obvious factors leading to location advantage are transportation costs, trade barriers (such as tariffs, wage differentials, and access to critical intermediate goods). However, factors, such as access to established distribution systems and business networks, also lead to location advantages.

Internalization is the most difficult of the OLI triumvirate to characterize. In its simplest form, internalization advantages are those factors that make internally managed coordination preferable to other forms of business organization. As noted by Rugman (1998), internalization encompasses much of the transaction cost and asset ownership frameworks.

Internalization also points to one of the fundamental challenges concerning the treatment of the firm in international trade models. There are many forms of managed coordination beyond pure internalization.[viii] The use of franchising, licensing, and other forms of strategic alliances are becoming increasingly common in international commerce. Indeed, Feenstra (1998) has argued the presence of such arrangements is evidence of a disintegration of the traditional vertically integrated firm. Implicit in this argument is the belief that internalization advantages are giving way to other forms of coordinated international economic activity.

3. TRADE MODELS AND THE FIRM

As aptly noted by Feenstra (1998), the importance of trade in terms of its contribution to GDP was not much different in 1990 than it was just prior to World War I. However, the structure of trade has changed dramatically. An increasing portion of international trade is occurring under the various forms of managed coordination of business firms rather than through arms-length market transactions that characterized the pre-war period. Late-nineteenth century firms were more similar to our neoclassical construct than are those of today.[ix]

This change in the structure of international trade has not gone unnoticed in the new trade theory literature. Prior to this literature, the HOS model dominated international trade theory. In this model, trade in homogenous goods with different factor intensities took place between countries with identical technology, homothetic demand, but with different relative factor endowments, and it was the difference in factor endowments that dictated the patterns of trade. While much empirical evidence (for example, Balassa, 1966; Gruebel, 1967; and Gruebel and Lloyd, 1975) suggested that a significant amount of trade among countries was intra-industry trade and trade by multinationals, the HOS model has little to say about these activities.

It was the attempt to explain theoretically the significant volumes of intra-industry trade that, in part, led to the emergence of new trade theory (Helpman, 1985). Early pioneers were Krugman (1979, 1980), and Lancaster (1980) who explained intra-industry trade in one-sector trade models by assuming Chamberlinian monopolistic competition in the production of differentiated products, the existence in all countries of a demand for a wide spectrum of varieties, each variety produced with internal economies of scale. Modeling demand for varieties is handled either by a Dixit-Stiglitz (1977) utility function that allows utility from consumption of varieties or a Lancaster-type utility function (1979) in which products are differentiated by characteristics, and utility is derived from the consumption of characteristics. In equilibrium, a finite number of varieties is produced, and each firm competes by specializing in the production of one variety. Since each country consumes all varieties, while each variety is produced in only one country, intra-industry trade is the obvious result.

Dixit and Norman (1980) and Helpman (1981) extend the analysis to a two-sector model and integrate Chamberlinian monopolistic competition with the traditional HOS model. One sector produces under perfect competition, while the other produces under imperfect competition. Relative-factor endowments determine trade in the perfectly competitive sector according to the HOS theory, while Chamberlin is used to explain intra-industry trade in consumer goods. Later, Lawrence and Spiller (1983) extend the Helpman model by introducing fixed cost of entry by a firm into a market.

Notably absent from these models is trade by multinational enterprises, or, for that matter, any real notion of the firm. Helpman (1984) was the first to develop a general-equilibrium model, which allowed vertical integration of the multinational firm. In his model, all firms were single-product producers. There were two sectors as before, but, now, the firm in the differentiated-product sector produced both differentiated-final goods and headquarter services. Firms could vertically integrate multinationally by locating headquarters in the high-wage country, exporting headquarter services to the production plants located in the low-wage country, and, then, export the differentiated product back to the high-wage country.

Later, Helpman (1985) extended the model to include production in finished goods, intermediate goods, and invisibles (headquarter services).

This allowed vertically integrated, multi-product multinational firms with production plants in more than one country. The country rich in headquarter services, or the shared input, would import the homogenous good as well as it would import varieties of the differentiated-finished goods that were produced by its foreign subsidiaries, or by corporations in the relatively labor-rich country. The shared input-rich country would export varieties of the differentiated product good, middle products, and headquarter services required by its multinational corporation located in the labor-rich country. The result was inter-sector, intra-industry, and inter-industry trade with the multi-product firm having production facilities in both countries. Markusen (1984) develops a model allowing for horizontal integration of multinational, multiplant firms that produce roughly the same product in different locations, that is, substituting international production for trade.

All the above, and most related works, basically treat the firm as a World Trade Organization (WTO) black box. Although Helpman (1984) and Markusen (1984) allow for internalization, they do so by essentially treating the process as exogenous. Still, they do introduce the notion by Williamson (1971) of firm-specific assets, headquarter services that were tied to the unit, and their work can be categorized under the OLI framework. Further, these firm-specific inputs could serve many plants, and were not location specific in their use. Inputs of this type could be, for example, management, distribution, and product-specific research and development (R&D).

Most of the new trade models generally explained the pattern and volume of trade, particularly intra-industry and intra-firm trade, in terms of relative factor endowments and relative sizes of countries. Strategic trade theory, along the lines of Brander (1981), include other variables such as firm differences and market concentration.

These descriptives, however, say little about the firm and why it becomes a multinational entity. Zoologists might classify the environmental habitat of pandas. They could correlate density of certain plants with the probability of a higher density of pandas. They might also correlate the size of the habitat to the density of pandas. However, this tells us little as to why pandas choose to live in a given habitat. It does not consider panda behavior in terms of food needs and shelter.

One paper that does address internalization of firm activities is Ethier (1986) in which he endogenizes the firm's decision to international transactions. Ethier exploits ownership advantages, locational consideration, and internalization (as opposed to arms-length use of markets) and places his model squarely in the OLI framework. He suggests internalization should be the focus of FDI and states that internalization is largely a matter of the international economies of information.

These models do have strengths, but, as noted by Markusen (1995), "these newer streams of literature are very limited in their treatment of firms." Perhaps the most promising are those that utilize elements of the OLI framework when trying to endogenize the emergence of multinational firms. The most comprehensive study to date is that of Markusen et al. (1996) who provide a unified approach to the multinational firm in that the model allows for the existence of national firms, horizontally integrated multinationals, vertically integrated firms, and FDI. While this and similar models successfully establish that the existence of multinational firms is consistent with general equilibrium, the substance of such firms is excluded. Indeed, these models often treat the multinational firm and FDI as synonymous.

Such models provide clean analytical frameworks such that resource allocation and price implications of policy alternatives can be systematically explored. However, it should not be forgotten that, despite the criticisms that neoclassical trade theory has received, it has provided the philosophical foundation for the almost universally held belief among economists that free (or freer) trade is good. In fact, Helpman (1999) concludes that the new explanations of trade from the new theory are complimentary to explanations provided by the HOS framework.

The treatment of firms in these models does, however, create problems that should not be overlooked. These models represent a very simplified economic order in which there is no scope for management and most transactions continue to occur at arms length. There is also no attempt to relate technological change and management functions. For example, while the existence of knowledge-based capital may explain the existence of multinational firms, it does not explain, by itself, why firms would prefer making large investments in foreign-production units instead of a contractual arrangement with agents of the foreign nation (Markusen, 1995). This may be particularly important since some of the most significant sources of innovation may come from management of information, organizational structures, and incentive regimes. These innovations are not easily captured with the neoclassical firm or with trade models in which it is utilized.

In contrast to the treatment of the firm in trade models as discussed above, are the models of multinational enterprises. Casson (1995) provides one of the most comprehensive treatments. The strengths of these models are that they provide a rich framework for understanding the forces that drive the growth and organizational boundaries of firms. As such, through the use of case studies (for example, Hummels, Rappaport and Yi (1998); Campa and Goldberg (1997); Abbott and Solana-Rosillo (1998)), the frameworks can provide rich detail on the mechanics and structure of international-business relationships that square with the evolution of the post-war structure of trade.

The business theories of the firm also have limitations. Perhaps most importantly, the transaction cost, asset ownership, and OLI explanations for the evolution of firm structure and boundaries provide general frameworks for analysis rather than providing analytical models. As such, it is difficult to explicitly incorporate these theories into either partial- or general-equilibrium frameworks. This limits the usefulness of the models for general policy analysis.

Another limitation of these models is that explanations for intermediate forms of coordination (such as licensing, franchising and other forms of strategic alliances) remain elusive. There are, of course, explanations for such forms of coordination, but they have not been adequately explained in a systematic fashion. Given the importance of such forms of coordination in international commerce, this issue remains an important topic of inquiry.

4. CONCLUSIONS

International trade occurs increasingly under the managed coordination of business firms as opposed to arms-length transactions subject to direct-market forces. Indeed, the world of international commerce is much different from that described in the models generally taught in international trade courses, and used as the basis for empirical trade-policy analyses. National boundaries and the policy rights they convey are perhaps losing significance in relation to the economic boundaries implicitly defined by multinational enterprises, and by other forms of coordinated international commerce in which the power of market forces is less apparent.

Our tools of analyses leave us with an unfortunate trade-off between explaining outcomes of change and mechanisms of change. Models, based on neoclassical or new trade theory, enable systematic analysis of outcomes. This is extremely important for informed debate on trade-policy issues. However, the current structure of international trade suggests that by ignoring the mechanisms driving trade, the policy implications obtained from these models may be misleading or may even be wrong.

The sugarcane industry provides an example. One of the largest sugarcane growers and sugar processors in Florida is also one of the largest producers of sugarcane in the Dominican Republic. In the past, an import quota on sugar imports into the United States provided the means of supporting a domestic sugar price far above the world price of sugar. Not only did domestic producers gain the higher price but so did the foreign producers with quota rights. If one were to analyze the situation from a U.S. firm perspective only, one would miss much that may actually be driving sugar policy. By realizing that the large sugar producer is a multinational firm, the policy situation seems more rational. For example, from a domestic firm's position, a quota of zero sugar imports would be the most advantageous quota size. For a foreign firm with no quota rights, a non-binding quota would be the best quota size. For the foreign firm with quota rights, there would be an optimal quota above zero but it would be binding that it would be most beneficial for that firm. What about the multinational firm? It would opt for a quota system that would give its foreign operation quota rights, but the optimal quota for the multinational firm would jointly consider both its domestic and foreign production. The multinational firm would have an incentive to lobby for a quota of a certain size, and for much of the quota rights going to the foreign portion of the firm. It is interesting to note that the Dominican Republic had 48 percent of the Caribbean Basin's sugar quota allocation in 1989 (Messina and Seale, 1993).

The literature on firm boundaries provides a rich framework for systematically understanding the mechanisms driving the evolution of much of the post-war structure of trade. However, this literature provides more of a framework for analysis than it does as a basis for developing analytical models. Thus, research, which is based upon these models is not easily generalized for purposes of policy analysis.

From this, it is clear that it would be very useful to develop a theoretical framework that would combine the analytical rigor and the equilibrium framework of international trade models with the richness of explaining the business mechanisms that drive international commerce, as provided by the literature on firm boundaries. It also seems clear that the central actor in doing so is the firm and its theoretical structure.

5. ENDNOTES

[i] There are, of course, multiproduct representations of the neoclassical firm. However, their use in trade theory appears to be limited. Markusen (1984) is a notable exception.

[ii] A similar picture emerges using U.S. data on imports for the manufacturing sectors

[iii] There are other equivalent representations of the firm such as cost or profit functions. While these dual functions have great popularity, the production function remains the primary firm representation in most international trade models.

[iv] Shephard (1970) gives substance to a production technology by defining the input requirement sets and nine properties imparted to these sets. To a large extent, these sets are the fundamental matter of the neoclassical firm.

[v] Other ways to get around this problem are to assume imperfect competition, or consider short-run models where some factors of production are quasi-fixed.

[vi] The issue of why firms exist is not addressed here. Demsetz reviews three arguments: 1) Coasian transaction costs; 2) Knight's risk-bearing; and 3) specialization.

[vii] The ensuing discussion draws heavily on Holmstrom and Roberts' (1998) excellent review of recent developments in research analyzing firm boundaries.

[viii] In the present discussion, the term internalization is synonymous with vertical or horizontal integration.

[ix] Chandler (1990) provides an insightful discourse on the evolution of business firms. He argues that the rise of absentee ownership changed the basic structure of business from that of owner-operators to those controlled by hired managers.

6. REFERENCES

Abbott, P and J. Solana-Rosillo. 1998. " International Firms in the Manufacture and Distribution of Processed Foods," in D. Pick, D. Henderson, J. Kinsey and I. Sheldon, eds., *Global Markets for Processed Foods: Theoretical and Practical Issues*. Boulder, CO: Westview Press.

Belassa, B. 1966. "Tariff Reductions and Trade in Manufacturers among the Industrial Countries." *The American Economic Review* 56: 466–473.

Brander, J.A. 1981. "Intra-industry Trade in Identical Products." *Journal of International Economics* 11: 1–14.

Campa, J. and L. Goldberg. 1997. "The Evolving External Orientation of Manufacturing Industries: Evidence from Four Countries." National Bureau of Economic Research (NBER) Working Paper 5919 (February). Cambridge, MA: NBER.

Casson, M. 1995. *Enterprize and Competitiveness: A Systems View of International Business*. Oxford, UK: Clarendon Press.

Coase, R. 1937."The Nature of the Firm." *Economica* 4: 386–405.

Chandler, A. 1990. *Scale and Scope: The Dynamics of Industrial Capitalism*. Cambridge, MA: Belknap Press.

Demsetz, H. 1995. *The Economics of the Business Firm: Seven Critical Commentaries*. Cambridge, MA: Cambridge University Press.

Dixit, A.K. and V. Norman. 1980. *Theory of International Trade*. Cambridge, MA: Cambridge University Press.

Dixit, A.K. and J.E. Stiglitz. 1977. "Monopolistic Competition and Optimum Product Diversity." *American Economic Review* 67: 297–308.

Dunning, J.H. 1981. *International Production and the Multinational Enterprise*. London, UK: George Allen and Unwin.

_____. 1977. "Trade, Location of Economic Activity and MNE: A Search for an Eclectic Approach," in B. Ohlin, P.O. Hesselborn, and P.M. Wijkman, eds., *The International Allocation of Economic Activity*. London, UK: Macmillan Press.

Ethier, W. 1986. "The Multinational Firm." *The Quarterly Journal of Economics* 101: 805–834.

_____. 1977."Trade, Location of Economic Activity and MNE: A search for an Eclectic Approach," in B. Ohlin, P. Hesselborn and P. Wijman, eds., *The International Allocation of Economic Activity*. London, UK: Macmillan.

Feenstra, R. 1998. "Integration of Trade and Disintegration of Production in the Global Economy." *Journal of Economic Perspectives* 12: 31–50.

Goldin, C. 1995. "Cliometrics and the Nobel." *Journal of Economic Perspectives* 9: 191–200.

Grossman, S. and O. Hart. 1986. "The Cost and Benefits of Ownership: A Theory of Lateral and Vertical Integration." *Journal of Political Economy* 94: 691–719.

Grubel, H. 1967. "Intra-industry Specialization and the Pattern of Trade." *Canadian Journal of Economics* 33: 374–388.

Grubel, H. and P. Lloyd. 1975. *Intra-industry Trade*. New York, NY: Wiley.

Helpman, E. 1999. "The Structure of Foreign Trade." *Journal of Economic Perspectives* 13: 121–144.

_____. 1985. "Multinational Corporations and Trade Structure." *Review of Economic Studies* 52: 443–457.

_____. 1984. "A Simple Theory of International Trade with Multinational Corporations." *Journal of Political Economy* 92: 451–471.

_____. 1981. "International Trade in the Presence of Product Differentiation, Economies to Scale and Monopolistic Competition." *Journal of International Economics* 11: 305–340.

Holmstrom, B. and J. Roberts. 1998. "The Boundaries of the Firm Revisited." *Journal of Economic Perspectives* 12: 73–94.

Hummels, D., D. Rappaprot, and K. Yi. 1998. "Vertical Specialization and the Changing Nature of World Trade." *Federal Reserve Bank of New York Economic Policy Review* 6: 79–98.

Krugman, P. 1991. *Geography and Trade*. Cambridge, MA: MIT Press.

_____. 1980. "Scale Economies, Product Differentiation, and the Pattern of Trade." *American Economic Review* 70: 950–959.

_____. 1979. "A Model of Innovation, Technology Transfer, and the World Distribution of Income." *Journal of Political Economy* 87: 253–266.

Lancaster, K.J. 1979. *Variety, Equity and Efficiency*. New York, NY: Columbia University Press.

Lawrence, C. and P.T. Spiller. 1980. "Intra-Industry Trade under Imperfect Monopolistic Competition. *Journal of International Economics* 10: 151–175.

_____. 1983. "Product Diversity, Economies of Scale, and International Trade." *Quarterly Journal of Economics* 98: 63–83.

Markusen, J. 1995. "The Boundaries of Multinational Enterprizes and the Theory of International Trade." *Journal of Economic Perspectives* 9: 169–190.

_____. 1984. "Multinationals, Multi-plant Economies, and the Gains from Trade." *Journal of International Economics* 16: 205–226.

Markusen, J.R., A.J. Venables, D.E. Konan, and K.H. Zhang. 1996. "A Unified Treatment of Horizontal Direct Investment, Vertical Direct Investment, and the Pattern of Trade in Goods and Services." National Bureau of Economic Research (NBER) Working Paper 5696. Washington, DC.

Messina, W.A., Jr., and J.L. Seale, Jr. 1993. "U.S. Sugar Policy and the Caribbean Basin Economic Recovery Act: Conflicts between Domestic and Foreign Policy Objectives." *Review of Agricultural Economics* 15: 167–180.

Penrose, E. 1995. *The Theory of the Growth of the Firm, 3rd Ed.* Oxford, UK: Oxford University Press.

Petzinger, T. 1999. "A New Model of the Nature of Business: It's Alive." *Wall Street Journal* (February 26: A-1).

Rugman, A. 1998. "Foreign Direct Investment and Internalization in Processed Foods," in D. Pick, D. Henderson, J. Kinsey and I. Sheldon, eds., *Global Markets for Processed Foods: Theoretical and Practical Issues*. Boulder, CO: Westview Press.

Shephard, R. 1970. *Theory of Cost and Production Functions*. Princeton, NJ: Princeton University Press.

Teece, D. 1982. "Toward an Economic Theory of the Multi-product Firm." *Journal of Economic Behavior and Organization* 3: 39–63.

Williamson, O. 1985. *The Economic Institutions of Capitalism*. New York, NY: The Free Press.

_____. 1979. "Transaction-Cost Economics: The Governance of Contractual Relations." *Journal of Law and Economics* 22: 233–271.

_____. 1971 "The Vertical Integration of Production: Market Failure Considerations (in Conglomerate and Vertical Responses to Market Imperfection)." *American Economic Review* 61: 112–123.

Zhang, K. and J. Markusen. 1997. "Vertical Multi-nationals and Host-Country Characteristics." National Bureau of Economic Research (NBER) Working Paper 6203 (September). Washington, DC.

PART IV

TRADE AND THE ENVIRONMENT

Chapter 17

International Trade with Price Supports and Environmental Constraints: The Canadian Hog Industry

Richard Gray[1], Robert Romain[2], and Hartley Furtan[1]
[1]*University of Saskatchewan and* [2]*Laval University*

1. INTRODUCTION

The national resource endowments specific to individual countries have long been thought of as one of the sources of comparative advantage in trade. As production processes have had more serious ecological impacts, and as wealthier consumers have demanded improved environmental quality, the ability of a country's ecological system to absorb pollutants has become more important as a source of comparative advantage. This phenomenon is especially important for agriculture and agricultural trade. For instance, groundwater nitrification has become a very important issue in dense livestock production areas of the European Union (E.U.) and North America. The environmental costs and abatement costs associated with livestock manure have become a significant economic factor in these regions.

The relationship between trade and the environment has become more important in trade agreements as the economic importance of environmental quality has increased. The environmental standards used to improve environmental quality often shift competitive advantage and, therefore, have become an important issue in trade. This has led to the development of rules restricting the ability of countries to use environmental policy to protect industries and to distort trade. For instance, the North American Free Trade Agreement (NAFTA) prohibits the use of an inferior set of environmental regulations for products destined for export. Many countries have also recognized that trade agreements, which can restrict the ability to support an industry, could make the implementation of higher environmental standards difficult without jeopardizing the competitive advantage of domestic firms. This concern may explain, in part, the favorable World Trade Organization (WTO) treatment of forms of support used to assist producers when meeting higher environmental standards. These WTO provisions influence domestic environmental policy, particularly the choice between public and private pollution-abatement measures.

In this chapter, we examine the relationships among trade rules, the choice of domestic environmental policy, economic welfare, and trade. We show that the existing WTO rules can favor domestic environmental policies that distort trade. The existing trade rules allow the subsidization of many public pollution-abatement policies. While these policies may improve domestic welfare, they can distort trade patterns and can harm other trading partners, or other regions within the same country. The overall effect is to distort comparative advantage and to maintain production in areas with high environmental costs.

We discuss the extent to which trade rules distort domestic-abatement incentives. We also discuss how comparative advantage is dependent on the nature of the technologies available and on the ability of the private sector to develop the lowest cost-abatement technologies in the absence of public expenditure initiatives. The Canadian hog industry is used here to illustrate these effects.

We present the relative importance of the hog sector across Canada in Section 2, as well as its importance within Canada's trading community. We also present a summary of the legislation of environmental protection in Québec and in Saskatchewan—the two representative hog regions of Canada. We provide a brief description of the basic conceptual framework of the chapter in Section 4. We propose different models to deal with groundwater-nitrification-abatement policies within Canada, and examine the welfare effects of alternative policy regimes. Our conclusions, and a call for further research, are offered in Section 5.

2. THE CANADIAN HOG INDUSTRY

2.1 Production and Exports

Over the last decade, Canadian hog production first declined from 16.3 million head in 1988 to 15.4 million head in 1991, but increased significantly to 18.8 million head in 1997. We show that this increase in production is due mainly to higher production levels in Québec and in the western provinces where the percentage increases between 1991 and 1997 were 25 percent and 27 percent, respectively (Figure 17.1). On the other hand, Ontario increased its production by 13 percent over the same period.

Québec's share of Canadian hog production has been relatively constant over the last decade at about 30 percent (Figure 17.2). Ontario's share, however, has declined steadily from 31 percent in 1988 to 25 percent in 1997, while the western provinces increased their relative production share from 13 percent in 1988 to 18 percent in 1997.

Since 1991, Québec's share of Canadian exports has fallen from 51 percent to 40 percent; the western provinces' share has increased from 28 percent to 40 percent; and Ontario's share has remained around 20 percent (Figure 17.3).

The changes in the structure of hog production and trade across the Canadian regions have been related to changes in the competitive advantage of the regions, which, in turn, have been caused by changes in agricultural policy. For example, the abolition of the Crowsnest subsidy gave a competitive advantage to the western provinces in hog production. How ever, very different regulations concerning production and the environment have also played an important role in the evolution of the pork sector across Canada.

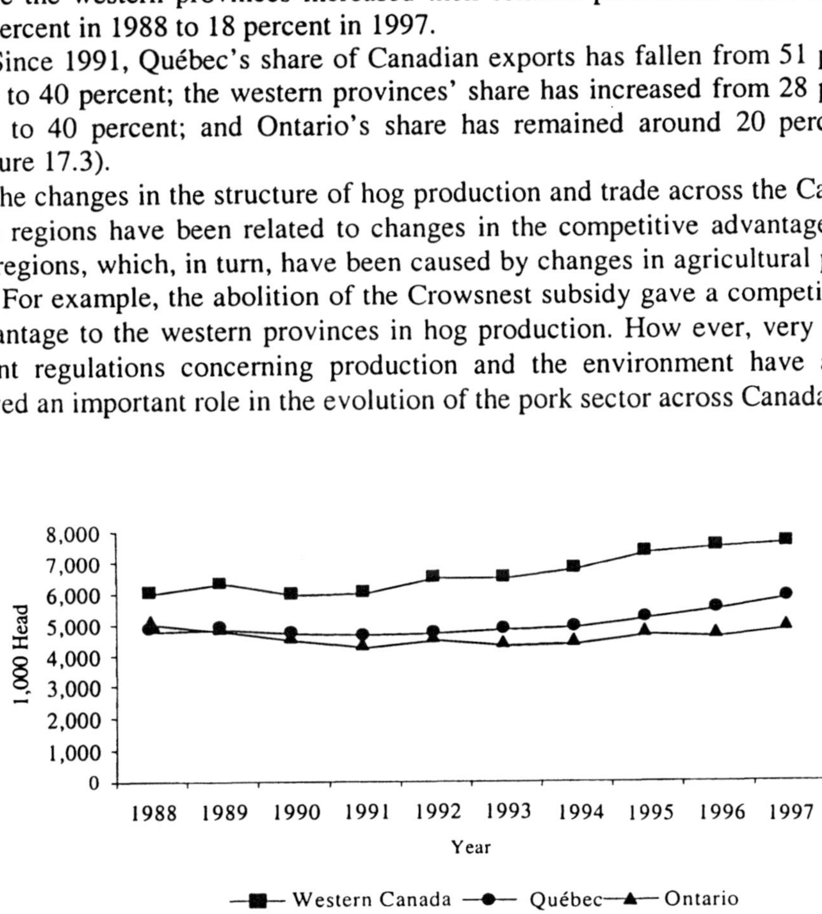

Figure 17.1 Hogs Marketed in Canada by Region, 1988-97

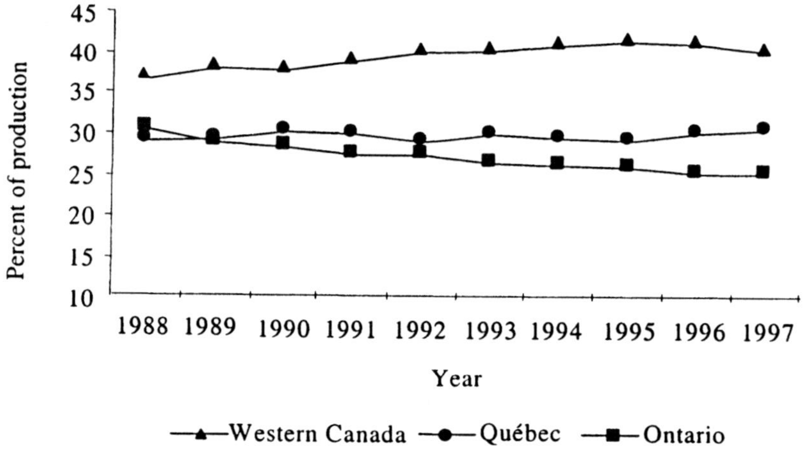

Figure 17.2 Share of Hog Production in Canada by Region, 1988-97

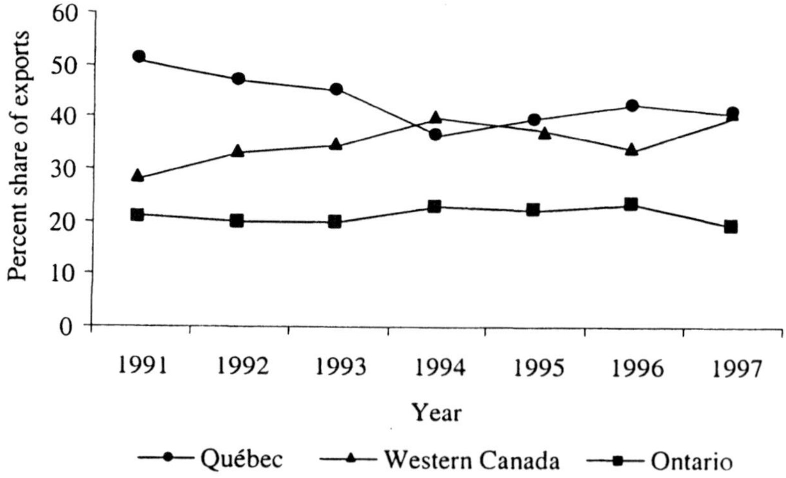

Figure 17.3 Export Share of Pork Meat in Canada by Region, 1988-97

2.2 Environmental Regulations

The environmental protection legislation in two provinces, Québec and Saskatchewan, represents regions with differing limits for expansion. On the one hand Québec, which markets 5.5 million hogs per year, has little or no room for expansion, while Saskatchewan, which markets 1.0 million hogs per year, has much greater expansion potential.

In Saskatchewan, the regulations for hog production and for protection of the environment from intensive livestock fall under the 1995 Agricultural Operations Act, which specifies the distance a new barn must be from existing structures. It does not state which technology must be used, but it lays out a process the farmer/investor must go through before getting a government issued permit for construction of a hog facility. Specifically, farmers must have a geological evaluation of the water system and of the waste-handling facility completed and made available for public scrutiny. Only when all the geological data have been collected and public meetings have been held can the farmer apply for a construction permit to build the facility. The fact that the farmer has applied for the permit is announced in the Local Chapter and individuals within the community have a period of time in which to register their complaints with the local government. If no complaints have been registered, the government will then issue the construction permit to the farmer. If complaints have been registered, the farmer must demonstrate to the local government that the criticisms have been recognized, and that they have been corrected. If the plaintiff is not satisfied that his or her complaints have been justified, he or she can take the farmer to court to block the production project, arguing that environmental damage will occur as a result of the proposed construction.

The advantage of this legislation is that, once farmers have gone through the review process and have been granted a construction permit, they cannot be sued in the future for nuisance or noise that could result from their hog production. If farmers maintain their production facilities in good working order, as permitted, they will remain free of the expense of court cases. Thus, the costs are known by the farmer/investor well in advance of his or her actual hog production.

The regulations for hog production in Québec are much more restrictive. There are several acts and regulations governing the quality of the environment and the protection of agricultural land, and several ministries (the Ministry of Agriculture, Fisheries and Food; the Ministry of Environment and Wildlife; and the Ministry of Municipal Affairs) are involved in their supervision.

There are regulations concerning odors, water pollution, and all other types of pollution that originate from agricultural production. These regulations stipulate several requirements, one of which is the minimum distance a barn must be from a neighbor in order to protect the neighbor from hog-production odors. These mandatory distances increase rapidly with the size of the farm unit. Regulations are also in place to prevent water pollution. For example, there are very strict controls to ensure that lagoons are waterproof. Also, the lagoon must be regulated in terms of its distance from a ditch, and the slope of the surrounding ground. Some floor and ventilation specifications are also in place for the barn. Still other regulations are associated with the storage of dry manure (for example, it is prohibited to spread manure between 1 October and 31 March).

A very important regulation is that hog producers must own, rent, or secure (with a four-year agreement) enough land to spread the manure their own animals produce, and that the land has to be actively in agriculture (that is, they cannot spread manure on idle land). The maximum amount of manure that can be spread over the land is determined by the crop that is grown. Producers have to prepare an annual agro-environmental plan of fertilization in which they identify where they will spread the manure, the quantity to be spread, and the ownership of the land. This plan has to be accepted by a professional agrologist. The land regulations associated with hog production in Québec have contributed to decreases in the hog industry in some regions.

Expansion of the hog industry has become very difficult in Québec. Moreover, there is pressure from the urban population against having large hog operations built in their municipalities. Therefore, even if a developer meets all the requirements of all governmental ministries for building a hog-operation unit, the municipality's issuance of building permits is often backlogged. These delays increase the producer's cost of hog production expansion.

Since 1976, agricultural land in Québec has also been protected by zoning laws, which make it very difficult for non-agricultural businesses to expand into a WTO-specified green zone. When they do expand into such a green zone, within the surroundings of a small rural town, for example, they have to comply with the same rules that agricultural producers do. [For example, if a garage is built in the green zone, it cannot be built within a predetermined distance (specified by law) from the closest farm operation.]

The expansion of hog production and its maintenance in Québec have therefore become difficult. The significant monetary and nonmonetary costs associated with environmental regulations have resulted in only moderate growth in hog production and a decline in the relative position of the province as a hog-exporting region.

2.3 Trade Rules and Environmental Subsidies

Many countries must operate within both the WTO and regional trade agreements. For North America both the WTO and NAFTA apply. The choice of domestic environmental standards has, to a large extent, been left out of the WTO and NAFTA. However, there are many parts of these agreements that deal with environmental issues. For instance, the extensive sanitary and phytosanitary (SPS) provisions are driven, at least in part, by environmental concerns.

Trade agreements deal with import protection, export subsidization, and the support given to domestic producers. In terms of import protection, the SPS agreement states that there must be an equal treatment of imported and domestically produced products, and that any restriction on the import of a good must be made based on a scientifically verifiable public risk. Given these rules, the environmental standards of the country in which a good is produced are only issues if the product produced presents some risk to a foreign country.

There are provisions to prevent an indirect production or export subsidy through discriminatory environmental policies. The side accord signed by the United States, Mexico, and Canada, entitled "North American Agreement on Environmental Cooperation," states:

> The Council shall cooperate with the NAFTA Free Trade Commission to achieve the environmental goals and objectives of the NAFTA by . . . (b) providing assistance in consultations under Article 1114 of the NAFTA where a Party considers that another Party is waiving or derogating from, or offering to waive or otherwise derogate from, an environmental measure as an encouragement to establish, acquire, expand, or retain an investment of an investor, with a view to avoiding any such encouragement. (Trade Compliance Center, 1993)

In contrast to these rules, which do not allow an indirect environmentally degrading subsidy, there are implicit and explicit provisions in the trade agreements to allow subsidies for pollution abatement. For example, the WTO states that non-actionable subsidies include assistance to help firms adopt environmental technology. Specifically,

> Assistance to adapt existing facilities to new environmental requirements. Such assistance must be on a one-time basis, be limited to 20 percent of adaptation costs, and be available to all firms which can adopt the new equipment and processes. (WTO, 1999)

This regulation appears to say that governments can subsidize the producers' purchase of new technology if that purchase is aimed at solving environmental problems. In such a case, the subsidy is not actionable. Indirectly, if a government publicly provides pollution-abatement measures (such as waste treatment), it is very unlikely that these measures would be actionable. Thus, there seems to be some asymmetry in the treatment of direct industry support and the support attached to policies for environmental enhancement.

3. THE CONCEPTUAL FRAMEWORK

3.1 Production in the Presence of Externalities

Some of the interesting welfare effects of various policy and technology scenarios are illustrated in Figure 17.4. Consider a small country facing a per-

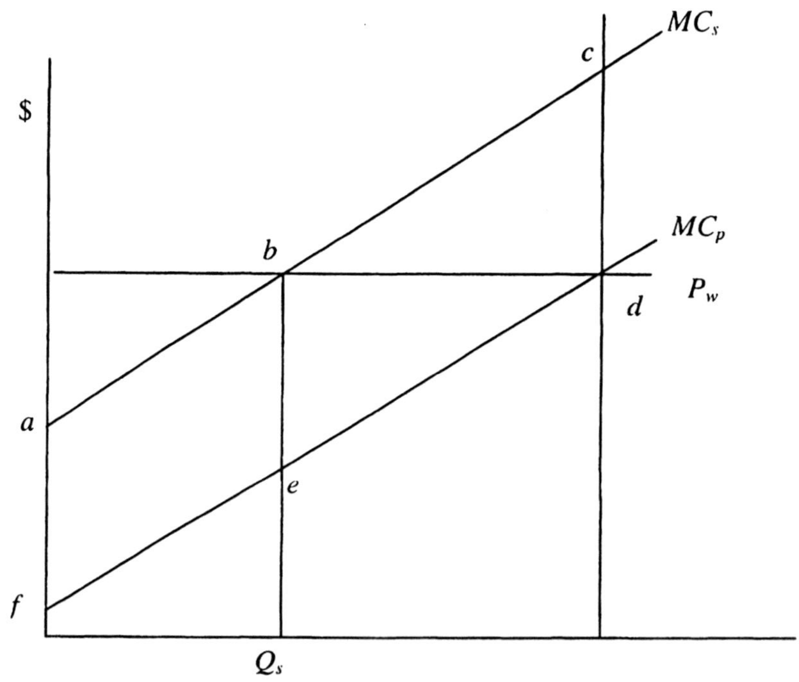

Figure 17.4 Competitive Equilibrium with Externality

fectly elastic world demand for output at price P_w. In this case, changes in the level of production will have no direct impact on consumer or foreign welfare. The supply of the product from the competitive industry is given by the private marginal cost of production, MC_p. In the absence of any regulation, the industry will produce a quantity, Q_p, where the world price is equal to the private marginal cost of production. Assuming there is a per-unit externality associated with the production of the good, then the marginal social cost, MC_s, lies above the private marginal cost by the per-unit externality associated with the output. The size of the per-unit externality will depend on the amount of pollutant produced, the ability of the ecosystem to absorb the pollutant, and the value society places on the quality of its ecosystem. Note that at Q_p, the social marginal cost of production exceeds the social benefit of production by the size of the per unit externality.

In the special case in which there are no other means of reducing the marginal social costs of pollution except through reducing output, and when there are no other externalities in the system, potential welfare is maximized at Q_s where the marginal social cost of production, MC_p, is equal to the marginal social benefit, P_w, of the last unit of production. In this case, the first-best solution can be achieved with either a regulation that restricts production to Q_s, or with a per-unit tax equal to the per-unit externality.

The comparative advantage of this region is determined by the social marginal cost curve, MC_s and competitive advantage is determined by the private marginal cost curve, MC_p (Segerson, 1988). In this case, ignoring the externality improves the competitive advantage of the industry while it imposes a cost on the rest of society. Ignoring the externality is not in the national interest since the marginal social cost of production, MC_s, exceeds the marginal social benefit, P_w. If the per-unit externality is large enough, then even the social marginal cost, MC_s, of the first unit of production will exceed the marginal benefit, P_w. In a less extreme case, if the average social cost of production exceeds the average social benefit, then that country is better off with no production at all, rather than maintaining production at Q_p. This is similar to the zero-gains-to-trade effect illustrated by Schmitz, Sigurson, and Doering (1986). Regardless of the size of a negative externality, if the social optimum is to the left of the private optimum, there is economic benefit from reducing production.

3.2 The Relationship Between Policy Choice and Marginal Cost

In general, there are ways, other than through strict reduction in industry output, to reduce an externality from production. For example, either the technologies or the level of input used can affect the per-unit externality associated with production. For instance, altering the protein content in a livestock feed ration can alter the amount of nitrogen in the manure. In this case, in which the input mix affects both the per-unit externality and the private marginal cost, a policy that affects the private action of the individual also affects the environment. Thus, environmental regulations and programs affect both the social and private marginal costs, and, as a result, they also affect the comparative and competitive advantage of the industry.

The socially optimal choice of environmental instruments will be a function of technology choices available, the transaction costs of implementing alternative policies, and, as we will show later, the trade rules under which the country operates. For example, suppose a producer were required to treat effluent before releasing it into a stream. If producers bear this extra cost, the private marginal cost, MC_p, shifts upward to MC'_p. Because of the lower level of pollution that would now occur, the marginal social cost, MC_s, shifts downward to MC'_s. The amount of production is reduced to Q'_p, where $P_w = MC'_p$. In this case, the producers must bear a marginal cost of *km* and a social cost of *jkdf* (Figure 17.5). This action forces an additional cost onto pro-

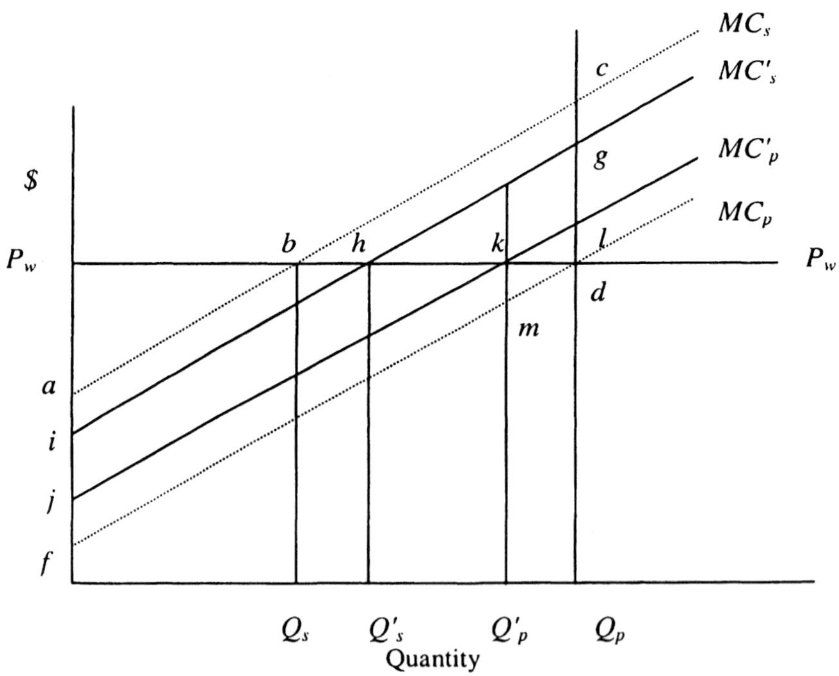

Figure 17.5 Effects of Environmental Policy on Private and Social Costs

ducers but it also reduces the social marginal cost from production from MC_s to MC'_s.

3.3 Compensating for the Private Costs of Regulation

Reducing the externalities associated with production in general will involve some policies other than those directly restricting output. The most common form of intervention is a set of regulations that requires firms to meet particular standards in order to avoid fines, or to be licensed to produce. Restrictions or taxation on input use or on emission levels are common. Each of these policies will reduce the cost of an externality associated with production, while at the same time increasing the private cost of production. Producers are often compensated for costly mitigation measures. For example, the adoption of some measures would be very difficult to implement and monitor, and are encouraged on a voluntary basis. In this case, producers are given an incentive to adopt the environmental practice. The incentive structure may include a subsidy, which would lower the private cost of adopting the new practice. In other cases, the government accepts that producers have a property right associated with the status quo. When the government is forcing producers to adhere to a costly regulation, it may feel obliged to partially compensate producers for their costs. The government may also wish to protect producers from the additional costs for competitive reasons. In this case, governments are not accepting the premise that environmental costs would affect the competitiveness of an industry. Finally, payments made to producers may be purely from rent-seeking activities, whereas the government is using compensation for environmental programs and other mechanisms to transfer resources to a particular group in society. For whatever reason, governments often make payments to compensate firms that are adopting practices, which, in turn, reduce the externalities associated with production.

3.4 Trade Effects of Externalities and Environmental Policies

Global economic surplus will be maximized when all countries face a common world price and when the cost of production in each country reflects a social-marginal cost of production. This suggests that non-market costs need to be considered when establishing trade rules so as to maximize total economic surplus.

Allowing a sector to produce without regard for externalities results in overproduction, which lowers the world price of the commodity and therefore reduces the economic surplus of other producers in the world. In effect, unregulated sectors that produce an untaxed externality are the recipients of an indirect subsidy from those harmed by the externality. If only the exporting firms are exempt from taxation, then it is an export subsidy.

If firms were compensated through either an output or an input subsidy for complying with environmental standards, then the subsidy would be considered a trade-distorting subsidy that would move the country away from production in which social-marginal cost is reflected in costs. The production-related, or input-related, subsidy would drive a wedge between the social marginal cost and the cost faced by the firm. The result would be a production level that is higher than the socially desirable level. The only policy for which this would not be the case would be one in which the subsidy is clearly decoupled from production. Given that even these decoupled payments are likely to affect entry to (or exit from) the industry, payments to compensate producers for environmental initiatives by themselves are very likely trade distorting.

4. TRADE RULES AND DOMESTIC-POLICY CHOICE

In Canada, agriculture is under the joint jurisdiction of the federal and provincial governments. Therefore, both levels of government can intervene in the sector, and provinces can develop specific policies and programs that fit their local needs. However, this also means that programs implemented in one province can affect the well being of the producers (and consumers) in other provinces if, for example, a provincial program is countervailed by an importing country. Hence, the choice of policy instruments to implement domestic environmental policy has to be made within the context of the operating trade rules. Thus, the optimal choice of policy instruments would be based on a consideration of the trade rules.

4.1 Modeling the Choice of Nitrification Abatement Policy in Québec

There are several environmental measures that the government of Québec could have used to address the problem of groundwater nitrification. They chose the requirement that hog producers acquire, or secure, enough agricultural land to safely use the nitrogen in the hog manure to prevent groundwater nitrification. This was a conscious policy choice. To illustrate how the choice of environmental policy may be influenced by trade rules, we compare this policy choice with two other possible policies: (1) the government could have chosen to do nothing about the problem, and the populous would have had to live with higher groundwater nitrates; alternatively, (2) the government could have built manure-processing facilities that would have utilized compost and would have dried the manure, which would have been sold at the commercial-fertilizer market.

In addition to the option of the acquisition of land, the government of Québec also offers a cost-of-production based, price-support program, which sometimes is binding. For this reason, we will also examine the effects of the alternate policies with and without a binding price support. In the case of a binding price support, it is assumed that additional price supports given to producers to offset manure-spreading costs will be countervailed by the United States. This action will have an impact on prices in Western Canada and on the countervail duty collected by the U.S. government.

To illustrate the welfare effects within the presence of trade rules, we first compare the alternative environmental measures when no price support exists. We then examine the policies in the presence of a binding price support. In order to isolate the effects of the policy set, we maintain the world-market conditions for all six scenarios.

4.2 Theoretical Model Without Binding Price Support

The model shown in Figure 17.6 depicts the hog-production sector in Québec when the price support is not binding. In this case, the price faced by producers is $P_{w,}$ which is the world price (that is, a U.S. price). The private marginal-cost curve MC_P is the cost before environmental costs are considered; thus the equilibrium output is Q_p. At output levels below Q_d there is neither social cost nor externality because hog production is so low that all the manure is demanded as fertilizer and is absorbed by the environment without difficulty. However, once the output level exceeds $Q_{d,}$ there is a pollution problem. This gives rise to social cost curve MC^0_s. As more hogs are produced, MC^0_s diverges from private cost curve, MC_p. The socially optimal level of hog production at the price P_w is Q_{s0}.

Consider scenario 2 wherein hog producers are regulated and cannot exceed a maximum manure application rate. In this scenario, it is assumed that farmers must pay for transporting the hog wastes and/or securing more land as more hogs are produced. This has the effect of rotating the private cost curve upward from MC_p to $MC^I_{s.}$ The policy also eliminates the environmental cost associated with manure, hence, the MC^0_s curve rotates downward to MC^I_s to coincide with the private cost curve. Note that because some waste has a positive fertilizer value, the cost curve facing the industry will be $k_1MC^I_s$. The difference between k_0 and k_1 reflects the net value of the manure

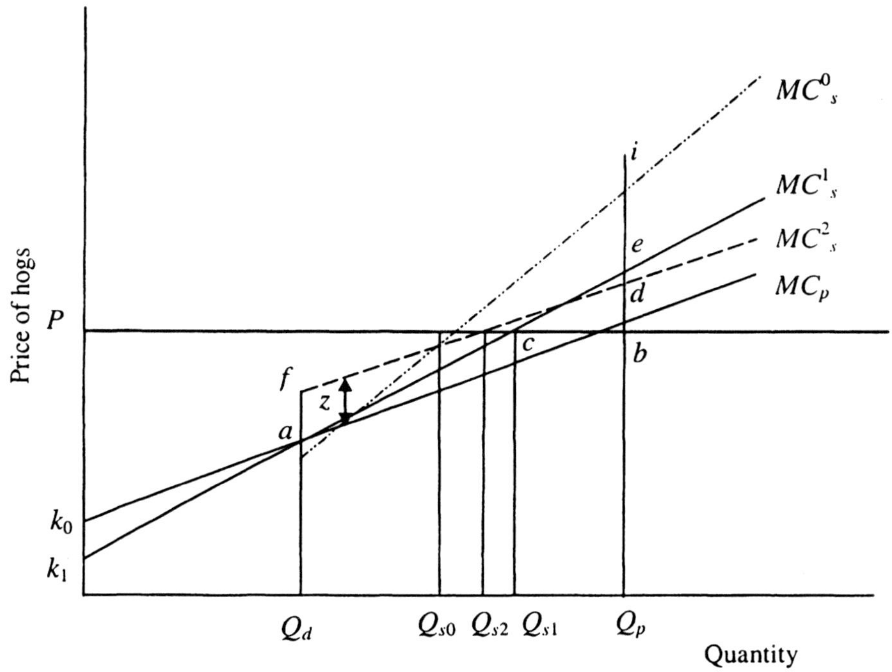

Figure 17.6 Welfare Effects without Binding Price Support

[that is, the fertilizer value net of the environmental costs (transport and land)], and this difference decreases with the output level. The distance between *e* and *b* reflects the net costs at Q_p. In this case, the new equilibrium level of output, Q_{s1}, is socially optimal.

Scenario 3 is a policy alternative to the regulation and represents a policy wherein the government builds a waste-disposal plant for hog waste at a cost of *z* per unit. Assuming that the cost for producers to transport the manure to the plant is the same as they incur to dispose of it under the present policy, this would leave them on private cost curve MC_P. The new social cost curve, MC^2_s, would represent the per-unit cost of the manure processing plus the cost of hog production. In this case, the socially optimal level of production would be Q_{s2}, which would lie somewhere between Q_{s0} and Q_p depending on the size of the per-unit cost *z*.

The welfare effects of the above scenarios can be summarized as follows. In the base case (scenario 1), producer surplus, *PS*, is the area P_wbak_1. The externality, or social cost, is area *aib* and there is no government cost. In scenario 2, *PS* decreases and is area P_wcak_1. The social cost is eliminated and there is no government cost since the price support is not binding. In scenario 3, *PS* is the same as in scenario 1, that is, there is no social cost, but the cost to the government is area *afdb*.

4.3 Theoretical Model With Binding Price Support

We now turn our attention to the case in which the price support is binding and for which there are trade implications (Figure 17.7). For the ease of comparison, we keep P_w at the same level, so the price that farmers are facing is now P_w+s. In the base case of scenario 4, optimal equilibrium output is increased from Q_p to Q_{ps}. With output level, Q_{ps}, *PS* increases to area $(P_w+s)gak_1$, and the taxpayers' transfer is area $P_w(P_w+s)gq$. As is presently the case, this price-support program is not countervailed, and the social cost is area *amg*.

We now assume that the regulation on the maximum application rate is implemented. In this case, the calculated cost of production will eventually increase by amount t, so the new price that producers face is P_w+s+t. This

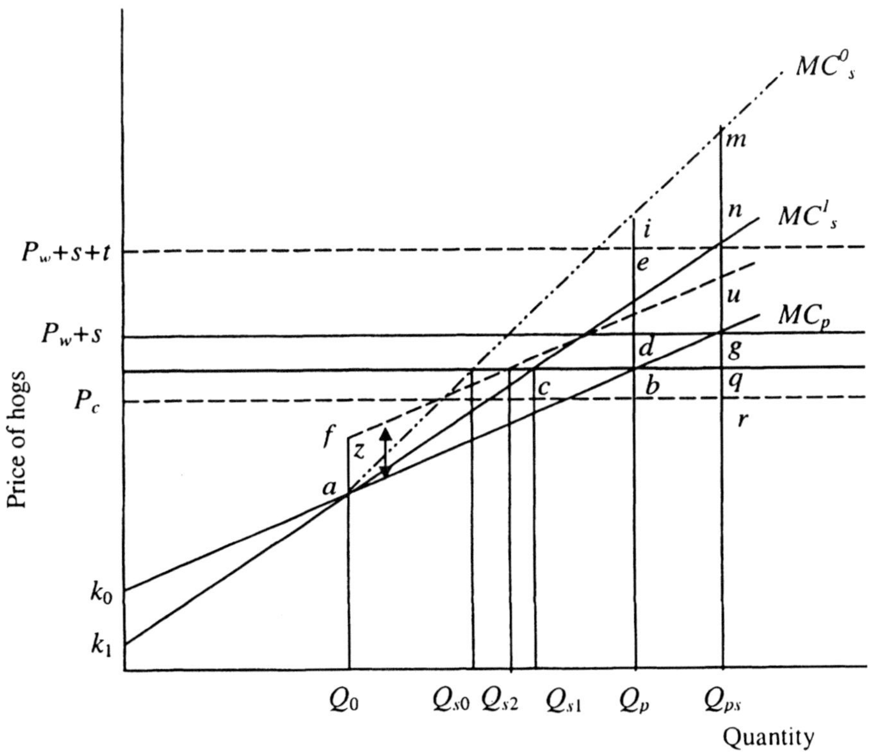

Figure 17.7 Welfare Effects with Binding Price Supports

situation represents more or less what is in place in Québec. We assume that the extra subsidy, t, is open for countervail action by the importing country (scenario 5). On the other hand, given the definitions discussed above, there is no countervail for the construction of the waste-disposal plants by the government (scenario 6).

When the extra subsidy is countervailed, the market price falls to P_c for those domestic producers outside Québec, thus reducing production. The tariff (P_w–P_c) is t/S, where S is Québec's production share of Canadian production. The subsidy to offset the cost plus the tariff is $t(1+1/S)$. Therefore, if the government is willing to completely protect the industry, then the price remains at P_w+s+t in Québec and production is maintained at Q_{ps}. Producer surplus is given by $(P_w+s+t)nak_1$ and the transfer to producers by taxpayers is $P_c(P_w+s+t)nr$. In this case, there is neither social cost nor government-abatement cost.

If subsidies that protect the environment are allowed, such as subsidized waste-disposal plants, they are not actionable under trade agreements. The marginal social cost curve is MC^2_s. In scenario 7, production remains at level Q_{ps} and producer surplus is given by area $(P_w+s)gak_1$. There is no social cost, but the government abatement cost is *afug*.

4.4 Estimating the Welfare Effects

4.4.1 The Base Parameters

We summarize the base parameters for the policy simulations that were conducted (Table 17.1). Two market-price levels were used in the policy analysis. In the no government-support scenarios, the price was set equal to the 1996 and 1997 Ontario average price of CDN $1.89 per kilogram (CPC, 1999). In the scenarios with a binding support price, a price of CDN $2.00 per kilogram was used, which is somewhat higher than the effective support price in Québec, but we decided to assume a higher support price in order to maintain the world price as constant in the comparative welfare analyses.

The quantities of carcass meat produced in Québec and the rest of Canada (ROC) were calculated on the 1997 basis of 85-kilogram carcass-weights and sales reported by the Canadian Pork Council (CPC, 1999). Consumption quantity was estimated from the 1997 per-capita consumption (carcass weight basis) of 25.2 kilograms (CPC, 1999), multiplied by the population in Québec (seven million) and in the rest of Canada (21 million). The ROC demand for Canadian hogs was calculated by subtracting estimated consumption from estimated production (CPC, 1999). We used the supply elasticity of 1.2, which is approximately the midpoint of the supply

Table 17.1 Base Parameters used for Policy Simulation

Base parameters	*Québec*	*SK*[1]	*ROC*[2]	*World*
Average price ($/kg)	1.89	1.89	1.89	
Farm support price ($/kg)	2.00	0.00	0.00	
Quantity produced (million kg) [no measures]	491.00	1,106.00	0.00	1,597
Domestic demand (million kg)	176.00	529.00	892.00	1,597
Elasticity of supply	1.50	1.5	0.00	
Elasticity of demand	-0.72	-0.72	-1,000.00	
Fertilizer value ($/kg of hog)	-0.08235	0.00	0.00	
Marginal land abatement cost ($/kg of hog)	0.08235	0.00	0.00	
Marginal treatment cost ($/kg of hog)	0.06	0.00	0.00	

[1]*Saskatchewan.*
[2]*Rest of Canada.*
Source: Authors' estimates and various sources (see text).

elasticity found in the literature. Short-run elasticities are likely to be less than this figure while long-run elasticities could be more. The domestic demand elasticity of –0.72 is equal to that reported by Chalfant et al. (1991).

The private and social costs of nitrification and nitrification abatement were difficult to quantify. As discussed earlier, producers are required to spread the manure on land over a minimum area per hog. We assumed that if the manure were to be spread according to the existing regulations, then the externality from hog production would be eliminated. (There is some market for spreading hog manure because crop producers close to the hog production areas are willing to pay for the nutrient content of hog manure, which can more than offset the hog producer's manure transportation and application costs.) Beyond a certain distance, the cost of the transport exceeds the fertilizer value of the manure and it becomes costly for hog producers to get rid of the manure.

Small quantities of manure have a value equal to the nutrient value minus the marginal application costs. At current prices for inorganic fertilizers and application costs, this was calculated to be CDN $7 per hog. The marginal cost for the last hog produced would be reflected in the price producers are currently willing to pay. Following a consultation with the industry, this figure also worked out to an amount of CDN $7 per hog.

Other externalities such as the smell of manure were assumed to be very small, or at least offset, by positive non-market benefits of hog production such as rural aesthetics. We also assumed that when the manure was put through a treatment facility, which separated the manure into useful plant nutrient, the externalities were zero. For the case in which there was no regulation for spreading the manure, we assumed the social externalities from groundwater nitrification were twice the abatement costs.

4.4.2 The Policy Simulation Results

The data in Table 17.2 show the welfare effects of three environmental policies in the absence of price supports as depicted in Figure 17.6. In scenario 1, no measures are used to address the manure externality. For the sake of comparison, this is the base case. The supply curve only reflects private

Table 17.2 Summary Results: Without Price Support

Scenario 1: No environmental measure				
Quantity produced (million kg)	491	1106		1,597
Producer surplus (CDN $ million)	319	695	0	1,014
Consumer surplus (CDN $ million)	231	693	1	925
Externality (CDN $ million)	20	0	0	20
Government abatement costs (CDN $ million)	0	0	0	0
Farm support expenditures (CDN $ million)	0	0	0	0
Countervail duties collected (CDN $ million)	0	0	0	0
Total economic surplus (CDN $ million)	529	1,388	1	1,918.1
Scenario 2: Land management regulation				
Quantity produced (million kg)	463	1,106		1,569
Producer surplus (CDN $ million)	310	695	0	1,005
Consumer surplus (CDN $ million)	231	693	1	924
Externality (CDN $ million)	0	0	0	0
Government abatement costs (CDN $ million)	0	0	0	0
Farm support expenditures (CDN $ million)	0	0	0	0
Countervail duties collected (CDN $ million)	0	0	0	0
Total economic surplus (CDN $ million)	541	1,388	1	1,929
Scenario 3: Manure treatment provision				
Quantity produced (million kg)	491	1,106		1,597
Producer surplus (CDN $ million)	319	695	0	1,014
Consumer surplus (CDN $ million)	231	693	1	925
Externality (CDN $ million)	0	0	0	0
Government abatement costs (CDN $ million)	15	0	0	15
Farm support expenditures (CDN $ million)	0	0	0	0
Countervail duties collected (CDN $ million)	0	0	0	0
Total economic surplus (CDN $ million)	535	1,388	1	1,923.6

Sources: Calculated by authors from simulation results.

production costs, and producer surplus is CDN $319 million. The social cost of hog-manure induced groundwater nitrification is CDN $20 million per year. The total economic surplus (TES) generated in this sector within Québec with this policy is CDN $529 million.

In scenario 2, a regulatory land-management policy requires hog producers to spread the manure at rates below the threshold at which groundwater nutrification occurs. This land-management policy is creating a TES of CDN $541 million per year. The policy reduces the social marginal cost and eliminates the externality by increasing the private marginal cost to reflect the abatement costs. Note that this regulatory solution reduces producer surplus relative to scenario 1. Producer surplus is decreased to CDN $310 million. Without price supports a second-best policy is the provision of manure treatment. This policy eliminates the pollution externality of CDN $20 million with a government abatement cost of CDN $15 million. The subsidy results in production beyond the socially optimal amount and a TES of CDN $535 million, which is higher than with no environmental measures but is less effective than the land-management regulation.

We show the impact of the same three policies in the presence of a producer price support, which maintains production at 536 million kilograms (Table 17.3). In scenario 4, no environmental measures are used. In this case, the support price increases production, and producer surplus increases to CDN $378 million at a cost to taxpayers of CDN $62 million per year. The larger production results also increase the pollution costs to CDN $24 million per year. The result is a decrease in TES to CDN $523 million for the Québec sector.

In Scenario 5, a regulatory land-management policy requires hog producers to spread the manure at rates below the threshold of where groundwater nitrification occurs. The policy reduces the social marginal cost and eliminates the externality by increasing the private marginal cost to reflect the abatement costs. However, the support policy means that in order to maintain production at 536 million kilograms, the price-support level is increased from CDN $2.00 to CDN $2.09 per kilogram. Making the assumption that this level of price support, which exceeds the CDN $2.00 minimum level, is countervailed with a $0.047 per kilogram U.S. countervail, the market price of hogs in Québec and the rest of Canada falls to CDN $1.837 per kilogram. This results in some increase in consumer surplus but a large overall impact on the cost of the price support, which increases the cost to CDN $165 million per year in Québec. Overall, the combination of price supports, land-use regulation, and countervail results in a loss in economic surplus to the Québec economy, with TES reduced to CDN $491 million.

Table 17.3 Summary Results: With Price Support

Scenario 4: No environmental measure				
Quantity produced (million kg)	536	1,106		1,642
Producer surplus (CDN $ million)	378	695	0	1,073
Consumer surplus (CDN $ million)	231	693	1	925
Externality (CDN $ million)	24	0	0	024
Government abatement costs (CDN $ million)	0	0	0	0
Farm support expenditures (CDN $ million)	62	0	0	62
Countervail duties collected (CDN $ million)	0	0	0	0
Total economic surplus (CDN $ million)	523	1,388	1	1,912
Scenario 5: Land management regulation				
Quantity produced (million kg)	536	1,065		1,601
Producer surplus (CDN $ million)	416	644	0	1,060
Consumer surplus (CDN $ million)	239	718	1	958
Externality (CDN $ million)	0	0	0	0
Government abatement costs (CDN $ million)	0	0	0	0
Farm support expenditures (CDN $ million)	165	0	0	165
Countervail duties collected (CDN $ million)	–	0	42	42
Total economic surplus (CDN $ million)	491	1,362	42	1,895
Scenario 6: Manure treatment provision				
Quantity produced (million kg)	536	1,106		1,642
Producer surplus (CDN $ million)	378	695	0	1,073
Consumer surplus (CDN $ million)	231	693	1	925
Externality (CDN $ million)	0	0	0	0
Government abatement costs (CDN $ million)	17	0	0	17
Farm support expenditures (CDN $ million)	62	0	0	62
Countervail duties collected (CDN $ million)	0	0	0	0
Total economic surplus (CDN $ million)	530	1,388	1	1,918

Source: Calculated by authors from simulation results.

The countervail duty also reduces economic surplus in the rest of Canada by CDN $16 million, reducing total economic surplus to CDN $1,362 million. In this scenario, producer surplus actually increases because producers receive the additional marginal cost of the land regulation over their entire output.

The pollution-abatement environmental measure that generates the largest TES with price supports is the subsidized public manure-treatment facilities. In this case, the policy does not affect producer costs or the level of the price support, and, as a generally available program, it does not attract countervail duties. Producer surplus remains at CDN $378 million and government income support costs remain at CDN $62 million. The publicly funded cost is CDN $17 million. The total economic surplus for the Québec hog sector is $530 million, which is the highest of the price support scenarios. Thus, with an income-support policy and countervail, the provision of manure treatment becomes the least costly policy.

Further analysis reveals that countervail duty has a direct effect on the optimal choice of policy. If no measures are countervailed, then the least-cost land-management option is chosen even with a price support. If the additional price support for this option is countervailed, then the second-best manure treatment option becomes preferred. Finally, if both environmental measures are countervailed, then Québec TES would be maximized with no measures in place. Thus, the choice of domestic instruments becomes driven by international trade rules, and in this case, the rules will tend to drive the sector away from compensated private abatement measures and toward policies of no abatement.

Thus, if private-abatement measures and, to some extent, public-abatement measures are subject to trade sanctions, then there may be a bias toward maintaining low environmental standards as an indirect form of protection. For agricultural commodities that tend to receive various forms of government support, domestic environmental measures will be influenced by international trade laws, and this influence may drive policy away from the most efficient set.

5. CONCLUSION

The example in this chapter clearly illustrates a linkage between international trade rules and the choice of domestic environmental policies. Given that lower standards will not be countervailed, whereas direct income support is countervailed, there will be some domestic bias toward lower standards. To the extent that the public provision of pollution abatement is less likely to be countervailed than is the income support for tighter regulation, there will be some bias toward the public provision of abatement.

The choice of environmental policy is potentially a very important issue for the implementation of international climate-change agreements. If these agreements significantly restrict the emission of greenhouse gases, then virtually every sector in the global economy will be affected. If a country chooses to subsidize the abatement of these gases for a domestic industry, then that industry will have a distorted competitive advantage over its trading partners who choose to use a Pigouvian tax to reduce emission levels.

Given the pervasiveness and the potential cost of these abatement policies, trade impacts could be substantial.

This chapter serves to illustrate the potential for intra-country spillovers when local policies result in trade actions that affect the whole country. This is especially important for environmental policies, which, in many parts of the world, are designed and implemented at the local government level. These issues will grow in importance over time as trade rules become more liberal and as environmental effects become more significant. For instance, the implementation of the Kyoto Accord is likely to generate a complex interaction between choice of local environmental policies, national welfare impacts, and international trade policies.

6. FURTHER WORK

The analysis presented here provides an example of how the choice of local environmental policy can be affected by international trade rules. If international trade rules can distort the choice of domestic environmental measures, this begs the question as to what international trade rules might result given an unbiased choice of environmental measures. This choice may also depend on whether the pollution is domestic or international in nature. The issues of compensation, property rights, and policy formation also become important. At the outset, international environmental policy and international trade policy will become intimately entwined. Anticipating the form of this relationship must begin with a better understanding of how trade policy can affect the choice of environmental policy and how environmental policy can affect trade. These issues will be particularly important in the design and implementation of international climate-change agreements.

7. REFERENCES

Abbott, P.C. and S.L. Haley. 1988. "International Trade Theory and Natural Resource Concepts," in J.D. Sutton, ed., pp. 35–62. *Agricultural Trade and Natural Resources—Discovering the Critical Linkages*. Boulder, CO and London, UK: Lynne Rienner Publishers.

CPC (Canadian Pork Council). 1999. *Pork Statistics*. Ottawa, ON: Canadian Pork Council.

_____. 1999. Internet website: www.canporc.ca/s

Chalfant, J.A., R.S. Gray, and K.J. White. 1991. "Evaluating Prior Beliefs in a Demand System: The Case of Meat Demand in Canada." *American Journal of Agricultural Economics* 73(2): 476–490.

Schmitz, A., D. Sigurdson, and O. Doering. 1986. "Domestic Farm Policy and the Gains from Trade." *American Journal of Agricultural Economics* 68(4): 820–827.

Segerson, K. 1988. "Natural Resource Concepts in Trade Analysis," in J.D. Sutton, ed., pp. 9–34. *Agricultural Trade and Natural Resources: Discovering the Critical Linkages*. Boulder, CO and London, UK: Lynne Rienner Publishers.

Trade Compliance Center. 1993. *North American Free Trade Side Agreement On Environmental Cooperation*. Internet website: http://199.88.185.106/tcc/data/commerce_html/TCC_2/Nafta/Naftaside2.html

WTO (World Trade Organization). 1999. Internet website: www.wto.ca

Chapter 18

Environmental Problems, Immigration, and Trade

Richard E. Just[1] and Sinaia Netanyahu[2]
[1] *University of Maryland and* [2] *Ben-Gurion University of the Negev*

1. INTRODUCTION

Issues between countries with common borders can be complex and involve non-trade issues as well as trade issues. The American-Mexican trading relationship is one of the most interesting. Recorded trade has increased from US $80 billion before the North American Free Trade Agreement (NAFTA) to US $200 billion in 1998, making Mexico the second largest U.S. trading partner. Simultaneously, the magnitude of American-Mexican non-trade issues has been growing as well. First, food safety and environmental quality have become major trade issues because of differences in standards between the two countries (Zepp, Kuchler, and Lucier, 1998). Second, Mexico is the primary source of illegal immigration into the United States, and much of it enters the agricultural labor force. Illegal Mexican immigration has far exceeded legal quotas for decades and has become a subject of agricultural economic research (Torok and Huffman, 1986). Third, Mexico is the primary source of illegal drugs for the United States. About 70 percent of the illegal drugs entering the United States in 1996 came from Mexico. Much of the increased smuggling has occurred with increased trade under NAFTA (Silvey and Brody, 1996).

In each of these issues, questions have been raised about inconsistencies in U.S. policy and enforcement. In the agricultural trade arena, for example, tariff barriers have begun to decline under NAFTA, leading to increased importation of winter fruits and vegetables to the United States (Calvin and Barrios, 1998). However, in view of growing concerns about environmental standards and food safety in 1997, President Clinton proposed Food and

Drug Administration inspections and the halting of imports of fruits and vegetables that do not meet U.S. standards. Because the NAFTA elimination of tariffs has a 15-year phase out for horticultural products, the Clinton proposal could curtail Mexican fruit and vegetable importation before NAFTA has been fully implemented. Thus, the United States may appear to be waffling on its agreement.

This chapter considers bargaining motivations that may develop as countries negotiate trade policy and environmental standards for it, and why compliance and enforcement of agreements may be intermittent. In addition, we consider how issues of illegal immigration and the drug trade, although typically considered to be largely independent of trade, may become useful issues to address in tandem, that is, to link to agricultural trade and related environmental safety. Illegal immigration and the drug trade are the subjects of considerable negotiation and imperfect enforcement. Illegal immigration is enforced imperfectly as evidenced by the relative ease of detecting violators on farms employing illegal immigrants. Similarly, after the United States has set up substantial political machinery to penalize countries not cooperating in eradicating illegal drug trade, President Clinton proposed that Mexico be designated a fully cooperating nation for political purposes even though drug enforcement officials strongly claim the contrary. In each case, many regard the United States as waffling on its positions.

Each of these issues has potential resolutions that are significantly one-sided. For example, profits from exporting illegal drugs or pesticide-laden fruits and vegetables bring economic gain to Mexico but can have high social costs in the United States. Alternatively, if Mexico controls illegal drug trade and immigration, much of the benefit accrues to the United States at the expense of Mexico. When issues have strong one-sided implications, a solution can be difficult to negotiate in isolation. On the other hand, if other issues, which have one-sided but counter-balancing effects, can be identified and linked, then an overall sustainable and self-enforcing agreement is more likely. This chapter examines properties that lead to useful linking of issues for negotiation. Interestingly, a side result is that many useful opportunities for linking result in partial cooperation, that is, intermittent cooperation that appears much like waffling.

Conventional wisdom implies that free trade should equate marginal values across economic activities. When trade has externalities, however, the marginal social impacts can be quite different. Pesticide- or disease-laden farm produce and illegal drugs are prime examples. When markets fail, trade and enforcement usually become a problem of bargaining between sovereign countries. Thus, the framework of game theory appears to be applicable. One of the fundamental insights of game theory is an explanation, under certain conditions, for non-cooperation on issues with prisoner's dilemma (PD) properties, that is, issues in which each side attains its most preferred outcome when one party takes an action contrary to the other party's incentives. The preferred outcome of a PD involves finding a solution where neither side gets its most desired outcome but both are better off than with noncooperation. When two such issues are negotiated in tandem, however, each party can sometimes get its most desired outcome on one of the issues by giving the other party what it desires on the other issue. Game theory considers such approaches in the context of linked games.

Linked games have been touted by Hauer and Runge (1997) for combining environmental and trade issues. Linking isolated games (negotiating multiple issues simultaneously) is claimed to enrich the set of Nash equilibria. Folmer, Mouche, and Ragland (1993) and Cesar and de Zeeuw (1994) make similar assertions regarding environmental problems. However, gains from linking are highly conditional (Netanyahu, 1998). Even where trade and environmental issues are linked with dramatically different issues like illegal immigration or drug trade, improved cooperation may not be possible.

This chapter explores conditions under which the feasible set of aggregated isolated games is dominated by the feasible set of associated linked games. In particular, this chapter investigates the characteristics of international trade policy games to determine whether gains accrue from linking. We begin by characterizing issues of trade and the environment in a market setting to evaluate potential welfare implications. This framework is used to demonstrate that a variety of potential game structures can result. Next, we examine potential gains from linking within the range of possibilities suggested by the market analysis. Gains are possible in some cases and are impossible in others. While game theory analyses have argued that these wide ranges of possibilities can explain wide ranges of observed behavior, we argue that the conditions that lead to gains from linking can be investigated empirically following typical econometric market analyses. The chapter concludes with a call for empirical work to narrow the range of explanations that are commonly attributed to game theory, and with an equity explanation for why international policies may appear to be unstable and intermittent.

2. STRUCTURES OF GAMES[1]

Before characterizing the economic effects of international policy and discussing the potential of linking trade issues, a delineation of game structures is useful as a framework in which to characterize welfare outcomes of countries' strategies in international negotiations. Although the principles of this chapter apply to multi-strategy games, for simplicity, the discussion here examines two 2-strategy, 2-player games, as in Figure 18.1. In reality, strategic possibilities may be continuous rather than discrete, but the basic implications of different shapes of the feasible sets (convexity, concavity, and slopes of the frontier in positive and negative spaces) carry through.

To establish the basic illustrative framework in the chapter consider Figure 18.1 where each cell in the first entry gives the payoff to Player A and the second entry gives the payoff to Player B if the strategies of both players correspond to that cell. Players A and B each have two strategies generically labeled "cooperate" and "defect" in each game. Without loss of generality, the payoffs of the cooperate-cooperate strategies in Game 1 can be represented by (1, 1) and the payoffs of defect-defect in both games can be represented by (0, 0). That is, any general set of payoffs can be re-scaled and translated to a different origin. For generality, payoffs of other various strategies in Game 1 are represented by q, r, s, and t while other payoffs in Game 2 are represented by u through z.

The equilibrium of a one-shot game can be determined according to the structure of the payoff set of the game. A useful, but incomplete, set of payoff structures is summarized in Table 18.1. To consider the practical problem where cooperation has merit, $u > 0$ and $v > 0$ are assumed. These four structures are selected to reflect empirical possibilities demonstrated in the next section. They are also interesting because each has potential defection equilibrium. (Games with equilibria consisting only of cooperation are not interesting because outcomes are clearly understood.)

To translate Table 18.1 into geometric terms consider first the case in which each of two isolated games is a PD game. Figure 18.2 is constructed using the payoffs in Figure 1 according to conditions in the first row of Table 18.1. The feasible sets of the isolated PD games are illustrated by the areas enclosed within solid lines. With PD games, both segments bounding the feasible set in the positive quadrant are negatively sloped. While consistent strategic behavior by both players must yield non-negative payoffs for both players, pure strategy partial cooperation, where one player cooperates and the other defects as in outcomes (q, r), (s, t), (w, x), or (y, z), results in a negative payoff for one of the players. Individual rationality restricts outcomes to the positive quadrant in isolated games because neither player is willing to accept being worse off than can be guaranteed by defecting. Thus, strategies outside the positive quadrant are not considered part of the feasible set in isolated strategic games. By comparison, the frontier outcomes where one player has a positive payoff and the other player has a zero payoff are feasible and are attained by convex (or probabilistic) combinations of pure strategies, which are called mixed strategies. For example, $(0, Y_1)$ is attained by following strategies (q, r) with probability p and strategies (1, 1) with probability $1 - p$ for some probability p.

(a) Game 1

		Player B	
		Cooperate	Defect
Player A	Cooperate	1 1	q r
	Defect	s t	0 0

(b) Game 2

		Player B	
		Cooperate	Defect
Player A	Cooperate	u v	w x
	Defect	y z	0 0

Figure 18.1 Two Generic 2-Player, 2-Strategy Games

Table 18.1 Alternative Game Structures

Game structure	*Conditions for Game 1*	*Conditions for Game 2* $(u > 0$ *and* $v > 0)$	*Nash Equilibrium*[a]
Prisoner's dilemma game	$q < 0$ $r > 1$ $s > 1$ $t < 0$	$w < 0$ $x > v$ $y > u$ $z < 0$	*dd*
Assurance game	$q < 0$ $r < 1$ $s < 1$ $t < 0$	$w < 0$ $x < v$ $y < u$ $z < 0$	*cc, dd*
Iterated dominance game	$q < 0$ $r > 1$ $s < 1$ $t < 0$	$w < 0$ $x > v$ $y < u$ $z < 0$	*dd*
Chicken game	$q > 0$ $r > 1$ $s > 1$ $t > 0$	$w > 0$ $x > v$ $y > u$ $z > 0$	*cd, dc*

[a]*The defect strategy is denoted by d and the cooperate strategy is denoted by c.*
Source: Authors.

The Nash equilibrium of PD games is for both players to defect. Figure 18.2 illustrates the feasible sets of isolated PD games by the area enclosed in solid lines. The convex hull of feasibility in Figure 18.2(a) is defined by points (0, 0), $(0, Y_1)$, (1, 1), $(X_1, 0)$ where $X_1 = (s - t)/(1 - t)$ and $Y_1 = (r - q)/(1 - q)$. The convex hull of feasibility in Figure 18.2(b) is defined by points (0, 0), $(0, Y_2)$, (u, v), $(X_2, 0)$ where $X_2 = u - v(y - u)/(z - v)$ and $Y_2 = v - u(x - v)/(w - u)$. The games in Figure 18.2 illustrate the case in which cooperation is preferred to convex combinations of partial cooperation, for example, (1, 1) lies above a segment connecting (q, r) and (s, t). The conditions of this case are $s_1 > s_2$ for Game 1 and $s_3 > s_4$ for Game 2, where s_1, s_2, s_3, and s_4 are slopes of the respective segments as shown.

The structures of Table 18.1 are illustrated geometrically in Figure 18.3. Figure 18.3(a) is the case of a PD game where cooperation is inferior to convex combinations of partial cooperation in Game 1, that is, $s_1 < s_2$. Figure 18.3(b) is the case where Game 1 is an assurance game, that is, both segments bounding the feasible set in the positive quadrant are positively sloped and partial cooperation results in a negative payoff for one of the players. Figure 18.3(c) is the case of an iterated dominance (ID) game. One of the segments bounding the feasible set in the positive quadrant is positively sloped and the other is negatively sloped. In this case also, partial cooperation results in a negative payoff for one of the players in each case. Finally, Figure 18.3(d) is the case of a chicken game where the entire feasible set is strictly contained in the positive quadrant. The full cooperation point may be either inferior to combinations of partial cooperation (as shown) or superior, in which case (1,1) would lie above the segment connecting (q, r) and (s, t). The following section shows that each of these structures is possible in the typical range of economic parameters of trade problems, but that the specific structure can be determined econometrically.

(a) Game 1

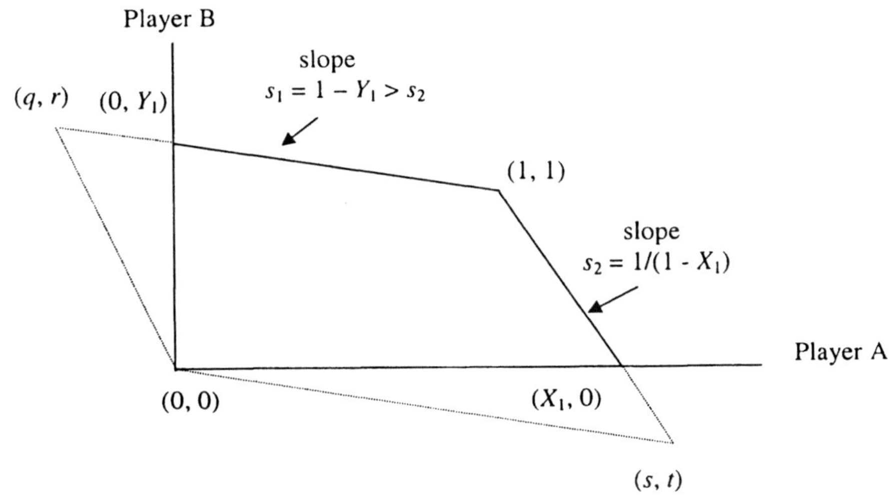

(b) Game 2

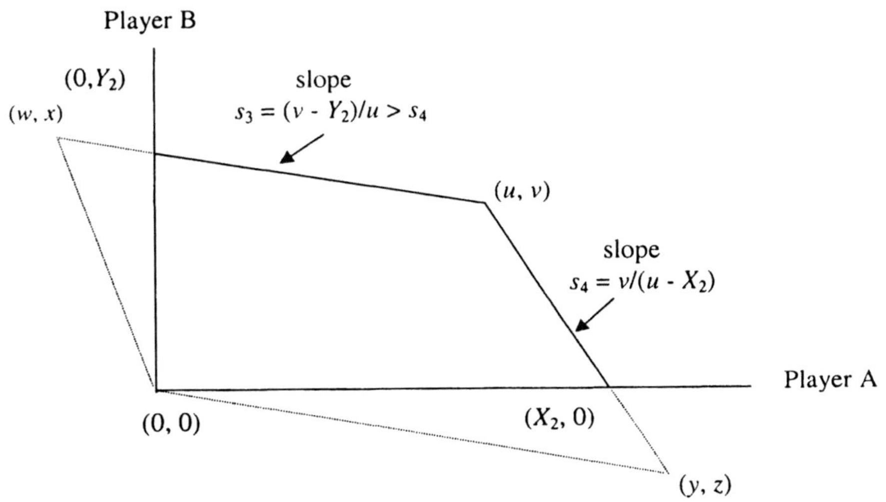

Figure 18.2 The Feasible Payoff Set in the Prisoner's Dilemma Case

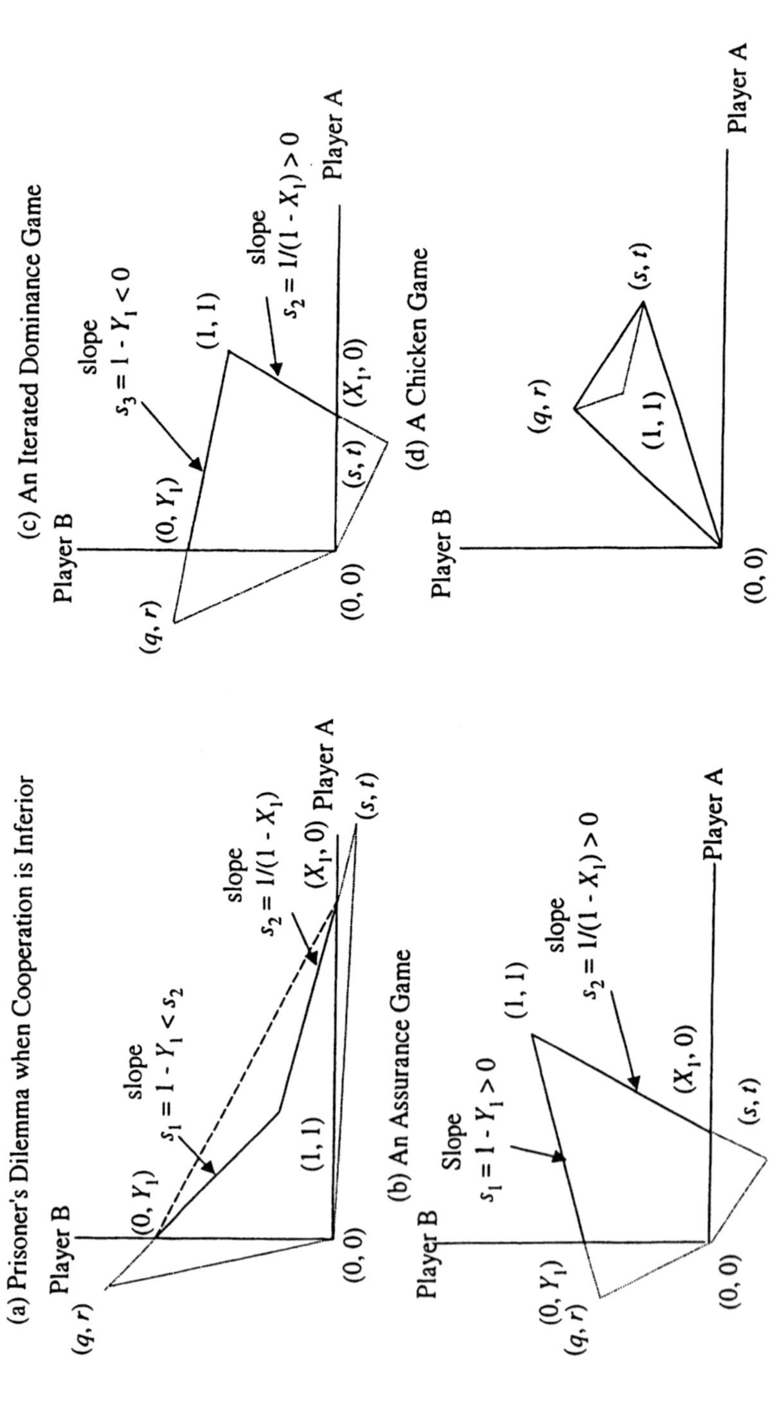

Figure 18.3 The Feasible Payoff Set of Various Game Structures

3. CONCEPTUAL ASSESSMENT OF IMPACTS OF COUNTRY ACTIONS

The potential for linking trade and environmental issues, as well as other issues, can be investigated conceptually based on standard welfare analyses. We undertake this discussion in the context of U.S.-Mexican trade, although the analysis can apply to many other trade relationships. To guide the discussion and to give a basis for applying game theoretic results regarding linking, consider, first, a conceptual analysis that serves to quantify the magnitudes of welfare impacts associated with practical strategy choices for each player (country). The policymakers of each country are assumed guided by standard economic welfare evaluations. More generally, consideration of weighted welfare criteria may be appropriate to represent policymaker concerns for equity, or response to rent seeking.

3.1 A Diagrammatic Analysis

Consider, first, the case in which an environmental or food-safety standard limits the quality of products that can be traded under a free-trade agreement. In Figure 18.4(a), suppose D_U represents the demand for fruits and vegetables in the United States, and that S_U represents supply of fruits and vegetables produced in the United States under Environmental Protection Agency standards on pesticide use. Suppose D_M and S_M represent the corresponding demand and supply in Mexico, except that Mexican production takes place under lower pesticide standards. Without free trade, restrictions prevent Mexican produce from entering the United States. With free trade, excess demand from the United States is represented by E_M, and total demand for Mexican production is D_T. With restricted trade, prices are P_U in the United States and P_M in Mexico. With free trade, the price is P_T in both countries, and U.S. consumption increases from Q_U to Q_T.

The market welfare effects of instituting free trade are a gain for U.S. consumers from area a to area $a + b + c$, a loss for U.S. producers from area $b + d$ to area d, a loss for Mexican consumers of area g, and a gain for Mexican producers of area $f + g$. However, if the higher pesticide levels are truly detrimental to consumers of the foods that carry them, then there is an external cost function, say C_M, associated with the consumption of Mexican produce. Thus, under free trade with no standards, the external cost incurred by U.S. consumers, who are unable to distinguish between pesticide-laden Mexican produce and safer U.S. produce, is area e.[ii] This external cost may negate most or all of the apparent U.S. market benefit of free trade, area c. After considering this impact, net country effects are a gain of area $c - e$ for

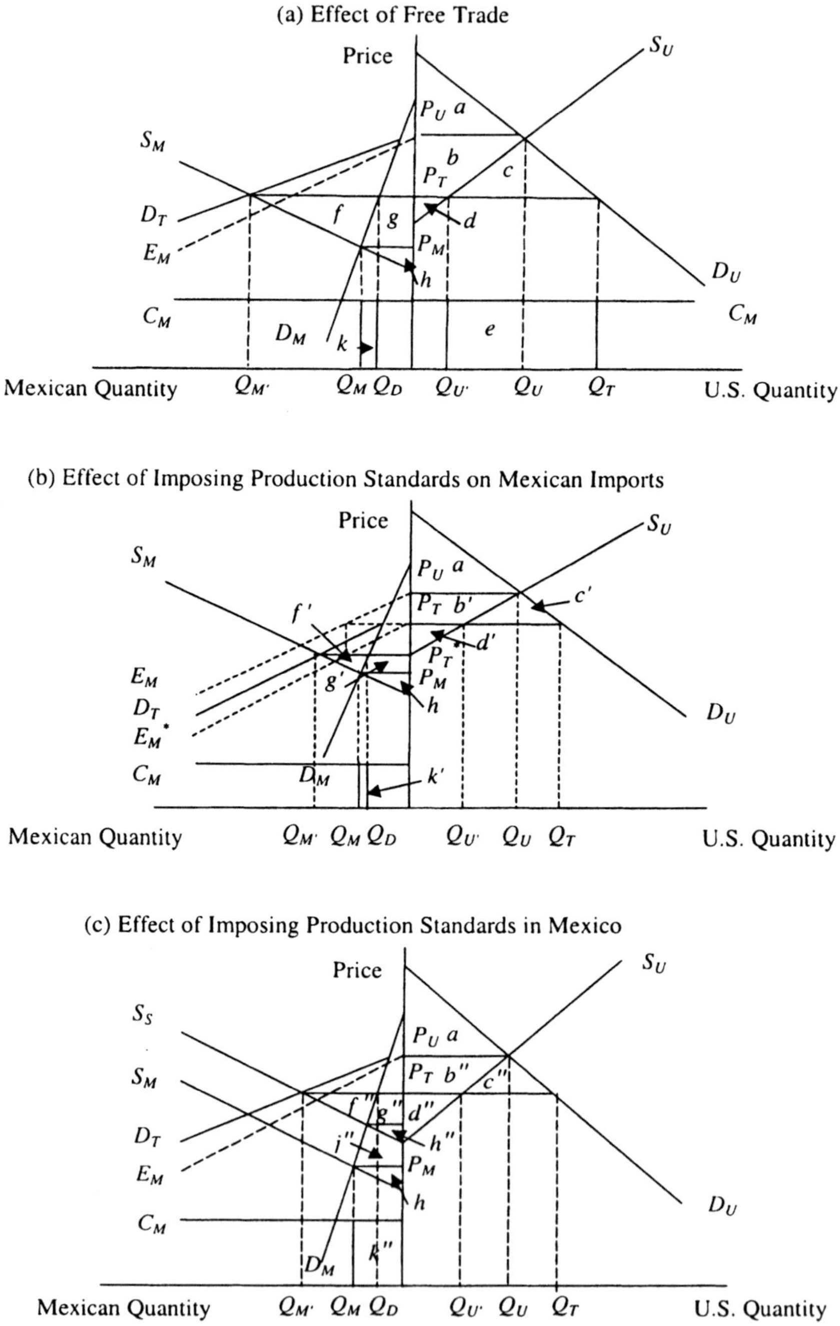

Figure 18.4 Effects of Sovereign Actions on Trade with an Environmental Externality

the United States and a gain of area $f + k$ for Mexico. (The external cost of Mexican consumption is reduced by area k, which may or may not be fully considered by Mexican policymakers.) This environmental externality shifts the gains-from-trade relationship to a highly unbalanced effect against the country whose external effect increases compared to the normal distribution of gains from trade.

One approach to addressing environmental and food-safety concerns is to impose or raise standards in the country where they are lower. Figure 18.4(b) considers the case of imposing or raising environmental production standards in Mexico only on produce bound for the United States (as under NAFTA). Again, E_M represents excess demand from the United States but now a higher production cost is required. To simplify graphical presentation, this production cost is subtracted from the E_M excess demand relationship obtaining E_M^*, which reflects the effective return to producers of U.S.-bound goods. (In the constant-cost case, E_M and E_M^* are parallel, but this need not be the case.) The total effective demand facing producers is D_T, which results in production $Q_{M'}$. Of this amount, Q_D is produced by conventional methods and is consumed domestically while the remainder, Q_T - $Q_{U'}$, is produced with a safer and more expensive production process and is exported to the United States. (Note that both safe and unsafe supply can be represented in the same diagram because of the constant cost difference in the two technologies.)

In this regulated-trade case relative to the no-trade case, U.S. consumers gain area $b' + c'$ while U.S. producers lose area b' for a net gain of area c'. Mexican consumers lose area $g' - k'$ while Mexican producers gain area $f' + g'$ for a net gain of area $f' + k'$ where, again, area k' may or may not be fully considered by Mexican policymakers.

Further possibilities are depicted in Figure 18.4(c). One possibility is that the environmental externality is a global externality; an example is ozone depletion caused by methyl bromide use.[iii] In this case, the United States may not be willing to settle for Mexico using safe production practices only on produce shipped to the United States, but may want to negotiate for safer production practices on all Mexican production. If so, all Mexican production is subject to the higher cost of production, which is represented by supply S_S in Figure 18.4(c) under safety standards. One possibility is to try to negotiate safe Mexican production practices even without trade. This outcome would be represented by equilibrium where S_S intersects D_M with a welfare loss for Mexican producers of area h'' - h, and a loss for Mexican consumers of area $h'' + j'' - k''$, for a net loss of $h + j'' - k''$.

In the event that Mexico does not initiate safe production practices independently, the United States may consider creating such incentives by making trade conditional on Mexico's adoption of safety standards on all its production. This case is also represented in Figure 18.4(c). With trade, total demand for Mexican production follows D_T, as in Figure 18.4(a), resulting in production $Q_{M'}$ and price P_T for safe product. Mexican consumption declines from Q_M to Q_D, and the U.S. imports the remainder, $Q_T - Q_{U'}$. In the case of country-wide internal Mexican-production controls relative to the no-trade case, U.S. consumers gain area $b'' + c''$ while U.S. producers lose area b'' for a net gain of area c''. Mexican consumers lose area $g'' + h'' + j'' - k''$ while Mexican producers gain area $f'' + g'' + h'' - h$ for a net gain of area $f'' - j'' - h + k''$, where again area k'' may or may not be fully considered by Mexican policymakers.

The various outcomes in Figure 18.4 present several alternatives that may be considered in negotiations between the two parties according to Figure 18.5. The United States basically controls the decision about whether or not to allow imports of Mexican products that do not conform to U.S. safety standards. Mexico controls the decision about whether or not to impose production standards in Mexico that will allow their products to meet U.S. standards. In each cell, the first entry gives the net welfare payoff relative to no trade and no Mexican standards for the United States; the second entry gives the net welfare payoff relative to no trade and no Mexican standards for Mexico. The environmental cost in Mexico is weighted by ϕ with $0 < \phi < 1$ to represent the case in which a poor country views the rich country's standard as too stringent, and, thus, views consumption to have less external cost than does the rich country. In other words, $\phi < 1$ corresponds to the plausible case where the income elasticity of environmental quality is positive.

		Mexico			
		Standards		No Standards	
United States	Trade	c'	$f' + \phi k'$	$c - e$	$f + \phi k$
	Restricted Trade	c''	$f'' - h - j'' + \phi k''$	0	0

Figure 18.5 Choices of Free Trade and Environmental Production Standards

3.2 Translating Welfare Quantities into Payoffs Guiding Negotiation

The welfare effects illustrated in Figure 18.4 can represent the payoffs to various outcomes of negotiation. For example, in Figure 18.5, the choice of free trade with standards (upper left cell) corresponds to the trade case of Figure 18.4(b) where standards are imposed on products brought into the United States. This outcome corresponds to the trade envisioned under NAFTA. By comparison, trade with no standards (upper right cell) corresponds to the trade case of Figure 18.4(a) where contaminated imports are brought into the United States and sold without discriminatory labeling. In other words, it corresponds to the case where the United States chooses free trade unilaterally, and Mexico chooses to export to the United States without safety standards. Restricted trade with standards (lower left cell) corresponds to the case where the United States does not allow imports that do not meet its safety standards and Mexico chooses to impose standards that qualify its production for the U.S. market. The base case (lower right cell) corresponds to a breakdown of cooperation where the United States chooses not to allow trade unless Mexico imposes standards and Mexico chooses not to impose any production standards.

Again, several generalities can be added to this analysis. For example, a weighting factor can be added to producer or consumer welfare in either country reflecting policymaker concerns for equity, or reflecting response to rent seeking. Additionally, the externality associated with trade may represent a case where both Mexican-bound and American-bound production in Mexico damages the U.S. environment, for example, because of a global externality as in the case of methyl-bromide use. For this case, one may add a factor multiplied by all Mexican production to U.S. welfare representing the degree of U.S. environmental concern with Mexican production standards. In addition, due to asymmetric information, the United States may need to undertake inspection of imports to verify Mexican safety enforcement. The implications of such inspection costs are ignored here, but obviously would entail some discounting of the benefits of trade. Even without these complications, however, the generality of this problem is sufficient to obtain all the game structures in Table 18.1.

3.3 Potential Games Structures

For purposes of further specific analysis, consider the variety of outcomes that can occur in Figure 18.5 under equal welfare weighting and where externalities occur only through trade. To exemplify the possibilities, suppose supplies and demands are linear. Let the price and Mexican production under free trade without standards be both normalized to 1, and let the share of Mexican production consumed domestically under these conditions be represented by λ. At this equilibrium, let the elasticity of Mexican supply be ε, the absolute elasticity of Mexican demand be η, and the absolute elasticity of U.S. excess demand be ν. Finally, let μ represent the marginal external cost (assumed constant) as a share of the free-trade equilibrium product price under no standards. In this model, the implications of a variety of elasticity and share assumptions are depicted in Figure 18.6.

Consider the rather plausible case where Mexico evaluates the marginal external cost at 50 percent of the U.S. evaluation ($\phi = .5$), the elasticity of Mexican supply is 1 ($\varepsilon = 1$), and the Mexican consumption share of its winter fruit and vegetable production is 30 percent ($\lambda = .3$). Then a Mexican demand elasticity of 5 ($\eta = 5$), a U.S. excess demand elasticity of 1 ($\nu = 1$), and a marginal external cost equal to 60 percent of the free-trade no-standards price ($\mu = .6$) yields the standard prisoner's dilemma in Figure 18.6(a) where cooperation is superior to mixed partial cooperation. Alternatively, a Mexican demand elasticity of 10, a U.S. excess demand elasticity of 1, and a marginal external cost equal to 90 percent of the free-trade, no-standards price yields the prisoner's dilemma in Figure 18.6(b) where cooperation is inferior to mixed partial cooperation. A Mexican demand elasticity of 1, a U.S. excess demand elasticity of .3, and a marginal external cost equal to 90 percent of the free-trade, no-standards price yields the chicken game in Figure 18.6(c). Finally, assuming a Mexican demand elasticity of 2, a U.S. excess demand elasticity of 1, and a marginal external cost equal to 60 percent of the free-trade, no-standards price yields the ID game of Figure 18.6(d). Similarly, an assurance game is obtained for the case with $\phi = .1$, $\varepsilon = .25$, $\eta = .3$, $\nu = .6$, $\lambda = .4$, $\mu = .9$.

Clearly, a wide variety of game structures are possible depending on elasticities and shares. The implication is that the determination of game structure is an empirical issue. Little empirical work has been done in the context of game theory. The analysis in this chapter suggests that empirical evaluation can narrow the set of possibilities that need to be considered under certain explanations. As a result, many potential explanations attributed to game theory underpinnings can be eliminated econometrically, yielding greater clarity of insight.

3.4 Welfare Analysis of Other Issues

Against the backdrop of bargaining possibilities associated with trade and environmental standards, consider next the possibility of linking to other issues to enhance the potential of negotiation. For example, suppose the issue of illegal immigration is considered as a possibility for enhancing joint

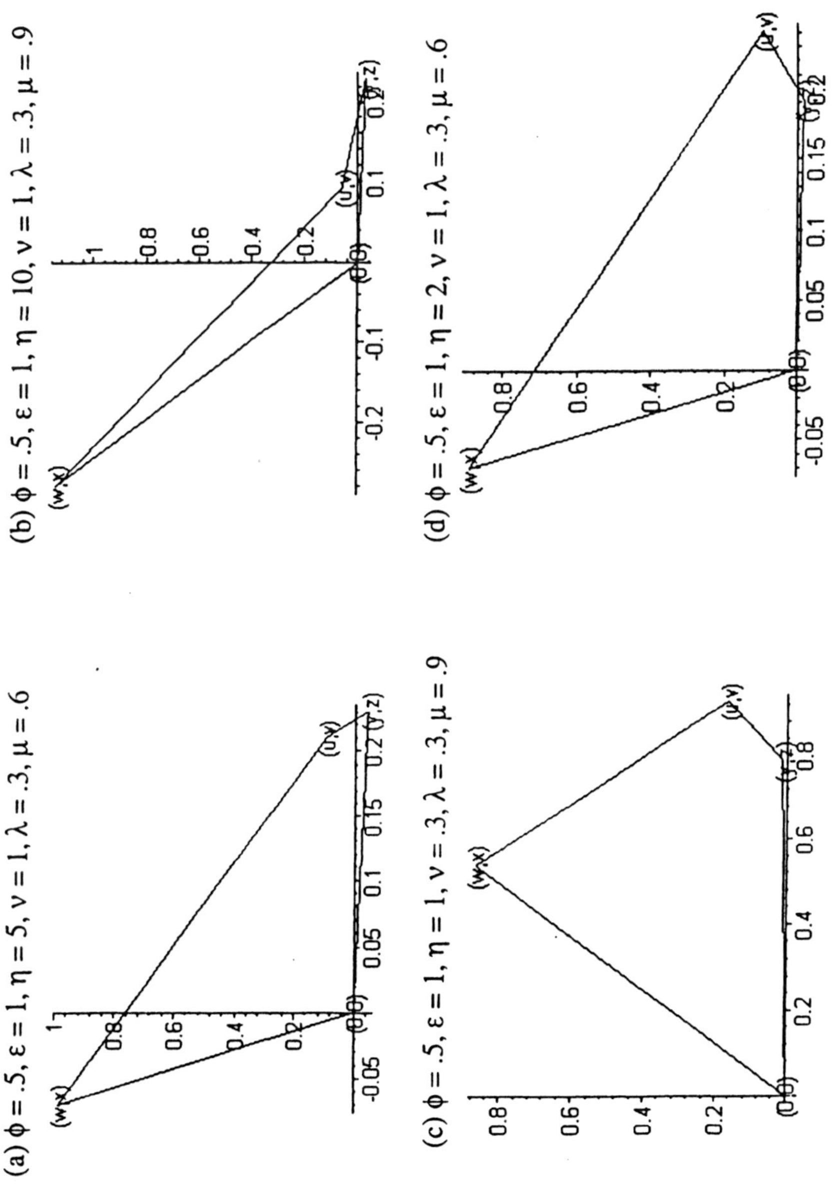

Figure 18.6 Practical Alternative Game Structure

cooperation. For the sake of brevity, these possibilities can be illustrated by analogy with the cases above. For example, the analysis of Figures 4 and 5 can be re-interpreted where Mexican supply and demand represent the domestic labor market, and U.S. supply and demand represent a market for menial farm labor in the United States conditional to equilibrium adjustments in other U.S. labor markets. The choice of restricted trade can correspond to stringent U.S. border enforcement and the choice of standards can correspond to Mexican cooperation in border enforcement. The cost of complying with *standards* can represent the risk premium of getting caught plus other costs of trans-boundary, labor-market participation. The externality from the U.S. standpoint can represent income-tax evasion associated with illegal labor or other social costs associated with illegal immigration. Admittedly, this analogy does not represent a complete analysis, but the parallel concepts suggest that a wide variety of game structures can apply to negotiation potential on labor issues. Additionally, further consideration of possibly asymmetric enforcement costs may be needed on both sides of the border. Note, however, that replacement of the lower left-hand cell of Figure 18.5 with the no-trade, uniform-standards case of Figure 18.4(c) (as might be more appropriate in this case) does not limit the range of game structures that are possible.

Similar analysis applies to the case of illegal drug traffic. In this case, the social cost of drug use may be quite high and borne mainly in the United States. As underscored by the February, 1999 Clinton talks in Merida, Mexico, levels of enforcement on both sides of the border are major issues of negotiation. Because of limited space, this application of the analysis will be left to implicit analogy. But again, the potential for a variety of game structures is apparent, making empirical reduction of the parameter space a useful approach. In the absence of such empirical work, one could claim to explain almost any outcome in these problems by game theory constructs.

4. LINKING GAMES (OR COMBINING ISSUES FOR NEGOTIATION)

We now turn to examining the potential benefits of linking games under a variety of game structures to illustrate the importance of narrowing possibilities by econometric means. Folmer, Mouche, and Ragland (1993), Ragland (1995), and Bennett, Ragland, and Yolles (1998) consider linking games of different structures. Folmer, Mouche, and Ragland (1993) link a pair of asymmetric PD games. Ragland (1995) links two isolated PD games, a PD and a chicken game, two chicken games, and a PD and an ID game. Bennett, Ragland, and Yolles link a PD game and an ID game. Each of these suggests that linking leads to improved cooperation. However, Ragland gives examples where strict dominance is not obtained. More recently, Netanyahu (1998) presents a general framework for linking games. She determines conditions that result in strict dominance as well as conditions that result in no dominance, and finds a wide variety of cases where linking offers no advantages.

To examine the potential benefits of linking, the procedure is to compare the convex hull of the feasible sets of the aggregated isolated games to the feasible set of the linked game. First, consider constructing the feasible set of the aggregated isolated games. Table 18.2 gives the points defining the feasible set, which is obtained by adding the outcomes of all strategies that satisfy individual rationality (non-negative payoffs). Figure 18.7 illustrates the exercise graphically where both games are PD games having cooperation superior to convex combinations of partial cooperation. The convex hull of the aggregate payoffs of all such combinations is illustrated by dotted lines in Figure 18.7. Points on the convex hull, other than vertices, correspond to mixed strategies, that is, playing each of two strategies with fixed probabilities in one or both of the games.

Next, consider the linked game obtained by considering all combinations of mixed and pure strategies in the two games. The aggregate outcomes from all pure strategy combinations are depicted in Figure 18.8 and define the feasible set of the linked game apart from non-negativity. The difference in the linked case from the aggregated case is that the strategies in the linked game need not be individually rational in the separate games, as long as the aggregate outcome is individually rational in the linked game. This difference is illustrated by comparing the representation of linked strategies in Figure 18.9(a) with the aggregated strategies in Figure 18.7. By considering strategies with negative payoffs for one of the players in an individual game, strategies with higher positive payoffs for the other player may be available

Table 18.2 Aggregation of Two Isolated Games

Point	*Payoffs in Game 1 Figure 18.2(a)*	*Payoffs in Game 2 Figure 18.2(b)*	*Aggregate payoffs*
1	$(0, Y_1)$	$(0, Y_2)$	$(0, Y_1 + Y_2)$
2	$(0, Y_1)$	(u, v)	$(u, Y_1 + v)$
3	$(0, Y_1)$	$(X_2, 0)$	(X_2, Y_1)
4	$(1, 1)$	$(0, Y_2)$	$(1, 1 + Y_2)$
5	$(1, 1)$	(u, v)	$(1 + u, 1 + v)$
6	$(1, 1)$	$(X_2, 0)$	$(1 + X_2, 1)$
7	$(X_1, 0)$	$(0, Y_2)$	(X_1, Y_2)
8	$(X_1, 0)$	(u, v)	$(X_1 + u, v)$
9	$(X_1, 0)$	$(X_2, 0)$	$(X_1 + X_2, 0)$

Source: Authors.

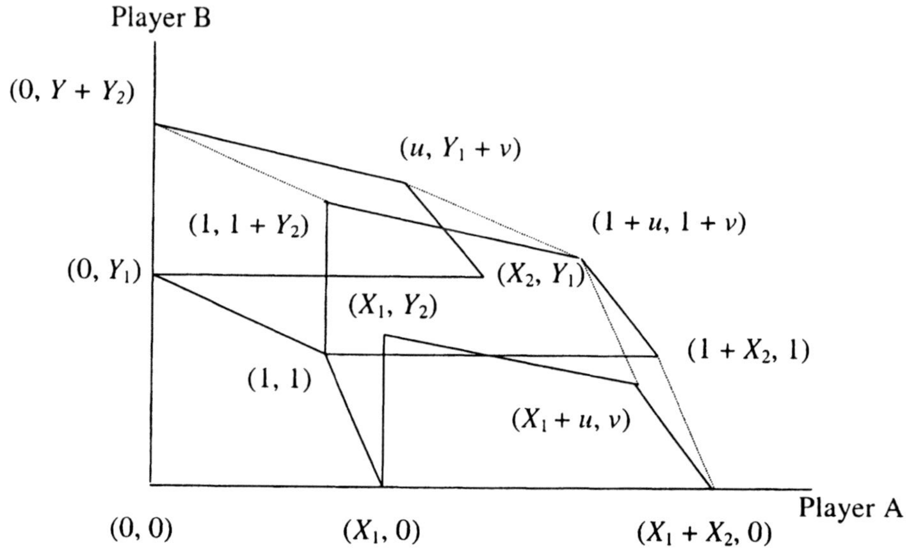

Figure 18.7 The Feasible Payoff Set of Aggregated Prisoner's Dilemmas

in the linked game. If this occurs with reciprocity for both players, then improved cooperation can be achieved.

Regardless of the game structure in Table 18.1, many of the strategies in Figure 18.8 are dominated by others (that is, they lie on the interior of the convex hull) and can be eliminated from consideration. For example, under the maintained assumption that $u > 0$ and $v > 0$, $(1 + u, 1 + v)$ dominates (u, v), $(1, 1)$, and $(0, 0)$. Also, $(1 + w, 1 + x)$ dominates (w, x), and $(1 + y, 1 + z)$ dominates (y, z). Strategies $(s + y, t + z)$ and $(q + w, r + x)$ can be eliminated from consideration in each case considered here (except where one of the games is a chicken game) because they do not satisfy individual rationality, that is, $t + z < 0$ and $q + w < 0$. Thus, only 8 of the 16 alternatives in Figure 18.8 are needed to define the linked game convex hull as depicted in Figure 18.9(a).

The comparison of feasible sets in the aggregated and linked games is illustrated in Figure 18.10. The solid frontier depicts the feasible set of the linked game, whereas the dotted line illustrates the somewhat smaller feasible set of the aggregated isolated games. Figure 18.10 depicts a case where linking weakly dominates the aggregated isolated games because the frontiers of the feasible sets overlap partially.

		Player B			
		c c	d c	c d	d d
Player A	c c	$1+u$ $1+v$	$q+u$ $r+v$	$1+w$ $1+x$	$q+w$ $r+x$
	d c	$s+u$ $t+v$	u v	$s+w$ $t+x$	w x
	c d	$1+y$ $1+z$	$q+y$ $r+z$	1 1	q r
	d d	$s+y$ $t+z$	y z	s t	0 0

Figure 18.8 The General Linked Game

Depending on conditions, some combinations of game types can yield strict dominance of the linked game over the aggregated isolated games while others yield no dominance. In general, linking issues is potentially advantageous (1) when it expands the feasible set of choices for the parties and (2) when it makes new strategies possible that are not permitted in the isolated games.

4.1 Linking Games: The Case with Prisoner's Dilemmas

When both games are PD games, the comparison of aggregated and linked games must consider a number of alternative possibilities. A detailed examination of these cases facilitates rather simple analysis of other alternative game structures.

4.1.1 Superiority of Cooperation to Partial Cooperation in Both Games

One issue is whether or not cooperation is preferred to combinations of partial cooperation in the separate games as in Figure 18.2(a) rather than Figure 18.3(a). The case with superiority of cooperation to partial cooperation is depicted in Figures 7, 9(a), and 10.

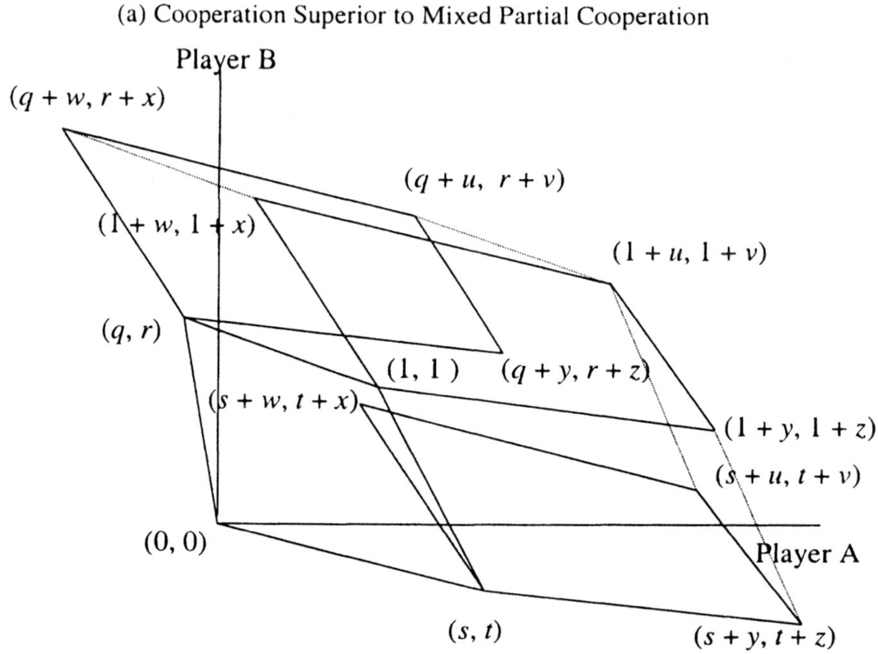

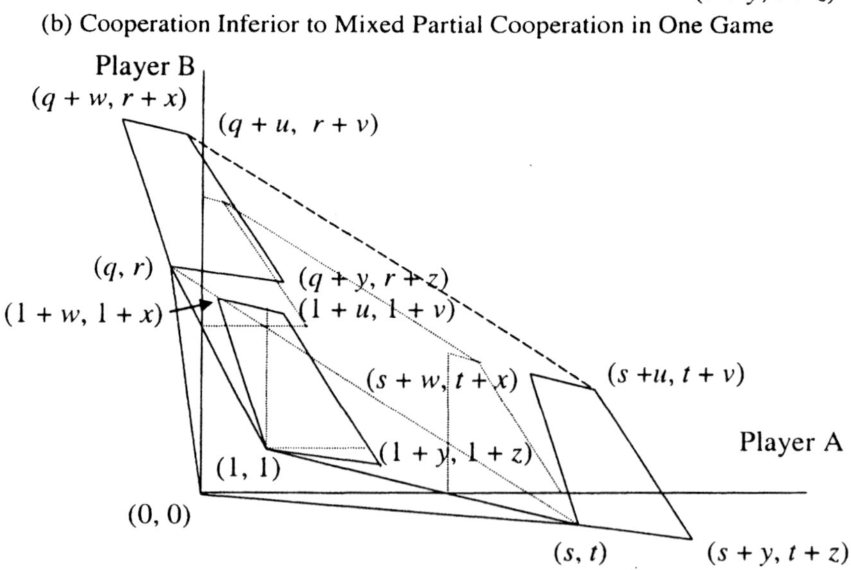

Figure 18.9 The Feasible Set of Linked Prisoner's Dilemmas

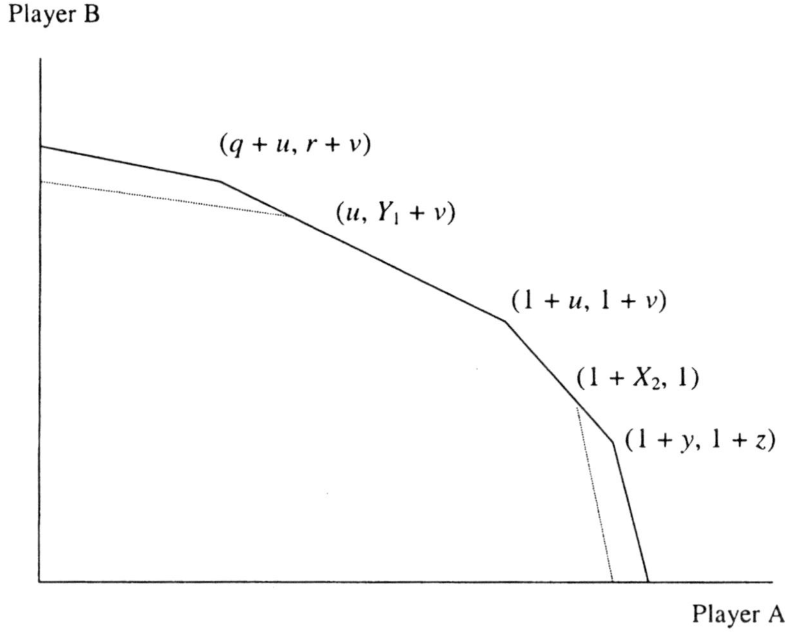

Figure 18.10 The Feasible Sets of the Aggregated and Linked Games

4.1.1.1 Comparison when Games are Symmetric

The results of aggregation depend on the relationship of the separate games. With PD structure, $s_1 > s_2$ and $s_3 > s_4$ leaving six possible orderings of the slopes in Figure 18.2: $s_1 > s_2 > s_3 > s_4$, $s_3 > s_4 > s_1 > s_2$, $s_1 > s_3 > s_4 > s_2$, $s_1 > s_3 > s_2 > s_4$, $s_3 > s_1 > s_2 > s_4$, and $s_3 > s_1 > s_4 > s_2$. We characterize the last four of these as *sufficiently symmetric* because the payoff sets of the isolated games have similar shapes in the positive quadrant of rationality. The convex hull of the feasible set includes points (0, 0), $(X_1 + X_2, 0)$, $(1 + u, 1 + v)$, and $(0, Y_1 + Y_2)$, and excludes points (X_1, Y_2) and (X_2, Y_1). Which of the other points will be on the frontier is determined as follows.[iv] If $s_1 < (>)\ s_3$, then $(u, Y_1 + v)$ is (is not) on the frontier and $(1, 1 + Y_2)$ is not (is). Similarly, $(1 + X_2, 1)$ is on the frontier when $s_2 \leq s_4$, and $(X_1 + u, v)$ is on the frontier when $s_2 \geq s_4$. Thus, the convex hull is completely characterized by conditions on the slopes of the frontiers in Figure 18.2.

To consider linking in this case of sufficient symmetry, manipulation of the points in Figure 18.8 reveals that the positive frontier of the convex hull includes $(1 + u, 1 + v)$. Which of the other points will be on the frontier can be determined as follows. The points $(q + w, r + x)$, $(1 + w, 1 + x)$, $(1 + u, 1$

+ v), and ($q + u$, $r + v$) form a parallelogram in which all the segments have slopes s_1 or s_3. If $s_1 < (>) s_3$, then ($q + u$, $r + v$) is (is not) on the frontier and ($1 + w$, $1 + x$) is not (is). Similarly, ($1 + y$, $1 + z$) is (is not) on the frontier when $s_2 \leq (\geq) s_4$, and ($s + u$, $t + v$) is (is not) on the frontier when $s_2 \geq (\leq) s_4$.

Qualitative comparisons reveal considerable overlap of the frontiers of the aggregated and linked cases as depicted in Figure 18.10. The point ($1 + u$, $1 + v$) lies on both frontiers, and the adjacent frontier segments have the same slope in both cases. The only difference is that the adjacent segments extend somewhat farther in the linked case producing weak dominance of the linked frontier. The most striking result is that the linked outcomes dominate the aggregated outcomes only for cases where payoff combinations substantially different than the full cooperation case are preferred. As demonstrated below, outcomes other than full cooperation are chosen in these cases when equity is a concern.[v]

4.1.1.2 Comparison when Games are Asymmetric

Consider next the case when slopes are ordered as $s_1 > s_2 > s_3 > s_4$ or $s_3 > s_4 > s_1 > s_2$. We characterize these cases as *sufficiently asymmetric* because the payoff sets of the isolated games have different shapes in the positive quadrant of rationality. That is, one mixed strategy highly favors one player in one game and another mixed strategy highly favors the other in the other game. Graphically, the entire frontier of one game is steeper than the entire frontier of the other game. As a result, the full cooperation point is in the interior of the feasible set of the aggregated games (not shown for brevity). For example, if $s_3 > s_4 > s_1 > s_2$, then the convex hull is defined by (0, 0), ($X_1 + X_2$, 0), ($1 + X_2$, 1), (X_2, Y_1), (u, $Y_1 + v$) and (0, $Y_1 + Y_2$). If $s_1 > s_2 > s_3 > s_4$, then the convex hull is defined by (0, 0), ($X_1 + X_2$, 0), ($X_1 + u$, v), (X_1, Y_2), (1, $1 + Y_2$) and (0, $Y_1 + Y_2$). These are interesting cases because full cooperation cannot be an equilibrium. This situation could not apply when full cooperation is an economically desirable outcome.

In this case of sufficient asymmetry, full cooperation is in the interior of the feasible payoffs of the linked game as well. For example, if $s_3 > s_4 > s_1 > s_2$, then the upper right frontier of the convex hull is defined by ($q + w$, $r + x$), ($q + u$, $r + v$), ($q + y$, $r + z$), ($1 + y$, $1 + z$), ($s + y$, $t + z$). If $s_1 > s_2 > s_3 > s_4$, then the upper right frontier of the convex hull is defined by ($s + y$, $t + z$), ($s + u$, $t + v$), ($s + w$, $t + x$), ($1 + w$, $1 + x$), and ($q + w$, $r + x$). Interestingly, both of these cases generate a convex hull for the linked game that strictly dominates the aggregated games (not shown). However, as indicated above, these are cases in which full cooperation is not economically desirable. For example, full cooperation cannot achieve economic efficiency in the traditional sense. Thus, the definition of cooperation seems strained or inappropriate for this case.

4.1.2 Inferiority of Cooperation to Partial Cooperation in One Game

Suppose cooperation is inferior to combinations of partial cooperation in Game 1 as in Figure 18.3(a). The dotted lines in Figure 18.9(b) depict the case of aggregating isolated games where $s_4 < -Y_1/X_1 < s_3$. The convex hull is defined by (0, 0), $(0, Y_1 + Y_2)$, $(u, Y_1 + v)$, $(X_1 + u, v)$ and $(X_1 + X_2, 0)$ with all other points from Table 18.2 in the interior. If $s_4 < s_3 < -Y_1/X_1$, then the convex hull is defined by (0, 0), $(0, Y_1 + Y_2)$, $(u, Y_1 + v)$, $(X_1 + u, v)$ and $(X_1 + X_2, 0)$ with all other points from Table 18.2 in the interior. If $-Y_1/X_1 < s_4 < s_3$, then the convex hull is defined by (0, 0), $(0, Y_1 + Y_2)$, $(u, Y_1 + v)$, (X_2, Y_1), $(X_1 + u, v)$ and $(X_1 + X_2, 0)$ with all other points from Table 18.2 in the interior. Intuitively, these outcomes correspond to adding the possibilities of Game 2 onto the combinations of partial cooperation in Game 1. The feasible set is depicted by dotted lines that correspond to adding strategies in the positive quadrant of Game 2 onto convex combinations of partial cooperation of Game 1.

The expanded feasibility set obtained with linking is depicted by solid and broken line segments in Figure 18.9(b). Figure 18.9(b) is drawn for the case with $s_4 < (t - r)/(s - q) < s_3$ where $(t - r)/(s - q)$ is the slope of the broken line. In this case, the upper right frontier of the convex hull is defined by $(q + w, r + x)$, $(q + u, r + v)$, $(s + u, t + v)$, and $(s + y, t + z)$. If $s_4 < s_3 < (t - r)/(s - q)$, then the upper right frontier of the convex hull is defined by $(q + w, r + x)$, $(s + w, t + x)$, $(s + u, t + v)$, and $(s + y, t + z)$. If $(t - r)/(s - q) < s_4 < s_3$, then the convex hull is defined by $(q + w, r + x)$, $(q + u, r + v)$, $(q + y, r + z)$, and $(s + y, t + z)$. Clearly, the convex hull of the linked game strictly dominates the convex hull of the aggregated games. Note that dominance is obtained by adding payoffs from one game onto the outcomes of the other game that have negative payoffs for one of the players but overcompensating benefits for the other player. As in other such cases, this is a case where the full cooperation outcome, $(1 + u, 1 + v)$, is in the interior of the feasible payoff set. As in other cases, this is a case in which cooperation is not attractive on the basis of economic efficiency alone, that is, according to a preference map with linear contours of slope -1.

$(X_1 + u, v)$ and $(1, 1 + Y_2)$. To determine which of the other two points is on the frontier, if $Y_1/X_1 > (<)\ Y_2/X_2$ then (X_2, Y_1) is (is not) on the frontier and (X_1, Y_2) is not (is). Intuitively, these conditions depend on how much one player benefits relative to the other player's loss in cases of partial cooperation compared to no cooperation.

4.1.3 Inferiority of Cooperation to Partial Cooperation in Both Games

Similar manipulation in the case where both games have cooperation inferior to mixed partial cooperation reveals that the convex hull includes (0, 0), $(X_1 + X_2, 0)$, and $(0, Y_1 + Y_2)$, and excludes $(1 + u, 1 + v)$, $(u, Y_1 + v)$, $(1 + X_2, 1)$, $(X_1 + u, v)$ and $(1, 1 + Y_2)$. To determine which of the other two points is on the frontier, if $Y_1/X_1 > (<) Y_2/X_2$, then (X_2, Y_1) is (is not) on the frontier and (X_1, Y_2) is not (is). Intuitively, these conditions depend on how much one-player benefits relative to the other player's loss in cases of partial cooperation compared to no cooperation.

By analogy with Figure 18.9(b), one can see that the feasible payoff set under linking for this case strictly dominates that for the aggregated isolated games. The upper right frontier of the convex hull of payoffs will be defined by $(q + w, r + x)$ and $(s + y, t + z)$ and either $(q + y, r + z)$ or $(s + w, t + x)$. The point $(q + y, r + z)$ will be on the frontier if $(z - x)/(y - w) \geq (t - r)/(s - q)$ and $(s + w, t + x)$ will be on the frontier if $(z - x)/(y - w) \leq (t - r)/(s - q)$. Again, full cooperation cannot represent an economically efficient outcome (even in the individual games).

4.1.4 Conclusions for Linking Two Prisoner's Dilemma Games

The linked game associated with two isolated PD games does not necessarily strictly dominate the aggregated isolated games. In many cases, at least one point is common to both frontiers. When the two frontiers have a common point, the full cooperation point $(1 + u, 1 + v)$ will lie on both frontiers, and the two frontiers will have common segments on each side of full cooperation. These common points other than full cooperation result from mixed strategies.

Interestingly, the entire frontier of the linked game can possibly be identical to the frontier of the aggregated games. These results run counter to the general assertions of Folmer, Mouche, and Ragland (1993) and Cesar and de Zeeuw (1995) who suggest that linking PD games enriches the set of subgame perfect Nash equilibria, or that linked games strictly dominate aggregated isolated games. Depending on circumstances, linking may or may not expand the feasible set, and may expand it only in ways that do not affect the equilibrium outcome.

4.2 Linking a Prisoner's Dilemma with an Assurance Game

A similar approach can be applied to each pair of game structures, which are suggested by Table 18.1. Suppose an assurance game is linked to a PD game. An assurance game follows the second case of Table 18.1 and is represented by Game 1, which starts from the origin of Figure 18.11(a). Both segments of the convex hull in the positive quadrant are positively sloped and terminate at (1,1). No payoff makes any player better off than at the equilibrium. If Game 2 is a PD game, the full cooperation point $(1 + u, 1 + v)$ will be on the convex hull if cooperation is preferred to combinations of partial cooperation in the PD game. The feasible set, thus, consists of the points circumscribed by dotted lines and solid lines in Figure 18.11(a).

The feasible set of the linked game is the area circumscribed by solid and broken lines in Figure 18.11(a). The sets outlined by solid lines are obtained by adding all possible payoffs in Game 2 to the points defining the convex hull of payoffs from Game 1. The feasible set has an upper right-hand frontier defined by $(q + w, r + x)$, $(1 + w, 1 + x)$, $(1 + u, 1 + v)$, $(1 + y, 1 + z)$, and $(s + y, t + z)$. The points $(1 + w. 1 + x)$ and $(1 + y, 1 + z)$ possibly lie in negative quadrants, in which case the entire positive frontier is defined by the two segments from $(1 + w, 1 + x)$ to $(1 + u, 1 + v)$ and from $(1 + u, 1 + v)$ to $(1 + y, 1 + z)$. If the PD game has cooperation inferior to combinations of partial cooperation, then the full cooperation point $(1 + u, 1 + v)$ will not be on the frontier and the frontier follows a straight line from $(1 + w, 1 + x)$ to $(1 + y, 1 + z)$.

4.2.1 Superiority of Cooperation to Partial Cooperation in the Prisoner's Dilemma Game

Comparing feasible sets of the linked and aggregated games, two cases are possible. If cooperation is preferred to partial cooperation in the PD game as in Figure 18.11(a), the linked game (represented by broken lines) weakly dominates the aggregated games (represented by dotted lines). However, the full cooperation point is on the frontier of both sets and adjacent segments are common to both frontiers. Thus, the linked game offers an advantage only if full cooperation is not selected.

4.2.2 Inferiority of Cooperation to Partial Cooperation in the Prisoner's Dilemma Game

If cooperation is inferior to combinations of partial cooperation in the PD game, then the linked game strictly dominates the aggregated isolated games (not shown). In this case, full cooperation is also not selected. The relevant

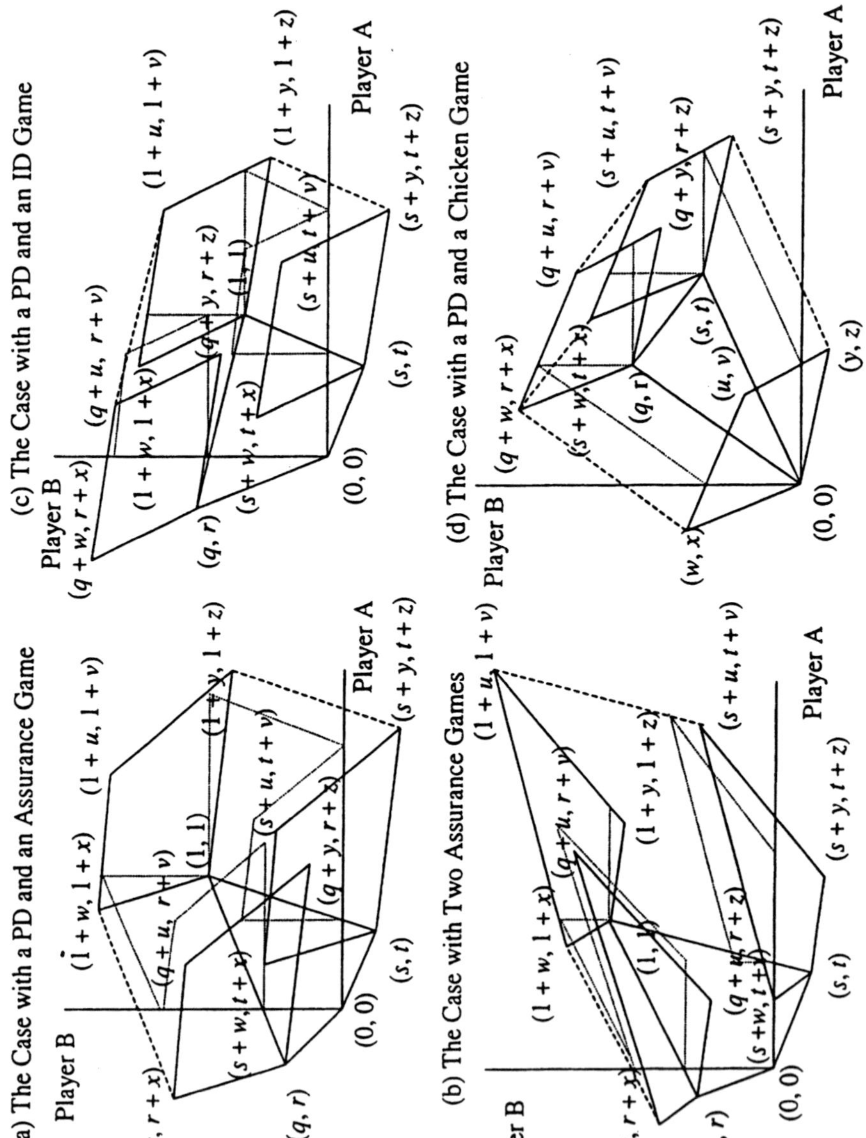

Figure 18.11 The Feasibility Set of the Linked Game

frontier of the linked payoff set will be a mix of the strategies with payoffs $(1 + w, 1 + x)$ and $(1 + y, 1 + z)$. The relevant frontier of the aggregated games will be a mix of strategies with payoffs $(1, 1 + Y_2)$ and $(1 + X_2, 1)$. In both cases, the result corresponds to cooperation in the assurance game and combinations of partial cooperation in the PD game. The difference is that partial cooperation strategies are constrained to the individually rational set in the individual games, but only to jointly rational strategies in the linked game.

4.2.3 Conclusions for Linking Assurance and PD Games

The results of this section sharply contrast with the assertions of Hauer and Runge (1997). When a PD game and assurance game are linked, Hauer and Runge (1997) claim that negotiations become more complex, decreasing the chances of an agreement relative to the single issue case. Figure 18.11(a) shows that a significant portion of the frontiers of the aggregated isolated games and the linked game overlap, suggesting that chances of agreeing to full cooperation are similar in both aggregated and linked games. Furthermore, when full cooperation is reached by isolated games, there is no advantage in linking. Finally, while Hauer and Runge (1997) focus on either full cooperation or full defection, the results here show that a mix of strategies is possible in the linked game, and that the linked game weakly dominates the aggregated isolated games only because of these mixed strategies.

4.3 Linking Two Assurance Games

Linking two assurance games leads to a trivial outcome of full cooperation. Suppose Game 1 is the same as in Figure 18.11(a) and that Game 2 follows a similar structure. The feasible set of the aggregated assurance games is represented by dotted lines in Figure 18.11(b).[vi] The feasible set of the linked game is the area circumscribed by solid and broken lines in Figure 18.11(b), and has a frontier defined by points including $(q + w, r + x)$, $(1 + u, 1 + v)$, and $(s + y, t + z)$. If $s_1 > s_3$, then $(q + u, r + v)$ is also on the frontier and otherwise $(1 + w, 1 + x)$ is on the frontier (because of the associated parallelogram with slopes s_1 and s_3). Similarly, if $s_2 \geq s_4$, then $(1 + y, 1 + z)$ is on the frontier and, otherwise, $(s + u, t + v)$ is on the frontier.

Comparing the linked feasible set with the aggregated feasible set, and noting that the dotted line coincides with the solid line between $(1 + w, 1 + x)$ and $(1 + u, 1 + v)$, the linked game weakly dominates the aggregated isolated games but the linked game never strictly dominates. While the feasible sets are not identical, they share the same full cooperation case as the upper right-hand extreme. Thus, linking provides no additional benefits. Both players choose the most preferred outcome of cooperation without negotiation.

4.4 Linking an Iterated Dominance Game with a Prisoner's Dilemma

Consider next linking a PD game with an ID game.[vii] Figure 18.11(c) illustrates the case in which Game 1 is an ID game following the third case of Table 18.1. The convex hull in the positive quadrant has one downward sloping segment, as in the PD game, and one upward sloping segment, as in the assurance game.

4.4.1.1 Comparison when Games are Symmetric

Where the PD game follows Figure 18.2(b) with cooperation preferred to mixed partial cooperation, the feasible set of the aggregated isolated games follows the dotted lines of Figure 18.11(c) when the two games are sufficiently symmetric, that is to say, when $s_1 > s_3 > s_2$ or $s_3 > s_1 > s_2$.[viii] The structure of the feasible set is similar to the case of two PD games on one side and to the case of one PD game and one assurance game on the other side.

The feasible set of the linked game is represented by the area circumscribed by solid and broken lines [Figure 18.11(c)] for the case where the two games are sufficiently symmetric and cooperation is preferred to mixed partial cooperation in the PD game. The feasible set is defined by points (0, 0), $(q + w, r + x)$, $(1 + u, 1 + v)$, $(1 + y, 1 + z)$, and $(q + u, r + v)$ if $s_3 \geq s_1$, or $(1 + w, 1 + x)$ if $s_1 \geq s_3$ (as implied by the parallelogram with slopes s_1 and s_3).

Comparing the feasible set of the aggregated games, the linked game does not strictly dominate.[ix] For example, the full cooperation point $(1 + u, 1 + v)$ is common to both frontiers. Furthermore, the slopes of the segments adjacent to $(1 + u, 1 + v)$ are the same in both cases. Thus, when $(1 + u, 1 + v)$ is on the frontier of the linked game, the frontiers have common segments. However, some strategies are clearly attainable in the linked game, that are not attainable in the aggregated isolated games, such as those along the segment from $(s + y, t + z)$ to $(1 + y, 1 + z)$ in the positive quadrant. Thus, the linked game weakly dominates.

4.4.1.2 Comparison when Games are Asymmetric

The case in which the two games are sufficiently asymmetric and cooperation is preferred to mixed partial cooperation in the isolated PD game is not shown for brevity. The condition of *sufficient asymmetry* is defined by $s_1 > s_2 > s_3$ (s_4 is irrelevant—it is positive). Other possibilities associated with sufficient asymmetry are excluded by the requirement that $s_1 > s_2$ when cooperation is preferred to mixed partial cooperation in the PD game. Notationally, the frontier of the aggregated isolated game is defined by the same set of points as with asymmetry in the case of two PD games. The only difference is that $X_1 < 1$ implies that $(X_1 + X_2, 0)$ will lie to the left of $(1 + X_2, 1)$. Thus, full cooperation will not lie on the frontier of the aggregated isolated games.[x] The linked game results also parallel the case of asymmetry with two PD games. Full cooperation will be in the interior of payoffs of the linked game.[xi] The linked game will strictly dominate the aggregated isolated games. For the case in which cooperation is inferior to mixed partial cooperation in the PD game, the linked game will strictly dominate the aggregated isolated games.[xii]

4.4.2 Linking a Chicken Game with a Prisoner's Dilemma

Finally, consider linking a PD game to a chicken game. Figure 18.11(d) shows the case where Game 1 is a chicken game corresponding to row 4 of Table 18.1. These conditions result in a convex hull in the positive quadrant with downward sloping segments as in the PD game and upward sloping segments as in the assurance game. However, the entire convex hull of payoffs is in the positive quadrant assuming positive payoffs for full cooperation so that cooperation is preferred to defection.[xiii] The full cooperation point (1, 1) can either be inside the frontier as in Figure 18.3(d) when cooperation is inferior to mixed partial cooperation, or it can be above and right of the (q, r)-(s, t) segment when cooperation is superior to mixed partial cooperation.

Graphically, possibilities generated by combining a PD game with a chicken game parallel those generated by combining a PD game with another PD game or a PD game with an ID game. The only difference is that the origin is shifted, in effect, to (q, t) because both players prefer moves from the origin to (s, t) and (q, r).[xiv] Figure 18.11(d) shows the feasible set of the aggregated PD and chicken games by dotted lines.[xv] The feasible set of the linked game is illustrated by broken lines.[xvi] The linked game does not strictly dominate the aggregated isolated games if cooperation is preferred to mixed partial cooperation in the PD game and if mixed partial cooperation is preferred to cooperation in the chicken game.

In the linked game, full cooperation is in the interior when it is inferior to mixed partial cooperation in either isolated game. The frontier of the linked game and the frontier of the aggregated isolated games have considerable overlap because mixed cooperation does not generate negative payoffs for either player in the chicken game. Thus, much of the opportunity for linking is lost, that is, opportunities to use strategies that are not feasible under rationality with isolated play. The only points added to the feasible set with linking are those associated with mixed cooperation strategies from the PD game.

4.4.3 Linking Two Chicken Games

The feasible sets of the isolated games of two chicken games are entirely contained in the positive quadrant. Thus, linking does not add strategies that are not feasible in the aggregated isolated games. The feasible set of the linked game is identical to aggregation, so linking offers no advantages. Visual examination also reveals that if either chicken game has cooperation dominated by mixed partial cooperation, as in Figure 18.11(d), then full cooperation cannot be on the frontier of either the linked or aggregated isolated games.

5. SUMMARY OF IMPLICATIONS OF GAME STRUCTURES

The various results of alternative game structures can be summarized in a few succinct conclusions. First, there are significant cases in which linking offers no improvements over aggregated isolated games where issues are negotiated independently. The conditions where this occurs are well defined and can be verified once the payoff structures of the games are estimated.

Second, strict domination of linking tends to occur only when the feasible set of one of the games is not convex or, specifically, when cooperation is inferior to mixed partial cooperation. Such cases of non-convexity may have a practical basis where economic externalities are present as illustrated in Figure 18.6(b). In these cases, strict preference for linking tends to be gained by alternating between mixed partial cooperation strategies that may appear to represent waffling or inconsistent behavior to casual observers. These are cases in which the pure strategy of full cooperation appears to be a mischaracterization because both parties are better off *cooperating* with mixes of non-cooperative strategies.

Finally, the cases contained in this chapter suggest that many reasonable cases in which linking is advantageous occur where linking dominates the aggregate games only weakly. In these cases, linking is preferred because of opportunities to expand the feasible set far from the point of full cooperation. These are likely the cases in which full cooperation does not serve equity concerns well. When these outcomes are preferred, behavior may be observed either in the form of pure strategies or mixed strategies but some strategies will result in negative payoffs for one of the parties with respect to one of the individual issues. That is, gains occur from linking because such possibilities become feasible when the resolution of other issues is used as compensation.

5.1 When Equity Considerations Are of Concern in Negotiation

Before concluding, a few additional points can help to clarify why optimal behavior may include less than full cooperation and even seemingly intermittent cooperation such as intermittent enforcement of immigration laws and drug trade restrictions, and waffling on originally envisioned operations of NAFTA. To consider equity considerations in bargaining, it is useful to consider the Nash bargaining solution (NBS) obtained by maximizing $\Pi = (u_A - \bar{u}_A)(u_B - \bar{u}_B)$ where u_A (u_B) corresponds to the utility derived by Player A (B) and $\bar{u}_A$ ($\bar{u}_B$) corresponds to the reservation utility of Player A (B). In the case of games, such as those represented in Figures 18.1 and 18.5, the reservation utilities are normalized to zero. Thus, contours that hold the product of utility, Π, constant are rectangular hyperbolas as represented by Π_L in Figure 18.12.

The illustration in Figure 18.12 demonstrates why outcomes substantially different than full cooperation are sometimes preferred. When the feasible set is highly asymmetric, the parties may be reluctant to pursue full cooperation because of inequity. But, yet, the parties may be able to identify partial cooperation strategies that obtain an outcome preferable to both noncooperation and full cooperation. That is, the tangency of the rectangular hyperbola contour of the NBS to the feasible set may represent a subgame perfect equilibrium.

Interestingly, these circumstances can arise when traditional economic efficiency analysis suggests full cooperation as the preferred outcome. Suppose, for example, that the payoffs in the games represent standard surplus measures as in Figure 18.5. Suppose the feasible payoffs of the aggregated and linked games were represented by dotted lines and solid lines, respectively, as in Figure 18.12. The standard criterion of economic efficiency corresponds to maximizing the sum of surpluses, that is, it is represented by linear welfare indifference contours with slope –1 as depicted by the broken line in Figure 18.12. The resulting *optimum* is highly inequitable. So also is the NBS of the aggregated isolated games. However, the NBS of the linked game is highly equitable, as represented by the tangency of the Π_L contour with the linked feasible set. This phenomenon may explain why international onlookers highly encourage full cooperation based on economic efficiency grounds, while the parties themselves are reluctant to accept full cooperation because of equity concerns. For such cases, linking may identify an equitable outcome that is nearly efficient but more acceptable to the parties.

Similar considerations can explain the choice of intermittent strategies that appear to represent waffling or inconsistency. To save space, we discuss this case by reference to the isolated games in Figure 18.9(b). Imposition of

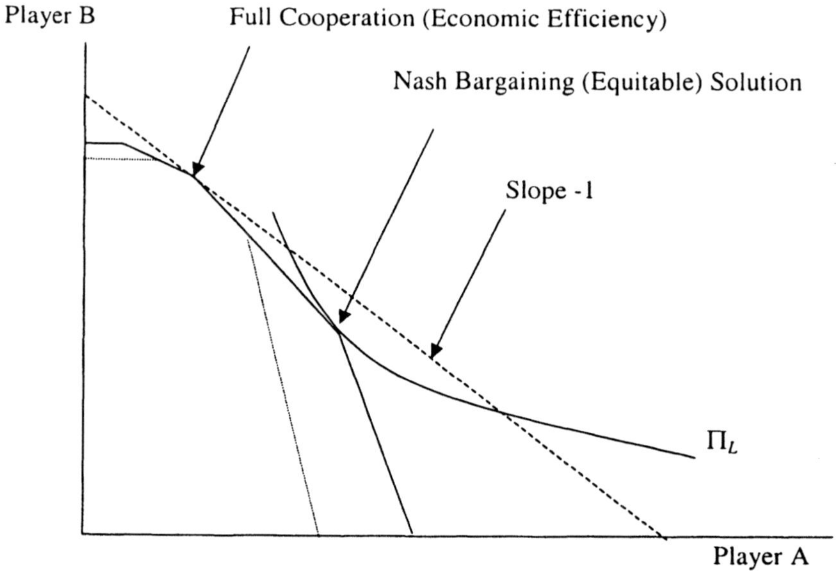

Figure 18.12 The Case In Which Equity is Preferred to Efficiency

a rectangular hyperbola tangent to the feasibility set (not shown) results in a tangency point other than full cooperation. If the tangency occurs along the broken line segment, then the preferred outcome corresponds to random variation (with a specified probability) between the strategies corresponding to the two endpoints of the segment. Thus, seemingly intermittent and inconsistent behavior would be observed simply as a result of a bargaining process that trades efficiency for equity.

While the framework illustrated in Figure 18.12 is discussed in the context of cooperative bargaining, similar constructs apply under reasonable conditions for noncooperative games as well. Such results have been demonstrated in the context of the axiomatic approach and the sequential approach to bargaining games due to Nash (1953) and Binmore, Rubinstein and Wolinsky (1986). For example, under risk of breakdown approaches suggested by Osborne and Rubinstein (1990), strategic bargaining with discounting in the alternating offer game proposed by Rubinstein (1982) induces players to reach an agreement by their impatience for outcomes. Osborne and Rubinstein show that time preferences and risk of breakdown yield solutions that can be represented by the NBS. For example, when players have symmetric preferences with respect to time and risk, a game of alternating offers has a unique subgame perfect equilibrium that converges to the NBS in which the reservation utilities are $[\lambda/(\lambda + r)](\bar{u}_A, \bar{u}_B)$ where λ represents the effect of time on the probability of breakdown and r is the logarithm of the discount factor. As r goes to zero (the discount factor goes to 1), the reservation utilities converge to $(\bar{u}_i, \bar{u}_p)$. If the probability of breakdown, λ, goes to zero, then the reservation utilities approach (0, 0). These various outcomes demonstrate the relevance of the NBS, and show that the principles of Figure 18.12 apply under reasonable non-cooperative bargaining cases as well although the reservation utilities may require some modification that causes some tilting or shifting of the contour illustrated in Figure 18.12.

6. POTENTIAL FOR LINKING TRADE AND ENVIRONMENTAL ISSUES

The above cases reveal a myriad of possibilities for linking games or, in other words, linking issues in trade negotiations. While conceptual analysis with game theoretic constructs admits many possibilities, the structure of individual games and the specific possibilities provided by linking is determined by the magnitudes of payoffs in Figure 18.1. Specifically, the cases illustrated here show that each type of game structure is potentially possible in international issues of agricultural trade and related safety standards. Based on the conceptual results in Figure 18.5, however, it appears that possibilities depend heavily on relatively few factors.

In general, unless one of the candidate games for linking has among the feasible set an outcome with a negative payoff for one of the parties coupled with a large payoff for the other party, linking generally does not achieve even weak dominance over aggregation of isolated games. As suggested by Figure 18.8, a strong negative possibility for w occurs when (w, x) is considerably left of the positive quadrant. One can show, under the market structure of Figure 18.4, that $w = (1 - \lambda)[(1/2\nu) - \mu]$. Thus, (w, x) is left of the positive quadrant if the marginal external cost is greater than half the price flexibility of U.S. excess demand. If excess U.S. demand is highly elastic, then this condition is likely. Estimation of this demand elasticity and marginal external effect are feasible empirically and render a statistical answer to the question.

The conditions under which z is negative, which causes (y, z) to lie below the positive quadrant, are less clear. However, at typical levels of other parameters as shown in Figure 18.8, z is more likely negative when Mexican consideration of the external cost (ϕ) is low, Mexican elasticity of supply (ε) is low, and the elasticity of excess U.S. demand (η) is high. Again, these are empirical questions that can be addressed by standard econometric analysis. A major point of this chapter is that such empirical work needs to be undertaken before offering serious game theoretic explanations of related outcomes, and before linking is offered as a serious suggestion for resolving trade issues.

7. CONCLUSION

Linking games has been proposed elsewhere as a way of improving international agreements on trade and environmental concerns. This chapter has investigated the gains from linking games of different and similar structures. Contrary to previous assertions, gains from linking may not be present in many cases. Specifically, linked games do not strictly dominate aggregated isolated games unless full cooperation is an inferior outcome. Also, linked outcomes do not improve on non-linked outcomes if full cooperation is the best outcome under linking. Thus, linking is not an approach that can be expected to bring about full cooperation.

Alternatively, a typical case is where linking expands the feasible set, which for lopsided issues may make more equitable opportunities available. In these cases, linking may facilitate partial cooperation and may result in intermittent and seeming inconsistent strategies. Furthermore, such outcomes may be inconsistent with standard criteria of economic efficiency even though it represents a trade-off of efficiency for equity among the trading countries.

We demonstrate that empirical work can help to determine where certain principles of game theory may be applicable. Specifically, we show how information from standard market econometric analyses can be useful when determining potential outcomes of linking, and that potential empirical outcomes span a wide list of possibilities.

The results of this chapter have interesting implications for U.S.-Mexican negotiations. Frustration has often been expressed about progress and cooperation on enforcement regarding illegal immigration and drug traffic. Proposals by the Clinton administration suggest the possibility that certain aspects of NAFTA will be undermined. For such problems, the results here raise a question about whether or not full cooperation is really a desired outcome. In general, both of the conditions that tend not to lead to full cooperation in the linked game may apply. First, payoffs may be highly asymmetric because of structural differences in the economies, and particularly because of the direction of flow of externalities. Second, there is likely a high concern for equity affecting the negotiations.

8. ENDNOTES

[1] Some of the material on game theory in this chapter, including some figures, overlaps that in Just and Netanyahu. 2000. "The Importance of Structure in Linking Games," *Agricultural Economics* 22: 87–100. (With permission from Elsevier Science).

[ii] For simplicity of presentation, C_M is depicted as a constant cost diagrammatically, but more general functions are allowed by the analysis of this chapter.

[iii] Methyl bromide is an important consideration in the U.S.-Mexican relationship. It was designated as a class I ozone depleter by the Montreal Protocol, which called for a complete phase out by 2010 (Fernandez-Cornejo, 1998). However, the U.S. Environmental Protection Agency, under the Clean Air Act, has initiated action to phase out methyl bromide in the United States by 2001 (Ferguson and Padula, 1994). Thus, environmental standards will apparently become incongruent between the two countries so that a consideration such as suggested by Figure 4(c) could become relevant.

[iv] In this case, as in many others throughout the paper, which points fall on the frontier can be determined by observing a parallelogram. For example, in this case the points $(0, Y_1 + Y_2)$, $(u, Y_1 + v)$, $(1 + u, 1 + v)$, and $(1, 1 + Y_2)$ form a parallelogram in which all the segments have slopes s_1 or s_3. Because two opposite points of the parallelogram are on the frontier, which of the other two points is on the frontier is determined by comparing slopes of the defining segments.

[v] Hereafter, situations where payoff combinations are preferred that are substantially different than the full cooperation case will be called cases where equity is a concern.

[vi] Using prior notation, $(1 + X_2, 1)$ is on the frontier if $s_2 \geq s_4$ and $(X_1 + u, v)$ is on the frontier if $s_2 \leq s_4$. Similarly, $(1, 1 + Y_2)$ is on the frontier if $s_1 \leq s_3$ and $(u, v + Y_1)$ is on the frontier if $s_1 \geq s_3$. These results follow from the associated parallelograms with slopes s_2 and s_4 (s_1 and s_3).

[vii] The payoff structure of the ID game is based on Bennett, Ragland, and Yolles (1998).

[viii] The convex hull is defined by $(0, 0)$, $(0, Y_1 + Y_2)$, $(1 + u, 1 + v)$, $(1 + X_2, 1)$, $(X_1 + X_2, 0)$, and $(u, v + Y_1)$ if $s_3 \geq s_1$ or $(1, 1 + Y_2)$ if $s_1 \geq s_3$ (note the associated parallelogram with slopes s_1 and s_3).

[ix] The feasible set of the aggregated games has $(u, v + Y_1)$ or $(1, 1 + Y_2)$ on the frontier as s_1 is less than or greater than s_3.

[x] The frontier is defined by $(0, Y_1 + Y_2)$, $(1 + X_2, 1)$, $(X_1 + X_2, 0)$, and (X_2, Y_1) if $-Y_2/X_2 \geq 1 - Y_1$ or $(1, 1 + Y_2)$ if $1 - Y_1 \geq -Y_2/X_2$ (as implied by the parallelogram with slopes $-Y_2/X_2$ and $1 - Y_1$).

[xi] The convex hull of the linked game is defined by points $(q + w, r + x)$, $(q + u, r + v)$, $(q + y, r + z)$, $(1 + y, 1 + z)$, and $(s + y, t + z)$.

[xii] The frontier of the linked game is defined by $(q + w, r + x)$, $(s + y, t + z)$, $(1 + y, 1 + z)$, and $(q + y, r + z)$ if $(z - x)/(y - w) \leq (r - 1)/(q - 1)$ or $(1 + w, 1 + x)$ if $(z - x)/(y - w) \geq (r - 1)/(q - 1)$.

[xiii] The convex hull is defined by points $(0, 0)$, (q, r), and (s, t).

[xiv] For more detail, see Netanyahu (1998).

[xv] The feasible convex hull of the aggregated game is defined by the points $(0, 0)$, $(0, Y_2)$, $(q, r + Y_2)$, $(s + X_2, t)$, $(X_2, 0)$ and by $(s, t + Y_2)$ if $(t - r)/(s - q) > s_3$ or $(q + u, r + v)$ if $s_3 > (t - r)/(s - q)$, and by $(q + X_2, r)$ if $s_4 > (t - r)/(s - q)$ or $(s + u, t + v)$ if $s_4 < (t - r)/(s - q)$.

[xvi] The convex hull of the linked game is defined by $(0, 0)$, (w, x), $(q + w, r + x)$, $(s + y, t + z)$, (y, z), and $(q + u, r + v)$ if $s_3 > (t - r)/(s - q)$ or $(s + w, t + x)$ if $s_3 > (t - r)/(s - q)$, and $(s + u, t + v)$ if $s_4 < (t - r)/(s - q)$ and $(q + y, r + z)$ if $s_4 > (t - r)/(s - q)$.

9. REFERENCES

Bennett, L., S. Ragland, and P. Yolles. 1998. "Facilitating International Agreements through an Interconnected Game Approach—The Case of River Basins." in *Conflict and Cooperation on Transboundary Water Resources*, R.E. Just and S. Netanyahu, eds., Boston, MA: Kluwer Academic Publishers.

Binmore, K., A. Rubinstein, A. Wolinsky. 1986. "The Nash Bargaining Solution in Economic Modeling." *Rand Journal of Economics* 17: 176–188.

Calvin, L. and V. Barrios. 1998. "Marketing Winter Vegetables from Mexico." *Vegetables and Specialties Situation and Outlook*, U.S. Department of Agriculture, Economic Research Service, Report No. VGS-274, April.

Cesar, H. and A. de Zeeuw. 1994. "Issue Linkage in Global Environmental Problems." Working Paper 56.94, Fondazione ENI Enrico Mattei.

Fernandez-Cornejo, J. 1998. "Short Run Input and Environmental Effects of Increased Agricultural Trade Between the U.S. and Mexico." Working Paper, Economic Research Service, U.S. Department of Agriculture.

Ferguson, W. and A. Padula. 1994. *Economic Effects of Banning Methyl Bromide for Soil Fumigation*. Economic Research Service, U.S. Department of Agriculture, Report No. AER-677 (March).

Folmer, H., P.V. Mouche, and S. Ragland. 1993. "Interconnected Games and International Environmental Problems. *Environmental and Resource Economics* 3: 313–335.

Hauer, G. and C.F. Runge. 1997. Trade-Environment Linkage in the Resolution of Transboundary Externalities. Center for International Food and Agricultural Policy, Working Paper WP97-1, University of Minnesota. St. Paul, MN (July).

Just, R. and S. Netanyahu. 2000. "The Importance of Structure in Linking Games." *Agricultural Economics* 22: 87–100.

Nash, J.F. 1953. Two-Person Cooperative Games. *Econometrica* 21: 128-140.

Netanyahu, S. 1998. "Bilateral Cooperation On Transboundary Water Resources: The Case Of The Israeli–Palestinian Mountain Aquifer," unpublished Ph.D. dissertation, Department of Agricultural and Resource Economics, University of Maryland, College Park, MD.

Osborne, M.J. and A. Rubinstein. 1990. *Bargaining and Markets*. San Diego, CA: Academic Press, Inc.

Ragland, S.E. 1995. "International Environmental Externalities and Interconnected Games." unpublished Ph.D. dissertation, Department of Economics, University of Colorado, Boulder.

Rubinstein, A. 1982. Perfect Equilibrium in a Bargaining Model. *Econometrica* 50: 97–110.

Silvey, H., and J. Brody. 1996. *Herculean Effort Needed Along U.S.-Mexican Border*. Council on Hemispheric Affairs, Washington, DC.

Torok, S.J., and W.E. Huffman. 1986. "U.S.-Mexican Trade in Winter Vegetables and Illegal Immigration." *American Journal of Agricultural Economics* 68: 246–260.

Zepp, G., F. Kuchler, and G. Lucier. 1998. "Food Safety and Fresh Fruits and Vegetables: Is There a Difference Between Imported and Domestically Produced Products?" *Vegetables and Specialties Situation and Outlook*, U.S. Department of Agriculture, Economic Research Service, Report No. VGS-274 (April).

Chapter 19

Welfare Gains under Tradable CO_2 Permits

Larry Karp and Xuemei Liu
University of California at Berkeley

1. INTRODUCTION

It is easy to understand the opposition to liberalized trade in established markets: Domestic producers lose from increased foreign competition. It is harder to understand the opposition to creating markets, including international markets, where they currently do not exist. Many economists and policymakers have proposed establishing tradable carbon permits to decrease the cost of reducing global carbon emissions. Since, as of 2001, there are no enforceable ceilings on emissions, the right to emit carbon has no market value. Emissions permits are not commodities. The usual forces that oppose market liberalization are obviously not present in this (proposed) market.

Environmentalists who favor reducing carbon emissions, frequently oppose international trade in emissions permits. It is puzzling that the group most in favor of a proposed change (reductions in emissions) is also the most opposed to a method of achieving that change cheaply (via trade). There may be a rational basis for this opposition. The theory of the second best alerts us to the possibility that, in a world with distortions, opening a new market may lower welfare. If there is a plausible second-best argument against trade in carbon permits, we have not found it. There is probably an emotional basis for environmentalists' opposition to tradable permits. The environmental problems we face are related to growth, which is related to the existence of liberal markets. A misunderstanding of the relation between markets and pollution may lead some environmentalists to incorrectly equate environmental deterioration with market liberalization of any kind. Certainly there is a deep skepticism among environmentalists regarding the merits of markets.

Economists advance the usual abstract arguments in favor of markets to explain why internationally tradable permits would be helpful in achieving reduced carbon emissions. These arguments are probably correct, but they are not convincing to people who are ill disposed toward markets in general. Economists' involvement in the debate over tradable permits differs from their involvement in previous debates over trade liberalization (for example, during the Uruguay Round of the General Agreement on Tariffs and Trade (GATT) negotiations). In both cases, the fundamental argument for market liberalization is theoretical. However, during previous trade negotiations, these theoretical arguments were backed by many empirical studies. Those studies attempted to measure, either econometrically or by means of simulation, the trade and welfare effects of various forms of liberalization. The validity of these empirical results is always debatable, but their concreteness sometimes makes them persuasive.

Although trade in carbon permits is potentially important, the possibility of this trade has led to little empirical work. The obvious explanation for the absence of empirical work is that the relevant market is missing. The existence of a world market for wheat makes it relatively straightforward to estimate supply and demand curves that can be used to study the effects of liberalized trade in wheat. We cannot use the same procedure to study the effects of creating a market where none exists.

Nevertheless, we do observe cross-country and inter-temporal variation in carbon emissions together with changes in inputs, such as capital and labor. This data can be used to estimate a relation between emissions and income to obtain an estimate of the marginal value of carbon emissions in countries.

These estimates enable us to compare the level of income when each country's emissions are restricted at a given level, with the level of income achieved when the country is able to trade permits. The difference in income is a measure of the welfare gains from trade. We also compare the amount of reductions in total emissions that could be achieved with and without trade when each country's income is held fixed at a given level. This comparison may be especially interesting to people who care more about reducing emissions than they do about increasing income.

The conclusions from this empirical exercise are speculative. However, they give us an estimate of the magnitude of the importance of trade in carbon permits. The simplicity and transparency of our model is appealing. More complicated models (for example, those based on optimization and engineering estimates) do a better job of describing some aspects of the world. However, the complexity of these models makes their conclusions difficult to assess.

In the following, we estimate a system of simultaneous equations in which national income and CO_2 emissions are endogenously determined by country-specific characteristics, including levels of capital, labor, and technology. We view pollution and gross domestic product (GDP) as joint outputs of a production function that depends on capital, labor, and technology (variables which we treat as exogenous). We estimate a national revenue function by regressing GDP on capital, labor, technology, and emissions. This function represents the efficiency frontier between income and emissions for given levels of the exogenous variables. A country's environmental policies and economic structure, which we proxy using per-capita energy consumption, determine the equilibrium level of GDP and of emissions.

We use the estimated model to simulate prices and efficiency gains under tradable emissions permits. We suppose that countries enter into an international agreement that allocates CO_2 emissions permits and that this agreement supersedes the mechanism that would otherwise determine the country's emissions (the point on its efficiency frontier). The joint production function (which depends on technology and factor endowments) has not been altered by the agreement. This fact means that we can use the estimated revenue function to determine the effect on GDP when there is a change in emissions. This function implies a demand for emissions permits, which we use to calculate the price of permits when trade is permitted. We simulate the efficiency gains resulting from trade in permits.

2. BACKGROUND

The Kyoto Protocol requires that by 2012 the industrialized countries reduce their collective emissions of greenhouse gasses by 5.2 percent of 1990 levels. The country-specific targets in the Kyoto Protocol may be difficult for some nations to achieve. There may be considerable cross-country variation in marginal-abatement costs, and the strength of environmental lobbies also differs. Emissions trading, which was proposed to enable signatories to achieve reductions efficiently, allows developed countries to trade emissions credits among themselves. This trade makes sense only among those countries that have agreed to quotas, predominately the Organization for Economic Development (OECD) countries; therefore, we include only these countries in our empirical model.

The U.S. Acid Rain Program, which allows trade in SO_2 emissions, is an important experiment in tradable pollution rights (Stavins, 1998). The United States' experience with the SO_2 program suggests that trade in CO_2 permits could have considerable benefits.

There have been many attempts to estimate the costs of reducing carbon emissions, and several attempts to synthesize the estimation results. If countries were allowed to trade emissions quotas, the equilibrium price would be determined by the costs of reducing emissions. We use the estimates from previous costs studies as a basis for comparison of the estimates of quota prices that we obtain from a simple econometric model.

Nordhaus (1991) collects estimates of marginal costs of abatement and then estimates a relation between these costs and the percentage reduction of emissions. Bohm and Larsen (1994) use this relation to estimate the price of tradable permits and the efficiency gain for intra-European trade. They estimate an equilibrium price of U.S. $240 per ton of carbon if only Western European countries trade. Including the remaining OECD countries (China, Eastern Europe, and the former Soviet Union) causes the estimated price to fall to U.S. $33.50 per ton of carbon. Larsen and Shah (1994) calculate the price of emissions if all countries participate in trade at U.S. $58 per ton of carbon and if only OECD countries participate at U.S. $181 per ton.

3. THE EMPIRICAL MODEL

We estimate a revenue function and an emissions function using 1975-90 panel data for 24 OECD countries.[i] We assume that GDP and CO_2 emissions are joint products produced by country-specific factors (capital, labor, and technology). This joint production function determines the trade-off between emissions and GDP for given levels of factors. We refer to this frontier as the revenue function. The second equation is the emissions function, which determines the equilibrium point on the efficiency frontier.

To conserve notation we suppress time and country subscripts in describing the model. The joint production function is $F(Y,E) = G(C,K,L,T,Pop)$, where Y = GDP (measured in constant 1987 U.S. dollars); E = Industrial CO_2 Emissions (in kt, that is, thousands of metric tons)[ii]; C is a country specific dummy; K = Physical Capital Stock (in constant 1987 U.S. dollars); L = Labor force; T = Patent applications (a proxy for technology[iii]); and Pop = Country Population. We invert the relation $F() = G()$ to obtain the revenue function $Y = f(C,K,L,T,Pop,E)$, which represents the feasible trade-off between income and emissions, for given levels of the other variables. We divide all variables (except the dummy) by Pop to obtain per-capita variables, and estimate a log-linear relation.

The estimation equation for the revenue function is:

$$(1) \quad y_{is} = c_i + \alpha_1 k_{is} + \alpha_2 l_{is} + \alpha_3 t_{is} + \alpha_4 e_{is} + \varepsilon_{1is}.$$

Lower case variables *y, k, l, t* and *e* are logarithm of the per-capita of the corresponding upper-case variables; c_i is the country specific dummy; ε_{1is} is the error associated with country *i* in period *s;* and the parameters α_j, $j=1, 4$ are to be estimated. We view *Y* and *E* as endogenous, and we treat *K, L, T* and *Pop* as exogenous. These explanatory variables are stock variables. Thus, we treat their levels as predetermined in a period. We include the country dummy to account for country-specific variables such as arable land and cultural factors. The revenue function describes the technological trade-off between emissions and income.

A second relation, the emissions function, describes the social trade-off between income and emissions. In principle, the emissions function should include variables that proxy political constraints (for example, membership in environmental groups and relative income of workers in dirty industries). Much of this kind of information is not available for our sample. In an effort to improve the specification of the emissions function and to maintain identification, we include commercial energy use (kt of oil equivalent), *N*, as a regressor in the emissions function. We view *N* as a proxy for the structure of the economy; that is, an indication of the opportunity cost of reducing emissions.

We estimate a log-linear specification of the emissions function:

$$(2) \quad e_{is} = d + \beta_1 y_{is} + \beta_2 n_{is} + \varepsilon_{2is} .$$

The variable n_{is} is the log of per-capita energy consumption in country *i*, year *s*; *d* is a constant; and ε_{2is} is the error term.

4. ESTIMATION RESULTS

In order to provide a basis for comparison, we first estimate equation (1) using ordinary least squares (OLS), and then jointly estimate equations (1) and (2) using three-stage least squares (3SLQ). In Tables 19.1 and 19.2 we report the OLS and 3SLQ results, respectively, with standard in parentheses.

Our parameter estimates for equation (1) are comparable to the augmented Solow growth model estimated by Nonneman and Vanhoudt (1996) for the OECD countries. Their estimated production function is $Y = K^{.33}L^{.4}\tilde{T}^{.08}H^{.15}$. Their measure of technology, $\tilde{T}$, uses research and development (R&D) expenditures and *H* is a measure of human capital. Our estimate of the elasticity with respect to capital is larger, and our estimate of the labor elasticity is smaller, relative to Nonneman and Vanhoudt (1996). Although we use a different variable to measure technology, our elasticity estimate is similar to theirs.

Table 19.1 OLS Estimates of Equation 1

Parameter name	*Estimate*
α_1 (*k = log of capital*)	0.534 (0.024)[1]
α_2 (*l = log of labor*)	0.3385 (0.022)
α_3 (*t = log of technology*)	0.0558 (0.006)
α_4 (*e = log of emissions*)	0.0452 (0.013)
R^2	0.99

[1]Numbers in parentheses denote standard errors.
Source: Authors' estimates.

Table 19.2 3SLQ Estimates

Parameter name	*Estimate*
α_1 (*k = log of capital*)	0.517 (0.025)[1]
α_2 (*l = log of labor*)	0.287 (0.031)
α_3 (*t = log of technology*)	0.0625 (0.007)
α_4 (*e= log of emissions*)	0.106 (0.027)
β_1 (*y = log of GDP*)	-0.216 (0.039)
β_2 (*n = log of energy consumption*)	1.179 (0.038)

[1]Numbers in parentheses denote standard errors.
Source: Authors' estimates.

5. SIMULATION RESULTS

We use the structural model—particularly the revenue function, equation (1)—to estimate the effect of trade in permits. It is convenient to rewrite this equation as:

$$
(3)\quad \begin{aligned} & Y_i = A_i E_i^{\alpha_4} \quad \text{with} \quad A_i \equiv F_i Pop_i^{\sigma};\ \sigma \equiv \left(1 - \sum\nolimits_j \alpha_j\right) \\ & F_i \equiv \exp(c_i) K_i^{\alpha_1} L_i^{\alpha_2} T_i^{\alpha_3} \end{aligned}
$$

The positively sloped solid curve in Figure 19.1 shows the graph of the revenue function for a particular country in the year 1990, and the negatively sloped solid curve shows the emissions function. The intersection of these curves, point *x*, represents the 1990 equilibrium. If factors of production and population increase [and, thus, variable *A* defined in equation (3)], then the revenue function shifts as shown by the upward-sloping, dashed-curve in Figure 19.1.

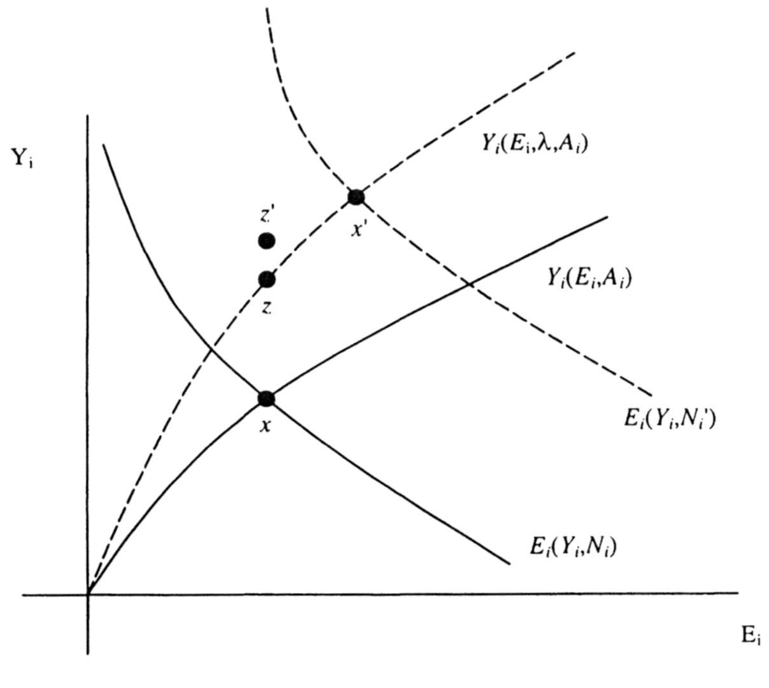

Figure 19.1 Equilibrium with and without Quotas

To reflect the view that emissions would increase in the absence of an agreement, the dashed curve labeled $E_i(Y_i N_i')$ represents the future (for example, year 2010) emissions function. The point x' is the equilibrium combination of emissions and income in the absence of an agreement to constrain emissions.

An international agreement changes the regime that determines the level of emissions. If the agreement restricts emissions in the year 2010 to its 1990 level, then a country's level of income without trade is given by point z. If a country receives an allocation equal to its 1990 emissions, but is able to trade permits, then it can achieve a higher level of income, such as point z'.[iv]

When we use the model to calculate the equilibrium price of tradable permits, we assume that the percentage increase in A_i over the 1991 to 2010 period is the same for all countries. That is, $A_{i,2010} = \lambda A_{i,1990}$ for $\lambda > 1$. There is a simple relation (described below) between the equilibrium price of permits and the value of λ. Therefore, in the next section we report the simulated equilibrium price under the (implausible) assumption that $\lambda = 1$. The reader can adjust these prices depending on the value of λ that seems reasonable. The efficiency gains due to trade are independent of the value of λ (that is, independent of the growth of factors of production).

6. ESTIMATES OF PRICES AND EFFICIENCY GAINS

With tradable emissions and perfect competition, the value of the marginal product of emissions in each country equals the world price of permits, denoted by P. Using equation (3), country i's value of marginal product (its equilibrium inverse demand) for emissions is $P = \alpha_4 A_i E_i^{\alpha_4 - 1}$, which implies demand:

$$(4) \quad E_i = \left(\frac{P}{\alpha_4 A_i} \right)^{\frac{1}{\alpha_4 - 1}} .$$

Using our 3SLQ point estimate (Table 19.2) $\alpha_4 = .106$, the elasticity of demand (both for a single country and for the aggregate of all countries) is 1.12. Summing equation (4) over I, and setting the result equal to the aggregate level of emissions $\bar{E}$ gives the equilibrium price $P^*(\bar{E})$ as the solution to:

$$(5) \quad \bar{E} = \Sigma_i E_i = \sum_i \left(\frac{P^*}{\alpha_4 A_i} \right)^{\frac{1}{\alpha_4 - 1}} .$$

We report the simulated price (1987 U.S. dollars per metric ton of CO_2) when OECD aggregate emissions and country *i*'s factors (and, thus, A_i) equal their 1990 levels using the OLS and the 3SLQ parameter estimates (Table 19.3).

(Here we set λ =1.) The second row of the table reports the simulated price for CO_2; and the third row converts this into the price of carbon.[v] The price estimates summarized in Section 2 refer to tons of carbon, so the third row of Table 19.3 should be used for comparison.

Figure 19.2 graphs each country's 1990 marginal product of emissions in the absence of trade and shows the equilibrium price (using the 3SLQ estimates). The 13 countries whose marginal product of emissions is higher than the price would buy permits. Switzerland and Sweden have the highest marginal product of emissions. Eleven countries, including the United States, gain from selling permits.

The estimated equilibrium price is primarily useful as a means of comparing our results with previous literature. Our price estimates (using the assumption λ =1) are substantially higher than those we summarized in Section 2. The more interesting economic question concerns the welfare effects of allowing trade in permits. Fortunately, the answer to this question is independent of the value of λ.

In order to estimate the efficiency gain due to tradable permits, we compare a country's estimated GDP with and without tradable permits, given a quota allocation equal to its 1990 emissions level. We denote Y_i^* as country *i*'s GDP when it uses the efficient level of emissions, E_i^* [that is, the value given by equation (4)] $\left(Y_i^* = A_i E_i^{*\alpha_4}\right)$. The value of its exports of permits, given an allocation equal to its actual 1990 emissions ($E_{i,1990}$) is $P^*\left(E_{i,1990} - E_i^*\right)$, where P^* is the equilibrium price from equation (5). Under tradable permits, country *i*'s total income is Y_i^{TP} :

$$(6) \quad Y_i^{TP} = Y_i^* + P^*\left(E_{i,1990} - E_i^*\right).$$

Table 19.3 Simulated Prices

Coefficients	*OLS (Table 19.1)*	*3SLS (Table 19.2)*
Price per metric ton of CO_2	U.S. \$57.50	U.S. \$156.80
Price per metric ton of carbon	U.S. \$210.00	U.S. \$574.94

Source: Authors' estimates.

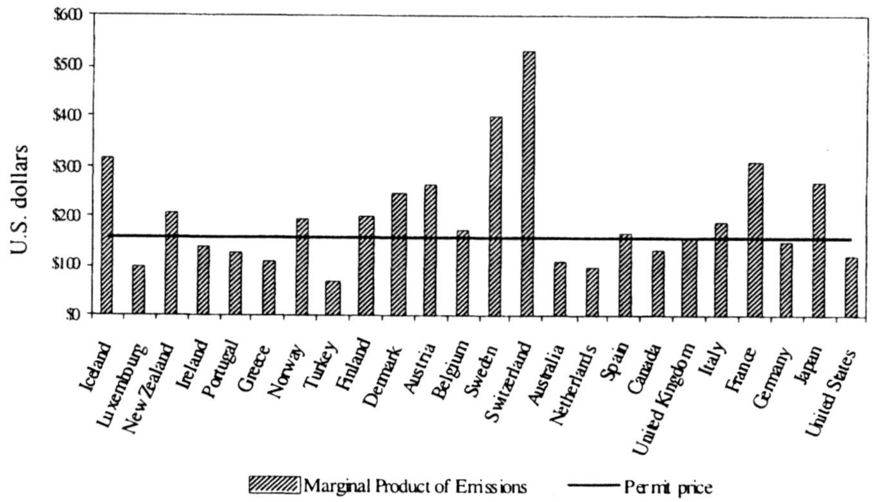

Figure 19.2 Marginal Product of Emissions and Permit Price

The estimated level of income without trade is $Y_{i,1990} = A_i\, E_{i,1990}^{\alpha_4}$. A measure of the efficiency gain due to trade is $\dfrac{Y_{i,1990}^{TP} - Y_{i,1990}}{Y_{i,1990}}$.

We show the efficiency gains for the countries in our sample using the 3SLQ parameter estimates of equation (1) (Figure 19.3). For most countries, the gains are below two percent of GDP; only three countries gain more than three percent. For some countries, for example, Germany, the gain is negligible; the United States gains 0.53 percent. The unweighted average of the gains for the 24 countries is 1.36 percent.

The results above held A_i at its (estimated) 1990 level. If A_i increases in the future (the time at which the quota becomes binding) the equilibrium price would be higher. For example, suppose that A_i is replaced by λA_i, $\lambda \geq 1$ to represent an increase in factors of production and population. Using equation (4) and the equilibrium condition $\overline{E} = \Sigma_i E_i$, it is easy to show that $\dfrac{dP^*}{P^*} = \dfrac{d\lambda}{\lambda}$.

The estimated equilibrium shares and efficiency gain are independent of λ (provided that the value of λ is the same for each country). In the absence of trade, income is $\lambda A_i E_{i,1990}^{\alpha_4}$. With equal proportional growth in A_i for all i,

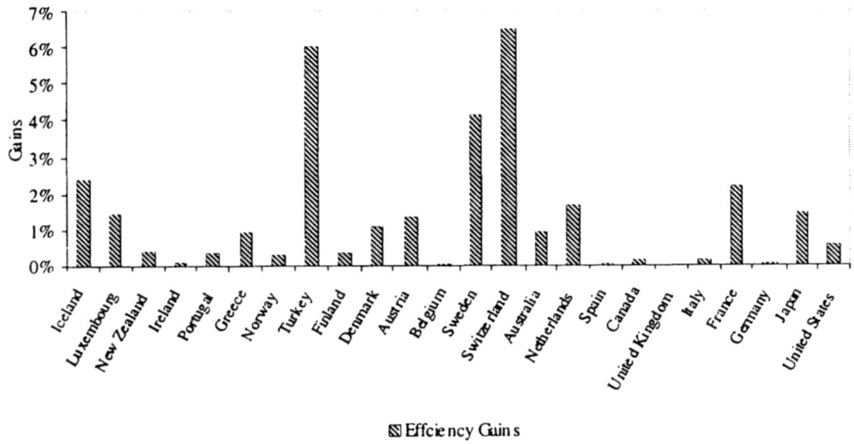

Figure 19.3 Efficiency Gains

each country's demand for emissions shifts up by the same amount, and its equilibrium share with trade remains the same. Since the percentage increase in price equals the percentage increase in λ, income under trade $\left(Y_i^{TP}\right)$ increases by the same proportion as income in the absence of trade: The efficiency gain due to tradable permits is independent of λ.

7. ESTIMATES OF POTENTIAL REDUCTIONS AND QUOTA SHARES

Another way to measure the efficiency gains of permit trading is to calculate the maximum additional reduction in emissions that can be achieved by allowing trade without reducing income. If countries were to agree to limit emissions to their 1990 level, then their estimated future income in the absence of trade is $\lambda A_i E_{i,1990}^{\alpha_4}$. The parameter λ >1 represents the increase in their factors of production relative to 1990 levels.

If the countries agree to allow trade in permits and attempt to reduce aggregate emissions, $\overline{E}$ below the 1990 level, then the constraint that no country is worse off can be written as:

$$(7) \quad \lambda A_i E_i^{*\alpha_4} + P^*\left(\overline{E};\lambda\right)\left(\mu_i \overline{E} - E_i^*\right) \geq \lambda A_i E_{i,1990}^{\alpha_4}.$$

Here μ_i is country i's share of aggregate emissions. The first term on the left side of (7) is the value of domestic production, given the efficient level of emissions (a function of P^*). The second term is the value of net exports of permit. The equilibrium price $P^*(\bar{E};\lambda)$ is proportional to λ and the equilibrium values E_i^* are independent of λ. Therefore, we can divide both sides of equation (7) by λ and write the constraint on income as independent of the growth parameter λ.

The optimization problem that determines the new agreement is:

$$\text{(8)} \quad \min_{\bar{E},\mu_i} \bar{E}, \text{ subject to } \Sigma_i \mu_i = 1, \text{ and equation (7).}$$

The equilibrium price and each country's equilibrium use of emissions depend on $\bar{E}$, but are independent of the allocation of quota rights. However, a country's income (and its willingness to sign an agreement) does depend on the allocation.

The solution to (8), that is, the minimal level of $\bar{E}$, is 8.06 percent lower than 1990 levels.[vi] Thus, tradable permits make it possible to achieve a significantly higher reduction in emissions without a loss in income. This model probably overstates the actual gain, because it ignores transaction costs and adjustment costs which would undoubtedly be associated with a reallocation of emissions. Thus, our estimates of gains should be viewed as plausible upper bounds, rather than unbiased estimates.

We show the 1990 shares of emissions; the equilibrium shares when aggregate emissions are fixed at 1990 levels and trade in permits is allowed (identified as Simulation 1); and the optimal shares, μ_i, implied by the solution to equation (8) (identified as Simulation 2) (Figure 19.4). We show the 1990 levels of emissions and when aggregate emissions are minimized (Figure 19.5).

Several countries (notably Japan and France) receive a share of quota rights (μ_i) less than their actual share in 1990. However, they emit more CO_2 under the equilibrium implied by the solution to equation (8) than they did in 1990. For the United States, on the other hand, the optimal quota share under the solution to (8) exceeds its historical 1990 share, but the equilibrium share of emissions is lower. Thus, Japan and France are net buyers of quota rights, and the United States is a net seller.

The Cobb Douglas functional form for income implies that a country's equilibrium share of emissions, $E_i^*/\Sigma_j E_j^*$, equals its equilibrium share of income from production, $Y_i^*/\Sigma_j Y_j^*$. Since the United States had roughly 35

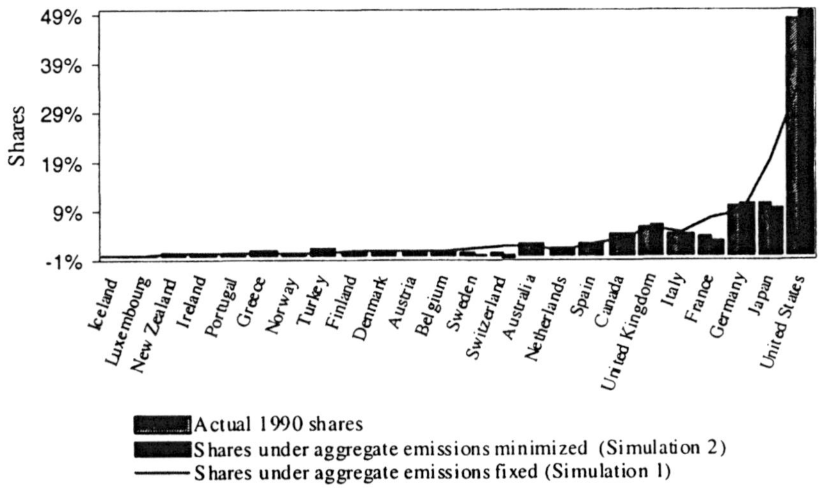

Figure 19.4 Emissions: Actual Shares, Efficient Shares, and Optimal Quota Allocations

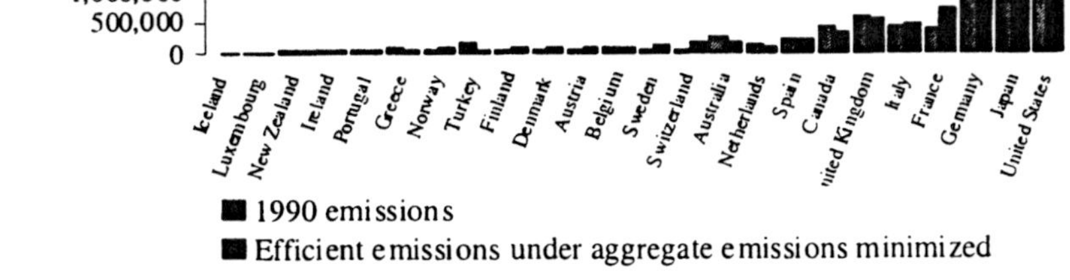

Figure 19.5 Emission Reduction

percent of OECD GDP in 1990 (World Bank, 1998), its share of emissions is approximately 35 percent for all the experiments.

8. CONCLUSION

We estimated a structural model to assess the effects of tradable permits for CO_2 emissions. One equation in our model describes the relationship between GDP and factors of production, including CO_2 emissions. We view these emissions as representing environmental services, the supply of which is endogenous. The second equation uses income and energy consumption (a proxy for the structure of the economy) to explain the equilibrium supply of these services.

We assumed that an international agreement supersedes the mechanism that would, in the absence of the agreement, determine the endogenous supply of environmental services (the level of emissions). We used our estimated revenue function to simulate the equilibrium price and efficiency gains of tradable permits, given a particular level of aggregate emissions. Our estimated carbon prices are two or three times as large as previous estimates without accounting for growth in demand (due to growth in factors of production).

Some proposals aim to reduce year 2010 aggregate emissions to 1990 levels. Our results suggest that an additional eight percent reduction in aggregate emissions could be achieved without income loss by appropriate distribution of emissions rights. This distribution gives the United States a larger share than its historic level, but the U.S. exports permits, leading to smaller U.S. emissions.

There are several limitations on our study. We ignored adjustment costs, which would require a dynamic model, and transaction costs. Therefore, our estimates of the gains from trade are more likely to represent plausible upper bounds, rather than estimates of expected gains. The data limitations mean that we have been unable to include all of the explanatory variables that should ideally be in the model. We used a systems estimator in an attempt to reduce the problem of endogeneity of regressors, but the lack of data (that is, the lack of better instrumental variables) means that this problem has probably not been eliminated.

We ignored other methods of reducing the costs of controlling stocks of greenhouse gasses, including the U.S. proposal to obtain credits by using carbon sinks. The acceptance of this proposal would reduce the U.S. marginal-abatement costs. Our estimates suggest that U.S. marginal costs are already relatively low and that the United States would be a net exporter. Allowing the United States to receive credits for carbon sinks would shift out its export supply function, reducing the price of permits and increasing the gains from trade.

Despite these limitations, our approach has the virtue of providing a simple and easily understood means of addressing an important policy issue. Our estimates suggest that tradable carbon permits could substantially reduce abatement costs. Consequently, proposals to allow this trade should be seriously considered in international negotiations aimed at reducing the dangers of global warming.

9. ENDNOTES

[i] Karp and Liu (1999) describe the data and provide a more complete report of the estimation results. That paper also discusses in detail the emissions function and its relation to the literature on the environmental Kuznets curve.

[ii] These include emissions arising from burning fossil fuels, from manufacturing cement, and contributions from other solid, liquid, and gas fuels and gas flaring. The data also includes emissions from commercial and residential sources, but not from changes in land-use (Roberts and Grimes, 1997; World Bank, 1998). This data accounts for approximately 94 percent of the measure of total anthropogenic emissions excluding land-use change and forestry found in the Buenos Aires press kit (1998).

[iii] Gardner and Joutz (1996) discuss the relative merits of using patent applications and research and development expenditures as proxies for technological innovation, and recommend the former.

[iv] The horizontal coordinate of z' represents the country's quota allocation, which differs from its actual emissions by the amount of trade. The vertical coordinate represents the value of production $Y_i(E_i,\lambda A_i)$ plus the value of sales of quota licenses.

[v] CO_2 has a molecular weight of 12 + 2(16) = 44. Thus the ratio of the weight of CO_2 to carbon is 44/12 = 3.6667.

[vi] Recall that in the 1992 Framework Convention on Climate Change, industrialized countries set a target for the year 2010 at 1990 levels. The Kyoto Protocol set a target at 5.2 percent of 1990 levels.

10. REFERENCES

Bohm, P. and B. Larsen. 1994. "Fairness in a Tradable-permit Treaty for Carbon Emissions Reductions in Europe and the Former Soviet Union." *Environmental and Resource Economics* 4(3): 219–239.

Buenos Aires Press Kit. 1998. Fourth Session of the Conference of the Parties U.N. Framework Convention on Climate Change. Internet website: http://www.unep.ch/iuc/submenu/press/climate.

Gardner, T. and F. Joutz. 1996. "Economic Growth, Energy Prices and Technological Innovation." *Southern Economic Journal* 62(3): 653–666.

Karp, L. and X. Liu. 1999. "Valuing Tradable CO_2 Permits and OECD Countries." Working Paper No. 872, Department of Agricultural and Resource Economics, University of California, Berkeley.

Larsen, B. and A. Shah. 1994. "Global Tradable Carbon Permits, Participation Incentives, and Transfers." *Oxford Economic Papers* 46: 841–856.

Nonneman, W. and P. Vanhoudt. 1996. "A Further Augmentation of the Solow Model and the Empirics of Economic Growth for OECD Countries." *Quarterly Journal of Economics* 111(3): 943–953.

Nordhaus, W.D. 1991. "The Cost of Slowing Climate Change: A Survey." *The Energy Journal* 12(1): 37–64.

Roberts, J.T. and P. Grimes. 1997. "Carbon Intensity and Economic Development 1962–91: A Brief Exploration of the Environmental Kuznets Curve." *World Development* 25(2): 191–198.

Stavins, R. 1998. "What Can We Learn from the Grand Policy Experiment? Lessons from SO_2 Allowance Trading." *Journal of Economic Perspectives* 12(3): 69–88.

World Bank. 1998. *World Bank World Development Indicator, 1998*. CD-Rom.

Chapter 20

The Impact of Climate Change on Agriculture: A Global Perspective

David Zilberman, Xuemei Liu, and David Sunding
University of California at Berkeley

1. INTRODUCTION

There is growing concern about the prospect of global warming caused by the accumulation of carbon dioxide and other greenhouse gases in the atmosphere. Several studies address the impact of climate change on the agricultural sector in various countries (Marland, McCarl, and Schneider, 1999). While many of these studies assess the impact of climate change on farm commodity yields and prices, only a few address the problems of rising seawater levels, water supplies, pest dynamics, and energy price increases. This chapter focuses on each of these issues. We present a comprehensive overview of the implications of climate change on agricultural productivity, including changes in yield from increased temperatures, variations in water supply, increased carbon concentration in the atmosphere, pest movement, and other factors. We also address the implications of climate change on water and land resources, in particular, the issues associated with rising water levels. We discuss strategies to slow the pace of global warming, the effects of incentives to conserve energy, and the benefits and institutional changes resulting from the implementation of soil carbon sequestration in farming and agro-forestry. We conclude with a presentation of policy implications for developed and developing countries.

While some economic studies (for example, Mendelsohn, Nordhaus, and Shaw, 1994) argue that, for the most part, the impacts of climate change are modest, we will show that such studies ignore some key considerations that require more concentrated attention. We suggest that climate change may have substantial impacts on agriculture, and may even contribute to the destabilization of agricultural and political systems. Dealing with climate change may require significant adjustment and investment, some of which may have to be coordinated globally.

2. THE IMPACT OF CLIMATE CHANGE ON YIELD, OUTPUT, AND PRICE

The main manifestation of climate change is an increase in average temperature of between 3° to 7° Fahrenheit. Climate change will also affect the variability of temperature and precipitation patterns. It is reasonable to anticipate that climate change will shift existing weather patterns two to three hundred miles northward in the Northern Hemisphere and an equal distance southward in the Southern Hemisphere. Thus, in the United States, central Oregon will have the climate that exists today in central California, southern Montana will have the current climate of Nebraska, and crop patterns will change accordingly. For example, we will see corn grown in Montana, and more fruits and vegetables grown in central Oregon (of course, the latter will require a more extensive water supply infrastructure).

Agronomists have calculated the impact on several crop yields under these varying conditions and some studies use an agronomic approach to compute the economic cost of climate change under current land-use patterns. Mendelsohn, Nordhaus, and Shaw (1994) develop a model to assess the economic impact of climate change, recognizing the importance of adaptation. They relate per-acre rent to weather conditions, since rent reflects net economic surplus to producers from their agricultural production activities; the authors use this rent function to estimate the distribution of per-acre rents under alternative scenarios of climate change. They use the difference in aggregate rents before and after climate change as an estimator of the economic impact on agriculture. Mendelsohn, Nordhaus, and Shaw (1994) expect farmers will adapt to the change in climate and modify their land-use patterns to the changing conditions. Their analysis estimates the economic impact of climate change on U.S. agriculture to be rather small: a total change of less than five percent in aggregate farm income. Of course, the distributional impact varies across regions. Some regions (states near the Canadian border, for example) may gain, while other, more southern states, may lose. But the overall effect seems to be modest.

The results of the Mendelsohn, Nordhaus, and Shaw (1994) model suggest that we should not be alarmed by the prospect of climate change, especially within the agricultural sector. We argue, however, that these results ignore several important phenomena that should be incorporated into the analysis of climate change.

2.1 Daylight Effect

Daylight density decreases as one moves northward or southward from the equator. This is especially true in areas north of the 23.5° latitude. Crop yield is positively related to daylight; therefore, moving production away from the equator may result in lower productivity.

2.2 Fertilization Effect

Increased carbon dioxide in the atmosphere will increase agricultural production, since higher carbon dioxide levels increase photosynthesis. Some expect this to increase yields by up to 30 percent, but Lincoln (1999) argues that, while the quantity of production will increase, the quality will decline because green plant matter will include less protein. This effect should be taken into account when computing the impact of climate change; if the gains associated with increased fertilization are greater than the losses associated with reduced quality, the fertilization effect will reduce the cost of climate change.

2.3 Pest Effect

In many cases, low temperatures control pest populations. Global warming may lead to the northward migration of many pests which could negatively affect crop yields. Furthermore, Gutierrez (1999) argues that movement of pest populations and changes in climatic conditions may reduce the effectiveness of current pest management schemes; this could affect productivity or require extra adjustment costs.

2.4 Settlement and Adjustment Costs

The adaptation and relocation associated with climate change will require significant fixed costs to farmers that may not be reflected in rent changes. Farmers may have to modify their capital costs and invest in redesigning production systems to accommodate changes in crop choices, pest control, and water management. Furthermore, farmers may migrate to new areas, which will entail significant settlement costs. The more extensive the climatic changes, the larger the areas that will potentially exit or enter agricultural production. Climate change may also require additional costs to agribusiness for adjustment and relocation of processing facilities.

The public sector will assume most of the adjustment costs. Safety-net programs will have to accommodate farmers and farm workers who lose their employment. Furthermore, agricultural research establishments will be challenged to develop new production strategies and technologies to accommodate the changes in the climatic conditions. As Olmstead and Rhode (1993) argue, much of the credit for the high yield of American agriculture is due to many years of research by crop breeders and other scholars in public research universities. Climate change will raise new challenges for new crop-production systems.

2.5 Price Effects

As noted above, several factors associated with climate change will contribute to reduced yields, thereby offsetting the effect of fertilization and reducing agricultural supplies. However, farmers may not necessarily be the group most affected by climate change. In cases with a relatively low elasticity of demand (which is true for most grains), the main welfare impact may be on consumers who will pay higher prices. The impact on producers is ambiguous since they may receive higher prices, but their levels of output could be lower. Within the agricultural economy, we expect livestock operations to fare less favorably than grain farmers because lower supplies, as a result of climate change, will raise the price of grains and thus the cost of producing meat.

2.6 The Impact of Geography

To a large extent, the overall effect of climate change on a country depends on its geographical location and shape. Countries like the United States that are rectangular will have northern regions that will benefit from climate change and southern regions that will become worse off. The acreage of both the gaining and losing regions may be quite similar. On the other hand, in a country with a triangular shape, the acreage of a region that loses may be much larger than the small number of regions that win, and the net effect of climate change may be quite negative.

2.7 Impact of Adjustment Speed

The capacity of regions to adjust to climatic changes and to develop new technologies is limited; therefore, the cost of adjusting to climate change depends on the speed of adjustment. If climate change occurs gradually over a period of time, the overall costs will be much lower than if climate change occurs quickly, leaving only a very short adjustment period. Furthermore, the ability to adjust to climate change will depend on the ability to correctly monitor climatic conditions and to adjust to changes as they occur. For example, if significant changes in conditions are expected within a short time, then developing policies that monitor climatic conditions should be emphasized to provide policymakers with as much information on the weather as possible. Unlike Mendelsohn, Nordhaus, and Shaw (1994), we expect climate change to result in significant costs to society, higher costs for producers, and significant adjustment costs to farmers and the government.

2.8 Government Expenditures on Public Goods

Government provides public goods in agriculture to address the under-investment arising from competitive structures (Cochrane, 1993). These government expenditures are of two types. One category aims to increase farm supplies by funding, for example, research and development (R&D) or investments in irrigation projects. The other category aims to facilitate adaptation. Government expenditures can aid both adaptation and crop switching in response to climate change. After climate change, the changes in crop yields and land rents will call for crop switching or for developing new areas. Government might find it beneficial to increase investment in crops likely to suffer from climate change, or to increase investment in the infrastructure crucial for crop switching.

Farmers' decisions on land allocation will be affected by changes in yields, commodity prices, and commodity support programs. In addition to direct subsidization, governments can invest in infrastructure and provide facilities that change the private cost of adaptation, and thus affect farmer decisions on crop switching and other land-use options.

Since farmers will reallocate land among crops as a result of climate change, government expenditures for each crop will be adjusted. If the temperature effect is dominant and the impact of public goods on cost savings is small, then the cutoff location moves northward. In this case, the expenditure for the cold crop falls while that for the hot crop increases. Climate change makes the hot crop more profitable around the cutoff location, and government expenditure cannot significantly increase the yield of the cold crop. Therefore, it is better for government to decrease investment in the cold crop and increase investment in crop switching.

Although expenditure will increase for one crop and decrease for the other crop, the magnitudes are not necessarily the same. Thus, the total expenditure is likely to change after warming occurs. Whether the total expenditure will increase depends on the relative magnitude of the change in each expenditure.

3. THE EFFECTS OF CLIMATE CHANGE ON WATER RESOURCES

Scientists suggest that some of the most significant effects of global warming will be on water resources. These effects include rising sea levels, changes in patterns of precipitation, changes in patterns of snowmelt, and the overall movement of water resources. These changes, which have not been extensively investigated, have a significant effect on food supply and agriculture; they are discussed below.

3.1 Rising Sea Levels

Sea levels could rise from 40 centimeters to one meter and even higher as global warming continues. This process may lead to complete flooding of some islands in the Atlantic Ocean, the Caribbean Sea, the Indian Ocean (for example, the Maldive Islands), and the Pacific Ocean. Moreover, rising sea levels will result in flooding of low-lying regions near oceans and seas (areas in the subcontinent of India, for example).

Where real estate values are high, especially in developed countries, the response to this phenomenon will involve construction of dikes and protective walls. For example, the dike systems in Holland and other low-lying countries will likely be expanded and strengthened, and resources will certainly be available to protect New York's Manhattan Island from the invasion of the sea. Some developing countries, such as the Maldive Islands, are already considering billion-dollar investments to protect their shores.

But what about agricultural regions throughout the world, particularly in countries such as Bangladesh and Pakistan, where millions of people live and farm near the ocean? These countries are already being threatened and harmed by floods. These countries (and the global community) will need to design strategies to build protective dikes and relocate masses of people.

The higher and more rapid the rise of the sea levels, the more severe the problems become. Dealing with rising sea levels, which result in flooding and relocation, will be a major challenge to international communities. It may put a million or perhaps even 10 million people in peril and could also require investment schemes and relocations of unparalleled dimensions, leaving some nations in dire straits.

Unless some of the food and agricultural production problems resulting from climate change are addressed, there may be significant political and military ramifications. This perception is based on an analysis of the relationship between weather patterns and historical development. According to Issar (1995), many of the most dramatic historical events were weather induced. The invasion of Rome by Germanic tribes from the north was associated with a mini-Ice Age around the year 500 AD. Global warming might have been a motivating factor in the movement of invading forces from the Arab Peninsula northward during the 8th and 9th centuries.

A major policy challenge resulting from climate change may be the development of resettlement strategies for protection of agricultural areas vulnerable to rising seawater invasion or relocation of affected individuals. In many cases, affected countries will not be able to deal with this problem by themselves. In addition, there are significant differences in population density and geographic locations that may make some countries (such as Bangladesh, Pakistan, and India) more vulnerable than others. This problem suggests rethinking the sharing of power and responsibility between sovereign nations and international communities when it comes to protecting the welfare of those in need.

3.2 Seawater Intrusion

The interaction of seawater intrusion with groundwater aquifers is a delicate balance of competing forces (the weight of the seawater versus the hydrologic weight of aquifers). In many situations, the over-extraction of groundwater has caused depletion of aquifers and has already resulted in the salinization of many precious groundwater resources. Examples include Australia, Hawaii, California, and the Gaza Strip. Protection of coastal aquifers requires significant investment in projects to divert sweet water and reduce pumping to preserve hydrologic heads. Rising seawater levels may drastically intensify the invasion and desalinization of some groundwater aquifers, and protection may require both a substantial reduction of groundwater pumping and diversion of resources to build up hydrologic pressure.

There are other effects of seawater intrusion that may counteract this phenomenon. For example, increased rainfall in some areas, a larger flow of rivers, and the buildup of groundwater aquifers may counter some seawater invasion of sweet groundwater. Nevertheless, in many regions where at least part of the agricultural production depends on groundwater pumping, rising seawater and the threat it poses to aquifers will greatly reduce the possibilities of agricultural production. Rising seawater levels may also lead to drastic changes in agricultural production practices, such as more fallowing of land, adoption of modern irrigation technologies, production of crops that are less water intensive, or introduction of varieties that are more salt tolerant. California was able to implement some of these practices in response to the 1987–91 drought (Zilberman et al., 1992). However, such a transition may be much more demanding for some developing countries.

3.3 Changes in Precipitation, Snowmelt, and Rainfall Patterns

Climate change will result in changes to precipitation patterns that will drastically affect yields, especially for rainfed agriculture. Some aspects of this problem were discussed in a previous section. However, climate change will also result in significant secondary effects that relate to the movement of water and the buildup of water reservoirs for irrigated agriculture. Snowmelt from the Sierras is essential for agricultural production in California. This is also the case for the Himalayas and Andes, for example. The snowmelt process is gradual and it sometimes takes several months for the water to be distributed. This process allows rivers to have a more or less constant flow, supplies water to agricultural regions in the spring and summer, and enriches groundwater resources.

Furthermore, a complex network of dams has been built throughout the world to capture snowmelt and other sources of water to stabilize water supply throughout the year. Dams that required immense investment were designed based on then-current precipitation patterns. Unfortunately, many existing dams may not be able to contain increased snowmelt and water movement caused by drastic changes in weather patterns. Dams could be swept away, resulting in a significant loss of water resources. New patterns of snowmelt and precipitation could also result in the redirection of major bodies of water. The cost of adjusting to such changes is immense. For example, the United States still has not completely recovered from the major flooding of the Mississippi River that occurred in the 1980s and 1990s. The cost of restoring dams is high (in the billions of dollars) and, there is often substantial debate on how to design new, more effective, and less vulnerable systems.

Many developing countries may also encounter these types of dilemmas, stuck with the knowledge that catastrophes may occur, but unable to afford suitable protective measures. Again, the India Peninsula, which benefits from the Himalayas, and regions in Asia and Latin America are examples of vulnerable regions. These highly populated areas with vulnerable agricultural regions may be at great risk from changes in water resource patterns, and the adjustment process may be costly and time consuming.

3.4 Impact on Fisheries and Fish Resources

Rising water levels may significantly impact fishing communities, because many are located in estuarine areas that can flood. In addition, there may be major changes in the dynamics of fish populations, resulting in exploitation of seafood. Although on the surface the impact on fisheries appears to be less devastating than in some agricultural regions near bodies of water, adjustment and relocation costs could be exorbitant. Fishing strategies will need to be redesigned, requiring investment in new equipment. In some areas, new forms of fishing may not be feasible, and populations may be at risk. Some relatively stable estuarine regions (Thailand and Ecuador, for example) have been developed as major aquaculture industries, and other regions are also considering these types of activities. Because these industries may be especially vulnerable to changing weather patterns, some may need to be redesigned. Aquaculture and mariculture, which were developed in estuaries near oceans, were designed to be more sustainable substitutes for depleted fisheries. With climate change, the modification of these strategies may be very costly.

Societies may respond to climatic changes both by relocation and the implementation of technological as well as institutional change. While it is desirable to design many of these changes in advance and implement them gradually, Rausser and Zusman (1991) suggest that institutional changes are often only induced by crisis situations. For example, California introduced water trading only in response to a severe drought. Part of the challenge in developing climate change policies is to overcome the tendency of political systems to wait for a crisis to introduce change. Thus societies that have the capacity to develop contingency plans to adjust to climate change by modifying their institutions in advance may fare better than societies that take a passive approach and adjust their institutions and policies only in response to climatic changes.

The possible impacts of climate change on the sea and other water resources suggest that populations and industries will require relocation and new sources of livelihood. Furthermore, particularly for developing countries, immense investments should be made to protect the environment. The impacts of climate change may force major communities to resort to desperate measures.

4. THE IMPACT OF STRATEGIES TO DELAY CLIMATE CHANGE IN AGRICULTURE

Some of the dire consequences of global warming have led the financial community to consider strategies that will slow the human contribution to climate change. Potential effects of global warming may even be averted by developing strategies to reduce carbon dioxide emissions and other global gases, and by sequestering gases from the atmosphere. Many of these strategies are contained in the Kyoto Protocol that aims to reduce developed countries' emissions of global warming gases by certain percentages below 1990 levels. Kyoto established institutions that enabled developed countries to assist developing nations in expanding their energy production without contributing to the process of climate change.

While the future of the Kyoto Protocol is uncertain at present and will likely be modified, it seems likely that if the threat of global warming increases, nations will reach an agreement to slow the buildup of greenhouse gases; this may help to reverse the process of climate change. A key element in any strategy involves buying time to enable transition from fossil fuels to alternative sources of energy. However, it is not clear when these technologies will be available and the extent to which they will be adopted (or affordable, especially by developing nations). Therefore, policies that aim to avert climate change may affect agriculture by increasing fuel prices and providing incentives for agricultural activities that sequester carbon.

An increase in the price of energy will affect agriculture in several ways. First, it will increase the price of inputs such as fertilizers (particularly nitrates which are energy intensive). In addition, the price of some agricultural chemicals will increase. However, as a result of higher energy prices, the cost of mechanical activities, including ploughing, may increase the use of chemicals, in particular, herbicides. Irrigated agriculture may be substantially affected by increased energy prices since the cost of pumping will rise, reducing the effectiveness of irrigated agriculture. With higher irrigation prices, farmers may have more incentive to adopt modern irrigation technologies, such as sprinkler and drip, and switch to water-conservation practices, such as the use of computers and other farm management tools.

As Lal et al. (1998) argue, most carbon in the atmosphere results from farming activities that release carbon through plowing and from the mismanagement of soils. In agriculture and forestry, planting trees and cover crops, and switching to practices such as no tillage and low tillage, can sequester substantial amounts of carbon from the atmosphere. The potential for sequestration is very large. It may meet or even surpass U.S. and international reduction and emission levels suggested by the Kyoto Protocol.

However, tapping this potential is costly and challenging. Furthermore, there is a limit to the amount of carbon absorption possible through agriculture and forestry since the capacity of soil to absorb carbon is finite. Nevertheless, Lal et al. (1998) argue that, in the next 50 years, agriculture will play a major role in reducing the buildup and eventually the absolute stock of carbon in the atmosphere, which will result in an economy that is less dependent on fossil fuel. Since agricultural sequestration can be executed in the short run, it will serve to buy time as new energy sources are developed. However, designing schemes that will enhance carbon sequestration in agriculture and forestry is a significant political challenge.

From an agricultural perspective, sequestering carbon can be an ecological service that may yield significant amounts of income. If an acre of land can sequester, say, 5 tons of carbon and 10 tons of carbon dioxide at a price of U.S.$25 a ton, the result will be an income of U.S.$250 per acre. However, such an activity may entail switching from full tillage to low or no tillage or the use of cover crops. As Lal et al. (1998) argue, this transition may be valuable and may also contribute to other socially desirable objectives such as reducing soil erosion and improving water quality. Furthermore, the profits of individual farmers may eventually increase during the transition to alternative soil management practices because relying on tillage-extensive activities will deplete the farmers' own soil and reduce its productivity. A transition to other systems will build up soil resources and that, in itself, is a source of higher yield. However, the argument for no-tillage is one in which the recommendations of scientists and the perceptions of farmers do not necessarily coincide. Yet we have evidence of high rates of adoption (sometimes more than 50 percent of low-tillage practices) when farmers obtain extra incentives through commodity program payments.

Designing programs that will encourage farmers to adopt resource-sequestering activities is challenging. On the one hand, this type of program will pay farmers to adjust their behavior toward low tillage; but, on the other hand, farmers need to be aware that they can be penalized for switching back to full tillage, which will release carbon into the atmosphere. Thus, implementation of this program requires both monitoring of behavior and enforcement of penalties. The implementation of the program may be even more difficult because the capacity to absorb the carbon varies across locations. So before programs are established, the soil situation and the capacity to sequester should be benchmarked.

There may be several mechanisms to induce a farmer to adopt soil carbon sequestration activities. They include government programs that will pay for sequestration, which will be especially pertinent if sequestration activities are credited to the Kyoto Protocol. Alternatively, farmers and farm groups may participate in markets for sequestration and emission rights. By sequestering soil, they will sell emission rights to individuals or organizations (utilities that will need them). Some experiments along these lines have been carried out by organizations (Zilberman and Sunding, 2001). Their future depends on the evolution of international agreements on climate change and whether the agricultural activities will be recognized as activities that count in meeting the emission reduction targets.

Overall, it seems that countries like the United States, Canada, Brazil, and Russia support the inclusion of soil carbon activities as legitimate means to reduce emissions, as currently defined by the Kyoto Protocol. Thus, sequestration activities may be a new source of income for farmers.

5. IMPACT OF CLIMATE CHANGE ON AGRICULTURAL POLICY

Climate change may lead to a reduction in agricultural productivity and the loss of agricultural resources. Furthermore, it may require a significant reallocation of resources and migration. The timing and magnitude of these impacts involve significant uncertainty, and the rate of change is also unknown.

The design of policies to address the possible impacts of climate change should occur within the context of other future developments. First, we expect global populations to increase and ultimately reach levels of 10 billion or more (IFPRI, 1995). As incomes of countries in Asia rise, so will the demand for meat and other products that require significant increases in agricultural production. Thus, the agricultural sector will be challenged to increase its output (maybe even double it) while at the same time its productive capacity in some regions may worsen. Further, there is evidence that the rates of increase in productivity have declined since the first three decades following World War II. Regions such as the subcontinent of India and the Middle East suffer from decreasing resources such as water and soil. In those regions, climate change problems may add to significant natural resource problems.

5.1 Research Policies

The work of Alston, Norton, and Pardey (1995), and Huffman and Evenson (1993) suggest that in the past, research, especially public research, has been a major contributor to increased productivity. The continuation and expansion of research should be major components of policies that address the risks and costs associated with climate change. Alston, Norton, and

Pardey (1995) express concern with the reduction in public support for research and argue that, while some public funding of research can be supplemented with private funding, these two types of research are not perfectly substitutable. Public research can develop new paradigms and breakthrough innovations, while private research is complementary and can refine these technologies. Furthermore, in the case of climate change, there is a need for more research on the capacity to grow various crops under varying weather conditions, the adaptive capacity of production systems to changes in temperature associated with pest migration, and increased productivity with declining water resources. The financial commitment needed for research will depend, of course, on the rate of climate change, but the development of an extended research infrastructure to tackle unexpected events is becoming increasingly more crucial.

The development of expanded research capacity is not sufficient, however. It should be associated with mechanisms to enhance technology transfer and the introduction of new technologies. While traditional breeding played a crucial role in increasing productivity in the past, it seems that other tools may play a major role in enabling adjustment to climatic changes. It is important to sort out regulatory and intellectual property rights issues on the use of biotechnology in agriculture (Zilberman, Yarkin, and Heiman, 1997). Access to biotechnology tools is essential, especially for developing countries and some less-developed and more vulnerable regions. The Consultative Group on International Agricultural Research (CGIAR) should work with the United States and other governments, and with private companies, to assure that new technologies, and the expertise to use them, are available to developing countries.

Climate change poses new challenges to the international research establishment in agriculture. Should the research agenda include issues on climate monitoring and responses to changes in climate? The CGIAR and its institution provide an organizational setup that will enable new research challenges that are introduced by the prospect of global warming to be carried out. But the capacity to take advantage of the opportunity depends on obtaining new funding, and on modifying research priorities to emphasize environmental and climatic aspects. Furthermore, to be effective, CGIAR has to learn to work better with the private sector and to develop the capacity to transfer technologies that go beyond new varieties toward resource management practices at various levels of organization.

5.2 Migration Policies

The expansion of international trade has been one of the major achievements of the second half of the 20th century, but more has to be achieved in the area of trade liberalization. One very important issue is freedom of movement of inputs. Climate change may cause some regions and countries to be worse off and incapable of supporting their population. The capacity to develop effective policies for population transfer and resettlement is a major international challenge.

Throughout the late 19th and early 20th centuries, we have seen major population movements that have reshaped global population distribution and contributed to the improvement in global welfare. Although freedom of movement among nations has increased, it is still very restricted. International communities can establish mechanisms to enable individuals to move within regions, provide incentives to absorb new immigrants in regions with economic potential and dense population, introduce new mechanisms to ease transition, set up an infrastructure to provide resources, and most importantly, build institutional setups that lead to peaceful relocation that will enhance human well-being. In spite of the large numbers of people moving across nations, we don't have the appropriate legal structure that explicitly recognizes an orderly managed transition. The establishment of such an arrangement is a major institutional and policy challenge.

5.3 Food Supply Policies

Climate change is likely to increase the variability of food supply. There may be periods of adjustment or crises when global production will be low, resulting in an inability to meet food demand. On the other hand, there may be periods of surplus. We have been fortunate that in the last 50 years, we have not had a global shortage of food, and actually the major problems of agricultural policy have been the management of surplus. One by-product of agricultural support policies has been an excess food capacity. However, in 1997, global food inventories were significantly sparse. In the 21st century, with climate change, population growth, and increased uncertainty, the risk of shortages is greater. And as we move toward an increased reliance on markets, we may reach a situation where this extra capacity, and the cushion that it provides, is not available. Obviously, this excess supply was provided at cost, and one aim of this agricultural policy is to provide chronic overproduction. Nevertheless, when it comes to the future, it is important to examine both sides of the supply and demand relationship. Mechanisms are needed to monitor global food situations and to prevent severe shortages. Furthermore, as the weather becomes more erratic, it is imperative that food and other relief supplies be readily available in the aftermath of extreme climatic events. The global community has already developed some mechanisms for intervention in crises situations, but in the future, these methods need to be rigorous and significantly expanded to deal with the risks of climate change.

Climate change is a global event that can be confronted if nations coordinate policies and establish international agreements to provide assistance to individual nations or regions. Two mechanisms, freer trade and globalization, can help to overcome some of the risks associated with climate change, but they should be intensified. They can be augmented by emphasizing greater expansion of global research and development, especially when it comes to better understanding of ecological and agricultural events in developing countries. More importantly, more policies should be designed to stabilize weather on a global rather than just on a national or regional level.

6. REFERENCES

Alston, J.M., G.W. Norton, and P.G. Pardey. 1995. *Science Under Scarcity: Principles and Practice for Agricultural Research Evaluation and Priority Setting*. Ithaca, NY: Cornell University Press.

Ayres, M. 1999. "Climate Change, Forests, and Pesticides." Paper presented at the Climate Change Impacts on Agriculture and Potential Mitigation Schemes Conference, Gainesville, FL (10 September).

Cochrane, W. 1993. *The Development of American Agriculture: A Historical Analysis (2nd Edition)*. Minneapolis, MN: University of Minnesota.

Gutierrez, A. 1999. "Climate Change and Pest Control." Paper presented at the Climate Change Impacts on Agriculture and Potential Mitigation Schemes Conference, Gainesville, FL (10 September).

Huffman, W.C. and R.E. Evenson. 1993. *Science for Agriculture: A Long Term Perspective*. Ames, IA: Iowa State University Press.

IFPRI (International Food Policy Research Institute). 1995. *A 2020 Vision for Food, Agriculture, and the Environment: The Vision, Challenge, and Recommended Action*. Washington, DC: IFPRI.

Issar, A.S. 1995. "Climatic Change and the History of the Middle East." *American Scientist* 83(4): 350–355.

Lal, R., J. Kimble, R. Follett, and C. Cole. 1998. *The Potential of U.S. Cropland to Sequester Carbon and Mitigate the Greenhouse Effect*. Chelsea, MI: Sleeping Bear Press.

Lincoln, D. 1999. "Climate Change and Pesticide." Paper presented at the Climate Change Impacts on Agriculture and Potential Mitigation Schemes Conference, Gainesville, FL (September 10).

Marland, G., B.A. McCarl, and U. Schneider. 1999. "Soil Carbon: Policy and Economics." in N.J. Rosenberg, R. Izaurralde, and E.L. Malone eds., pp. 151–178, *Carbon Sequestration in Soils: Science, Monitoring, and Beyond*, Proceedings of the St. Michaels' Workshop. Columbus, OH: Battelle Press.

Mendelsohn, R., W. Nordhaus, and D. Shaw. 1994. "The Impact of Global Warming on Agriculture: A Ricardian Analysis." *American Economic Review* 84(4): 753–771.

Olmstead, A.L. and P. Rhode. 1993. "Induced Innovation in American Agriculture: A Reconsideration." *Journal of Political Economy* 101(1): 100–118.

Rausser, G.C. and P. Zusman. 1991. "Organizational Failure and the Political Economy of Water Resources Management," in *The Economics and Management of Water and Drainage in Agriculture*, pp. 735–758. Norwell, MA: Kluwer Academic Publishers.

Zilberman, D. and D. Sunding. 2001. "Climate Change Policy and the Agricultural Sector," in R. Lal, J.M. Kimble, R.F. Follett and B.A. Stewart, eds., pp. 629–644, *Assessment Methods for Soil Carbon*. Boca Raton, FL: Lewis Publishers/CRC Press.

Zilberman, D., C. Yarkin, and A. Heiman. 1997. "Agricultural Biotechnology: Economic and International Implications." Paper presented at the XXIII International Conference of Agricultural Economists, Food Security, Diversification, and Resource Management: Refocusing the Role of Agriculture, Sacramento, CA (10–16 August).

Zilberman, D., A. Dinar, N. MacDougall, M. Khanna, C. Brown, and F. Castillo. 1992. "How California Responded to the Drought." Giannini Foundation working paper, Department of Agricultural and Resource Economics, University of California, Berkeley.

Chapter 21

Tradable Permits and Agricultural Sequestration of Carbon

Chris Dumas[1], Charles B. Moss[2] and Andrew Schmitz[2]
[1]*University of North Carolina at Wilmington and* [2]*University of Florida*

1. INTRODUCTION

Mounting scientific evidence supports a relationship between increasing global temperatures and the concentration of greenhouse gases in the atmosphere. While gases, such as methane, chlorofluorocarbons (CFCs) and nitrous oxide, are believed to contribute to increasing global temperature, the Intergovernmental Panel on Climate Change (IPCC) attributes 60 percent of this greenhouse effect to carbon dioxide (IPCC, 1996). IPCC estimates of the potential increase in global mean surface temperature over the next hundred years range from 1° centigrade (C) to 3.5° C, with a best estimate increase of 2.0° C. Beyond the year 2100, temperatures may climb an additional 0.5°C to 2.0° C as the thermal inertia of the world's oceans is overcome. Despite uncertainty concerning the eventual magnitude of the warming effect, there is broad acceptance that global temperature is rising. The IPCC projects that the rate of warming will likely exceed any experienced in the last 10,000 years.

Economic studies have examined agriculture's contribution to global warming (for example, Schmitz et al., 1994) and the potential impact of warming on agricultural productivity and income (for example, Cline, 1992; Fankhauser, 1993; Mendelsohn, Nordhaus, and Shaw, 1994; and Nordhaus 1991, 1994). Other studies have examined the ability of agriculture to adapt to climate change by altering cultural practices or international trade flows (for example, Rosenzweig and Parry, 1994). More recently, researchers have begun to investigate the role of agriculture in any future market for carbon reduction, or sequestration, services (Lal et al., 1999; Luciuk et al., 1999). This chapter investigates carbon sequestration as a new market for agriculture by analyzing the interaction of agricultural carbon sequestration activities and international trade in carbon-reduction permits.

2. POLICY IMPLICATIONS OF GLOBAL WARMING FOR AGRICULTURE: TAXES AND SEQUESTRATION

Some points of agreement exist in several models that have been developed to analyze the economic effects of global warming. First, it is agreed that increasing temperatures may impose significant costs on the global economy; however, these costs will not be fully realized for some time. Second, society can reduce the level of future costs by using either Pigouvian taxes or quotas on carbon emissions. While economic models often favor carbon taxes over quotas, Pearce (1991) suggests that quotas may be more viable for international agreements. Third, a disparity exists between industrialized and developing countries regarding the economic costs and benefits of reducing carbon emissions. Carbon fuels are of great importance in developing nations, and any international tax will significantly impinge upon growth. Further, developing nations have little incentive to participate in any international agreement unless their agriculture would be significantly degraded by changing climate. These points undoubtedly influenced the form of carbon emissions limits set forth in the Kyoto accord. Specifically, the Kyoto Accord sets country-specific carbon emission quotas based on 1990 carbon emissions levels. In general, developing nations were given less stringent quotas.

Economists anticipate that most countries will use domestic carbon taxes, or other market mechanisms, to meet their emissions reduction requirements. Pearce (1991) surveys the literature on such taxes and finds a range of U.S. \$3 per ton of carbon to U.S. \$623 per ton of carbon. Poterba (1993) concludes that a carbon tax of U.S. \$5 per ton will add 13 percent to the price of coal, four percent to the price of oil, and five percent to the price of natural gas. Given agriculture's dependence on fossil fuels, a carbon tax would shift the supply curve for agricultural products to the left, from S^0 to S^1, reducing the equilibrium quantity produced (Figure 21.1). Assuming that at least some of the carbon-tax revenue is dissipated by the government, a tax will decrease aggregate income, shifting the demand for agricultural products to the left, from D^0 to D^1. The sum of these effects could be a significant loss of agricultural producer surplus equivalent to area B – area A.

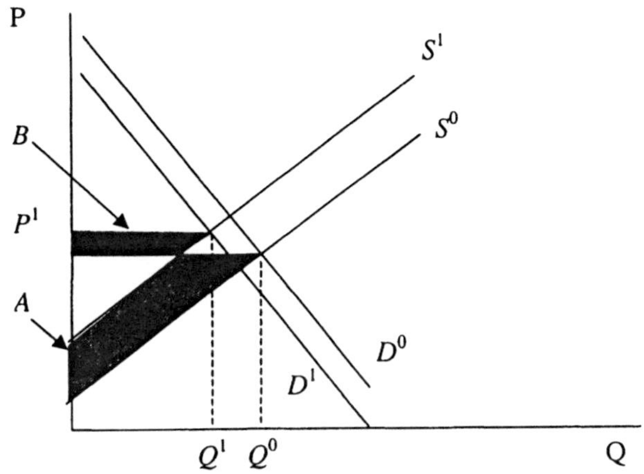

Figure 21.1 Effect of Pigouvian Tax on Agriculture

Carbon taxes may also impact agricultural-trade flows. Given that developing countries are assigned less strict carbon limits under the Kyoto Accord, the excess demands of developing countries for agricultural outputs on world markets may become excess supplies. Consider a domestic country that exported agricultural products before the Kyoto Accord. Suppose carbon taxes cause both domestic supply and demand to shift inward, as depicted in Figure 21.1. In this case, the domestic excess supply function of agricultural commodities shifts upward from ES^0 to ES^1 (Figure 21.2). Assuming that a smaller carbon tax is levied in the importing (that is, foreign) country, the foreign excess demand for agricultural commodities shifts upward from ED^0 to ED^1. Although several outcomes are possible, in some cases differential carbon taxes may be sufficient to shift the trade position of the domestic country from net exporter to net importer.

While the carbon tax has been the most widely discussed method of meeting Kyoto Accord commitments, other alternatives have been suggested. From an agricultural perspective, the most significant alternative is the biological sequestration of carbon. While the most frequently cited method of biological sequestration is increased forestry, Lal et al. (1999) investigate carbon sequestration through agriculture. Although 5,000 million tonnes of carbon have been released into the atmosphere as a result of cultivation in the United States, Lal et el. estimate that U.S. agriculture has the potential to partially reverse this situation and sequester 2,500 to 3,500 million tonnes of carbon over a 25- to 50-year period.

Lal et al. (1999) point to four possible ways that domestic agriculture can be used to reduce atmospheric carbon. First, soil erosion can be slowed to

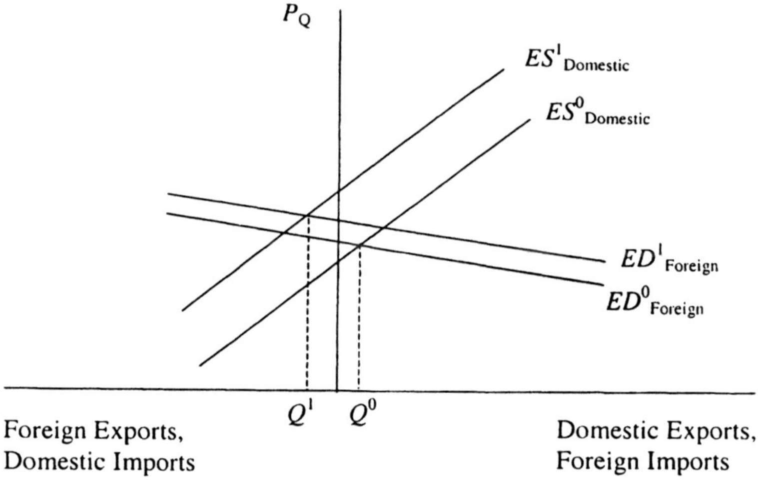

Figure 21.2 A Pigouvian Carbon Tax Can Alter Agricultural Trade Flows

reduce the amount of carbon lost into the atmosphere. Second, fragile or economically marginal land can be converted to permanent vegetation. Third, farmers can produce bio-fuels that would displace some fossil fuels, recycling carbon already in the atmosphere. Fourth, agricultural production could be concentrated on prime agricultural land. Under the last alternative, no-till or low-till technologies could be used to sequester carbon more effectively. Further arguments can be made that organic fertilizer use, winter crop rotation, and summer fallow elimination would increase the level of carbon sequestered on each acre of farmland.

For the purposes of this chapter, we are primarily interested in the second alternative listed above: winter crop rotation. While the first and fourth alternatives may provide significant reductions in atmospheric carbon, it is unlikely that these alternatives will lead to direct compensation to farmers. It is more likely that these alternatives will be coupled with tax breaks to offset the carbon tax. Similarly, the production of bio-fuels raises a host of market issues beyond the scope of the current analysis.

The remainder of this chapter analyzes the potential economic consequences of a new agricultural reserve, created for the purpose of sequestering carbon, similar to the Conservation Reserve Program (CRP) in the United States. In our formulation, a farmer would enter into a long-term contract with the government to convert currently tilled farmland to a permanent cover crop for a stated period. This policy would be a short- to intermediate-term strategy for the reduction of carbon. As outlined in Jorgenson and Wilcoxen (1999), we anticipate that the cost of the carbon tax coupled with increased scarcity of carbon-based fuels will lead to technological innovations that reduce the emission of carbon over time. Agriculture may provide a means of lessening the economic cost of carbon reduction in the short- to intermediate-term.

3. CARBON REMEDIATION WITHOUT CARBON TRADE

Under the Kyoto Accord, each signatory country is assigned greenhouse gas emission targets to be met by 2008 to 2012. Following the above discussion, we assume that each country will adopt a carbon tax to help meet the assigned targets. However, we also assume that each government augments the carbon tax by creating a domestic market for carbon sinks. In this section of our chapter, international carbon trade is not allowed.

3.1 The Domestic Market for Carbon Remediation

Under the Kyoto Accord, the United States has agreed to reduce carbon emissions to 1,225 million tonnes of carbon by 2008 to 2012 (Table 21.1). Given that 1996 emissions levels were 1,447 million tonnes, the United States must reduce carbon emissions by 222 million tonnes. In the absence of international carbon trade, the United States has several ways to achieve this reduction, including improved industrial efficiency and the sequestration of carbon through agriculture and forestry (see Chakrovorty, Raomasset, and Tse, 1997, for a discussion of alternative technologies). We interpret emission reductions from improved industrial efficiency and carbon sequestration in agriculture or forestry as the production of carbon sequestration, or remediation, R (Figure 21.3). Total supply of carbon remediation available in the United States is, S_T^{US}, as the sum of remediation available from industry adopting new technology, S_I^{US}, and the supply of carbon dioxide sequestra tion available from forestry and agriculture, S_{Ag}^{US}. The demand for carbon reduction, D^{US}, is equal to the fixed level of reduction that is required in the clears at carbon remediation price P_R^0. At this price, industry contributes R_I units of carbon remediation, while agriculture contributes R_{Ag}.

We focus next on the farm-level supply of carbon remediation. Figure 21.4 depicts isoprofit contours and a stylized acreage constraint for two agricultural activities, wheat production and grass production. Assuming competitive markets, the shapes of the isoprofit contours imply that marginal products are decreasing. We assume that conversion of wheat acreage, W, to grass acreage, G, results in the net removal of carbon from the atmosphere.

Mathematically, the farmer's profit-maximizing choice of production activities can be depicted as:

Table 21.1 Carbon Limits Under the Kyoto Accord, Agricultural Land Available for Sequestration, and Value of Agricultural Output

	Kyoto Accord target (per-cent of 1990 emissions)[a]	*Excess carbon production (1,000 tonnes)*[b]	*1991 Arable land (1,000 hectares)*[c]	*Value of agricultural output (U.S. $ /hectare)*[c]
Australia	108	5,279	48,991	136
Austria	92	1,774	1,533	114
Belgium and Luxembourg	92	4,320	819	1,454
Bulgaria	92	-3,832	4,146	562
Canada	94	6,633	45,960	206
Denmark	92	2,701	2,555	722
Finland	92	3,326	2,453	294
France and Monaco	92	10,068	19,126	854
Germany	92	-11,156	7,478	1,030
Greece	92	3,873	3,910	1,091
Hungary	94	-219	5,287	722
Iceland	110	-7	8	125
Ireland	92	2,033	3,171	121
Italy	92	9,904	11,967	1,105
Japan	94	44,007	4,637	1,723
Netherlands	92	7,474	905	2,488
New Zealand	100	1,680	412	119
Norway	101	5,150	882	315
Poland	94	8,202	14,759	738
Portugal	92	2,454	3,167	459
Romania	92	-6,382	10,053	511
Russia	100	-102,327	229,930	254
Spain	92	9,292	20,324	561
Sweden	92	2,589	2,853	422
Switzerland	92	1,359	412	1,396
United Kingdom	92	10,580	6,738	846
United States	93	222,349	187,776	394

[a]*Kyoto Protocol to the United Nations Framework Convention on Climate Change.*
[b]*Excess carbon production is derived by subtracting the percentage of carbon emissions allowed for each country under the Kyoto Accord times the estimated carbon emissions in 1990 from the estimated carbon emissions in 1996. Estimated carbon emissions are taken from Marland, Boden and Brenkert (2000).*
[c]*United States Department of Agriculture, Economic Research Service (1994).*

$$\text{(1)} \quad \max_{W,G} \quad \pi = \pi_W\left(P_W, W\right)\cdot W + \pi_G\left(P_G, P_R, G\right)\cdot G$$
$$\textit{subject to:} \quad L = W + G$$

where $\pi_W\left(P_W, W\right)$ is profit per hectare of wheat as a function of the price of wheat, P_W, $\pi_G\left(P_G, P_R, G\right)$ is profit per hectare of grass as a function of grass

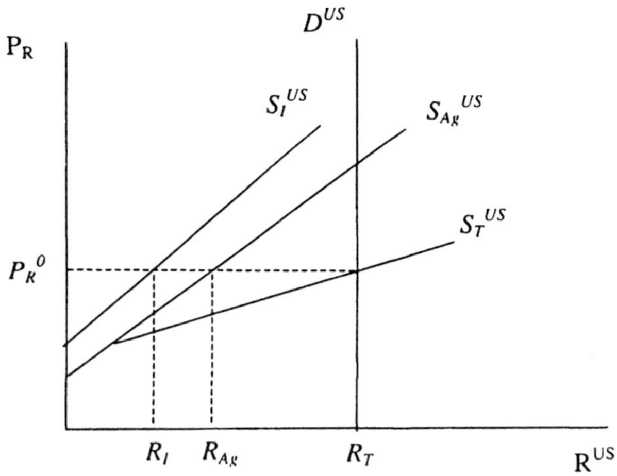

Figure 21.3 Carbon Remediation Equilibrium in the United States without Trade

price, P_G, and carbon remediation price, P_R. L is the acreage constraint. The farmer's profit-maximizing levels of W and G equate the slope of the iso-profit contour (that is, the ratio of marginal profits) with the slope of the acreage constraint (equal to –1):

$$\frac{\partial \pi / \partial G}{\partial \pi / \partial W} = -1. \qquad (2)$$

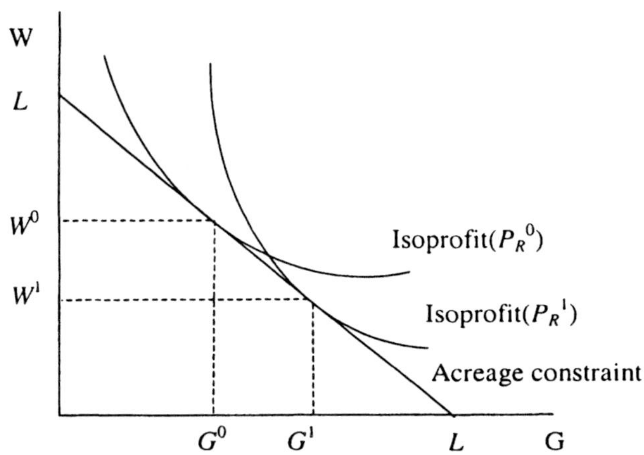

Figure 21.4 Allocation Between Wheat and Grass as the Carbon Price Increases

An increase in carbon remediation price from P_R^0 to P_R^1 shifts acreage from wheat production to grass (Figure 21.4). If the price of wheat declines, the tangency depicted in Figure 21.4 shifts toward more production of grass, and the supply of agricultural carbon remediation, depicted in Figure 21.3, shifts outward.

The substitution of grass for wheat could be increased by the imposition of a carbon tax on fuel. Although a carbon tax would increase the cost of producing annual crops such as wheat, for permanent cover crops like grasses, additional fuel would not be required after the crop is established. Figure 21.5 depicts the combined effect of a carbon tax on fuel and a demand for carbon sequestration on the market for wheat. A carbon tax shifts the supply of wheat upward from S_W^0 to S_W^1. At the same time, the carbon tax shifts consumers' demand curve for wheat inward from D_W^0 to D_W^1. Given that the demand for wheat is generally inelastic in price and income, the price of wheat, P_W, will probably increase. Nonetheless, agriculture will likely suffer a net economic loss.

However, given that agriculture can provide an economically viable carbon sink, the demand for agricultural carbon sequestration activities shifts outward. The right-hand panel of Figure 21.5 depicts the market for agricultural carbon sequestration. We assume that before the Kyoto Accord, the country possesses a relatively small demand for carbon reduction, D_{Ag}^0, due to the free-rider problem implicit in global warming. Figure 21.5 presents the demand as a downward sloping function of price. Implicitly, this assumes that other technologies may be available to reduce carbon emissions or to

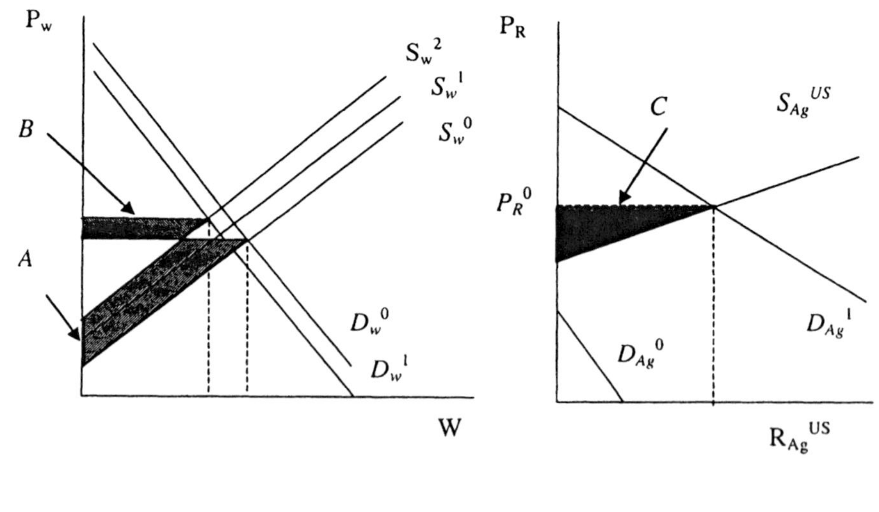

Figure 21.5 Effect of Carbon Tax and Sequestration on Domestic Wheat Market

sequester carbon. After the Kyoto Accord, the demand for carbon sequestration shifted from D^0_{Ag} to D^1_{Ag} as each country agreed to meet stated carbon limits. As a result, the equilibrium price for agricultural carbon remediation increased from zero to P^0_R.

The emergence of carbon sequestration as a profitable agricultural activity shifts resources (land) from the production of wheat into carbon sequestering grass production (Figure 21.4). The reallocation of land into grass production results in a further decrease in wheat supply to S^2_W, which implies a further increase in wheat price (Figure 21.5). Farmers lose economic rents equivalent to area *A* minus area *B* in fuel taxes and reduced wheat sales, but gain area *C* in carbon remediation payments.

A rigorous parameterization of this model is beyond the scope of the current analysis. An important determinant of net agricultural rents is the quality of land used to sequester carbon. If marginal land is used to sequester carbon, the reduction in wheat marketed may be small. In this case, agriculture would profit by the rent from carbon sequestration less the direct cost of the carbon tax. Furthermore, it may be possible to reduce the cost of the carbon tax through tillage alternatives.

3.2 Cross-country Divergence of Carbon Price Under Autarky

An important implication of the no-trade scenario is the divergence in the price of carbon sequestration across countries under autarky (Figure 21.6). The left panel depicts the no-trade equilibrium in Russia, while the right

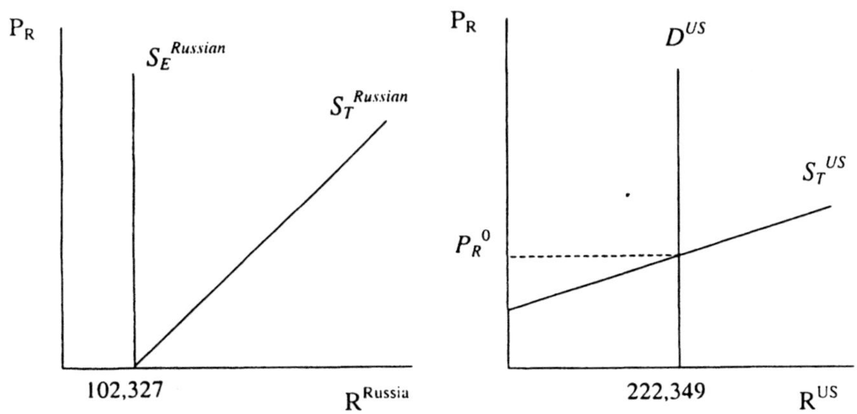

Figure 21.6 Carbon Sequestration Markets under Autarky

panel depicts the no-trade equilibrium in the United States (note that under autarky we assume perfectly inelastic demand for carbon sequestration at the national level). Under the Kyoto Accord, Russia is allowed to emit up to 108 percent of its 1990 carbon emissions level. This fact, along with the overall decline in economic activity in the country over the past decade, results in an excess carbon credit of 102 million tonnes of carbon (Table 21.1). Figure 21.6 represents this excess carbon credit as a fixed supply of carbon, $S_E^{Russian}$. Adding the potential supply of carbon sequestration from Russian agriculture results in total Russian carbon supply, $S_T^{Russian}$. On the other hand, the demand for carbon remediation in the United States is 222 million tonnes. Without trade, a positive price for carbon remediation will only exist in the United States. Similar conclusions about the divergence of carbon taxes across countries have been reached by other researchers, such as Poterba (1993).

4. CARBON REMEDIATION WITH CARBON PERMIT TRADE

Figure 21.7 depicts the derivation of the United States' excess demand for carbon sequestration on the world market. U.S. excess demand, ED^{US}, is simply the fixed level of U.S. carbon reduction requirements, D^{US}, less the supply of carbon reduction from agricultural and industrial sources located

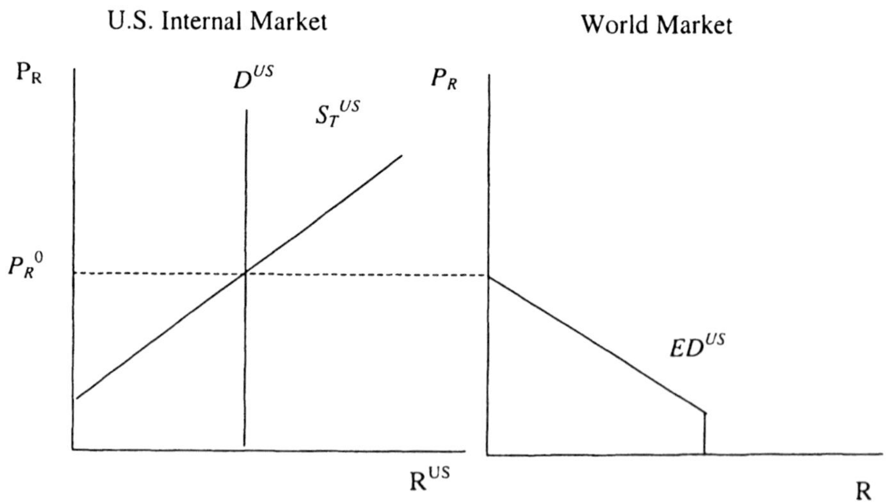

Figure 21.7 Derivation of U.S. Excess Demand for Carbon Remediation

within the United States. Due to upward-sloping remediation supply in the United States, the U.S. excess demand curve for remediation is downward sloping, despite perfectly inelastic demand within the United States. Russian excess supply of carbon remediation on the world market, ES^{Russia}, can be developed in a similar manner.

The excess demand curve for carbon reduction in the United States and the Russian excess supply curve for carbon are combined to depict the world market in carbon sequestration permits (Figure 21.8). Russian excess supply equals excess demand from the United States at an equilibrium carbon permit price of P_R^*. At this price, R^* units of carbon are traded. Without trade, the economic cost of carbon reduction in the United States is equal to area $J + K + L$. However, under the tradable permits regime, the cost of carbon reduction in the United States is simply L, with the additional cost in Russia of N. Moving from autarky to trade, U.S. producers of carbon sequestration lose M, while Russian producers gain O units of surplus. Trading carbon permits results in a lower world cost of carbon remediation since $J + K$ is less than N. Introduction of tradable carbon permits leads to an overall reduction in the level of carbon sequestration compared to the no-trade scenario. This result follows from the initial allocation of emissions targets across countries under the Kyoto agreement.

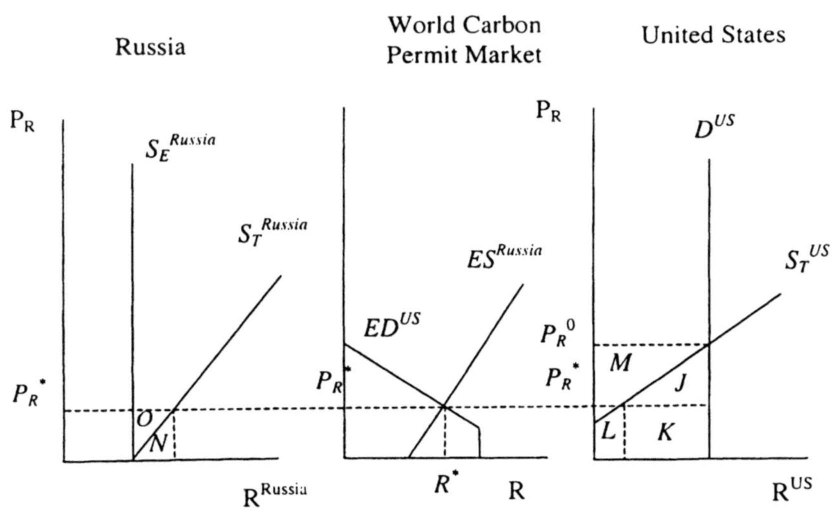

Figure 21.8 International Trade in Carbon Permits

5. EMPIRICAL EVIDENCE

The Kyoto Accord restricts carbon emissions to some fraction of 1990 levels, typically 92 percent (Table 21.1). Given that signatory countries will require 174.7 million tonnes in additional carbon remediation (beyond the Russian surplus supply) to meet Kyoto Accord obligations, tradable carbon permits will have a positive impact on international prices.

What is the potential significance of agriculture in an international market for carbon permits? One approach to this question is to determine the relationship between carbon permit price and the quantity of carbon sequestered by agriculture. Table 21.1 presents arable hectares (excluding permanent pasture), and mean, per-hectare returns to traditional agricultural activities by country. We assume that per-hectare carbon sequestration payments (carbon-permit price multiplied by carbon-removal rate) above mean per-hectare returns to traditional activities would shift 25 percent of arable hectares into carbon sequestering activities. Agricultural sequestration is assumed to remove 10.43 tonnes of carbon per hectare; potential carbon sequestration by country is presented in Table 21.2. Table 21.3 presents results for an alternative carbon removal rate of 4.22 tonnes per hectare.

World agricultural supply of carbon remediation is illustrated in Figure 21.9 for a carbon-removal rate of 10.43 tonnes per hectare. If agricultural sequestration alone is used to meet the world requirement of 174.7 million tonnes of carbon removal, the equilibrium price of carbon permits is U.S. $106.82 per tonne. At this price, all countries except Switzerland, Belgium, Japan, and the Netherlands will devote up to 25 percent of their tillable acreage to carbon sequestration. However, under the assumption that each hectare of agricultural sequestration removes only 4.22 tonnes of carbon, agriculture alone cannot remove an adequate quantity of carbon to meet Kyoto Accord obligations.

6. DISCUSSION AND IMPLICATIONS

The Kyoto Accord on climate change requires that signatory countries reduce carbon-dioxide emissions. Most economists anticipate that carbon taxes and other market mechanisms will be used to reduce the cost of achieving the required emissions reductions. While there is debate, scientific evidence supports the ability of agricultural activities to sequester carbon. This sequestration would come in many forms, including the adoption of conservation tillage, and the retirement of tillable farmland through a carbon reserve. Highlighting the latter approach, this chapter develops an economic

Table 21.2 Agricultural Carbon Sequestration at 10.43 tonnes per Hectare

	Arable land (hectares)	*Carbon sequestration supplied (1,000 tonnes)*	*Total supply of sequestration (1,000 tonnes)*	*Price of sequestration (U.S. $/tonnes)*
Free Resource		123,923	123,923	0.00
Austria	1,533	3,965	127,888	11.05
New Zealand	412	1,066	128,953	11.52
Ireland	3,171	8,201	137,154	11.66
Iceland	8	21	137,175	12.08
Australia	48,991	126,701	263,876	13.15
Canada	45,960	118,862	382,738	19.91
Russia	229,930	594,647	977,384	24.55
Finland	2,453	6,344	983,728	28.42
Norway	882	2,281	986,009	30.45
United States	187,776	485,628	1,471,637	38.09
Sweden	2,853	7,378	1,479,015	40.79
Portugal	3,167	8,191	1,487,206	44.37
Romania	10,053	25,999	1,513,205	49.40
Spain	20,324	52,562	1,565,767	54.23
Bulgaria	4,146	10,722	1,576,489	54.33
Denmark	2,555	6,608	1,583,097	69.79
Hungary	5,287	13,673	1,596,771	69.79
Poland	14,759	38,170	1,634,940	71.34
United Kingdom	6,738	17,426	1,652,366	81.78
France and Monaco	19,126	49,464	1,701,830	82.55
Germany	7,478	19,340	1,721,170	99.57
Greece	3,910	10,112	1,731,282	105.46
Italy	11,967	30,949	1,762,231	106.82
Switzerland	412	1,066	1,763,296	134.95
Belgium and Luxembourg	819	2,118	1,765,414	140.55
Japan	4,637	11,992	1,777,407	166.56
Netherlands	905	2,341	1,779,747	240.51

Source: USDA/ERS (1994), and authors' calculations.

model of the choice between raising wheat and allocating acreage to carbon sequestration.

As a major user of fossil fuels, agriculture will be affected by a carbon tax. The increased price of fossil fuels will make the production of commercial crops relatively more expensive, increasing the relative profitability of carbon sequestration. One implication of a carbon tax will be an increased supply of land available for a carbon reserve.

Additionally, changes in traditional agricultural trade flows may occur as developed countries implement higher carbon taxes to achieve more stringent reduction requirements. Our study illustrates the connection between international trade and agricultural carbon sequestration. Under the Kyoto

Table 21.3 Agricultural Carbon Sequestration at 4.22 tonnes per Hectare

	1991 Arable land (hectares)	*Carbon sequestration supplied (1,000 tonnes)*	*Total supply of sequestration (1,000 tonnes)*	*Price of sequestration (U.S. $/tonnes)*
Free Resource		123,923	123,923	0.00
Austria	1,533	1,619	125,542	27.06
New Zealand	412	435	125,977	28.22
Ireland	3,171	3,348	129,325	28.55
Iceland	8	8	129,334	29.59
Australia	48,991	51,731	181,064	32.20
Canada	45,960	48,530	229,595	48.77
Russia	229,930	242,789	472,384	60.14
Finland	2,453	2,590	474,974	69.61
Norway	882	931	475,905	74.58
United States	187,776	198,277	674,183	93.28
Sweden	2,853	3,013	677,195	99.91
Portugal	3,167	3,344	680,539	108.67
Romania	10,053	10,615	691,154	120.98
Spain	20,324	21,461	712,615	132.82
Bulgaria	4,146	4,378	716,993	133.06
Denmark	2,555	2,698	719,691	170.94
Hungary	5,287	5,583	725,273	170.94
Poland	14,759	15,584	740,858	174.73
United Kingdom	6,738	7,115	747,973	200.30
France and Monaco	19,126	20,196	768,168	202.19
Germany	7,478	7,896	776,064	243.86
Greece	3,910	4,129	780,193	258.30
Italy	11,967	12,636	792,829	261.62
Switzerland	412	435	793,264	330.52
Belgium and Luxembourg	819	865	794,129	344.25
Japan	4,637	4,896	799,026	407.94
Netherlands	905	956	799,981	589.06

Source: USDA/ERS (1994), and authors' calculations.

Accord, developing countries are required to make proportionately smaller reductions in carbon emissions than are developed nations. The divergence of carbon prices across countries under autarky implies that tradable carbon credits would reduce the economic cost of meeting the Kyoto carbon standards. Under plausible assumptions, we derive the supply curve of carbon sequestration by agriculture on the world market. Results suggest that Russia will be a dominant player in a world market for agricultural carbon credits.

The limited scope of this study requires several caveats. First, we have not attempted to reconcile alternative estimates of agriculture's technical ability to sequester carbon (for example, Lal et al. 1999; Luciuk et al., 1999; and Schmitz et al., 1994). Second, the analysis considers neither the empirical relationship between agricultural and forest sequestration of carbon nor the

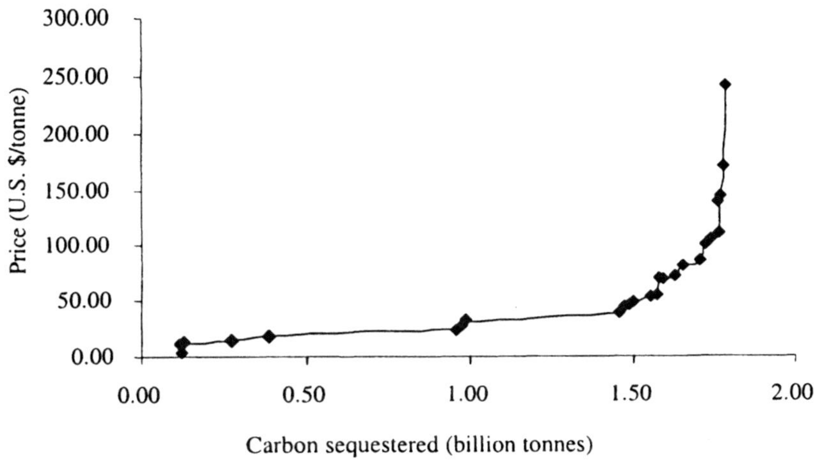

Figure 21.9 World Agricultural Supply of Carbon Remediation

adoption of new technology by industry. For example, Osborne (1999) found that for forest management in Guyana, U.S. $2.50 per tonne of carbon is required for carbon sequestration to compete with commercial logging. At the estimated equilibrium price of U.S. $106 per tonne of carbon, the supply of sequestration from forestry may be sufficient to satisfy the carbon targets. Finally, we do not address the potential role of less-developed countries in a world carbon market. However, participation by these countries may be limited due to food-security issues. It may be difficult for these countries to reduce food-crop plantings for added carbon sequestration in the midst of food scarcity.

7. REFERENCES

Chakrovorty, U., J. Roumasset, and K. Tse. 1997. "Endogenous Substitution among Energy Resources and Global Warming." *Journal of Political Economy* 105(6): 1201–1234.

Cline, W.R. 1992. *The Economics of Global Warming*. Washington, DC: Institute of International Economics.

Fankhauser, S. 1993. "The Economic Costs of Global Warming: Some Monetary Estimates," in Y. Kaya, N. Nakicenovic, W.D. Nordhaus, and F.L. Toth, eds., pp. 83–105. *Costs, Impacts and Benefits of CO_2 Mitigation*. International Institute for Applied Systems Analysis CP-93-2, Laxenburg, Austria.

IPCC (Intergovernmental Panel on Climate Change). 1996. *IPCC Second Assessment: Climate Change 1995*. R. Watson, M. Zinyowera, R. Moss and D. Dokken, eds, pp. 4–10. Cambridge, MA: Cambridge University Press.

Jorgenson, D.W. and P.J. Wilcoxen. 1999."Energy, the Environment, and Economic Growth," in A.V. Knesse and J.L. Seeney, eds, 1267–1349. *Handbook of Natural Resource and Energy Economics, Volume 3*. Amsterdam, Netherlands: North-Holland Publishers.

Kyoto Protocol to the United Nations Framework Convention on Climate Change, Article 3, Annex B (United Nations, New York, 1997). Internet website: http://www.unfccc.de

Lal, R., R.F. Follett, J. Kimble, and C.V. Cole. 1999. "Managing U.S. Cropland to Sequester Carbon in Soil." *Journal of Soil and Water Conservation* 54: 374–381.

Luciuk, G.M., M.A. Bonneau, D.M. Boyle, and E. Viberg. 1999."Carbon Sequestration-Additional Environmental Benefits of Forages in the PFRA Permanent Cover Progam." Mimeographed Manuscript No. 1967, PFRA, Agriculture and Agri-Food Canada, Ottawa.

Marland, G., T.A. Boden, and A.L. Brenkert. 2000. "Global, Regional, and Natural Fossil Fuel CO_2 Emissions," in *Trends: A Compendium of Data on Global Change*. Environmental Sciences Division, Carbon Dioxide Information Analysis Center, Oak Ridge National Laboratory, Oak Ridge, TN.

Mendelsohn, R., W.D. Nordhaus, and D. Shaw. 1994. "The Impact of Global Warming on Agriculture: A Ricardian Analysis." *American Economic Review* 84(4): 753–771.

Nordhaus, W.D. 1994. *Managing the Global Commons: The Economics of Climate Change*. Cambridge, MA: MIT Press.

Nordhaus, W.D. 1991. "A Sketch of the Economics of the Greenhouse Effect." *American Economic Review* 81(2)(May): 146–150.

Pearce, D. 1991. "The Role of Carbon Taxes in Adjusting to Global Warming." *Economic Journal* 101(July): 938–948.

Poterba, J.M. 1993. "Global Warming Policy: A Public Finance Perspective." *Journal of Economic Perspectives* 7(4): 47–63.

Osborne, T. 1999. "Carbon Offsets as an Alternative to Large-scale Logging: A Case Study in Guyana's Forest." Unpublished M.S. Thesis, University of Florida, Gainesville FL.

Rosenzweig, C. and M. Parry. 1994. "Potential Impact of Climate Change on World Food Supply." *Nature* 367:133–138.

Schmitz, A., H. Furtan, P. Kuch, and A. Guzel. 1994. "North American Agriculture: Its Contribution to the Greenhouse Effect." *International Journal of Environment and Pollution* 4(1/2): 75–96.

USDA/ERS (United States Department of Agriculture, Economic Research Service). 1994. *World Agriculture: Trends & Indicators, 1961–1991*. Washington, D.C. Internet website: http://usda.mannlib.cornell.edu/data-sets/international/89024

PART V

TRADE AND BIOTECHNOLOGY

Chapter 22

The Timing of Evaluation of Genebank Materials and the Effects of Biotechnology

Bonwoo Koo and Brian D. Wright
International Food Policy Research Institute and University of California at Berkeley

1. INTRODUCTION

Plant genetic resources have provided basic building blocks for the improvement of plant varieties, and recent advances in biotechnology have opened up more possibilities for the use of these resources in crop improvement. Collections of germplasm (the "material that controls heredity," Witt, 1985: 8) for conservation purposes in *ex situ* genebanks have been greatly expanded over the past decades. However, materials in genebanks are not used extensively by plant breeders (Wright, 1997). One frequently cited obstacle to greater utilization is the lack of information useful to breeders regarding the samples of germplasm held as accessions in *ex situ* collections.

Evaluation data[i] are of the greatest value to plant breeders seeking particular genetic traits for crop improvement, but only a small fraction of samples in the genebank network has evaluation data.[ii] Even among the collections with evaluation data, the information currently provided by genebanks is often perceived by plant breeders as inadequate. While this situation has led many plant breeders to demand more complete evaluation of genebank materials for crop improvement (Duvick, 1984; Goodman, 1990; NRC, 1993; Peeters and Galwey, 1988), this viewpoint is not universally shared (see for example Frankel, 1989).

Existing studies on the use and management of collections of genetic resources deal largely with the optimal size of *ex situ* genebank collections (Brown, 1989; Chang, 1989), the value of genetic resources (Evenson, 1996; Simpson, Sedjo, and Reid, 1996), or the optimal search strategy for useful traits (Gollin, Smale, and Skovmand, 2000). However, the neglected issue of the timing of evaluation and utilization of genebank materials has become an important consideration for genebank managers, especially in the current rapidly changing technological environment.

The timing of identifying and isolating useful traits has been an important factor in crop improvement. For example, when the Russian wheat aphid began to affect the United States in the late 1980s, the damage might have been mitigated if the sources of resistance had already been identified. Searches for the resistance trait took several years, and lack of preparation for this disease might have contributed to the estimated economic damage of U.S. $670 million by 1991 (Russian Wheat Aphid Task Force, 1991). On the other hand, when barley stripe rust fungus devastated barley crops in South America after its arrival from Europe in 1975, plant breeders in the United States worked to identify sources of resistance to the disease and were already breeding varieties resistant to it when it reached the United States in 1991 (Leonard, 1999).

In this chapter, we focus on cases similar to those just mentioned, in which a trait for resistance to a pest or disease is already known but has not yet infested the relevant crop. We address the following questions on evaluation and utilization of genebank collections. When is it optimal to evaluate genebank material for a trait, that is, *ex ante* in anticipation of an infestation, or *ex post* after infestation of the disease in the relevant crop? When is it optimal to develop a new variety incorporating the trait? What is the effect of advances in biotechnology on the timing of evaluation and development? And how is the timing of evaluation and development changed when there is a possibility of a further technological breakthrough in either of these functions in the future?

Resistance is often achieved by transferring resistance traits found either in genebank collections or breeders' genetic stocks into existing varieties, by conventional breeding or other means such as wide crossing or genetic transformation. Identification of resistance before infestation of a disease can be difficult and costly. If infestation of a disease occurs, the development of a new disease resistant variety will be faster if the gene for the resistance trait has already been located. However, there is some chance that the disease never occurs or occurs in the remote future, in which case the money spent for evaluation is, in hindsight (*ex post*), wasted. If evaluation is initiated after the disease occurs, excess *ex ante* evaluation is avoided. However, social losses due to damages from the disease will be incurred during the delay until a new variety is developed.

In a simple model of evaluation and development, we find that if a disease has sufficiently low likelihood of occurrence, it is cost effective to delay evaluation of the trait until the disease occurs. In addition, the benefit from *ex ante* evaluation turns out to be the largest when the hazard rate of the disease is at an intermediate, rather than at the maximum, level. Several recent innovations in tools and methods have made evaluation for resistance traits and development of useful cultivars incorporating these traits cheaper and faster.[iii] Such a cheaper and faster response might seem to favor *ex post* evaluation. But we show that in economic terms advances in biotechnology that reduce the cost or increase the speed of evaluation or development tend to favor *ex ante* evaluation more than *ex post*.

Current advances in biotechnology imply that further innovations are likely. A well-known result of real option theory is that uncertainty about costs tends to favor delay in an investment (Arrow and Fisher, 1974; Dixit and Pindyck, 1994; McDonald and Siegel, 1986). Does this imply that a further anticipated breakthrough in the technology of evaluation and development tends to favor delay, in contrast to the effect of a current breakthrough? The usual real option effect is observed in our model for the case of innovation in evaluation. But we show that anticipation of the possibility of a technological breakthrough in development may speed up, rather than delay, the timing of evaluation and development. We conclude that both the level and rate of technological progress in development may justify advancing timing of the evaluation of genebank accessions as well as the development of cultivars incorporating newly identified traits.

2. THE MODEL OF EVALUATION AND DEVELOPMENT

2.1 Assumptions

We propose a highly stylized model that abstracts from many biological details. We consider the search for a single qualitative trait[iv] (say, a resistance to a certain disease) for a specific environment from a given *ex situ* collection of germplasm, either *ex ante* or *ex post*.[v] For simplicity, we assume resistance is conferred by a single gene, and (contrary to common experience) that single-gene resistance retains its effectiveness forever. The evaluation process ceases as soon as a variety with the targeted gene is identified.[vi]

We postulate a two-stage process for the development of a new disease-resistant variety: the evaluation (or search) process in which a genebank conducts a search in its collection for a disease resistant trait, and the development process in which a genebank develops a new variety, expressing the evaluated gene, for release to farmers.[vii] We assume that the search process must be completed prior to the initiation of the development process.

A disease breaks out at a random future time t, according to a Poisson disease-hazard rate λ. By spending a constant flow search cost c,[viii] the genebank finds the targeted gene at time s after the initiation of evaluation, according to a Poisson discovery rate ϕ. Once a disease occurs, society suffers losses from the disease until the development of a new disease resistant variety.[ix] It takes l periods with a constant development cost m per period to develop a new variety,[x] and the disease cost due to the delayed introduction of a new variety is d per period of infestation. Diffusion of the disease and of the new variety after development are, for simplicity, assumed immediate. The genebank is risk neutral and the discount rate is r.

Figure 22.1 shows the decision tree of a genebank manager on the timing of evaluation and development.[xi] For simplicity, we ignore the option of "no action" by assuming that the cost from a disease is high enough to justify the development of a new variety once a disease occurs. The first decision that a genebank manager should make at time zero, prior to an infestation, concerns the timing of evaluation; that is, whether to evaluate immediately (*ex ante*) or delay evaluation till the infestation of a disease (*ex post*). If he

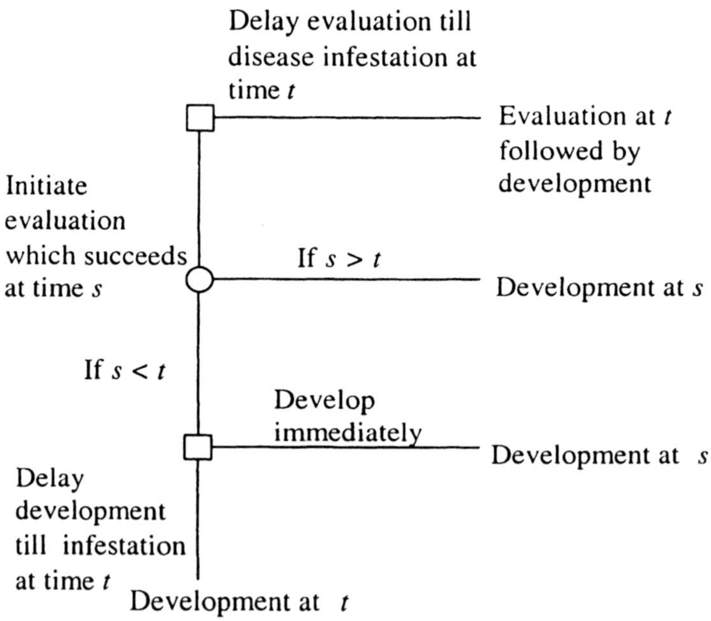

Figure 22.1 Decisions on the Timing of Evaluation and Development

chooses to delay evaluation, evaluation and development should (by assumption) follow in sequence immediately after infestation. If he chooses *ex ante* evaluation at time zero and the search for the trait succeeds at time s, he must decide the timing of development if the infestation has not yet occurred ($s < t$). His decision then is whether to develop immediately, before the disease infestation, or to delay till infestation occurs. Therefore, we consider three alternatives for the timing of evaluation and development: (1) *ex post* evaluation/development (P_t^t), (2) *ex ante* evaluation/*ex post* development (A_0^t), and (3) *ex ante* evaluation/*ex ante* development (A_0^0).[xiii]

2.2 The Expected Costs of Evaluation and Development

2.2.1 *Ex post* Evaluation/Development

One option a genebank manager can choose is to delay evaluation (that is, the search for the gene) until infestation of a disease. This is the current standard practice. (Of course, if the disease is unknown prior to infestation, there is no alternative.) If the genebank starts to search for a resistant gene after the disease infestation, society will suffer losses during the evaluation period $[t, t + s]$ as well as during the development period $[t + s, t + s + l]$. The expected cost contingent on a search of duration s, evaluated at time zero, is:

$$C_1(s) = \int_0^{\infty} \left[\frac{c+d}{r}\left(1 - e^{-rs}\right) + \frac{(m+d)L}{r} e^{-rs} \right] \lambda e^{-(r+\lambda)t} dt$$

where $L \equiv 1 - e^{-rl}$. The expected cost of *ex post* evaluation/development is:

$$P_t^t = \int_0^{\infty} C_1(s)\phi e^{-\phi s} ds = \frac{c\lambda}{(r+\phi)(r+\lambda)} + \frac{mL\phi\lambda}{r(r+\phi)(r+\lambda)} + \frac{d(r+L\phi)\lambda}{r(r+\phi)(r+\lambda)}.$$

The terms on the right-hand side are the expected search cost, development cost, and disease cost, respectively. Because all of these costs are incurred only if the disease breaks out, the expected cost approaches zero as the disease-hazard rate λ approaches zero.

2.2.2 *Ex ante* Evaluation/*Ex post* Development

The genebank manager can also consider the option of starting before the infestation ("outbreak") of a disease in the relevant cropping area to search for a targeted gene that has resistance to the disease. If it takes time to evaluate a trait and the disease cost is sufficiently high, it can be optimal to initiate the evaluation process before the disease infestation occurs. By initiating the evaluation process in advance, development can begin earlier, reducing the time over which an infestation can cause damage. If it is optimal to start evaluation before a disease infestation, the Poisson assumptions of this model imply that it is optimal to start at time zero. An example is the anticipatory search for barley stripe rust resistance in germplasm relevant to the United States. When the search results in discovery *after* the disease occurs ($s > t$), development follows immediately by assumption. The expected cost when $s > t$, evaluated at time zero, is:

$$C_0(s) = \int_0^s \left[\frac{c}{r}\left(1 - e^{-rs}\right) + \frac{d}{r}\left(e^{-rt} - e^{-rs}\right) + \frac{(m+d)L}{r} e^{-rs} \right] \lambda e^{-\lambda t} dt .$$

If the gene is found *before* the disease occurs ($s < t$) and the genebank manager chooses to delay development till the disease infestation, the expected cost is:

$$C_2(s) = \int_s^\infty \left[\frac{c}{r}\left(1 - e^{-rs}\right) + \frac{(m+d)L}{r} e^{-rt} \right] \lambda e^{-\lambda t} dt .$$

The expected cost of *ex ante* evaluation/*ex post* development is derived using $C_0(s)$ and $C_2(s)$.

$$A_0' = \frac{c}{r+\phi} + \left[\frac{mL\phi\lambda}{r(r+\phi)(r+\lambda)} + \frac{mL\phi\lambda}{(r+\phi)(r+\lambda)(r+\phi+\lambda)} \right] + \left[\frac{d(r+L\phi)\lambda}{r(r+\phi)(r+\lambda)} - \frac{d(1-L)\phi\lambda}{(r+\phi)(r+\lambda)(r+\phi+\lambda)} \right]$$

Unlike the case of *ex post* evaluation/development, the expected search cost (the first term on the right-hand side) is incurred regardless of the disease infestation. The expected development cost is larger (in present value) than the cost under *ex post* evaluation/development because it begins earlier and so is less discounted. However, the expected disease cost is smaller because the resistant variety is, in expectation, developed earlier.

2.2.3 *Ex ante* Evaluation/*Ex ante* development

Another alternative is to develop a new variety even before the disease breaks out. There exists a practice among plant breeders called "anticipatory breeding" in which breeders try to incorporate resistance in cultivars for possible infestation of diseases. Examples of *ex ante* development are Australia's development of locally adapted cultivars resistant to Russian wheat aphids (Skovmand, 1999) and to wheat stem rust (McIntosh and Brown, 1997) in anticipation of infestations in the future. The expected cost of *ex ante* development when $s < t$ is:

$$C_3(s) = \int_s^{\infty}\left[\frac{c}{r}\left(1-e^{-rs}\right)+\frac{mL}{r}e^{-rs}\right]\lambda e^{-\lambda t}dt + \int_s^{s+l}\frac{d}{r}\left(e^{-rt}-e^{-r(s+l)}\right)\lambda e^{-\lambda t}dt$$

The expected cost of *ex ante* evaluation/*ex ante* development is derived using $C_0(s)$ and $C_3(s)$.

$$A_0^0 = \frac{c}{r+\phi}+\frac{mL\phi}{r(r+\phi)}+\left[\frac{d(r+L\phi)\lambda}{r(r+\phi)(r+\lambda)}-\frac{d(1-L)\phi[\lambda+(r+\phi)(1-e^{-\lambda l})]}{(r+\phi)(r+\lambda)(r+\phi+\lambda)}\right].$$

While the expected development cost is higher than for other alternatives, the expected disease cost is the lowest. If the disease cost d is sufficiently large or the likelihood of disease occurrence is high, *ex ante* development can be the most cost effective.

2.3 The Timing of Evaluation

Given *ex ante* evaluation at the first stage, the decision on the timing of development at the second stage is determined by comparing the expected cost of *ex ante* evaluation/*ex ante* development (A_0^0) with that of *ex ante* evaluation/*ex post* development (A_0^l). *Ex post* development is preferred given *ex ante* evaluation if A_0^l is less than A_0^0; that is, if condition (1) is satisfied.

(1) $mL > d(1-L)(1-e^{-\lambda l})$.

If *ex post* development is chosen at the second stage, the decision at the first stage is determined by considering the following cost difference function:

$$G \equiv P_t^t - A_0^t$$

$$(2) \quad = -\frac{cr}{(r+\phi)(r+\lambda)} - \frac{mL\phi\lambda}{(r+\phi)(r+\lambda)(r+\phi+\lambda)} + . \frac{d(1-L)\phi\lambda}{(r+\phi)(r+\lambda)(r+\phi+\lambda)}$$

If $G < 0$, *ex post* evaluation/development incurs lower expected cost and delaying evaluation is optimal. The first term on the right-hand side of (2) is the difference in the expected search cost, which is non-positive because the search cost is always incurred, and incurred earlier, under *ex ante* evaluation. The second term, the difference in the expected development cost, is also non-positive for a positive discount rate because development occurs earlier when evaluation is *ex ante*. The third term is the difference in the expected disease cost, which is non-negative due to the longer period of delay under *ex post* evaluation.

For a rare disease (small λ), the cost difference is negative and *ex post* evaluation is more cost effective. The expected development cost and disease cost under both alternatives vanish when the disease-hazard rate λ approaches zero, and only the expected search cost remains as an important consideration. As the disease-hazard rate becomes high, the cost difference function G approaches zero, implying that there is a vanishing difference in the expected costs. Because the disease is expected to occur very soon, the expected timing of the evaluation process as well as the development process is almost the same for both alternatives and thus the cost difference is negligible.

Proposition 1 (Optimal timing of evaluation and disease-hazard rate) Given ex post *development at the second stage,* ex post *evaluation is optimal if disease-hazard rate is below a critical level λ_1 and search cost is less than a critical level c_1.*

Proof: From equation (2), we can derive the range of the disease-hazard rate in which *ex post* evaluation is preferred.

$$(3) \quad 0 < \lambda < \lambda_1 \equiv \frac{cr(r+\phi)}{K\phi - cr}$$

where $K \equiv d(1 - L) - mL > 0$. If λ is less than the cutoff level λ_1, *ex post* evaluation brings lower expected cost. In addition, the search cost should be less than c_1 to make λ_1 positive.

$$(4) \quad c < c_1 \equiv \frac{K\phi}{r} \quad \text{Q.E.D.}$$

From the genebank manager's point of view, proposition 1 implies that it is better not to evaluate accessions *ex ante* for a gene (or trait) which is expected to be used rarely in the future. For a rare disease, the use of the evaluated resistant gene will be long delayed and the search cost incurred at the current time is large relative to the expected present value of the benefits captured at a later time. For a disease that is more likely to cause an infestation soon, *ex ante* evaluation may be preferred if it is cheap enough, because it reduces the expected disease cost. But the advantage of *ex ante* evaluation is not monotonic in the disease-hazard rate.

Proposition 2 (Ex ante evaluation and disease-hazard rate) The benefit from ex ante *evaluation is maximized when the disease-hazard rate is at an intermediate level.*

Proof: The first-order condition for maximization of function G is:

$$(5) \quad \frac{\partial G}{\partial \lambda} = \frac{-(K\phi - cr)\lambda^2 + 2cr(r+\phi)\lambda + r(r+\phi)[K\phi + c(r+\phi)]}{(r+\phi)(r+\lambda)^2(r+\phi+\lambda)^2} = 0 .$$

The rate λ^* which is associated with the maximum cost difference G (or, maximum net benefit from *ex ante* evaluation) is calculated as:

$$(6) \quad \lambda^* = \frac{cr(r+\phi) + \sqrt{J}}{K\phi - cr} > 0$$

where $J \equiv K(K+c)r(r+\phi)\phi^2 > 0$. By condition (4), only one root of λ that satisfies the condition of the above equation is positive and finite.[xiii] For *ex ante* evaluation to be optimal, the value of λ^* should be greater than λ_1. The second-order condition is satisfied around λ^*.

$$\frac{\partial^2 G}{\partial \lambda^2} = \frac{-2[(K\phi - cr)\lambda - cr(r+\phi)]}{(r+\phi)(r+\lambda)^2(r+\phi+\lambda)^2} < 0 . \quad \text{Q.E.D.}$$

If, in addition to the costs accounted for above, there is a fixed cost associated with the decision to evaluate *ex ante* (for example, in terms of the genebank manager's time), then the value λ^* gives guidance to a genebank manager in forming priorities regarding evaluation for resistance to various potential diseases. If a disease rarely occurs (that is, low λ), the advantage of early evaluation is less attractive as shown in proposition 1, and thus *ex post* evaluation is optimal. If a disease is expected to occur soon, on the other hand, the trait will likely be evaluated soon under either approach. Making the optimal decision is thus less crucial. *Ceteris paribus*, diseases with hazard rates near λ^* are the best candidates for *ex ante* evaluation. Therefore, if a genebank manager is to allocate scarce managerial resources to *ex ante* evaluation, he should pay more attention to potential diseases with intermediate hazard rates.

Figure 22.2 shows the graph of a typical cost difference as a function of the disease-hazard rate. The value λ_1 defined in (3) is the cutoff disease-hazard rate below which *ex post* evaluation is preferred. The size of the cost difference G for $\lambda > \lambda_1$ indicates the degree to which *ex ante* evaluation is preferred, that is, the larger the value, the better the *ex ante* evaluation approach in terms of cost effectiveness. The benefit from *ex ante* evaluation is maximized at λ^*, and then approaches zero as λ is further increased.

3. THE EFFECT OF ADVANCES IN BIOTECHNOLOGY

Recently, innovations in biotechnology have occurred that can reduce the cost of the search or development process and/or speed up these processes. Here we consider how such innovations affect the decision on the timing of evaluation and development. We identify two types of technological changes for our analysis. The first type includes the technologies which primarily affect the evaluation process such as molecular genetic techniques, while the other type includes those that mainly affect the development process, such as transformation technology, cell culture techniques, and the use of molecular markers to identify transformed cultivars in the breeding process.

3.1 Advances in Biotechnology Affecting the Evaluation Process

The effects of the changes in the flow search cost (c) and the discovery rate (ϕ) on the cutoff rate λ_1 are:

$$\text{(7a)} \quad \frac{\partial \lambda_1}{\partial c} = \frac{K\phi r(r+\phi)}{(K\phi - cr)^2} > 0$$

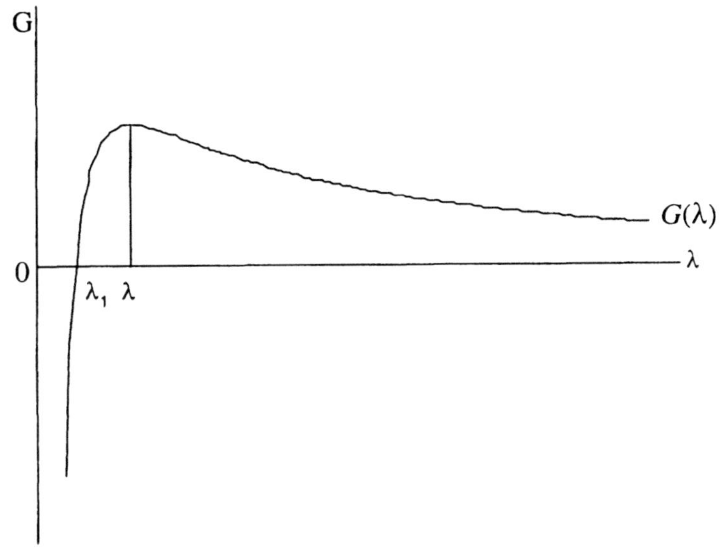

Figure 22.2 A Cost Difference Function *G*

$$\text{(7b)} \quad \frac{\partial \lambda_1}{\partial \phi} = -\frac{cr^2(c+K)}{(K\phi - cr)^2} < 0$$

A higher search cost c favors *ex post* evaluation [equation (7a)]. The search cost differs by the types of traits to be evaluated. For example, Peeters and Williams (1984) report that the per unit evaluation cost for *rhizoctonia* resistance in sugarbeet was U.S. $175, while the cost of evaluating nematode resistance was only U.S. $60 per unit. If the cost and probability of an outbreak of each disease were the same and if there were no economies of scope in evaluation (ignored here), then *ex ante* evaluation becomes relatively more attractive for nematode resistance.

A high discovery rate ϕ reduces the expected present value of the search cost more when evaluation is started earlier, and thus favors *ex ante* evaluation. However, it also favors *ex post* evaluation because it is less costly to delay evaluation when discovery is fast. Equation (7b) shows that the first effect dominates the second effect around the cutoff rate λ_1. If modern tools of biotechnology such as genetic-marker techniques and new genomic information reduce the time spent for the evaluation process of certain resistance traits sufficiently, without increasing the flow search cost, this should tend to favor *ex ante* evaluation of germplasm for such traits.

3.2 Advances in Biotechnology Affecting the Development Process

The effects of the changes in the development period (l) and the development cost (m) on the cutoff rate are:

$$\text{(8a)} \quad \frac{\partial \lambda_1}{\partial m} = \frac{crL\phi(r+\phi)}{(K\phi - cr)^2} > 0$$

$$\text{(8b)} \quad \frac{\partial \lambda_1}{\partial l} = \frac{cr^2\phi(r+\phi)(d+m)(1-L)}{(K\phi - cr)^2} > 0 .$$

A low development cost favors *ex ante* evaluation [equation (8a)]. Because timing of the development process under *ex ante* evaluation is earlier than that under *ex post* evaluation, a reduction in development cost has a larger impact when evaluation is made *ex ante.* Modern technologies which reduce the development cost will favor *ex ante* evaluation. On the other hand, if strict government regulations on the use of biotechnology increase the development cost and delay the introduction of a new variety, *ex post* evaluation will be more encouraged.

A shorter development period l decreases the expected development and disease costs under both alternatives. However, the rate of decrease in the expected costs under *ex ante* evaluation is higher than that under *ex post* evaluation, because the saved costs are less discounted for *ex ante* evaluation. Modern genetic engineering techniques which speed up the development process favor *ex ante* evaluation. We can summarize these implications of advances in biotechnology in the following proposition.

Proposition 3 (Optimal timing of evaluation and advances in biotechnology) Advances in biotechnology which reduce the cost of the search or development process and/or speed up these processes increase range of disease-hazard rates over which ex ante *evaluation dominates* ex post *evaluation.*

4. THE IMPLICATIONS OF DYNAMIC TECHNOLOGICAL CHANGE

The fast pace of innovations in biotechnology that has made the search for traits and the development of new varieties incorporating them cheaper and faster is likely to continue. Does anticipation of further advances in technology reinforce or offset the effects of cheaper and faster search and development on the timing of evaluation and development?

If a technological breakthrough could occur in the evaluation process, standard real option theory clearly implies that anticipation of this possibility should tend to delay evaluation (Dixit and Pindyck, 1994: 140). If the breakthrough possibility occurs in the development process, however, the effect is ambiguous because the benefit from the technological breakthrough can be realized only after a successful search process. In this section, we focus on this interesting case by assuming that a technological breakthrough occurs only in the development process, and ask whether the potential for such further technological changes tends to delay or advance the timing of evaluation.

4.1 The Expected Costs with a Technological Breakthrough

We now assume that a technological breakthrough reduces the flow development cost by x from m to n (that is, $n = m - x$). The breakthrough is assumed to be discrete; it happens only once in the future at time u with a Poisson breakthrough rate β. Thus, while the flow development cost at time zero is m, the cost at time t has fallen to n with probability $(1 - e^{-\beta t})$. Another simplifying assumption is that the breakthrough can be utilized only if it occurs before the initiation of the development process. That is, we rule out converting the development process to the new technology if a breakthrough occurs in the middle of the process.

Figure 22.3 illustrates the decision tree of a genebank manager when there exists an anticipated technological breakthrough in development. The genebank manager now has additional decision nodes at the time of a technological breakthrough, u. The decision at time zero is whether to evaluate immediately (*ex ante* evaluation), or to delay, either till a technological breakthrough ("post-breakthrough" evaluation) or till disease infestation ("post-infestation" evaluation). Given *ex ante* evaluation (that is, the left branch in Figure 22.3), if a disease breaks out before finishing the evaluation ($t < s$), development must follow immediately after evaluation, by assumption. If a technological breakthrough happens before finishing the evaluation and the disease has not yet broken out ($u < s < t$), he must decide whether to

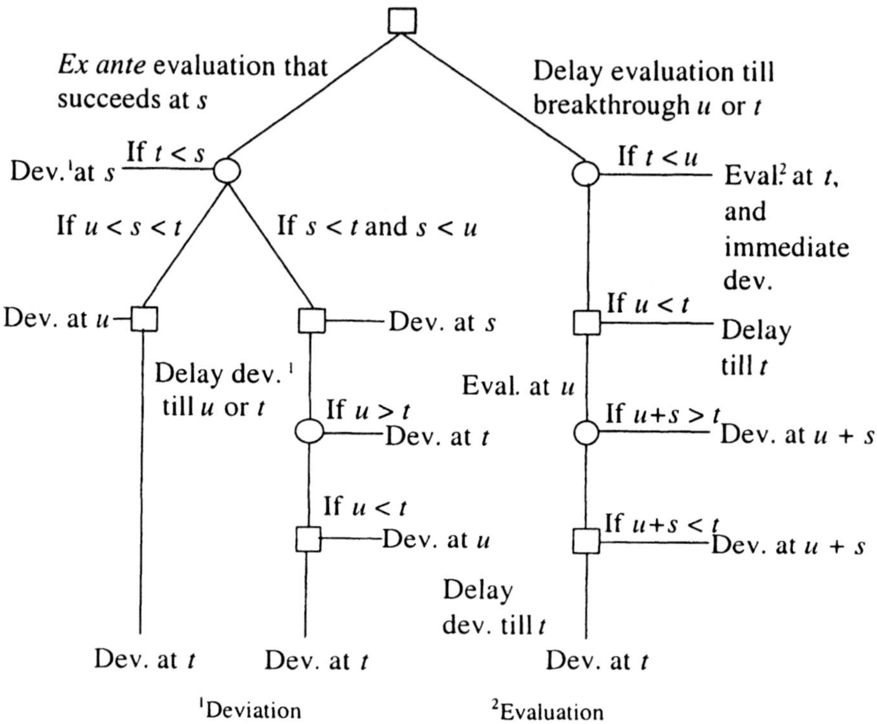

Figure 22.3 Decisions on Timing with a Technological Breakthrough

develop immediately, A_0^0, or to delay till disease infestation, A_0^t. If neither happens before finishing the evaluation ($s < t$ and $s < u$), he can (1) delay development until disease infestation regardless of the technological breakthrough, A_0^t, (2) delay till a technological breakthrough, A_0^u, or (3) develop immediately, A_0^0.

On the other hand, if the genebank manager decides to delay evaluation (that is, the right branch in Figure 22.3), his decision depends on the relative timing of infestation and breakthrough. If infestation occurs before a breakthrough ($t < u$), he begins evaluation immediately, followed by development. Otherwise, the decision environment is the same as the previous case without a breakthrough possibility except for the decrease in the size of development cost. One alternative is that the manager delays both evaluation and development until infestation even if the technological breakthrough has happened before the disease, in which case the expected cost is P_t^t. The second alternative is to evaluate after the breakthrough but to delay development till the disease infestation, P_u^t, and the third is to develop immediately after the evaluation, even if the disease infestation has not yet occurred, P_u^u. The expected costs of each alternative are summarized in Table 22.1.[xiv]

Table 22.1 Expected Costs with a Possibility of a Technological Breakthrough

Schemes	Search cost	Development cost	Disease cost
A_0^t	S	$N-N_1+X-X_1$	$D-D_0$
A_0^u	S	$N-N_2+X-X_1$	$D-D_0-D_1+D_2$
A_0^0	S	$N+X$	$D-D_0-D_1$
P_t^t	$S\left(\frac{\lambda}{r+\lambda}\right)$	$N\left(\frac{\lambda}{r+\lambda}\right)+X_2$	D
P_u^t	$S\left(\frac{\beta+\lambda}{r+\beta+\lambda}\right)$	$N\left(\frac{\lambda}{r+\lambda}\right)+N_1\left(\frac{\lambda\beta}{(r+\phi)(r+\beta+\lambda)}\right)+X_2$	$D-D_0\left(\frac{\beta}{r+\beta+\lambda}\right)$
P_u^u	$S\left(\frac{\beta+\lambda}{r+\beta+\lambda}\right)$	$N\left(\frac{\beta+\lambda}{r+\beta+\lambda}\right)+X_2$	$D-(D_0+D_1)\left(\frac{\beta}{r+\beta+\lambda}\right)$

where $S\equiv\frac{c}{r+\phi}$, $N\equiv\frac{nL\phi}{r(r+\phi)}$, $N_1\equiv\frac{nL\phi}{(r+\lambda)(r+\phi+\lambda)}$, $N_2\equiv\frac{nL\phi}{(r+\beta+\lambda)(r+\beta+\phi+\lambda)}$, $X\equiv\frac{xL\phi}{r(r+\beta+\phi)}$,

$X_1\equiv\frac{xL\phi(r+\beta)}{r(r+\beta+\lambda)(r+\beta+\phi+\lambda)}$, $X_2\equiv\frac{\lambda X}{r+\beta+\lambda}$, $D\equiv\frac{d(r+L\phi)\lambda}{r(r+\phi)(r+\lambda)}$, $D_0\equiv\frac{d(1-L)\phi\lambda}{(r+\phi)(r+\lambda)(r+\phi+\lambda)}$,

$D_1\equiv\frac{d(1-L)\phi(1-e^{-\lambda l})}{(r+\lambda)(r+\phi+\lambda)}$, and $D_2\equiv\frac{d(1-L)\phi(1-e^{-\lambda l})}{(r+\beta+\lambda)(r+\beta+\phi+\lambda)}$

4.2 The Effect of a Breakthrough Possibility on the Timing of Evaluation and Development

We now analyze how the possibility of a technological breakthrough in the future affects the current decision on the timing of evaluation and development. From the previous section, the conditions that *ex post* development is preferred given *ex ante* evaluation [equation (1)] and *ex post* evaluation is preferred to *ex ante* evaluation given *ex post* development [equation (2)] can be rewritten as follows.

$$(9)\quad V \equiv mL\phi - d(1-L)\phi(1-e^{-\lambda t}) > 0$$

$$(10)\quad W \equiv cr(r+\phi+\lambda) + mL\phi\lambda - d(1-L)\phi\lambda > 0\,.$$

These equations specify the parameter space in which *ex post* evaluation and development is preferred without a breakthrough possibility. When there exists an anticipated technological breakthrough, *ex ante* development before a breakthrough (and before an infestation occurs) is always (conditional on the chosen parameter space and *ex ante* evaluation) dominated by the development at the time of a breakthrough, A_0^u, for $V > 0$. Because the disease cost is low relative to the development cost, delaying development till the breakthrough will bring a lower cost to a genebank manager. In addition, if the size of a technological breakthrough is large, delaying development further till the disease infestation is not optimal. That is, $A_0^u < A_0^t$ if:

$$(11)\quad xL\phi > V\,.$$

If the reduction in the expected development cost due to a breakthrough is large enough to dominate the increase in the present value of the cost due to early development, it is optimal to develop at the time of a breakthrough rather than to delay till the disease infestation. Thus, given conditions (9), (10), and (11), *ex ante* evaluation/post-breakthrough development, A_0^u, is optimal if evaluation is made *ex ante.*

If, on the other hand, evaluation is initiated at the time of a technological breakthrough and the disease has not occurred after the evaluation ($u + s < t$), development follows immediately after evaluation if condition (11) holds, that is, $P_u^u < P_u^t$. To find the optimal *ex post* evaluation alternative, we need to compare P_u^u with the expected cost from delaying evaluation till the disease infestation, P_t^t :

$$P_u^u - P_t^t = \theta\left[W + V(r+\phi) - xL\phi(r+\phi+\lambda)\right]$$

where $\theta \equiv \beta / \left[(r+\phi)(r+\lambda)(r+\phi+\lambda)(r+\beta+\lambda)\right]$. If, given *ex post* development, there is negligible difference in expected cost between *ex ante* evaluation and *ex post* evaluation without a breakthrough possibility (that is, W, defined in (10), is close to zero), the sign of the above equation is negative by condition (11). Therefore, post-breakthrough evaluation and development, P_u^u, is optimal if evaluation is delayed.

Proposition 4 (Optimal timing of evaluation and a breakthrough possibility) An anticipated technological breakthrough in development can advance the timing of evaluation from ex post to ex ante, which in turn advances the timing of development.

Proof: If condition (11) holds and W is close to zero, A_0^u is the optimal *ex ante* evaluation alternative and P_u^u is the optimal delayed evaluation alternative.

$$A_0^u - P_u^u = \eta\left\{\left(V - xL\phi\right)r(r+\phi)(r+\beta+\phi) + xL\phi\lambda\left[\phi(\phi+\lambda) - r(r+\beta)\right]\right\}$$

where $\eta \equiv \beta / \left[r(r+\phi)(r+\beta+\phi)(r+\beta+\lambda)(r+\beta+\phi+\lambda)\right]$. If the expression on the right-hand side of the above equation is less than zero, *ex ante* evaluation is optimal. The first term in the bracket is negative by condition (11), and the second term is negative if $\phi(\phi + \lambda) < r(r + \beta)$ [and if $\lambda > 0$, assumed in equation (4)]. Thus, the sufficient condition where *ex ante* evaluation is preferred is $\phi(\phi + \lambda) < r(r + \beta)$. Q.E.D.

Proposition 4 implies that the possibility of a technological breakthrough advances the timing of evaluation (and development) for $\phi(\phi + \lambda) < r(r + \beta)$. If a technological breakthrough is expected to happen soon (that is, β is high) and is likely to happen before a discovery occurs (that is, ϕ is low), this will favor early evaluation because the breakthrough then increases the value of immediate development. If λ is high, on the other hand, immediate development is likely even without a breakthrough, so the latter has less expected effect on timing. Similarly, if the breakthrough happens after discovery (that is, low β and high ϕ), there is no marginal advantage to advancing evaluation to exploit the breakthrough.

Figure 22.4 illustrates an example of the optimal evaluation alternatives under different parameter spaces. When there is no possibility of a technological breakthrough ($x = 0$), *ex ante* evaluation/*ex post* development, (A_0^t), is preferred for $\phi > 1.18$ and *ex post* evaluation (P_t^t) for $0.25 < \phi < 1.18$. Within this range where *ex post* evaluation is preferred, as the size of a

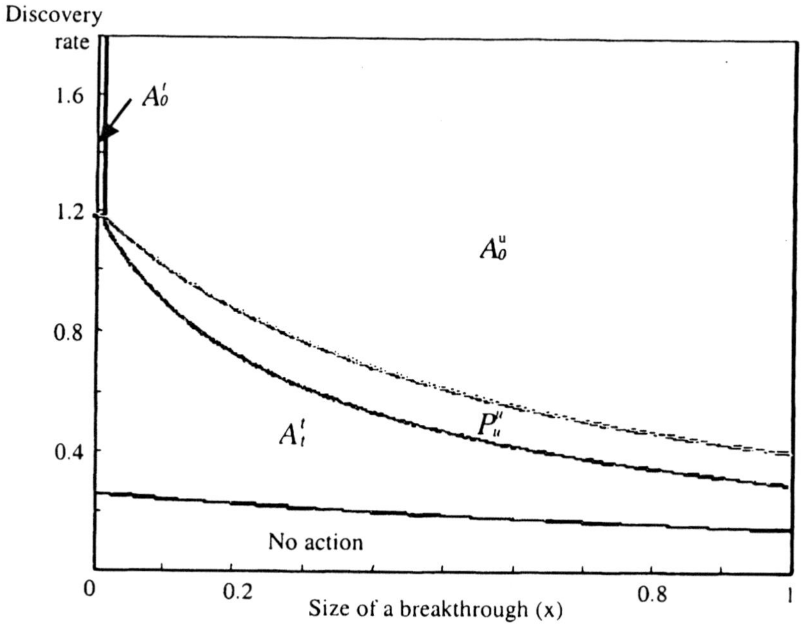

Figure 22.4 Optimal Timing of Evaluation with a Technological Breakthrough

breakthrough increases, the set of optimal alternatives expands to include post-breakthrough evaluation/development (P_u^u) and *ex ante* evaluation/post-breakthrough development (A_0^u). The benefit from a large technological breakthrough can be substantial if it is utilized earlier. This will induce *ex ante* evaluation to capture the benefit of earlier utilization if condition $\phi(\phi + \lambda) < r(r + \beta)$ holds. Thus, continuous advances in biotechnology in the future may induce early evaluation of genebank accessions and consequently early development of cultivars.

This result is contrary to the common inference from real option theory in which uncertain future environment delays the timing of a decision. A crucial feature of this model is that evaluation must be successfully completed before utilizing the breakthrough in development. If the evaluation process is instantaneous or the breakthrough affects the cost of evaluation rather than development, our model favors delaying as in standard real option theory. For example, Dixit and Pindyck (1994: 327) show that sequential investment is no different than ordinary investment if investment in each stage is instantaneous.

However, for situations in which it takes time to finish the first stage of an investment process, an increase in uncertainty may make early investment more attractive. Bar-Ilan and Strange (1996) argue that when there are "investment lags," investment may be hastened by an increase in uncertainty. With investment lags and the option to abandon a project, the opportunity cost of waiting, which is independent of uncertainty in the standard model, increases with uncertainty, resulting in earlier investment. In our model where it takes time to finish the first stage of a two-stage investment process, the ability to delay the second stage (development), if the breakthrough has not occurred when the evaluation is completed, makes early investment in the first stage (evaluation) more attractive. The real option value due to technological uncertainty actually can advance investment in both stages relative to the case of no technological uncertainty, in contrast to the results in other sequential investment models such as Majd and Pindyck (1987) and Bar-Ilan and Strange (1998).

5. CONCLUSION

The agricultural environment is continuously changing, and so is the demand for germplasm by plant breeders. Predicting future use of germplasm in *ex situ* genebanks is increasingly difficult. One of the most important policy issues regarding management of genebank collections concerns the evaluation of their collections prior to utilization. A commonly expressed view is that all traits likely to be relevant in crop improvement should be completely evaluated *ex ante* to facilitate and encourage the utilization by plant breeders.

This chapter examines the optimal timing of evaluation of germplasm for disease resistance traits and of development of cultivars incorporating those traits, from the social point of view. We consider the case where the disease-causing agent is known, but the date of crop infestation is stochastic. We find that for a trait which has low probability of being needed soon, *ex ante* evaluation tends to be dominated by delayed evaluation. This result is especially important for the management of genebanks, which suffer chronic funding problems. Instead of spending scarce financial resources for the expensive evaluation of rarely used genes, it might well be more efficient to focus on other activities such as provision of basic information and construction of an information network for better information flow (Frankel, 1989).

Technological progress has an important influence on the timing of evaluation of accessions and of the development of cultivars incorporating traits identified in the evaluation process. Innovations that reduce search and development costs and/or speed up search and development rates turn out to favor *ex ante* evaluation. On the other hand, the possibility of a future technological breakthrough in the cost of evaluation tends to delay evaluation, as one might expect from the common general intuition about real option value in which prospective technological progress affecting the cost of an investment tends to delay the timing of an investment.

But the possibility of a breakthrough in the cost of development may advance the timing of the whole process, contrary to the standard result (Majd and Pindyck, 1987) in a time-to-build model of real option theory. We showed that, for the parameter space where *ex post* evaluation is preferred, the breakthrough possibility may advance the timing of evaluation from *ex post* to *ex ante*, thus advancing the timing of development. The reason is that the marginal benefit from the technological breakthrough is larger when the development process is started earlier. If the initial situation favors *ex ante* evaluation, the possibility of a breakthrough reinforces the relative merits of earlier evaluation. In this Poisson model, the effect of the mean breakthrough and the variance cannot be distinguished. But in a model in which only variance increased, the option associated with the new technology would similarly increase in value. We can conclude that both currently realized technological progress and anticipation of future progress may advance the optimal timing of the evaluation process and consequently the development process, except when the anticipated progress refers to the evaluation stage.

This study covers only one small aspect of the problems faced by genebank managers, that is, evaluation for single-gene, disease-resistance traits. Depending on the specific cost conditions, evaluation for multiple traits simultaneously might often be optimal, complicating the above analysis. The case of a sequence of technological breakthroughs is another interesting extension. The question of evaluation for more complex quantitative traits associated with yield is the subject of ongoing research in functional genomics. As the science becomes better understood, exploration of the economics of different managerial strategies regarding evaluation of genebank accessions should continue to be a very interesting topic.

6. ENDNOTES

[i] The types of information regarding germplasm can be broadly categorized as passport data, characterization data, and evaluation data (OTA, 1987). Passport data include information on the origin and environmental conditions of the material. Characterization data include environment-insensitive traits such as morphological, biochemical and molecular information, while evaluation data include environment-sensitive traits such as yield potential and disease resistance.

[ii] Although this point is frequently made, we must go back to Peeters and Williams (1984) for hard data. They report that 65 percent of the samples in the genebank network have no passport data, 80 percent have no characterization data, 95 percent have no evaluation data, and only 1 percent have extensive evaluation data.

[iii] Molecular genetic techniques such as restriction fragment length polymorphism (RFLP), random amplified polymorphic DNA (RAPD), and amplified fragment length polymorphism (AFLP), are used to identify specific genotypes and agronomic traits of interest and to "fingerprint" individual accessions. These techniques together with cell culture techniques and transformation techniques involving recombinant DNA (rDNA) are also used for the development of new plant varieties by facilitating the transfer of the desired genes and the development of new cultivars in a fast, reliable, and cost-effective way (Rao and Riley, 1994).

[iv] Frankel (1989) categorizes two kinds of traits of concern to plant breeders: qualitative traits and quantitative traits. Disease resistance traits are typically qualitative, and are among the most sought-after traits in germplasm collections.

[v] *Ex post* evaluation is often performed by plant breeders, but for simplicity of exposition we assume the genebank manager conducts the search.

[vi] Our search process is close to the process of screening assumed by Simpson, Sedjo, and Reid (1996). In practice, some sources of a trait may be more desirable as breeding materials than others, so evaluation may continue after the first positive result of the search (Evenson and Kislev, 1976; Gollin, Smale, and Skovmand, 2000). We ignore this consideration here.

[vii] Gollin, Smale, and Skovmand (2000) explicitly consider the pre-breeding process, by which resistance genes found in the search process are transferred into a breeding program before the development of a new variety. Our model assumes that pre-breeding is a part of the development process.

[viii] This implies that the search intensity during the evaluation process does not change at all. In general, if there is some capacity to rank the prospects of samples according to their *ex ante* probability of success, search costs may increase with sample size during the process. Furthermore, per unit search cost can be changed by capital investment (Simpson and Sedjo, 1996).

[ix] We ignore the possibility of other types of disease control such as chemical pesticides. Some soil-borne diseases may naturally decline in severity after 7 to 10 years of continuous infestation, and in other cases crop rotation can be used as a means of disease control (Leonard, 1999). Finally, substitution of a non-susceptible crop can reduce the losses caused by the disease. The flow cost of infestation could be modified to recognize these mitigating factors.

[x] Similar qualitative results are derived by assuming a stochastic development process.

[xi] Squares in Figure 22.1 indicate the decision nodes for the genebank manager, and the circle indicates the chance node with exogenously determined probabilities.

[xii] In the following analysis, A_i^j (P_i^j) denotes the expected cost of *ex ante* (*ex post*) evaluation, where i and j are the timings of evaluation and development, respectively. For example, P_t^t implies the expected cost of *ex post* evaluation where evaluation is done at time t and development follows immediately.

[xiii] The sign of the other root of λ is checked with

$$c^2r^2(r+\phi)^2 - K(K+c)r(r+\phi)\phi^2 = r(r+\phi)(cr - K\phi)\left[c(r+\phi)+K\phi\right]$$

which is negative for $cr < K\phi$ of equation (4).

[xiv] See Koo and Wright (2000) for details of the derivations. This chapter also appears as "The Optimal Timing of Evaluation of Genebank Accessions and the Effects of Biotechnology." *American Journal of Agricultural Economics* 82(4): 797–811. With permission from the *American Journal of Agricultural Economics*.

7. REFERENCES

Arrow, K. and A. Fisher. 1974. "Environmental Preservation, Uncertainty, and Irreversibility." *Quarterly Journal of Economics* 88(May): 312–319.

Bar-Ilan, A. and W. Strange. 1998. "A Model of Sequential Investment." *Journal of Economic Dynamics and Control* 22(March): 437–463.

_____. 1996. "Investment Lags." *American Economic Review* 86(June): 610–622.

Brown, A.H. 1989. "The Case for Core Collections," in A.H. Brown, O. Frankel, D. Marshall, and J. Williams, eds. *The Use of Plant Genetic Resources*. Cambridge, MA: Cambridge University Press.

Chang, T.T. 1989. "The Case for Large Collections," in A.H. Brown, O. Frankel, D. Marshall, and J. Williams, eds., pp. 123–135. *The Use of Plant Genetic Resources*. Cambridge, MA: Cambridge University Press.

Dixit, A. and R. Pindyck. 1994. *Investment under Uncertainty*. Princeton, NJ: Princeton University Press.

Duvick, D. 1984. "Genetic Diversity in Major Farm Crops on the Farm and in Reserve." *Economic Botany* 38: 161–178.

Evenson, R. 1996. "Valuing Genetic Resources for Plant Breeding: Hedonic Trait Value, and Breeding Functions Methods." Paper presented at the Economics of Valuation and Conservation of Genetic Resources for Agriculture. Rome, Italy (May).

Evenson, R. and Y. Kislev. 1976. "A Stochastic Model of Applied Research." *Journal of Political Economy* 84(April): 265–281.

Frankel, O. 1989. "Principles and Strategies of Evaluation," in A.H. Brown, O. Frankel, D. Marshall, and J. Williams, eds., pp. 245–260. *The Use of Plant Genetic Resources*. Cambridge, MA: Cambridge University Press.

Gollin, D., M. Smale, and B. Skovmand. 2000. "Searching an *Ex Situ* Collection of Wheat Genetic Resources." *American Journal of Agricultural Economics* 82(4): 812–827.

Goodman, M. 1990. "Genetic and Germplasm Stocks Worth Conserving." *Journal of Heredity* 81: 11–16.

Koo, B. and B. Wright. 2000. "The Optimal Timing of Evaluation of Genebank Accessions and the Effects of Biotechnology." *American Journal of Agricultural Economics* 82(4): 797–811.

Leonard, K. 1999. Personal communication. University of Minnesota. St. Paul, MN.

Majd, S. and R. Pindyck. 1987. "Time to Build, Option Value, and Investment Decisions." *Journal of Finanancial Economics* 18(March): 7–27.

McDonald, R. and D. Siegel. 1986. "The Value of Waiting to Invest." *Quarterly Journal of Economics* 101(November): 707–727.

McIntosh, R. and G. Brown. 1997. "Anticipatory Breeding for Resistance to Rust Disease in Wheat." *Annual Review of Phytopathology* 35: 311–326.

NRC (National Research Council). 1993. *Managing Global Genetic Resources: Agricultural Crop Issues and Policies*. Committee on Managing Global Genetic Resources: Agricultural Imperatives. Washington, DC: National Academy Press.

OTA (Office of Technology Assessment). 1987. *Technologies to Maintain Biological Diversity*. OTA F-330. Washington, DC: OTA.

Peeters, J. and J. Williams. 1984. "Toward Better Use of Genebanks With Special Reference to Information." *Plant Genetic Resources Newsletter* 60: 22–32.

Peeters, J. and N. Galwey. 1988. "Germplasm Collections and Breeding Needs in Europe." *Economic Botany* 42: 503–521.

Rao, R. and K. Riley. 1994. "The Use of Biotechnology for Conservation and Utilization of Plant Genetic Resources." *Plant Genetic Resources Newsletter* 97: 3–19.

Russian Wheat Aphid Task Force. 1991. *Economic Impact of the Russian Wheat Aphid in the Western United States*. Manhattan, KS: Kansas State University.

Skovmand, B. 1999. Personal communication. International Wheat and Maize Improvement Center (CIMMYT), Mexico.

Simpson, R. and R. Sedjo. 1996. "Investments in Biodiversity Prospecting and Incentives for Conservation." Discussion paper 96-14, Resources for the Future Washington, DC (April).

Simpson, R., R. Sedjo, and J. Reid. 1996. "Valuing Biodiversity for Use in Pharmaceutical Research." *Journal of Political Economy* 104(February): 163–185.

Witt, S.C. 1985. *Biotechnology and Genetic Diversity*. California Agricultural Lands Project. San Francisco, CA.

Wright, B. 1997. "Crop Genetic Resource Policy: The Role of *Ex Situ* Genebanks." *Australian Journal of Agricultural and Resource Economics* 41(March): 81–115.

Chapter 23

The Identification and Classification of Genetically Modified Organisms: Implications for Trade

Richard Gray, Jill E. Hobbs, and Faycal Haggui
University of Saskatchewan

1. INTRODUCTION

The development, licensing, and production of products containing genetically modified organisms (GMOs) have recently become the center of some controversy. Those strongly opposed to genetically modified (GM) products argue that the genetic modification process is unnatural and could result in life forms that are dangerous to the ecosystem and to human health. At the other end of the spectrum there are those who believe that GMOs pose no additional risk and offer a great deal of potential for improvements in the production of food. This controversy has spilled over into the international trade arena, leading to different approaches to the development, licensing, and regulation of GMOs.

While the two extreme points of view continue to have voice in the media, there is a growing middle ground in which people argue that consumers and producers should be the ultimate judges of the benefits of GMOs. And further, consumers and producers should be fully informed about the traits of GMOs, and should also be informed when GMOs are used in the production of a product. GMOs should be clearly labeled, and the benefits of each GM product should be based on its ability to survive in the marketplace. Several recent polls have shown that many North American consumers are in favor of the labeling of GM products. Consumers International (1999) reports that the results of a survey sponsored by Industry Canada show that 83 percent to 94 percent of Canadians want foods to be specifically labeled when they are produced using biotechnology. In their final report, a Canadian citizens' panel on food biotechnology reported a high level of anxiety within the Canadian public with respect to consumer health and safety, and recommended that labeling issues with respect to biotechnology be resolved (Citizens' Panel, 1999). A national survey of U.S. consumers in 1997 found that more than 75 percent of them supported the Food and Drug Administration (FDA) labeling policy, which requires that food be labeled as a product of biotechnology only if it has been changed in a significant way (Hoban, 1998). Consumer reaction has tended to be stronger in Europe. In July 1998, a survey conducted by the U.K. National Federation of Women's Institutes (an organization with 265,000 members in 8,000 regional chapters) found that 93 percent of women surveyed wanted GM foods to be labeled (Perdikis, Kerr and Hobbs, 2000).

Taking the market test for GMOs at face value, it appears to be well founded in logic. If a consumer who has full information decides to purchase a GM product, then that individual must be better off or he or she would choose another product. Similarly, as long as producers have the choice of GM and non-GM products, then the choice to grow a GM product should reflect an expectation of a higher return. Finally, a marketer (defined as any downstream food processing or food-handling firm) who decides to sell, or to use, a GM product must be better off or he or she, too, would use another product. If GM products survive in the marketplace, then consumers, producers, and marketers are all better off. This simple argument suggests a limited role for regulation.

Using a market test to determine whether a GMO is desirable can be a fundamentally flawed test, however, if there are non-market externalities. Non-market externalities are costs and benefits related to economic goods that are external to the market, and, as such, these effects cannot be captured using a market viability test. If the non-market externalities are large and negative, market viability will overestimate the net benefits of the good and the market will overproduce this good. When there are positive externalities, the market test will not reveal these benefits, and the market may underproduce this good. As a result, when making an evaluation of a GM product, it is important to identify non-market externalities so that they can be considered in the assessment of the good's desirability.

The existence of externalities also creates a potential role for government policy. Optimal government intervention could take many forms. It may include the provision of public goods in the areas of consumer information, product testing, or basic research. The optimal intervention might also include regulation, taxation, or subsidization of GMO production. Whether government intervention is desirable will depend on the nature and size of the externality and on the costs, which are required to address the externality, that are associated with the government intervention. Diverging approaches to the regulation of GMOs are creating international trade tensions. Against the backdrop of evolving national regulatory policies toward GMOs, a thorough analysis of their potential non-market externalities is timely.

The objectives of this chapter are twofold. First, we identify and classify the types of non-market externalities that are potentially present within the production, marketing, processing, and consumption of GM products. Second, we describe how institutions and other factors will influence the magnitude of these non-market costs and benefits.

The following section of this chapter contains a brief review of the body of economic theory that is relevant to GMO externalities and to the underlying causes of market failure. We then review a large number of potential negative and positive externalities that have been identified in both the academic and popular literatures, and present a framework for classifying these externalities. Wherever possible, we provide a qualitative assessment of the likely magnitude for each form of externality. This assessment includes a discussion of the factors that would increase or decrease the magnitude of each form of externality and the research required to accurately quantify these effects. We close this chapter with concluding comments that include a discussion of the implications for further research and the likely importance of each category of externality in the international trade arena.

2. THE ECONOMICS OF EXTERNALITIES

2.1 Private versus Social Costs and Benefits

Non-market externalities, hereafter referred to as externalities, are costs or benefits flowing among economic agents that have incomplete compensation or remuneration.[1] Most goods and services are bought and sold in the marketplace. Generally, the individual or firm that produces a good is paid by those who benefit from the purchase of that good. A positive externality exists when a good generates benefits that are not being fully paid for in the marketplace. These are often third-party benefits or costs that flow to those not involved in the sale or purchase of the good. For instance, an individual may pay to have a tree planted in his or her front yard, which provides an aesthetic benefit for the community, yet he or she is not paid for creating this benefit. A negative externality exists when a cost is imposed on other members of society without compensation. For example, a pulp mill may pollute a stream, causing hardship for those downstream. If those individuals were negatively affected and were not adequately compensated by the pulp mill, then the mill would be creating a negative externality. (Both negative and positive externalities are sometimes referred to as spillover effects.)

Indirect market impacts, such as changes in prices or in levels of employment, are sometimes referred to as pecuniary externalities. For instance, a GMO canola might expand acreage of canola and reduce the supply of peas. The resulting higher price of canola, and the economic loss to pea consumers, would be pecuniary externalities. These indirect market impacts are internal to the market and do not affect market efficiency.

Externalities are closely associated with the concept of market failure. Many economists have shown that when markets are perfectly competitive, such that individuals must pay for the costs they impose on others and are paid for the benefits for which they create, markets will maximize benefits to society.[ii] One of the assumptions of perfect competition is that there are no externalities present. When there are externalities, the market will not, except through chance, allocate resources efficiently.

How externalities affect decisions and overall economic welfare can be illustrated in a simple graph containing both private and social marginal costs and benefits. The market demand for a good represents the relationship between the private willingness to pay and the quantity available for sale. The market supply for a good represents the marginal cost of producing various quantities of the good. The market equilibrium, illustrated in Figure 23.1, occurs at a price at which the private quantity supplied is just equal to the private quantity demanded. If the private market demand reflects all marginal benefits of the good and the supply curve represents all the marginal costs of producing the good, then, at the market equilibrium, total net benefits from production and consumption are maximized. At quantities greater than Q_p, the cost of producing an additional unit exceeds its benefits, and, at quantities less than Q_p, the benefit from consuming another unit of the good exceeds the cost of producing the unit. In the absence of externalities, the market acts like an invisible hand to produce the socially desirable quantity of the good. This relationship holds only when both the cost and benefits of good provision are fully reflected in the private supply and demand curves.

When the production of a good creates a benefit that is not reflected in the private demand for the good, then the social, or total, marginal benefits exceed those reflected in the private demand curve. In this case, a positive externality exists, and the market undersupplies these goods. This is illustrated in Figure 23.2, where the social demand curve lies above the private demand curve. In this case, the optimal production of the good is Q_s, whereas the market, which only reflects private demand, would undervalue the good and supply quantity Q_p.

When the production of a good imposes some costs on others that are not reflected in the marketplace, a negative externality exists. The social, or total

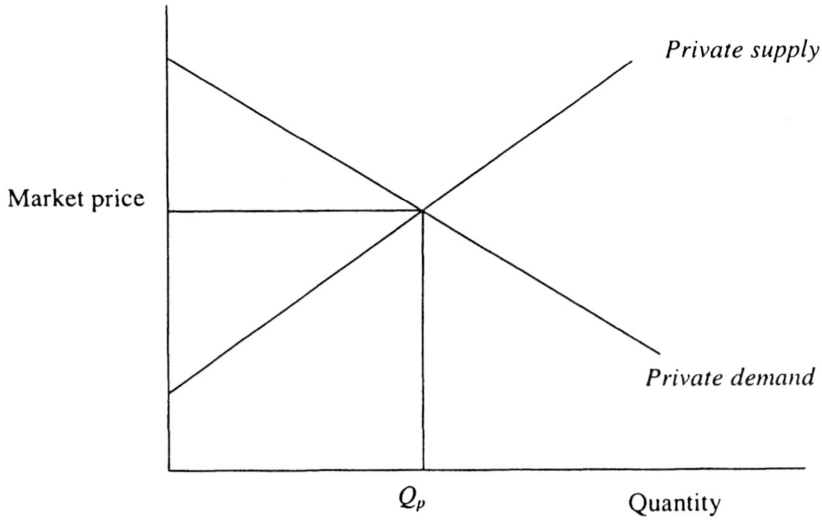

Figure 23.1 Private Supply and Demand for a Good

marginal costs exceed those reflected in the private supply curve. As illustrated in Figure 23.3, the social supply curve lies above the private supply curve. The market tends to understate the costs of production, and the good is oversupplied at level Q_p rather than at the social optimum level of Q_s. Externalities result in market failure by driving a wedge between the private supply and demand and between the social supply or demand.

2.2 The Nature of Consumer Choice

To understand the conditions under which market failure can occur, it is first necessary to outline some key economic concepts regarding the nature of consumer choice and relate these to the case of GMOs.

Market failure can occur when we relax the traditional neoclassical assumption of fully rational individuals operating in an environment of perfect information. Information asymmetry is a form of imperfect information that arises when there is an unequal distribution of information between transacting parties. For example, there might be unequal information about the attributes of a product, or about the characteristics of a buyer or seller. Adverse selection can be a consequence of hidden information prior to a transaction. For example, the seller may have more information about the true quality of a product than does the buyer. In Akerlof's classic discussion of the market for used cars (Akerlof, 1970) the buyers offer only poor-quality cars for sale. The poor-quality products (lemons) drive the high-quality products from the

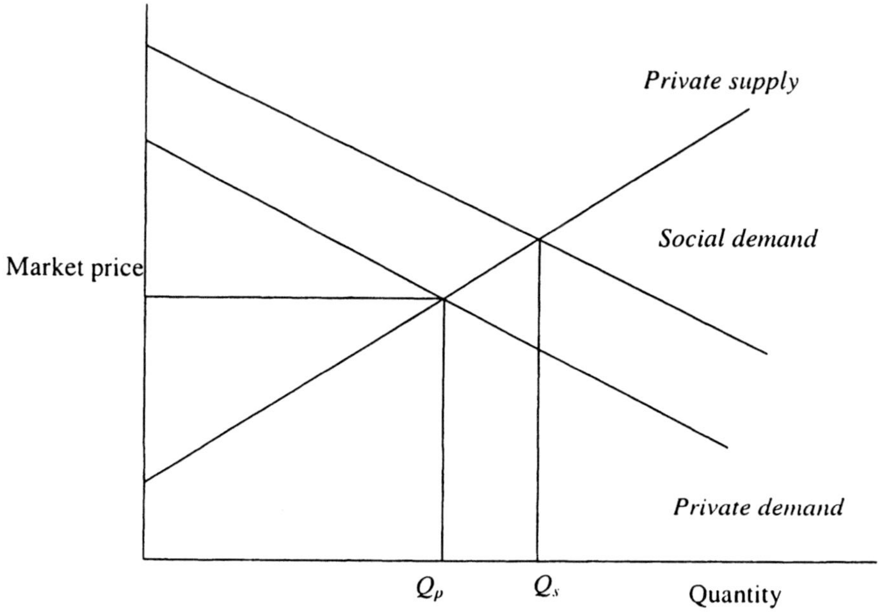

Figure 23.2 Private Versus Social Cost for a Good with a Positive Externality

marketplace. Consumers with ethical, or food-safety objections to GM foods will consider these foods to be of poorer quality. With information asymmetry consumers cannot tell which products contain GMOs (hence, which are lemons) without further information. They must, therefore, assume that all relevant foods are genetically modified, and they will discount these foods accordingly. The result is a welfare loss for those consumers wishing to consume non-GM foods. As in the Akerlof lemon case, GM foods could drive non-GM foods from the market (Gaisford and Plunkett, 2000).

The nature of the product determines the extent of information asymmetry, hence, the risk of adverse selection. We can distinguish between search, experience, and credence product attributes (Nelson, 1970; Darby and Karni, 1973). Attributes whose quality can be determined prior to the transaction through visual inspection [for example, the color of a car, the material in a shirt, the (uncooked) weight of a steak], are known as search attributes. Attributes whose quality can be determined only after they are purchased (for example, the juiciness of an orange, the tenderness of a steak, or the mechanical performance of a car), are experience attributes. Credence attributes are those whose quality cannot be determined accurately even after consumption, because the consumer lacks expert knowledge, or because there is insufficient information. Examples include the services of a doctor, or whether a beef product is contaminated with Bovine Spongiform Encephal-

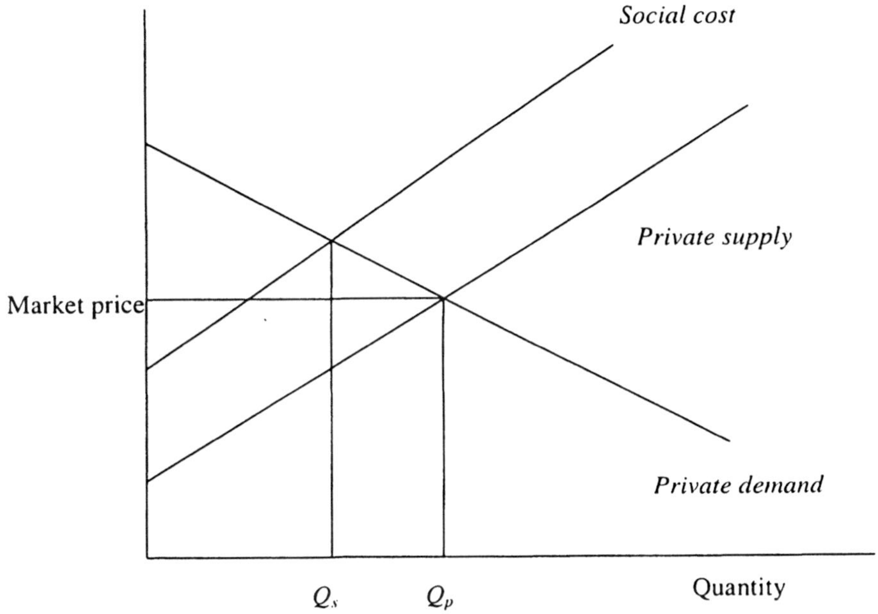

Figure 23.3 Private Versus Social Cost for a Good with a Negative Externality

opathy (BSE). Clearly, a GMO in food is a credence attribute. Due to information asymmetry, even after having consumed the food, the consumer does not know whether he or she has consumed a GMO. Adverse selection will be higher when goods are characterized by credence attributes. A method is required to signal the presence of the GMO credence attribute to consumers. Labeling of GM foods or GM-free foods is one such method.

Bounded rationality and opportunism compound the problem of information asymmetry. Bounded rationality recognizes that individuals have limited cognitive abilities; although they intend to act rationally, their ability to accurately evaluate all possible alternatives is physically limited (Simon, 1961). Opportunism is "self interest seeking with guile" (Williamson, 1979). Given bounded rationality and information asymmetry, it is not possible for economic agents to determine whether or not opportunism will occur, and who will act opportunistically. As suggested above, one solution to information asymmetry in the market for GM food is for companies to voluntarily label their food as a GM product. The extent to which this solves the information asymmetry problem depends on the credibility of the labeling system. Firms expecting a negative consumer backlash against GM foods do not have an incentive to label these foods accurately or display labels prominently. Bounded rationality means that it is not possible for consumers to determine which firms will act opportunistically to misrepresent their products. Market failure results. This is one argument in favor of mandatory labeling of GM food products to correct the informational asymmetry market failure. Of course, other non-market externalities may still result, as discussed later in this chapter.

3. THE QUANTIFICATION OF GMO EXTERNALITIES

There is significant and rapidly growing literature that describes many externalities associated with the production, marketing, and consumption of GM products. To assess whether these externalities are important, it is necessary to classify these externalities to ensure that the list is comprehensive, and to avoid double counting.

Choosing a system to classify GMO externalities is difficult. In standard environmental textbooks, externalities are often classified by naming both the activity of origin and the activity of incidence. A producer-consumer externality, for example, occurs when some effect originating from production has an external impact on a consumer. We rejected this approach because, for many externalities both the origin and the incidence are difficult to define. For instance, if a GMO affects one's health negatively, did the externality originate in the decision to create the organism, to grow the organism, or to consume the GMO? Similarly, if a GM product were to destroy the market for organic products, the producers, consumers, and marketers of the organic product would all be affected.

Instead, a classification system based on the type of direct impact provides the most workable system for GMOs. This classification, which contains six categories, or types, of direct impact for GMO externalities, could easily be expanded if new types of externalities were subsequently identified. By focusing on the direct impact, the system avoids the double counting that could occur if indirect impacts were also classified. The six categories, used in this section to review the externalities in the literature, are:

- Human-health externalities
- Ecosystem externalities
- Reduced demand for undifferentiated non-GM products
- Marketing/segregation externalities
- Production-cost externalities
- Information-cost externalities

3.1 Human-health Externalities

This category includes both the direct-health impacts on those consuming the GM product and the human-health impact from having the GMO in the environment. The components of the direct-health impact that are external to the consumer are the impacts on public-health expenditures (Gray and Malla, 2001). All of the environmental health impacts would be external to the GM market.

3.1.1 Negative Human-health Externalities

There could be a direct, negative impact on consumer health if, in the long run, GM input traits were to prove harmful to human health or to the environment. This is very much akin to the traditional notion of a negative externality, and implies that a possible side effect of consuming the product causes the social marginal cost of producing GM products to exceed consumers' private marginal cost.

Assessing the magnitude of this externality is difficult because long-run health concerns essentially are a fear of the unknown. As such, it may not be possible to assign probabilities to the negative-occurring event. In general, there would be a very small probability of a potentially very large external cost. Nevertheless, factors affecting the impact of this externality would include the extent to which GM foods are used in the diet (for example, the percent of dietary intake), the extent to which any negative effects could be constrained or reversed, and the size of the population potentially affected. Sensitivity analysis around these key assumptions would be essential.

It is important to note that the effects of consumers' perceptions (that is, that GMOs could represent a threat to health) are captured elsewhere (for example, the externality this imposes on producers and consumers of non-GM foods because segregation is required). The externality discussed here represents only actual negative-health impacts that lead to increased personal health costs for consumers, such as the loss of income and reduced quality of life. If one has faith in the regulatory-approval process through which food products must pass before being released onto the marketplace, and if one accepts the assurances of the majority of the scientific community that GM foods are safe, then the magnitude of this externality should be small. If one does not have faith in the regulatory-approval process, then this externality could be larger. Examples of these types of externalities can be found in Butler and Reichhardt (1999), Flint and Chesson (1999), Gendel (1998), Gilbert (1999), Hobbs and Plunkett (1999), Kappeli and Auberson (1998), Persley and Siedow (1999), and Walsh (1999).

3.1.2 Direct Positive Human-health Externalities from Output-trait Products

Output-trait genetically modified (OT-GM) products are part of the second wave of genetically engineered products that could offer positive health benefits for consumers and cost savings for taxpayers who support health care. In 2000, there were few OT-GM-enhanced products on the market. However, nutriceutical/functional foods are a growing market segment and offer insights into the value that consumers might place on future OT-GM products. Further research is required to determine the extent to which products of this nature are waiting in the wings in the final stages of pre-commercialization, or the extent to which products are awaiting regulatory approval. Based on currently available OT-GM products, the magnitude of this externality is likely to be small. However, it may be important in a longer-run consideration of externality effects. For examples and a discussion of these types of externalities, see Cavadini, Hertel and Hammes (1998), Dunwell (1998), Liu (1999), Persley and Siedow (1999), and Symonds (1999).

3.1.3 Indirect Positive-health Impacts from Reduced Use of Agricultural Chemicals

Many input-trait genetically modified (IT-GM) products are specifically designed to require less agricultural chemical use. This should have a positive effect on the environment and the general health of the population. The magnitude of this externality is determined by the extent to which chemical residues affect food safety and quality (hence, the efficacy of regulatory standards); by the extent to which significant reductions in pesticide and/or herbicide residues are forthcoming; and by consumers' risk perceptions and their willingness to pay for foods that have fewer agricultural chemicals. Existing in the literature are several examples of attempts to measure consumers' risk perceptions and their willingness to pay for residue-free foods (Caswell, 1996). These examples would provide a starting point for attempts to measure this externality. If current regulatory standards are sufficient to ensure that residues from the use of agricultural chemicals are not detrimental to human health, then the magnitude of this externality may not be large. For a discussion of these types of effects see also Froyd (1997), Persley and Siedow (1999), and Rissler (1999).

3.2 Ecosystem Externalities

Ecosystem externalities include the impacts of GMOs on the environment. Specifically, these externalities result in changes in the value of the natural environment. They include changes in social-use value (such as recreation or hunting) and in non-use values (such as aesthetics or existence values). To avoid double counting, these externalities do not include the indirect human-health impacts that are measured under the first category, nor do they include the external changes in the non-GM cost of agricultural products that are captured within the production-cost-externality category.

3.2.1 Positive Ecosystem Externalities from Reduced Pesticide Use

The reduced pesticide use within IT-GM products may have positive effects on the ecosystem. The magnitude of the effects depends on several factors: (1) how much individuals value ecosystem health, (2) the extent to which non-GM agricultural practices are creating long-term sustainability problems, and (3) the extent to which IT-GM crops can reverse this situation. The magnitude of the benefits also depends on whether there are substitute means of achieving the same benefits (examples include improved production practices, rotation, and zero-till). For discussion and examples of these types of effects see USDA/ERS (1999), Mooney and Klein (1999), and *Pesticide and Toxic Chemical News* (1998).

3.2.2 Negative Ecosystem Externalities

Potential negative ecosystem externalities originate either from the impact of the IT-GM crop, (for example, alleged impact of GM corn on the Monarch butterfly) or the out-crossing of new genetics which could change the ecosystem function. A negative externality would occur if the development of herbicide-resistant GM crops were to encourage the use of a single herbicide, which would result in an increase in the herbicide resistance of weed populations. A further negative externality would occur if GM crops were to cross-pollinate with their wild species, and would thereby transfer genes that would be resistant to weeds (Mooney and Klein, 1999).

This externality incorporates environmental effects that impact a broader section of society. The complex nature of ecosystems makes these effects difficult to quantify. For example, there has been controversy over the potential impact of GM corn on the Monarch butterfly, with contradictory scientific studies seeming to prove, or to disprove, the hypothesis that the GM crop could harm this butterfly species. Unforeseen impacts on other plant and animal species are significant concerns among environmental groups.

There is evidence that more than 200 weeds have become resistant to herbicides worldwide (Heap, 1999). Further evidence suggests that some insects are becoming resistant to Bacillus thuringiensis (Bt) sprays (Tabashnik, 1994).

The magnitude of this externality depends on scientific assessment of the probability of a negative impact, society's level of risk aversion, and the value that society places on potentially threatened environmental amenities and/or species. Additional important factors are the proximity of the GM crop to their wild and weedy species and the ability of the new gene to provide the weedy species with an advantage (Mooney and Klein, 1999). There appears to have been little or no research attempted that measures the magnitude of this potential negative externality with respect to GMOs. However, there is a rich body of literature that addresses these questions relative to other environmental externalities. Given the level of attention this has received in the public sphere, it will be important to include it in any future assessment of GMO externalities. Accepted methods of valuing non-market goods such as contingent valuation have a role to play in this regard. A number of studies discuss these issues: Adam and Köhler (1996), Clark (1999), Cook et al. (1996), De-Leij et al. (1998), Hauser, Jorgenson and Ostergard (1998), Heap (1999), Holt and LeBaron (1990), Kappeli and Auberson (1997), Mooney and Klein (1999), Nickson and McKee (1999), Nielson et al. (1999), Rissler and Mellon (1996), Saat (1996), and Tabashnik (1994).

3.3 Reduced Demand for Undifferentiated Non-GM Products

The existence of a GM product can reduce consumers' willingness to pay for a non-GM product. This consumer externality arises from the uncertainty that non-GM products may contain GMOs because the GMO is a credence attribute. The credence nature of the GM attribute means that it is not possible for consumers to determine whether GMOs are present and whether segregation or labeling of GM/non-GM products is accurate and credible. If the consumer is uncertain about the characteristics of a non-GM product, then the demand for these products will fall (that is, the poor-quality GM "lemons" chase the high-quality non-GM products from the market). This reduced demand, although originating at the consumer level, affects the marketing channel and the producers of these non-GM products. The cost of this impact can be measured as the reduction in consumer surplus of these products. The case in which there are positive or negative externalities from the consumption of these non-GM products, the change in health costs associated with the change in consumption also needs to be calculated.

The size of this externality will be determined by the size of the market affected, by the number of consumers that discount the value of GM products, by the extent to which they discount these products, and by the perceived probability that contamination has taken place. The size of this effect is difficult to measure even after the fact. Separating changes in consumer demand due to fears from other demand factors is very difficult. When consumer pressure leads governments to impose an import ban on both GM and non-GM products, it could be a very important source of externality. In response to this fear, there are also resources spent on market segregation. For further discussion see Biodemocracy and Organic Consumers Association (1999), Consumers International (1999), and Hobbs and Plunkett (1999).

3.4 Marketing and Segregating Externalities

The existence of GM products may force non-GM marketers to incur costs to segregate and distinguish their products from GM products. The credence nature of the GMO attribute requires a signaling mechanism, such as segregation and labeling. These are all the costs associated with maintaining separation between non-GM and GM products and distinguishing non-GM products from GM products in the marketplace. They are externalities because the costs are not born within the GM market. These additional marketing costs are reflected in lower producer prices and/or higher consumer prices.

Key determinants of the magnitude of this externality include the products affected, the importance of these crops to the agri-food sector (such as the number of producers, the volume and the value of sales), and the ability of the current supply-chain infrastructure (such as grain-handling and transportation facilities) to segregate GM and non-GM crops. The regulatory approaches in different markets, and the value of market opportunities foregone, will determine whether or not segregation is necessary. A qualitative assessment suggests that this will be a major externality and that it should be a priority for future research. If the production and transaction costs of segregating and labeling non-GM products from GM products are too high, then segregation is uneconomic and does not occur. In this case, the lemon-type case would manifest itself. In the presence of information asymmetry, perceived low-quality GM products would drive high-quality non-GM products from the market. In essence, the externalities described in the section above would occur, and the market for undifferentiated non-GM products would contract. For a further discussion of labeling and segregation issues see Caswell (1999), Herrman and Boland (1999), Hoef et al. (1998), Hobbs and

Plunkett (1999), Hurburgh (1994), Smyth and Phillips (1999), and Vandeburg et al. (2000).

3.5 Production-cost Externalities

These externalities exist when GM products affect the production costs of non-GM producers. For instance, a neighbor to a GM producer may have additional herbicide costs, or may have to maintain an additional margin around his or her field. These are separate from the segregation costs discussed above because the externality results from inadvertent pollution between farms rather than from being an issue of segregation through the marketing system.

3.5.1 Field-to-field Pollution from GM Crops

Field-to-field pollution is an externality imposed on the non-GM producer by the GM producer if, for example, GM pollen drift were to cause contamination of non-GM crops with GM material. The result can be increased on-farm production costs for producers of non-GM crops.

This externality is potentially important when crops are grown in close proximity, and when the onus is put on producers to prove that contamination cannot occur. Loss of markets, therefore, could result. A useful analogy is the loss an export market in Germany experienced as a result of the actions of a group of Western Canadian honey producers. The honey producers were unable to provide their German retail customer with an assurance that their honeybees did not collect pollen from GM crops (Tjaden, 1997). The German firm would no longer accept the producers' honey without this assurance and, given the credence nature of the GM attribute, it was not possible for the retailer to detect the presence of GMOs in the food product. This example also demonstrates that externalities can extend beyond crops and into other agri-food products.

The size of this externality would depend on the probability of field-to-field pollution occurring, the proportion of agricultural production potentially affected (that is, the extent of GM crops grown proximal to non-GM crops), adoption rates for GM crops; regulatory approaches in affected markets, and the potential value of markets foregone. The probability of detection may also be a determinant; that is, whether or not downstream firms will be able to test for the presence of GMOs in a non-GM crop. The cost of the measures required to prevent contamination—that is, the size of the buffer zone required to protect non-GM crops—will also be a factor. For further discussion see Giddings, Sackville-Hamilton, and Hayward (1997a and 1997b), Klein, Kerr and Hobbs (1998), and Tjaden (1997).

3.5.2 Higher Weed Control Costs and/or Lower Yields

This negative externality arises if an herbicide-resistant trait creates harder-to-kill weeds in neighboring crops. It is imposed by the specific GM-product markets' effect on other crop markets. If the herbicide-resistant trait were to encourage the development of herbicide resistance in other plants (for example, weeds), neighboring producers would face higher production costs. This externality is considered separately from the wider ecosystem externalities discussed earlier because it is specifically a production effect.

The magnitude of this effect depends on the amount of inter-field movement of genetic material that takes place, the type of crop grown in the neighboring fields, and the weed-control systems used in neighboring fields. This has the potential of becoming a dynamic problem because these volunteer herbicide-tolerant crops may reproduce. A recent study focussed on a triple herbicide-resistant strain of canola (*Western Producer*, 2000); this resistance means that weeds may emerge that, in the long run, could impose a major cost on GM and non-GM crop-production systems. The emergence of these weeds would also dramatically affect segregation costs, especially if low tolerance GM levels were adopted. For a discussion of these potential effects see Persley and Siedow (1999); Heap (1999); Holt and LeBaron (1990); Mann (1999); Mooney and Klein (1999); Saat (1996); and the *Western Producer* (2000).

3.6 Information-cost Externalities

When GM products are introduced, consumers are faced with a new consumption decision. This decision requires effort on behalf of the consumer so he or she can obtain and process new information; as well, he or she will need public and private resources to produce and communicate the information. Many of these costs will be outside the GM market.

It is important to emphasize that many of the costs and benefits associated with GM products are not externalities. In most cases, the externalities occur outside the GM market in non-GM products, or in goods in the public domain, such as the environment or public health care.

The credence nature of the GMO attribute imposes higher information and search costs on consumers, downstream food processors, and distributors. First, these costs occur when determining whether or not GMOs are present. Second, for consumers, costs occur when evaluating the potential impact, or risk, of the presence (or, in the case of the output trait, GMO absence) of the GMO. Consumer-information costs are exacerbated by the plethora of contradictory information, which raise the costs of determining the credibility of an information source, from various sources about GM foods.

The magnitude of this cost depends on the amount of unbiased information provided from the participants in the GM-product market. The more polarized the availability of information, the greater is the cost to the consumers who must invest time and effort to make consumption decisions. Measurement of this externality will require conducting survey(s) of consumers, or, where available, it will require using existing survey data to evaluate consumer perceptions of GM foods and the information sources at their disposal. The extent to which consumers have access to accurate sources of information, and whether or not they regard these information sources as credible, will be important factors. Conjoint analysis (which allows one to measure the relative importance of different product attributes in a consumer's purchase decision) or similar stated-preference methodologies may usefully be applied to this question. An evaluation should be made of private- and public-sector strategies that are used to reduce consumers' information costs (including labeling and certification). In particular, the extent to which these strategies reduce information costs and correct market failures should be evaluated. For further discussion see Burros (1997), Chess (1998), Frewer, Howard and Aaron (1998), Gath and Reimar (1998), Hoban (1998), Hobbs (2000), and Perdikis, Kerr, and Hobbs (2000).

4. CONCLUSIONS

There is great debate over the desirability of the development, production, marketing, and consumption of GM products. Many have argued that the fate of these products should ultimately be left up to the marketplace. Using the market test to determine the desirability of these new products may ignore the presence of externalities, which are costs and benefits not reflected in the market for GM products.

A large number of potential negative and positive externalities associated with GMOs have been identified in both the academic and popular literatures. This chapter has attempted to classify these externalities and to provide a qualitative assessment of the likely magnitude of each form of externality. This assessment included a discussion of the factors that would increase or decrease the magnitude of each form of externality. In some cases, this discussion has suggestions for the research that would be required to more accurately quantify these effects.

A review of the externalities found in the literature revealed some important differences in the form of externalities created by OT-GM and IT-GM technologies. OT-GM products are altered to produce a desirable trait and, as such, can offer health benefits or can become the basis of lower-cost pharmaceuticals. The desirable trait in these products means that they will generally sell for a premium and will self-segregate. This means that consumers of non-GM products have little risk of contamination from these products. The potential negative externalities from the OT-GM products include some unwanted producer-to-producer contamination, and some unanticipated health risks from consumption. The potential for ecosystem damage is likely to be limited and is likely to be similar to that created by non-GM crops.

The IT-GM crops may produce positive environmental externalities through reduced pesticide use, but they may also have a far greater potential for negative externalities of a number of types. The lower cost of production will entice producers to grow these GM varieties, as already evident in corn, soybeans, and canola. If GM products are discounted in the marketplace, then there will be an incentive to misrepresent these crops as non-GM products. This means that the non-GM markets will have to bear the cost of segregation, and will have to face a reduction in demand if they are associated with GM varieties of the same crop. Consumers of non-GM products will suffer because they will not be certain that their products are GM free. The prevalence of these GM crops will also increase the segregation and the potential production costs for non-GM producers.

There are many externalities—some positive and some negative—associated with the development, production, marketing, and consumption of GM products. The potential magnitude of these externalities suggests a need for further quantification and research. Each GM product will have different externalities. It is important that several case studies be examined in detail to determine whether or not further policy intervention is desirable.

While the current seed regulatory and licensing systems in Canada and the United States examine some of the health and environmental impacts of releasing GM crops, the effects on other markets are not considered. Given the large potential benefits and costs of GM products, it is important to consider the impacts on the non-GM product markets and the additional costs that may have to be borne in these markets.

The long-run potential for damage to non-GM crop production systems appears to be significant for IT-GM crops. The potential magnitude of many of these external effects suggests a great need to study these effects as they may have very detrimental long-term impacts on production and marketing costs. The irreversible nature of the release of these IT-GM products into the environment suggests that it may be prudent to carefully assess each of these potential impacts before any more of these products are made available.

A qualitative assessment suggests that the most important externalities affecting international trade relations are likely marketing/segregation externalities and reduced demand for undifferentiated non-GM products. For export-dependent crops, the size of these potential externalities is affected directly by GMO market access, and by the labeling policies of the importing countries. In the case of seed-crop exports, ecosystem externalities may also be important. The greater the externalities, the more likely it is that importing countries will impose restrictions on exports from GMO-producing countries. Potential negative human-health externalities and information-cost externalities have led a number of countries to introduce mandatory labeling of food containing GMOs in an attempt to correct the market failure that arises from information asymmetry. These policies differ in their coverage and in the threshold GM content allowed before labeling is required. This divergence in policy approaches raises transaction costs for firms engaged in international trade and has contributed to trade tensions over food safety and quality issues. Finally, production-cost externalities are more likely to be important domestic-policy issues than they are international-trade issues.

5. ENDNOTES

[i] This definition is consistent with the more extensive definition provided by Baumol and Oates (1988).

[ii] This notion was central to Adam Smith's "invisible hand" of the marketplace. The notion that competitive markets are Pareto optimal is one of the fundamental theorems of welfare economics (Varian, 1984).

6. REFERENCES

Adam, K.D. and W.H. Köhler. 1996. "Evolutionary Genetic Considerations on the Goals and Risks in Releasing Transgenic Crops," in Wöhrmann, K., Tomiuk, J. and Sentker, A., eds. *Transgenic Organisms—Biological and Social Implications,* Basel, Switzerland: Birkhaüser Verlag.

Akerlof, G.A. 1970. "The Market for 'Lemons': Quality, Uncertainty and the Market Mechanism." *Quarterly Journal of Economics* 84: 488–500.

Baumol, W.J. and W.E. Oates. 1988. *The Theory of Environmental Policy, 2nd Ed.* Cambridge, MA: University Press.

BioDemocracy and Organic Consumers Association. 1999. Internet website: http://www.purefood.org/gelink.html <http://www.purefood.org/gelink.html>

Blythman, J. 1999. "We don't believe our food is safe." Internet website: http://www.millennium-debate.org/indsun6june6.htm.

Burros, M. 1997. *The New York Times.* Internet website: http://www.purefood.org/whole.html.

Butler, D. and T. Reichhardt. 1999. "Long-term Effect of GM Crops Serves up Food for Thought." *Nature* 398 (6729, April 22): 651–653.

Caswell, J. 1999. "An Evaluation of Risk Analysis as Applied to Agricultural Biotechnology (with a Case Study of GMO Labeling)," in Lesser, W.H., ed., pp. 665–674, *Transitions in Agbiotech: Economics of Strategy and Policy.* Food Marketing Policy Center, University of Connecticut, Storrs, CT.

_____. 1996. *Valuing Food Safety and Nutrition.* Boulder, CO: Westview Press.

Cavadini, C., C. Hertel, and W.P. Hammes. 1998. "Application of lysostaphin-producing Lactobacilli to Control Staphylococcal Food Poisoning in Meat Products." *Journal of Food Protein* 61 (4, April): 419–424.

Chess, C. 1998. "Fearing Fear: Communication About Agricultural Biotechnology." *Agbioforum* 1 (1, Summer). Available online: http://www.agbioforum.missouri.edu

Citizens Panel. 1999. *Designer Genes at the Dinner Table: Citizens' Conference on Food Biotechnology.* Final report. Calgary, AB: University of Calgary (7 March).

Clark, E.A. 1999. "Debunking the Myths of Genetic Engineering in Field Crops." Internet website: http://www.oac.uoguelph.ca/CRCS/faculty/eac.myths.htm

Coase, R.H. 1960. "The Problem of Social Cost." *Journal of Law and Economics* 3 (October): 1–44.

Consumers International. 1999. Summary of Consumer Surveys Related to Labeling of Foods Produced Using Biotechnology. Internet website: http://www.oneworld.org/consumers/campaigns/biotech/surveys.html

Cook, R.J., W.L. Bruckart, J.R. Coulson, M.S. Goettel, R.A. Humber, R.D. Lumsden, J.V. Maddox, M.L. McManus, L. Moore, and S.F. Meyer. 1996. "Safety of Microorganisms intended for Pest and Plant Disease Control: A Framework for Scientific Evaluation." *Biological Control* 7 (3 December): 333–351.

Darby, M.R. and E. Karni. 1973. "Free Competition and the Optimal Amount of Fraud." *Journal of Law and Economics* 16 (1): 67–88.

De-Leij, F.A.A.M., M.J. Bailey, J.M. Whipps, I.P. Thompson, P.A. Bramwell, and J.M. Lynch. 1998. "Release of Genetically Modified Micro-organisms and Biomonitoring" *(Biotechnology Research Series;* Volume 7), in *Environmental Biomonitoring, the Biotechnology Ecotoxicology Interface,* J.M. Lynch, ed., pp. 70–100. Cambridge, England: Cambridge University Press.

Dunwell, J.M. 1998. "Novel Food Products from Genetically Modified Crop Plants: Methods and Future Prospects." *International Journal of Food Science Technology* 33 (3 June): 205–213.

Economic Research Service (ERS). 1999. *Genetically Engineered Crops for Pest Management.* United States Department of Agriculture, Washington, DC (25 June). Internet website: http://www.econ.ag.gov.whatsnew/issues/biotech

Flint,-H.J.; A. Chesson. 1999. "Safety Implications of Genetically Modified Organisms." *Feed-mix, Doetinchem* 7 (3): 3235.

Frewer, L.J., C. Howard, and J.I. Aaron. 1998. "Consumer Acceptance of Transgenic Crops." *Pesticide Science* 52 (April): 388–393.

Froyd, J.D. 1997. "Can Synthetic Pesticides be Replaced with Biologically Based Alternatives?: An Industry Perspective." *Journal of Industrial Microbiology and Biotechnology* 19 (3): 192–195.

Gaisford, J. and M. Plunkett. 2000. "Limiting Biotechnology? Information Problems and Policy Responses." *Current Agriculture, Food and Resource Issues* 1: 21–28.

Gath, M. and V.A. Reimar. 1998. "The Potential Effects of Labeling GM Foods on the Consumer Decisions–Preliminary Results of Conjoint Measurement Experiments in Germany." *Ar Norwegen* 4 (3): 18–28.

Gendel, S.M. 1998. "The Use of Amino Acid Sequence Alignments to Assess Potential Allergenicity of Proteins Used in Genetically Modified Foods." *Advanced Food Nutrition Research* 42: 45–62.

Giddings, G.D., N.R. Sackville-Hamilton, and M.D. Hayward. 1997a. "The Release of Genetically Modified Grasses, 2, The influence of Wind Direction on Pollen Dispersal." *Theoretical and Applied Genetics* 94 (8): 1007–1014 (June).

_____. (1997b). "The Release of Genetically Modified Grasses, 1, Pollen Dispersal to Traps in Lolium perenne." *Theoretical and Applied Genetics* 94 (8 June): 1000–1006.

Gilbert, S. 1999. *The New York Times*. Internet website: http://www.purefood.org/monlink.htm

Gray, R. and S. Malla. 2001. "The Evaluation of the Economic and External Health Benefits from Canola Research," in J.M. Alston, P.G. Pardey, and M.J. Taylor, eds., pp. 211–240, *Agricultural Science Policy*. Baltimore, MD: Johns Hopkins University Press.

Hauser, T.P., R.B. Jorgensen, and H. Ostergard. 1998. "Fitness of Backcross and F2 Hybrids Between Weedy Brassica rapa and Oilseed Rape (B. napus)." *Heredity*. 81 (4): 436–443 (October).

Heap, I.M. 1999. "International Survey of Herbicide Resistant Weeds." Internet website (February): http://www.weedscience.com.

Herrman, T. and M. Boland. 1999. "Segregation Strategies." *Feed and Grain*. February/March: 20–24.

Hoban, T. 1998. "Trends in Consumer Attitudes About Biotechnology." *Agbioforum* 1(1) (Summer). Internet website: http://www.agbioforum.missouri.edu

Hobbs, J.E. 2000. "Consumer Issues and Labeling in International Trade." Paper presented at the *Globalization and New Agricultural Trade Rules for the 21st Century* Conference, Canadian Agri-Food Trade Research Network, Saskatoon, Canada (February).

Hobbs, J.E. and M.D. Plunkett. 1999. "Genetically Modified Foods: Consumer Issues and the Role of Information Asymmetry." *Canadian Journal of Agricultural Economics* 47 (4): 445–455.

Hoef, A.M.A. van, E.J. Kok, E. Bouw, H.A. Kuiper, and J. Keijer. 1998. "Development and Application of a Selective Detection Method for Genetically Modified Soy and Soy-derived Products." *Food Additive Contamination* 15 (7): 767–774.

Holt, J.S. and H.M. LeBaron. 1990. "Significance and Distribution of Herbicide Resistance." *Weed Technology* 4: 141–149.

Hurburgh, C. R., Jr. 1994. "Identification and Segregation of High-Value Soybeans at a Country Elevator." *Journal of American Oil Chemists Society* 71 (10): 1073–1078.

Kappeli, O. and L. Auberson. 1998. "How Safe is Safe Enough in Plant Genetic Engineering." *Trends in Plant Science* 3(7): 276–281.

_____. 1997. "The Science and Intricacy of Environmental Safety Evaluations." *Trends in Biotechnology* 9 (September 15): 342–349.

Klein, K.K., W.A. Kerr, and J.E. Hobbs. 1998. "The Impact of Biotechnology on Agricultural Markets." *Canadian Journal of Agricultural Economics* 46(4): 441–453 (December).

Liu, K. 1999. "Biotech Crops: Products, Properties, and Prospects." *Food Technology* 53 (5): 42–49.

Mann, C.C. 1999. "Biotech. Goes Wild." *Technology Review* July/August: 37–43.

Mooney, S. and K.K. Klein. 1999. "Environmental Concerns and Risks of Genetically Modified Crops: Economic Contributions to the Debate." *Canadian Journal of Agricultural Economics* 47 (4): 437–444.

Nelson, P. 1970. "Information and Consumer Research." *Journal of Political Economy* 78(2): 311–329.
Nickson, T.E. and M.J. McKee. 1999. "Risk Assessment of Genetically Modified Plant Products." *Proceedings of the Western Society of Weed Science* 52: 6–9.
Nielsen, K.M., A.M. Bones, K. Smalla, and J.D. van Elsas. 1999. "Horizontal Gene Transfer from Transgenic Plants to Terrestrial Bacteria—A Rare Event." *FEMS-Microbiology Review* 22 (2): 79–103.
Perdikis, N, W.A. Kerr, and J.E. Hobbs. 2000. "Can the WTO/GATT Agreements on Sanitary and Phyto-Sanitary Measures and Technical Barriers to Trade Be Renegotiated to Accommodate Agricultural Biotechnology?" in W.H. Lesser, ed., pp. 692–707, *Transitions in Agbiotech: Economics of Strategy and Policy*. Food Marketing Policy Center, University of Connecticut, Storrs, CT.
Persley, G. and J.N. Siedow. 1999. "Applications of Biotechnology to Crops: Benefits and Risks." Internet website:
http://www.cast-science.org/biotc_ip.htm <http://www.cast-science.org/biotc_ip.htm.
Pesticide and Toxic Chemical News. 1998. Internet website:
http://www.purefood.org/ge/aaasge.htm.
Rissler, J. 1999. "Biotechnology and Pest Control: Quick Fix versus Sustainable Control." Internet website: http://www.netspeed.com.au/cogs/gen2.htm.
Rissler, J. and M. Mellon. 1996. *The Ecological Risks of Transgenic Crops*. Cambridge, MA: MIT Press.
Saat, T.A.W.M. 1996. "Out in the Open: Field Trials with Genetically Modified Crop Plants in Four European Countries, Chile and the United States," in E.R. Schmidt and T. Hankeln, eds. *Transgenic Organisms and Biosafety: Horizontal Gene Transfer, Stability of DNA, and Expression of Transgenes*. New York, NY: Springer-Verlag
Simon, H. 1961. *Administrative Behavior. 2nd Ed.* New York, NY: Macmillan, Inc.
Smyth, S. and P. Phillips. 1999. "Competitors Cooperating: Establishing a Supply Chain to Manage Genetically Modified Canola." Paper presented at the *Economics of Quality Control in Agriculture* Conference, University of Saskatchewan, Saskatoon, Canada (December).
Symonds, B.K. 1999. "Healthier, Nutritious Foods to Come From Biotech Grains." *Feedstuffs* 71 (16): 12–13.
Tabashnik, B.E. 1994. "Evolution of Resistance to Bacillus thuringiensis." *Annual Review of Entomology* 39: 47–49.
Tjaden, T. 1997. "Beekeepers Buzzing Over Pollen Fears." *The Western Producer* (29 May: 23).
USDA/ERS (U.S. Department of Agriculture, Economic Research Service). 1999. Internet website: http://www.ers.usda.gov/
Vandeburg, J.M., J.R. Fulton, F.J. Dooley, and P.V. Preckel. 2000. "Impact of Identity Preservation of Non-GMO Crops on the Grain Market System." *Current Agriculture, Food, and Resource Issues* 1: 29–36.
Varian, H.R. 1984. *Microeconomic Analysis, 2nd Ed.* New York, NY: W.W. Norton & Company.
Walsh, J. 1999. "Ready or Not, Genetically Modified Foods are Here to Stay." *Environmental Nutrition* 22 (4): 1–6.
Western Producer. 1999. "Senate Considers Hearings on GMO Issue." Internet website: http://www.producer/articles/1991007/ news/19991007news04.html
Western Producer. 2000. "Triple-resistant Canola Weeds Found in Alberta" (10 February: 1).
Williamson, O.E. 1979. "Transaction Cost Economics: The Governance of Contractual Relations." *Journal of Law and Economics* 22: 233–262.

Chapter 24

International Trade in Genetically Modified Agri-Food Products

Peter W.B. Phillips
University of Saskatchewan

1. INTRODUCTION

Commercialisation of the products of agricultural biotechnology promises to exacerbate the already difficult international trade situation in the agri-food system. By 2000, genetically modified (GM) agri-food products were produced in 13 countries and available in most markets through international trade. National regulatory systems of most countries attempt to manage any risks or concerns about biotechnology within their national borders, but find that they need to manage international flows to fully address their concerns. In response, a number of bilateral and multilateral efforts are underway to find mechanisms to both manage the risks and to maintain open access to international markets. So far none of the systems provide a comprehensive base for managing the concerns related to GM agri-food products. As a result, private supply chains are evolving to manage the risks and to capture the returns from this technology.

In this chapter we provide an overview of the 2001 international trade situation and agenda. Section 2 examines the technology, its uses, and current diffusion within global production and trade. Section 3 discusses the domestic regulatory capacity and efforts to manage the introduction and uses of biotechnologically based agri-food products. Section 4 discusses the current efforts to develop an international system to manage international trade in new products. Section 5 examines how the private sector is responding to the uncertain market access for these new products. Section 6 offers some concluding comments.

2. BIOTECHNOLOGY, GENETICALLY MODIFIED FOODS, AND INTERNATIONAL TRADE

Modern biotechnology, which dates from the early 1970s, is the latest and perhaps most fundamental innovative technology to be applied to the agri-food sector. The array of new tools called biotechnology—including genomics, tissue culture, micropropagation, cloning, marker-assisted breeding, gene splicing, and transgenes—now allows breeders to selectively modify plants or animals at the molecular level. The first products of this technology entered agri-food markets in the late 1980s.

Since then, agricultural research has developed a vast array of crops, animals, and microbes using new biotechnology methods that involve input or output attributes involving genetic modification. By January 2001 more than 40 genetic modifications related to 13 different crops were approved and produced in one of 13 countries, and to varying degrees were made available to other countries through international trade. A number of countries have approved the release of one or more varieties of GM fish (for example, salmon), trees (for example, poplar), microbes, drugs [for example, recombinant Bovine Somatotropin (rBST)], and various vaccines for animals. More than 40 crops and a range of animal species and microbes have been genetically modified and await regulatory approval in various countries involved in the international food trade (OECD, 2000a).

The introduction of these products into the global agricultural and food system has generated significant debate about two major issues: (1) the safety of the foods made from these products and their affect on the environment; and (2) a host of other issues involving ethical, legal, social, or economic consequences of the use of this technology. While there are a number of animal and microbial products now entering the market, the primary focus of this analysis is on GM-commodity food crops. So far almost 99 percent of the trials for new products have involved those plants, and they are the only extensively produced and traded GM products to date. Furthermore, all of the current trade disputes concerning biotechnology involve market access for such crops.

Although the first unconfined release of a GM crop was tobacco in China in the late 1980s, the main effort to commercialize such crops has been in the past five years. During 1995, a small number of acres of GM varieties of canola, corn, cotton, soybeans, and tomatoes were produced in Canada and the United States. Since then, seven more GM crops have entered the market—flax, melons, papaya, potato, rice, squash, and sugar beet. James (2000) estimates that global production of these crops, scattered among 13 countries, increased to more than 109 million acres in 2000. We show which countries have approved which crops as of June 2000 (Table 24.1).

Table 24.1 Countries Approved and Producing GM Crops, 1999

					European Union									
	Argentina	*Australia*	*Canada*	*China*	*France*	*Portugal*	*Spain*	*UK*	*Japan*	*Mexico*	*Romania*	*S. Africa*	*Ukraine*	*USA*
Recorded production	Y	Y	Y	Y	Y	Y	Y	N	N	Y	Y	Y	Y	Y
Canola			P					A	A					P
Cotton		P	A	P					A					P
Flax			A											A
Maize	P		P		P	P	P		A			P		P
Melon														P
Papaya														P
Potato			P						A				P	P
Rice														P
Soybeans	P		P						A		P			P
Squash			A											P
Sugar beets									A					P
Tobacco				P	A									
Tomato			A	A			A		A	A				P

A = Approved varieties, but no record of production.
P = Approved varieties, with a record of production.
Source: James (2000); OECD (2001) Biotechnology Regulatory Affairs Website.

Although there are more than 40 different genetic modifications involving 13 different crops and 13 countries, the commercial activity is concentrated. James (2000) estimates that the United States planted 74 percent of the global area for transgenic crops in 1999, Argentina 23 percent, Canada 7 percent, China 1 percent and the other 8 countries together accounted for less than 1 percent of the total planted area. Soybeans accounted for 59 percent of the global acreage, corn 23 percent, cotton 12 percent, canola 6 percent and the other crops accounted for less than one percent of the global acreage total. Herbicide-tolerant crops accounted for 74 percent of the acreage, insect-resistant crops (Bt) for 19 percent, stacked herbicide tolerance-Bt crops for 7 percent and all the other traits combined (for example, viral resistance, nutrition and delayed ripening) accounted for less than one percent of total acreage.

Output from that limited number of countries is consumed around the world through international trade (Table 24.2). Given that none of the countries producing GM varieties has completely effective crop-segregation systems in place for those crops commercially produced, even small amounts of GM production could co-mingle with the general commodity stream and lead to diffusing the GM seed through international exports. Although a maximum of eight countries produce and export any one of the GM commodities, this production is concentrated in countries that dominate both

Table 24.2 Distribution of Production and Trade in Genetically Modified Crops, 1998

	Total world production		*Production in countries producing approved GM varieties*		*Exports*	*Imports*
Commodity	*Number of countries*	*Volume (M t)*	*Number of countries*	*Percent total*	*Percent total world exports originating in countries with GM production*	*Total number of countries importing commodity*
Canola	51	35.9	2	23	50	68
Flax	46	2.7	2	46	81	74
Maize	158	613.3	7	53	85	168
Melon	17	0.7	1	0	0	13
Papaya	47	5.1	1	0	5	56
Potato	148	298.5	3	14	12	177
Rice	114	575.9	1	1	80	82
Soybeans	83	157.7	4	81	88	114
Squash	83	14.8	1	0	0	32
Sugar beets	52	259.6	1	11	19	26
Tobacco	128	6.9	1	36	9	154
Tomato	159	90.9	5[a]	37	54	140

[a]*There are inadequate records to determine whether or not GM tomatoes are being grown in four of these countries.*
Source: Table 24.1 and FAO (1998).

world production and world trade in those products. For example, although GM canola is produced commercially in only two countries (Canada and United States), they together produce approximately 23 percent of global production and account for 50 percent of global exports. GM corn (maize), soybeans, and flax are produced in countries that account for more than 80 percent of world trade in those products. In contrast, a number of the GM crops are either not extensively traded or are produced in countries that are not major exporters (for example, beets, papaya, potatoes, and tobacco).

Given that most of these GM products are commodities subject to significant blending and shipping, it is often hard to determine the exact originating country of all parts of any shipment. As a result, any GM production that enters the international market could potentially be imported by any of the importing countries. As many as 177 countries import some quantities of these 13 modified crops (Table 24.2). As a result, domestic and international trade rules, and, ultimately, private companies are being forced to adapt to consumer and citizen concerns.

Economic theory suggests a link between technology and its potential trade impact. The new, endogenous growth theory suggests that if comparative advantage is now based on knowledge rather than on land, labor, or capital, then it may be possible to create comparative advantage through research policy and industrial incentives. There is some evidence that biotechnology research exhibits economies of scale and scope that cause clustering of research, and that much of the research embodied in new plant and animal varieties has somewhat limited international transferability. If research is a non-transferable intermediate factor of production, then the fact that innovation begins in one jurisdiction could forever put that site on a higher trajectory of research and development (R&D) and of new product development. Grossman and Helpman (1991) argue that, as a result, the high-technology share of gross domestic product (GDP) and exports will be higher than otherwise, which excites many governments around the world (Phillips and Khachatourians, 2001). If this were to happen, then production of GM foods would tend to be relatively concentrated in a few exporting countries, which would increase pressure for international market access. So far, all of the GM varieties increase productivity, which, in a traditional trade model, would increase exportable surpluses in some countries and depress international prices. Without trade restrictions, production should decline in importing countries and trade should increase. As a result, biotechnology will add pressure for renewed rules for international trade in the agri-food sector.

3. DOMESTIC REGULATION OF GENETICALLY MODIFIED FOOD

Domestic-regulatory systems currently provide the foundation for regulating GM agri-food products. In the past, new products reviewed and approved in the country of production have generally been granted unrestricted access to global markets. Increasing consumer concerns about food safety and widely varying citizen concerns about the environmental, social, economic, and ethical aspects of GM foods has led many countries to change their practices such that a domestic review of the food products' safety be conducted before they are allowed to be imported, or produced, for sale to local consumers. Although the evidence is inconclusive, one could hypothesize that the health, safety, and ethical issues are luxury goods, and hence have a greater impact in developing countries.

This divergence of views is reflected in the array of domestic regulatory systems that are in operation. Most countries, whether they have specific concerns about GM foods, have some form of domestic regulation in the agri-food sector that predates the arrival of biotechnology in the global production and trade systems. Most countries have at least a basic structure for regulating the introduction of new plant varieties into their countries. The FAO (1998) reports that 75 percent of the countries in the world, and almost all of those that are likely targets for deliberate commercial release of foreign or GM plants or animals, have national quarantine rules. Almost 75 percent of those countries (103) are members of the International Plant Protection Convention and their quarantine rules conform to the Convention. Furthermore, more than 68 percent of the countries in the world have some form of controls on the seeds industry. In spite of regulations, though, many of these countries have limited technical capacity to undertake the scientific work to assess risks (FAO, 1998). As a result, most of the countries importing GM foods continue to rely on the regulatory oversight provided by the exporting country.

A handful of mostly developed countries have modified existing or have built new regulatory systems to handle their specific concerns about biotechnology (OECD, 2000b). Two divergent regulatory approaches have evolved (Table 24.3). The North American regulatory approach follows a rules- and science-based, legalistic approach, while the European model pursues a political-control approach. Isaac and Phillips (1999) argue that with respect to modern biotechnology, the North American model of regulations supports rapid commercialization of biotechnology products. While these systems are often triggered by the use of biotechnology processes, they evaluate the risks of the resulting varieties based on the specific resulting novel traits. By basing the review on product attributes, these countries can, for the most part, use existing or modified vertical (or sectoral) regulatory agencies or authorities. The focus on products also facilitates the review of new varieties, with those that are essentially equivalent subject to lower levels of investigation.

Table 24.3 Competing Regulatory Approaches for Agricultural Biotechnology

	North American approach	*E.U. approach*
Orientation	Supply push	Demand pull
Tradition	Legalistic	Accountability
Features of regulatory system:		
Trigger	Novel product attributes	Use of biotechnology processes
Regulatory base	Vertical through existing regulations and agencies	Horizontal through new regulations and agencies
Precautionary principle	Essential equivalency	Certainty
Access to system	Closed to interest groups	Open to interest groups
Labeling	Discretionary	Mandatory

Source: Isaac and Phillips (1999).

This approach forms the basis for the biotechnology regulatory systems in use in the United States, Canada, Australia, New Zealand, Japan and Mexico. The European approach to government regulatory intervention, in contrast, is dominated by concerns over the democratic accountability of the discretionary decision-making power of regulators (Majone, 1994). This has led to the development of new horizontal regulatory systems used to handle the perceived risks of this new technology. The regulatory system, reflecting consumer and citizen concerns, demands certainty about the safety of new products, withholding approval from those where all risks (that is, actual, hypothetical, and speculative) have not been resolved. Regulatory decision-making resides with elected public officials to ensure accountability, with the result that all regulatory interventions are subject to the day-to-day public interests, which dominate the concerns of elected officials. Although 18 products were reviewed and approved for release in the European Union be fore 1997, since then a succession of national and E.U. moratoriums on the release of new varieties has stopped approvals.

Since 1998, there has been increasing pressure from consumers for governments to develop and to impose mandatory labeling systems for GM foods in order to allow consumers to make their own choices about GM products. In response, 16 countries plus the European Union have proposed some form of labeling. Most of the countries propose to require mandatory, positive labels of GM elements in food, while the United States and Canada have chosen to develop voluntary rules for labels that would identify food products that are free of GM elements.

We show in Table 24.4 that there is no convergence yet on whether the presence of GM elements should be labeled, which products should be labeled (all products, only those products with detectable GM elements, or a prescribed list?), tolerances for GM ingredients (one percent or five percent, by ingredient or volume?), or where GM labels should be applied (stores, restaurants, take-aways?). It is hard to be conclusive given that, so far, only the United Kingdom and Japan have formally adopted and implemented their systems. All the rest of the systems are pending. The wide array of requirements and applications of labels, while perhaps justified on a domestic basis, have real potential to become technical barriers to trade (TBTs) that could be judged to contravene existing international trade agreements.

As a result of the incomplete, and often conflicting, domestic regulatory systems in exporting and importing countries, regulation of GM foods has become an international issue. There have been extensive efforts in recent years to resolve the impasse over the regulation and trade of GM products.

Table 24.4 National Rules for Labeling GM Foods

Countries	*Coverage*	*Tolerance*
Mandatory rules:		
E.U. (adopted; must be implemented by member states)	All foods in stores; no regulation for additives or feeds	1 percent GM per ingredient
UK (effective September 1998)	All whole foods, processed food, additives and flavorings in stores and restaurants	1 percent GM per ingredient
Japan (effective April 2000)	Any one of 28 listed products (excludes oils, additives and animal feed) if it is one of the three top ingredients and composes at least 5 percent total weight	5 percent by volume
Australia/New Zealand (pending)	Processed foods, fruits, vegetables in stores, take-outs, and restaurants	1 percent advantageous
Argentina, China, Czech Republic, Ethiopia, Hungary, Indonesia, Mexico, Poland, Russia, South Africa, South Korea, Switzerland, and Thailand	Pending; limited details	Pending; limited details
Voluntary Systems:		
Canada/USA	Being developed	Being developed

Source: Phillips and Foster (2000).

4. INTERNATIONAL REGULATION OF GENETICALLY MODIFIED FOOD

Currently, there are seven international bodies vying to coordinate and regulate products of biotechnology (Table 24.5). Conceptually they represent a progression from largely science-based institutions [International Plant Protection Convention (IPPC), International Office of Epizootics (OIE), and Codex Alimentarius Commission (Codex)] to ones with broader objectives such as trade facilitation, environmental protection and other social and political goals [Organization for Economic Cooperation and Development (OECD), regional initiatives, World Trade Organization (WTO) and Bio-Safety Protocol (BSP)]. This section offers a thumbnail sketch of each institution involved in regulating trade in GM foods and a brief synopsis of where the system is going (Buckingham et al., 1999; OECD, 2000c).

Table 24.5 Current Institutions Regulating International Trade in GM Crops

Institution	Date	Coverage	Member States	Dispute Settlement	Orientation
International Plant Protection Convention (IPPC)	1952	Pests and pathogens of plants and plant products	107	Non-binding; sets WTO standards via SPS S.3.4	International standard for plant measures involving quarantines
International Office of Epizootics (OIE)	1924	Infectious animal diseases	152	Non-binding; sets WTO standards via SPS S.3.4	Harmonize import and export regulations for animals and animal products through International Animal Health Code
The Codex Alimentarius Commision (Codex)	1962	Food labeling and safety standards	160	Non-binding; sets WTO standards via SPS S.3.4	International standards to provide guidance to the food industry and protection to consumer health
OECD	1969	Harmonization of international regulatory requirements, standards and policies	29	None	Consensus documents
Regional initiatives (TEP; ECTI)	1990s	Harmonization of the science of regulation	Bilateral	None	Regional side agreements, MOU, MRA, formal dialogues, and joint research projects
WTO/GATT	1947	Trade in all goods and most services	139	Binding	Establish rules for transparency and dispute settlement through TBT and SPS agreements
BioSafety Protocol	2000	Transboundary movements of GMOs	Na	None	Requires 50 countries to ratify before operation

Source: Modified from Buckingham et al. (1999).

The IPPC and OIE are multilateral treaties that seek to protect plants and animals from the spread of pathogens through international trade. The IPPC protects natural flora, cultivated plants, and plant products; the OIE protects animals and fish. In collaboration with both regional and national plant and animal protection organizations, they provide a forum for international co-operation, harmonization and technical exchange of plant- and animal-protection information. The IPPC has addressed the international regulation of GM crops through several International Standards for Phytosanitary Measures (ISPMs) while the OIE has developed standards for diagnostic reagents, sera, and vaccines for animals in the International Animal Health Code. Both institutions have their own non-binding dispute avoidance and settlement systems, but their most important role in international trade is through the WTO Sanitary and Phytosanitary (SPS) Agreement, which uses the IPPC and OIE standards as the basis for evaluating SPS disputes. National measures based on international standards from either of these institutions will generally not be open to challenge under the WTO dispute-resolution process. Furthermore, both organizations nominate experts for WTO/SPS dispute panels, and provide technical background information to the panels based on their standards. As such, they can have far-reaching economic and political consequences on trade in GM products.

Codex provides a similar service related to processed foods. Codex develops international food standards, which identify the product and its essential composition and quality factors, identify additives and potential contaminants, set hygiene requirements, provide labeling requirements, and establish the scientific procedures used to sample and analyze the product. Each standard normally takes six or more years to develop. Determination of the safety of the food product is based on scientific risk analysis and toxicological studies. Once a Codex standard is adopted, member countries are encouraged to incorporate it into any relevant domestic rules and legislation, but they may unilaterally impose more stringent food-safety regulations for consumer protection, provided the standards are scientifically justifiable. Codex plays an important role in agri-food trade because its standards, guidelines, and recommendations are acknowledged in the SPS and Technical Barriers to Trade Agreements of WTO. As with the IPPC and OIE standards, during consideration of trade disputes, domestic standards based on Codex standards are usually sustained.

While there are currently no Codex standards in place for products of biotechnology, there has been significant effort by Codex to develop a standard for labeling food products derived from biotechnology; these negotiations are ongoing.

The OECD has actively assisted in the harmonization of international regulatory requirements, standards, and policies related to the discipline of biotechnology, since 1995. It has undertaken a number of projects to make regulatory processes more transparent and more efficient, to facilitate trade in products derived through biotechnology, and to provide information exchange and dialogue within non-OECD countries. The OECD leads efforts to develop Consensus Documents, which are scientific background documents, mutually recognized by member states, that set out the biology of the crop planted and the introduced trait; these documents provide the base to be used in regulatory assessment of an agricultural, or food, product derived through modern biotechnology. By August 2000, 14 consensus documents were published by the OECD, that were related to viral resistance, to herbicide tolerance, to applications of pseudomonas, and to the biology of canola, potato, rice, bread, wheat, and spruce (OECD, 2000a). In addition, in June 1999 the Group of Eight Heads of Government Ministerial (G8) requested the OECD to undertake a study of the implications of biotechnology and other aspects of food safety. This involved, among other actions, an international food-safety conference in Edinburgh, Scotland in February 2000, which concluded that the regulatory systems around the world have to date only approved GM foods that are safe. The OECD reported its findings to the G8 in Tokyo in July 2000, and maintains an ongoing role in seeking a consensus of opinion.

A number of bilateral or multilateral regional initiatives have played, and will increasingly play, an important role in the regulation of trade in goods and services. The Trans-Atlantic Economic Partnership (TEP) between the United States and the European Union, for example, has undertaken talks to improve regulatory processes and scientific cooperation through several means: (1) mutual recognition of testing and approval procedures; (2) progressive realignment or adoption of the same standards, regulatory requirements and procedures; (3) adopting internationally agreed on standards; and (4) dialogue between scientific and other expert advisers in standard-setting bodies and regulatory agencies.

The European Union has similar trade liberalization initiatives with Canada and Japan. Since 1998, the Canadian Food Inspection Agency, the United States Department of Agriculture, Animal and Plant Health Inspection Service have also been studying and comparing the molecular genetic characterization of transgenic plants in search of ways to harmonize their regulatory review processes. Some agreement has already been achieved, although no formal binding bilateral agreement has yet been concluded. Meanwhile, the United States, Canada, and the European Union all offer training and support for regulators in key import markets (usually developing countries) in an effort to export their regulatory models to other countries. These bilateral processes could be an important way to resolve technically based trade disputes. Regional agreements, memoranda of understanding, mutual recognition agreements, formal dialogues, and joint-research projects are mechanisms that can be used to decrease bilateral regulatory barriers to the GM food trade. Using these mechanisms may help countries achieve the greatest trade liberalization that is possible at a given point in time. Then, if a critical mass of regions reach sufficient levels of liberalization, regional agreements could be incorporated with relative ease into the multilateral international trade regime. In this way, regional trade agreements may be complementary, if not preliminary, to the WTO/GATT process.

WTO is likely to be called on soon to consider American, Canadian and Argentinean concerns that national regulatory rules in the European Union unduly restrict access for GM products into Europe. While the WTO agreement permits environmental and health regulations, it sets some rules on when and how they may be used. Specifically, the WTO (Art. XI) permits "national standards or regulations for the classification, grading or marketing of commodities in international trade," and (WTO, Art. XX(b)) "the adoption or enforcement of measures necessary to protect human, animal or plant life or health." The SPS Agreement offers four rules on how they can be used: (1) SPS measures cannot offend either branch of the principle of non-discrimination; (2) SPS standards that conform to international standards developed by international organizations (that is, Codex, OIE, and IPPC) are presumed to be consistent with the obligations outlined in the SPS Agreement; (3) where a country implements national standards that are in excess of established international standards or where no international standards exist, those standards must be based on scientific principles and on the completion of a risk assessment study; and (4) SPS measures shall not be applied in a manner that would constitute a disguised restriction on international trade.

If a country believes another party has unjustly restricted market access through an SPS measure, it can use the WTO dispute-settlement system. Once a case has been presented, heard, and adjudicated, the General Council approves the final dispute-panel report, unless there is unanimous consent not to adopt it. To date, WTO panels have decided three cases concerning the validity of national SPS measures: the E.U. Hormones case, the Australian Salmon case and the Japanese Agricultural-products case. In all three cases the contested domestic SPS measure was struck down on the basis that there was no risk assessment completed to support the SPS measure, or that the risk assessment was improperly done. The Salmon-case decision set out three criteria for a proper risk assessment: (1) it must identify the diseases whose entry, establishment, or spread a member wants to prevent within its territory, as well as the potential biological and economic consequences associated with the entry, or spread, of these diseases; (2) it must evaluate the likelihood of the entry, establishment, or spread, of these diseases, as well as the potential biological and economic consequences of failing to prevent introduction; and (3) it must evaluate the likelihood of the entry, or spread, of these diseases according to the SPS measures to which it might be applied (WTO, 1998). If the risk assessment does not refer to the SPS measure, it is doomed to failure before a WTO Panel.

While the WTO is an attractive site for dispute resolution for many countries, it has some problems. As currently interpreted, the SPS Agreement does not permit non-science concerns (such as consumer preference, animal welfare, or non-measurable environmental risks) to be considered in the determination of whether an SPS measure is based on science and is supported by a proper risk assessment. WTO dispute cases show that the organization is likely to value unrestricted trade and scientific proof above other factors, such as environmental protection or socio-economic considerations. There is a very real risk that WTO decisions that are contrary to domestic concerns may be ignored, which could topple the WTO Panel process and the WTO as a whole, and jeopardize the other operations of the WTO in trade liberalization.

The BSP represents a recent effort to provide a comprehensive international structure to ensure the protection of biodiversity and to facilitate consideration of non-scientific concerns. The Cartagena Protocol (hereafter referred to as the protocol), concluded in Montreal in January 2000, provides rules for trans-boundary movements of GM organisms intended for environmental release and for those destined for the food chain. For living GM organisms (for example, seeds for propagation, seedlings, and fish for release), exporters will be required to obtain approval from importing countries. Within 15 days of approving a new GM variety, a country will notify a BioSafety Clearing House with information about the traits and evaluations. The first time that a new GM variety is to be exported as seed, the exporting country will notify the importing country. The importing country will then decide whether to approve the shipment because of risks identified through a science-based risk assessment. This process is called advanced informed agreement (AIA). Although this seems straightforward, the protocol includes two features that may raise conflict in coming years. First, the text indicates that countries may, in their reviews of GM foods, consider socio-economic factors (for example, the impact on local farmers), provided they respect their other international obligations. Second, the protocol includes a so-called precautionary principle, whereby countries do not have to have complete scientific certainty to block imports of a GM product that they fear could be harmful to biological diversity. Although it is unclear how the negotiating parties expect the two exemptions to operate, it is likely, given the reference in the preamble to other international obligations, that any import bans not based on scientific-risk assessments will be constrained. As with the SPS Agreement under the WTO, temporary bans may be permitted, but it is likely that countries will need to make real efforts to undertake the scientific research to validate (or refute) the concern. Framers of the protocol have attempted to focus it tightly on environmental risks. To that end, transboundary movements of GM organisms intended for food, feed and processing (for example, commodities) will be exempt from the advanced informed-agreement provisions. Nevertheless, exporters must label shipments with GM varieties, and countries can to decide whether to import those commodities based on a scientific-risk assessment. GM varieties intended for contained use (for example, national-breeding programs and research) and those in transit through other countries will not require AIAs.

It is perhaps too early to make a confident evaluation of the protocol. Almost all participants in the talks—developed and developing country governments, agricultural producers, biotechnology companies and public interest groups—have expressed optimism that the protocol will protect the environment without unduly impeding international trade. In part, its impact will depend on its membership. It requires 50 countries to sign and ratify the protocol for it to come into force. The United States, which is the single largest producer of GM crops, has not ratified the enabling 1992 Convention on Biological Diversity, which means that, although it may abide by the protocol, it will not be a party to it. Ultimately, the economic and trade impact of the protocol depends on how it is implemented. Isaac and Phillips (1999) examined the operation of the BSP on the Canadian canola trade and concluded that it might affect as little as 0.5 percent of total exports (that is, first time shipments of GM organisms intended for deliberate release).

This review of the current international institutions does not yield an obvious candidate to resolve the international trade difficulties in the GM agri-food market. Thus, the GM agri-food community will continue to face a patchwork of regulations, with national systems providing basic, but different, standards and international bodies setting minimum, but potentially unenforceable, standards for different aspects of the regulation of products of biotechnology. At their root, national governments do not have enough confidence in each other to accept their trading partners' systems. For the time being, the agri-food market will continue to face that uncertainty.

5. THE PRIVATE SECTOR RESPONSE

The agri-food market has begun to divide into two distinct markets, and, to date, there has been only limited interaction between the two. Some markets, such as the European Union, and some food processors, have decided to forgo the technology for now, and are devoting increasing effort to securing adequate volumes of GM-free foodstuffs to satisfy their customers. Consumers in those markets, for the most part, do not have the opportunity to consume GM foods, as they simply are not available. Other markets, such as those in North America, have rapidly adopted the technology, and, for the most part, do not offer a choice of GM-free food to their consumers. Trade between these two blocks has slowed dramatically in all those product markets in which GM varieties are being used. Some firms have begun investing to develop quality-assured supply chains that will be used to bridge the gap between these two markets.

Although European rules do not prohibit trade in GM products, they have had that effect. As noted, the E.U. regulatory system has approved 18 GM products for use in the European Union. But as new varieties, not yet approved in the European Union, are developed, adopted, and co-mingled in the United States and Canada, trade in key product areas has been sharply curtailed. For example, U.S. corn exports to the European Union dropped to U.S. $175 million in 1999 from U.S. $574 million in 1995 and soybean exports dropped to U.S. $1.1 billion from U.S. $2.1 billion, while Canada's canola exports to the European Union dropped to U.S. $8 million from U.S. $204 million (Strategis, 2000). Meanwhile, E.U. food processors have diverted their purchases to markets where GM varieties are not produced, such as those markets in northern Brazil (for soybeans) and in Australia (for canola).

This trend has been exacerbated by the new rules in many countries for the labeling of GM food elements. Processors, retailers, and restaurateurs, not wanting to label that their products contain GM elements, have announced that they are going GM-free. A recent survey shows that as of May 2000, 29 manufacturers, 21 retailers, and six restaurant chains around the world have signalled intentions to adopt labels for GM foods (Phillips and Foster, 2000). Thirteen of the processors only produce for the E.U. market, nine are in Japan, four are targeting world markets, four are only in North America, one is in Mexico, and one is in Thailand. E.U. retailers also lead the way on labeling. Even though the E.U. labeling law is not yet in force in most member states, 15 supermarket chains have moved to eliminate GM ingredients in their own-label products. In addition, two U.S. retailers, two in South Africa and one in Hong Kong have adopted GM-free policies. Finally, six international fast-food chains in the United Kingdom, responding to British labeling laws, have removed GM foods from their system. One chain in the United States has announced it will remove GM potatoes from its products.

Labeling is one thing, but actually delivering the promised product is quite another. As GM technology spreads to new markets, it will become more difficult for importing countries to find supplies that one can confidently state are GM-free. The main difficulty of buying from markets with both GM and non-GM varieties is that, as of June 2000, tests available for determining whether products are GM-free are costly, often inaccurate, and always slow. As a result, companies are being forced to develop identity-preserved production and marketing systems to secure GM-free varieties to serve these markets. These systems are costly, and markets, at least in early 2000, did not appear to offer a large enough premium to compensate for producer costs. In 1999, Archer-Daniels Midland (ADM), a major grain wholesaler, informed U.S. producers that it would be segregating GM from non-GM corn, and gave indications that it might not buy GM varieties in the 2000 crop year. Early in 2000, ADM reported that only five percent of sales (by volume) from July to December 1999 demanded GM-free products. As a result, ADM indicated it would continue to buy both GM and GM-free crops. There are some firms in Canada and the United States offering small premiums for producers to deliver GM-free soybeans and corn, but early indications are that the cost of the systems may offset any gains in market share. There are a number of examples in Canada in which producers in the canola industry have, for limited periods, adopted identity-preserved production and identity-preserved marketing systems for GM and non-GM novel trait varieties, but the resulting systems have imposed incremental costs ranging from 10 percent to 15 percent of farm-gate prices (Phillips and Smyth, 1999; Smyth and Phillips, 2000).

While industry remains concerned about the costs of adopting segregation systems, many firms in North America and Europe are testing the waters. It would be reasonable to hypothesize, however, that some crops and some regions will be more difficult and costly to segregate. Those crops with significant risk of cross-pollination (for example, canola) will be more difficult and expensive to segregate. Furthermore, many parts of the industry lack the necessary infrastructure to assure the quality of their products. Those crops, regions, or countries (mostly least-developed countries) that lack an effective pedigreed seeds industry, that have weak or nonexistent quality regulations, or that have rudimentary distribution systems, are likely to face insurmountable costs of segregating their crops for the diverging markets. Thus, market uncertainty will translate into unequal opportunities and impacts.

6. CONCLUDING OBSERVATIONS

The diffusion of agricultural biotech products is exacerbating the historically difficult international agri-food trade system. As new GM varieties are introduced into new product lines and new markets, public concern and regulatory oversight of production and trade is increasing. As time passes, the yield-enhancing effect of this new technology will expand production into key export markets, depress world prices, and increase demand for a freer flow of GM food products. In the absence of international consensus on how to jointly regulate the wide array of consumer and citizen concerns about GM technology, national regulators will continue to enact rules. These rules will effectively reduce international trade, which, if sustained long enough, will dampen private investments in research and development.

The simple conclusion is that the scale of diffusion of GM crops is so large that governments throughout the world must consider how to more effectively regulate production and trade in GM crops. A portfolio of approaches will be needed to comprehensively address the challenges of managing the varied effects of GM foods. This will likely include continued science-based consensus building through institutions such as the IPPC, OIE and Codex, bilateral consensus building through regional efforts (such as TEP or OECD), and multilateral negotiations and precedent setting through dispute settlements (WTO and BSP). Even so, producers and agri-food firms will continue to be challenged to meet differentiating standards in segmenting markets.

The stakes are high. This new technology has the potential to forestall, or to solve, many of the constraints to increased production both in developed and in developing countries. The biggest benefit to world welfare would be greater, and more secure supplies would lower the real price of food. Sustaining, or increasing, the global per-capita food supply would go a long way to help to improve the welfare of the 20 percent of the world's population living on less than one dollar per day. Nevertheless, technology poses a number of major challenges for industry and for governments. Given that many of the technologies are not universally applicable, their use will be uneven. Some producers and regions may gain and others may lose. Based on the evidence to date, it is conceivable that those parts of the agri-food sector that will benefit from the new technologies could be in both the developed and developing nations. The winners will probably self-select by making investments to adopt, adapt, and manage the technology. As a result, the production benefits will be unevenly distributed, which will challenge the global agri-food sector to allow or to facilitate adjustments of labour and capital (currently many agricultural and trade policies are designed to maintain farm production and to impede adjustment), while, at the same time, they provide more secure access to global markets.

7. REFERENCES

Buckingham, D., R. Gray, P. Phillips, T. Roberts, J. Bryce, B. Morris, D. Stovin, G. Isaac, and B. Anderson. 1999. *The International Co-ordination of Regulatory Approaches to*

Products of Biotechnology. Study prepared for Agriculture and Agri-Food Canada, Ottawa, Ontario, Canada (June).

FAO (Food and Agriculture Organization). 1998. *State of the world's plant genetic resources for food and agriculture*, Appendix 1. Rome, Italy: FAO.

_____. 2000. Data Collections. Internet website: http://apps.fao.org

Grossman, G. and E. Helpman. 1991. *Innovation and Growth in the Global Economy*. London, UK: The MIT Press.

Isaac, G. and P.W.B. Phillips. 1999. *Market Access and Market Acceptance for Agricultural Biotechnology Products*. Proceedings of the ICABR conference on "The Shape of the Coming, Agricultural Biotechnology Transformation," Ravello, Italy (June 17–19).

James, C. 2000. "Global Status of Commercialized Transgenic Crops." International Service for the Acquisition of Agri-biotech Applications (ISAAA) Brief No. 21-2000. Internet website: http://www.ISAAA.org/

Majone, G. 1994. *Interdependence vs. Accountability: Non-Majoritarian Institutions and Democratic Government in Europe*. European University Institute Working Paper SPS 94/3 (San Domenico di Fiesole, Italy : European University Institute).

OECD (Organization for Economic Cooperation and Development). 2001. *Regulatory Developments in Biotechnology in OECD Member Countries*. Internet website: http://www.oecd.org/ehs/country.htm.

_____. 2000a. *Biotrack Online: Consensus Documents*. Internet website: http://www.oecd.org/ehs/cd.htm.

_____.2000b. *Compendium of National Food Safety Systems and Activities*. Internet site: http://www.SG/ADHOC/FS(2000)5/ANN/FINAL (June 7).

_____.2000c. *Overview and Compendium of International Organizations with Food Safety Activities*. Internet site: http://www.SG/ADHOC/FS(2000)4/FINAL, (11 May).

Phillips, P. and D. Buckingham. 2001. "Agricultural Biotechnology, the Environment and International Trade Regulation: Your Place or Mine?" in H. Michelmann et al., eds., *Globalization and New Agricultural Trade Rules in the 21st Century*. Boulder, CO: Lynne Rienner Publishers.

Phillips, P. and G. Khachatourians. 2001. *The Biotechnology Revolution in Global Agriculture: Invention, Innovation and Investment in the Canola Sector*. Biotechnology in Agriculture Series, No. 24. Oxford: CABI Publishing.

Phillips, P. and H. Foster. 2000. *Labeling for GM Foods: Theory and Practice*. Proceedings of the International Consortium on Agricultural Biotechnology (ICABR) Conference, Ravello, Italy (24–28 August).

Phillips, P. and S. Smyth. 1999. *Competitors co-operating: Establishing a supply chain to manage GM canola*. Proceedings of the Economics of Quality Conference, Saskatoon, Saskatchewan, Canada (December).

Smyth, S. and P.W.B. Phillips. 2000. *Managing the value of new-trait varieties in the canola supply chain in Canada*. Proceedings of the 4th Wageningen Supply Chain Conference, Wageningen, Netherlands (May).

Strategis. 2000. Data search of Canadian exports by product and market. Industry Canada Internet website: http://strategis.ic.gc.ca/engdoc/main.html.

WTO (World Trade Organization). 1998. *Australia - Measures Affecting Importation of Salmon: Complaint by Canada*. Appellate Decision AB-1998-5. (Appeal from WT/DS18). Internet website: http://wto.org

Index

Printed in the United States
79124LV00001BB/10

9 780792 374725